Arthropod Management in Vineyards

Noubar J. Bostanian • Charles Vincent
Rufus Isaacs
Editors

Arthropod Management in Vineyards:

Pests, Approaches, and Future Directions

Springer

Editors
Dr. Noubar J. Bostanian
Agriculture and Agri-Food Canada
Horticultural Research and
Development Center
430 Gouin Blvd.
Saint-Jean-sur-Richelieu, QC, Canada

Dr. Charles Vincent
Agriculture and Agri-Food Canada
Horticultural Research and
Development Center
430 Gouin Blvd.
Saint-Jean-sur-Richelieu, QC, Canada

Dr. Rufus Isaacs
Department of Entomology
Michigan State University
East Lansing, MI, USA

ISBN 978-94-007-4031-0 ISBN 978-94-007-4032-7 (eBook)
DOI 10.1007/978-94-007-4032-7
Springer Dordrecht Heidelberg New York London

Library of Congress Control Number: 2012939840

Printed on acid-free paper

Springer is part of Springer Science+Business Media (www.springer.com)

Noubar J. Bostanian and Charles Vincent dedicate this volume to Charles-Henri de Coussergues and the late Victor Dietrich. Their queries and enthusiasm encouraged us to initiate research projects with them in viticultural entomology. Rufus Isaacs dedicates this volume to G. Stanley Howell, a pioneering viticulturist with a great interest in insects and their interactions with his beloved grapevines.

Foreword

I consider myself an old school IPM'er. That is because as a graduate student at the University of California Berkeley, I had the privilege of knowing and studying with some of the pioneering people who helped develop the IPM paradigm. Robert van den Bosch was on my PhD oral exam committee and Ken Hagen was on my thesis committee; two of the three authors who wrote the seminal 1959 Hilgardia paper that proposed the concept of integrated control which would become IPM. I was taught that an IPM program was based on knowledge of the ecology of the crop, the pest, its natural enemies, timely monitoring, adherence to an economic threshold and choosing a control action designed to minimize economic, environmental and health risks. I learned that IPM was an ecosystem approach to pest management. Furthermore, it is not static, but one that changes over time as we gain more knowledge about all of the above. Over the years many definitions of IPM have been proposed, with arguments ensuing about what is 'real' IPM. Arthropod Management in Vineyards has come along at the right time to present the most recent information and discussions on the basic tenets of IPM as they apply to modern vineyard management, including IPM principles, discussion of economic threshold and action thresholds, monitoring and arthropod population modeling.

In an ideal world, IPM decision-making is objective, based on sound science, quantitative pest monitoring, and experience. However, once I moved to being a private IPM practitioner out on commercial farms, I realized that arthropod pest management decisions are an outcome of a fascinating combination of knowledge, monitoring, time management, price of the crop, the perception of risk from pest damage, one's mood at the time the decision of what to do takes place, the growers willingness to take risks, what the neighbors are doing, and several other things that I am forgetting to mention. In the real world of pest management, time is money, and there is never enough of either one. Moreover, at least in the United States, many pest management consultants still derive much of their income from the input products they sell, creating an inherent conflict of interest in pest management decision-making. The challenge in implementing IPM in vineyards becomes one of taking the information presented in this book and using it to push back against the non-objective issues that interfere with science based pest management decision-making.

After many years of wrestling with the question of why there was not more IPM practiced in vineyards, I came to the conclusion that the goal in sound pest management decision-making was to match perceived risk of pest damage occurring with that of real risk. When a grower or a pest management practitioner makes a decision to implement a control tactic they do so because they perceive that the risk of pest damage occurring is unacceptable. The challenge is to determine that the perceived risk is in fact real. If perceived risk is high and real risk is low, management actions are taken un-necessarily. If perceived risk is low and real risk is high then no action is taken and economic levels of damage can occur. Risk of pest damage can be short term. For example, it might occur next week or the week after, or it can be long term and measured in years in perennial crops such as grapes. Poor planning in location and establishment of the vineyard and/or poor management of the landscape in which the vineyard occurs can increase long term risk. In either case, the information presented in this book will help grape growers and IPM practitioners determine if perceived risk is real risk.

In conclusion, I think it is very helpful to look at vineyard pest management as a continuum from no IPM being implemented on one end, to high-level IPM being implemented on the other end. IPM is not a static list of things to do or a recipe, like in a cookbook, that when followed always ends up with the same result. It is a paradigm. In the real world, grape growers are distributed all along the pest management continuum, some using no IPM, some implementing some aspects of IPM, and some practicing high level, landscape-based IPM. The reasons for their location on the continuum are many and varied. Nevertheless, the goal of everyone should be to move along the pest management continuum enhancing their IPM programs over time. The information presented in this book will be of great help to grape growers and pest management practitioners all along the pest management continuum. For those growers, consultants, or researchers just beginning to develop IPM programs for their own regional pest challenges, it will provide basic, well-established reference information highlighting approaches and success stories that will provide a great foundation on which to build. For those with sophisticated IPM programs already in place, it will provide the cutting edge information and theories that will allow them to push the envelope of IPM as they move into the future.

VP Professional Services
SureHarvest
Soquel, CA 95073

Clifford P. Ohmart

Preface

Wherever there are vineyards, there are insects and mites. Arthropods have inhabited vineyards for as long as the grapevine has been cultivated in the pursuit of fresh fruit, juice, raisins, or wine. Domestication of wild grapevines across the globe has provided a habitat of great suitability for specialist grapevine herbivores, and has opened up new possibilities for some generalist insects and mites with a penchant for the vine. Additionally, invasive insects transported to new grape production regions are finding their second homes most agreeable, disrupting established IPM programs and requiring rapid responses. The changing distribution of the grape industries coupled with the dynamic nature of pest and predator populations ensures that arthropod management in vineyards will remain an essential component of viticulture.

Vineyard managers have been battling unwanted six and eight legged creatures for thousands of years, and while human management of vineyards can exert great control over the system, at times arthropod pests can gain the upper hand. Whether phylloxera in the 1800s, glassy wing sharpshooter in the later 1900s or stink bugs in present day eastern US viticulture, vineyard managers must remain informed, prepared, and vigilant to ensure economical production of the highest quality grapes without succumbing to new pest arthropods. Failure to implement effective arthropod management practices can result in complete loss of this high value crop or the inability to make quality value-added products, and so it is essential that arthropods are managed using the latest technologies.

This collection of chapters by experts in their fields has been assembled to take a snapshot of the science of arthropod management in vineyards. Broader big-picture themes are discussed in the chapters towards the front of the book, followed by pest-specific chapters that provide 'state-of-the-science' information on how applied entomologists and grower educators are tackling pest challenges in vineyards around the world. The major grape-growing regions of the world are well represented among the chapters, along with some smaller and more recently developed production areas. Throughout this diversity some common themes have appeared, highlighting the challenges inherent in managing arthropods within a high value crop where consumer tolerance of infestation is essentially zero (fresh grapes) or where very low levels of insects present at harvest may be a concern not for their feeding on the vine but for their chemical secretions potentially causing contamination of the resulting wine.

Most of this collection of chapters has a focus on vineyard pests, the damage they cause, and the range of tactics that have been developed, or are being developed, for preventing economic injury to vines. The observant reader will notice a significant emphasis throughout this book on biological control, whether classical, inundative, or conservation in its nature. Vineyard managers in some regions of the world have been leaders in adopting biological control methods for pest management, through the conservation of predatory mites, provision of habitat for natural enemies, or importation of effective parasitoids and predators. There is a significant movement towards wider adoption of sustainable viticulture, and these biological approaches for arthropod management are an essential component of such efforts.

To minimize the injury to grapevines from pest arthropods, vineyard managers need access to timely information that can help them make sound decisions. This requires up-to-date research conducted by people trained in viticultural pest management. In most grape producing regions, there are agricultural universities or government research stations with a focus on grape management, often including researchers with an emphasis on insect biology and management. This book provides a review of the major themes in vineyard pest management and highlights some of the most recent scientific advances of viticultural entomologists, to present the current status of viticultural arthropod pest management science. This collection is broadly international, covering the primary insect pest groups, and we hope it will remain a relevant text for academic and technical students of viticulture and pest management as well as for practical vineyard managers.

We would like to thank all the contributors who supported the aims and the goals of this book, and our families who supported our completion of this project. The editors express special appreciation to Gaétan Racette, Pierre Lemoyne and Michel Brouillard for their generous assistance and cooperation in proof reading and formatting this book.

Saint-Jean-sur-Richelieu and East Lansing

Noubar J. Bostanian
Charles Vincent
Rufus Isaacs

Contents

Contributors

Pia Addison Department of Conservation Ecology and Entomology, Faculty of AgriSciences, University of Stellenbosch, Matieland, South Africa

Rodrigo P.P. Almeida Department of Environmental Science, Policy and Management, University of California, Berkeley, CA, USA

Miguel Altieri Department of Environmental Science, Policy and Management, University of California, Berkeley, CA, USA

Vaughn A. Bell New Zealand Institute for Plant & Food Research Limited, Havelock North, New Zealand

Walter J. Bentley Division of Agriculture and Natural Resources, Kearney Agricultural Research and Extension Center, University of California, Parlier, CA, USA

J. Christopher Bergh Alson H. Smith, Jr. Agricultural Research and Extension Center, Virginia Polytechnic Institute and State University, Winchester, VA, USA

Noubar J. Bostanian Horticultural Research and Development Centre, Agriculture and Agri-Food Canada, Saint-Jean-sur-Richelieu, QC, Canada

Natalia Botero Neerdaels Driscoll's Strawberry Associates Inc., Watsonville, CA, USA

Marcos Botton Embrapa Grape and Wine, Bento Gonçalves, RS, Brazil

Hannah J. Burrack Department of Entomology, North Carolina State University, Research Annex West, Raleigh, NC , USA

Marco Candolfi Innovative Environmental Services (IES) Ltd, Witterswil, Switzerland

Vanessa J. Connick National Wine & Grape Industry Centre, Faculty of Science, School of Agricultural & Wine Sciences, Charles Sturt University, Wagga Wagga, NSW, Australia

Richard L. Coviello Division of Agriculture and Natural Resources, University of California, Fresno, CA, USA

Kent M. Daane Department of Environmental Science, Policy and Management, University of California, Berkeley, CA, USA

Carlo Duso Department of Environmental Agronomy and Crop Science, University of Padova, Legnaro (PD), Italy

Majid Fallahzadeh Department of Entomology, Jahrom Branch, Islamic Azad University, Jahrom, Iran

Brian Galka Saskatoon Research Centre, Agriculture and Agri-Food Canada, Saskatoon, SK, Canada

Yann Guisard National Wine & Grape Industry Centre, Faculty of Science, School of Agricultural & Wine Sciences, Charles Sturt University, Wagga Wagga, NSW, Australia

Geoff M. Gurr EH Graham Centre for Agricultural Innovation (NSW Department of Primary Industries and Charles Sturt University), Faculty of Science, School of Agricultural & Wine Sciences, Charles Sturt University, Wagga Wagga, NSW, Australia

John Michael Hardman Atlantic Food and Horticulture Research Centre, Agriculture and Agri-Food Canada, Kentville, NS, Canada

Claudio Ioriatti Centre for Technology Transfer, FEM-IASMA, San Michele all'Adige, TN, Italy

Rufus Isaacs Department of Entomology, Michigan State University, East Lansing, MI, USA

Paul E. Jenkins Department of Entomology, Michigan State University, East Lansing, MI, USA

Serge Kreiter Montpellier SupAgro, UMR CBGP, Campus International de Baillarguet, Montferrier sur Lez cedex, France

Jacques Lasnier Co-Lab R&D, Division de Ag-Cord. Inc., Granby, QC, Canada

Tracy C. Leskey USDA-ARS, Appalachian Fruit Research Station, Kearneysville, WV, USA

Greg M. Loeb Department of Entomology, New York State Agricultural Experiment Station, Cornell University, Geneva, NY, USA

Andrea Lucchi Department CDSL, Section of Agricultural Entomology, University of Pisa, Pisa, Italy

Michael Maixner Institute for Plant Protection in Fruit Crops and Viticulture, Julius Kühn-Institute, Bernkastel-Kues, Germany

M. Mani Division of Entomology and Nematology, Indian Institute of Horticultural Research, Bangalore, India

Jose Luis Miano INTA EEA Mendoza, Luján de Cuyo, Mendoza, Argentina

Albie Miles Department of Environmental Science, Policy and Management, University of California, Berkeley, CA, USA

Clara Nicholls Department of Environmental Science, Policy and Management, University of California, Berkeley, CA, USA

Chrystel Olivier Saskatoon Research Centre, Agriculture and Agri-Food Canada, Saskatoon, SK, Canada

Douglas G. Pfeiffer Department of Entomology, Virginia Tech., Blacksburg, VA, USA

Kevin S. Powell Department of Primary Industries, Biosciences Research Division, Rutherglen Centre, Rutherglen, VIC, Australia

Alberto Pozzebon Department of Environmental Agronomy and Crop Science, University of Padova, Legnaro (PD), Italy

Olivia L. Reynolds (née Kvedaras) EH Graham Centre for Agricultural Innovation (NSW Department of Primary Industries and Charles Sturt University), Elizabeth Macarthur Agricultural Institute, Woodbridge Road, Menangle, NSW 2568, Australia

Julien Saguez CoLab R&D, Division de Ag-Cord. Inc., Granby, QC, Canada

Anthony Saliba National Wine & Grape Industry Centre, Faculty of Science, School of Agricultural & Wine Sciences, Charles Sturt University, Wagga Wagga, NSW, Australia

Michael J. Samways Department of Conservation Ecology and Entomology, Faculty of AgriSciences, University of Stellenbosch, Matieland, South Africa

Michael C. Saunders Department of Entomology, Pennsylvania State University, University Park, PA, USA

René Sforza USDA-ARS-European Biological Control Laboratory, Campus International de Baillarguet, St-Gely du Fesc, France

Marja Simpson EH Graham Centre for Agricultural Innovation (NSW Department of Primary Industries and Charles Sturt University), Charles Sturt University, Leeds Parade, Orange, NSW, Australia

Luis A.F. Teixeira Product Support and Renewal, DuPont Crop Protection, Wilmington, DE, USA

Marie-Stéphane Tixier Montpellier SupAgro, UMR CBGP, Campus International de Baillarguet, Montferrier sur Lez cedex, France

Jean-Marie Tompkins Department of Ecology, Bio-Protection Research Centre, Lincoln University, Lincoln, New Zealand

Lucia G. Varela UC Cooperative Extension, Division of Agriculture and Natural Resources, University of California, Santa Rosa, CA, USA

Charles Vincent Horticultural Research and Development Centre, Agriculture and Agri-Food Canada, Saint-Jean-sur-Richelieu, QC, Canada

James T.S. Walker New Zealand Institute for Plant & Food Research Limited, Havelock North, New Zealand

Vaughn M. Walton Department of Horticulture, Oregon State University, Corvallis, OR, USA

Phyllis G. Weintraub Agricultural Research Organization, Gilat Research Center, D.N. Negev, Israel

Houston Wilson Department of Environmental Science, Policy and Management, University of California, Berkeley, CA, USA

John C. Wise Applied Insecticide Toxicology Laboratory, Michigan State University, East Lansing, MI, USA

Steve D. Wratten Department of Ecology, Bio-Protection Research Centre, Lincoln University, Lincoln, New Zealand

Tania Zaviezo Departamento de Fruticultura y Enología, Universidad Católica de Chile, Santiago, Chile

Chapter 1
Principles of Arthropod Pest Management in Vineyards

Charles Vincent, Rufus Isaacs, Noubar J. Bostanian, and Jacques Lasnier

1.1 Introduction

The first appearance of *Vitis vinifera* L. has been dated to between 130 and 200 million years ago, with the human relationship to this plant dating from the Neolithic period. Wild grapes were harvested by foragers and early farmers. For thousands of years, the berry has been harvested for both medicinal and nutritional value and its history is intimately entwined with the history of wine. Domestication of the Eurasian grape (*V. vinifera* ssp. *sativa* Hegi) from its wild ancestor (*V. vinifera* L. ssp. *sylvestris* (C. C. Gmelin) Hegi) occurred in Transcaucasia where the greatest genetic diversity is found today. Other evidence based on the study of chloroplast DNA polymorphisms indicates there has also been early domestication in the Iberian Peninsula (Arroyo-García et al. 2006).

Changes in the shapes of pips, or seeds (narrower in domesticated forms), and the distribution of grapevines, points to domestication in 4,500–5,000 BC in Armenia and Georgia. A complete archaeological picture of wine production, 6,100 years old, was unearthed recently in a cave in Armenia. This included a wine press for stomping grapes, fermentation equipment, storage vessels and drinking

C. Vincent (✉) • N.J. Bostanian
Horticultural Research and Development Centre, Agriculture and Agri-Food Canada, 430 Gouin Blvd, Saint-Jean-sur-Richelieu, QC, Canada J3B 3E6
e-mail: charles.vincent@agr.gc.ca; noubar.bostanian@agr.gc.ca

R. Isaacs
Department of Entomology, Michigan State University, East Lansing, MI 48824, USA
e-mail: isaacsr@msu.edu

J. Lasnier
Co-Lab R&D, Division de Ag-Cord. Inc., 655 Delorme, Granby, QC, Canada J2J 2H4
e-mail: jlasnier@translog.ca

N.J. Bostanian et al. (eds.), *Arthropod Management in Vineyards: Pests, Approaches, and Future Directions*, DOI 10.1007/978-94-007-4032-7_1,

cups, as well as withered grape vines and skins. This discovery was supported by chemical studies revealing the presence of the anthocyanin malvidin that confers the red color to grapes and pomegranates (Barnard et al. 2011). The Egyptians and Romans both valued grapes and grape-based products, which they spread across their empires in Europe and Africa.

With movement to the New World, Europeans brought their winemaking skills and began the expansion of viticulture across the globe. In the Americas, early attempts to establish *V. vinifera* in the colonies were hampered by pests and diseases as well as by the unsuitability of the cold climate, until Spanish missionaries planted vines in California. Relying on the wild *Vitis riparia* Michaux and *Vitis rotundifolia* Michaux species, eastern US settlers gradually adapted local genotypes to viticultural production such as *V. riparia* that is now produced commercially for grape juice. Later they developed hybrid crosses between *V. vinifera* and *Vitis labrusca* L. that combine pest resistance traits with adaptation to the climate and berry suitability for winemaking.

Nowadays, grape production is aimed at two markets, namely fresh (table) grapes and processed grapes that are produced for raisins, juice or wine. Wine making is based on two complementary activities, namely viticulture (cultivation, protection and harvesting of grapes- the 'outdoor' occupations) and oenology (fermentation of grapes into wine- the 'indoor' occupations). Over the centuries viticulture and oenology have developed into a multi-billion dollar industry (Brostrom and Brostrom 2009), historically based in Europe but now spread across many continents. American prohibition in the 1920s exerted a strong negative influence on the viticultural industry in North America. As a consequence, little research work occurred for decades in all fields related to viticulture in this continent. In recent decades, wine making has undergone tremendous growth in several parts of the world, i.e. North America (history reviewed by Pinney 2005), South America, South Africa, Australia and New Zealand. There is an ancient history of wine production in China and although this was not popular for many years, Chinese grape and wine production has increased rapidly in the past few decades.

There are a number of textbooks addressing agronomic issues of viticulture (e.g. Reynier 1997; Jackson 2008; Keller 2010). Recently, it has been shown that different management systems affect the ecological sustainability of vineyards (Abbona et al. 2007), and this is a developing issue for viticulture, as will be evident in this book.

Grapes are grown in a variety of climates and agricultural situations ranging from extreme hot and dry (e.g. Israel, Greece, Arizona) to cool-climate conditions (e.g. Canada, New Zealand, Moldavia) that confer a specificity to the final product (Dominé 2010). In each of these situations, appropriate and relevant information must be acquired in order to develop sustainable pest management programs adapted to given wine-producing areas, and the local complex of diseases and arthropod pests.

From a crop protection point of view, fungal diseases, namely powdery mildew (*Erysiphe necator* (Schweinitz)), bunch rot (*Botrytis cinerea* Persoon ex Fries) and downy mildew (*Plasmopara viticola* (Berkely & M. A. Curtis) Berlandier & de Toni) are serious concerns and major drivers of pest management programs. In several regions of the world, a suite of diseases require several fungicide sprays per season to achieve optimal vine health. However, arthropod pests also pose serious threats that must be addressed. As stated in Bentley et al. (2005), the absolute and relative importance of insects in grape production depends on the crop market (fresh versus processed grapes) and environmental conditions. Hereafter, our aim is to review the main principles related to arthropod management in vineyards, whether those are for production of fresh grapes, wine, or juice. The conceptual framework of integrated pest management (IPM) (Kogan and Hilton 2009) and most of the principles and tactics relevant to pest management of perennial crops such as apples and peaches (Aluja et al. 2009) also apply to vineyards. After briefly discussing the plant itself, we will address entomological issues encountered in vineyards.

1.2 The Vine

Grape vines have phenological stages that offer cues for the timing of arthropod activity and the timing of pesticide treatments or other interventions (Fig. 1.1). In Europe these phenological stages are designated with different systems, notably the Baggiolini (letters from A to P), Eichhorn (22 stages denoted by numbers from 01 to 47) and BBCH (stages denoted by numbers from 01 to 100) (Bloesch and Viret 2008). In contrast to most tree crops, vines have indeterminate growth (i.e., continuous production of meristems during the growing season). This feature results in continuous availability of tender tissues during the growing season, which may favor the residence of some insect populations. Flower buds are coated with pectin that may provide food for some insects, e.g. the tarnished plant bug, *Lygus lineolaris* Palisot de Beauvois (Fleury et al. 2006). Likewise, depending on cultivars, phloem exudate is abundant in sugars (sucrose, glucose and fructose) and amino acids that may provide food for other arthropods (Gholami et al. 2004). Finally, in the course of berry development cell division and cell expansion occur before the veraison stage and, at the inception of ripening (after veraison), cell expansion is accompanied by increases in sugar content, fruit softness, color and flavor (Fig. 1.2). These changes in metabolism are coordinated within each individual berry, but are not synchronized within a given cluster, which results in uneven levels of maturity and size within berry clusters, providing berry resources for a rather long period during the growing season. Altogether, these features create opportunities for some arthropods to thrive.

1.3 Arthropod Biodiversity in Vineyards

The literature concerning arthropods associated with vineyards consists of ca. 1,000 scientific papers from 1972 to 2010. Historically, research in vineyards focused on arthropod pests, and consequently, relatively little is known about the biodiversity

Fig. 1.1 Main phenological stages of the vine plant. The numbers and the letters refer respectively to the Eichhorn and Lorenz, and the Baggiolini systems (After Carisse et al. (2009), with permission of the author)

of arthropods in unmanaged or lightly managed vineyards. Notable exceptions are the biodiversity studies conducted systematically in the cool-climate vineyards of Quebec (Vincent et al. 2009). Earlier, Bostanian et al. (2003) reported 60 cicadellid species, several of which are present in relatively low numerical importance. However, as some of these species are vectors of phytoplasma diseases (Olivier et al., Chap. 11), they may have an important economic impact despite their low numbers. In the same vineyards, Goulet et al. (2004) found 124 carabid species,

Fig. 1.1 continued

while Bolduc et al. (2005) found 97 spider species, Bouchard et al. (2005) reported 73 species of curculionids, Lucas et al. (2007a) reported 20 species of coccinellids, and Lesage et al. (2008) reported 59 species of chrysomelids. This high level of arthropod biodiversity demonstrates the potentially rich array of natural enemies that may be conserved in vineyard systems for biological control of pests. Such information should prove to be useful for the development of strategies to manage vineyards with relatively little use of broad-spectrum insecticides, such as in organic or biodynamic vineyards.

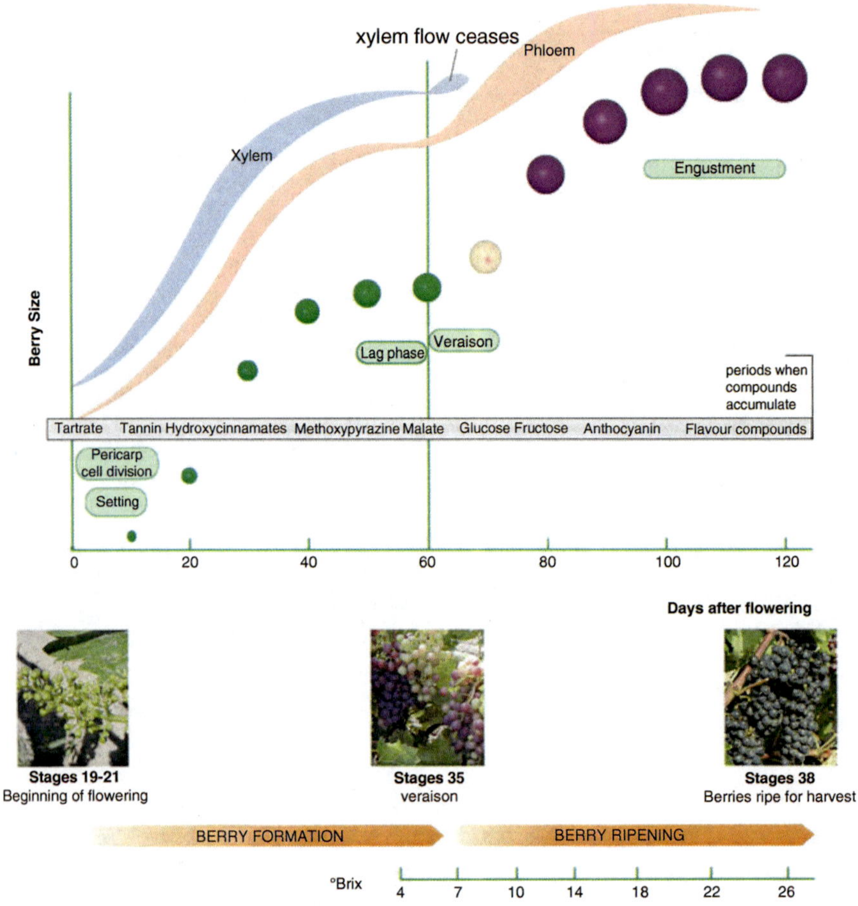

Fig. 1.2 Grape berry development and major physiological events (Redrawn after Kennedy (2002), with kind permission of Wine and Viticulture Journal and the author)

1.4 Arthropod Pests

In his review of grape insects, Bournier (1976) produced a thorough list of insects related to the parts of grape vines attacked, i.e., roots, wood trunk, shoots, buds, very young shoots, berries and leaves. He also included miscellaneous classes, such as gall makers, honeydew producers, aerial polyphagous insects and soil insects. That list, which also reported the main geographical regions where the pests are found, has been reprinted in Bentley et al. (2005) who stated that approximately 150 arthropod species are considered pests of vineyards worldwide. Arthropods of vineyards, with particular focus on France, have also been reviewed by Esmenjaud et al. (2008).

Damage caused by arthropods fall into two categories: irreversible and reversible. An example of an insect causing irreversible damage is the glassy-winged sharp-shooter, *Homalodisca vitripennis* (German) (formerly known as *H. coagulata* (Say)), vector of the bacterium *Xylella fastidiosa* Wells et al. which is the causal agent of Pierce's disease. Another important group is the cicadellids that vector phytoplasmas. Once phytoplasmas have entered the plant, only thermotherapy (Mannini 2007) can appropriately manage them. As this is impractical in the field, phytoplasma-positive vines are generally up-rooted and destroyed so as to avoid further spread of phytoplasmas in the vineyard (Olivier et al., Chap. 11).

As most other damage caused by arthropods is reversible, management techniques can be implemented at levels depending on the severity of the threat relative to the cost of treatments (Isaacs et al., Chap. 2). For example, direct damage caused by insects or mites feeding on the foliage can be addressed with insecticides or acaricides, respectively. In this book, major arthropod pests of vineyards are addressed: phytophagous mites (Duso et al., Chap. 9), phylloxera (Powell, Chap. 10), leafhoppers (Olivier et al., Chap. 11), mealybugs (Daane et al., Chap. 12), and grape berry moths (Ioriatti et al., Chap. 14; Isaacs et al., Chap. 15).

1.5 New Arthropod Issues

Since the review by Bournier (1976), several new entomological problems have arisen in vineyards, notably because of increases in acreage and commercial exchanges worldwide. For example, several cicadellid species are now considered economically important pests because they are vectors of phytoplasma diseases (Olivier et al., Chap. 11). Likewise, in California, the importance of hopperburn, a noncontagious symptom of plants caused by the direct feeding damage of certain leafhoppers and planthoppers (Backus et al. 2005), has increased dramatically.

As viticultural activities are steadily developing worldwide, new entomological problems arise in vineyards from time to time. The multicolored Asian lady-beetle, *Harmonia axyridis* (Pallas), is a case in point (Lucas et al. 2007a, b; Pfeiffer et al., Chap. 19). Originally introduced as a biocontrol agent in southern USA, it became problematic in vineyards in the late 1990s as large populations were found at harvest time in vineyards of North America, notably Ohio, Ontario and Quebec (Lucas et al. 2007b). Upon harvest of grapes, crushed adult beetles release a number of methoxypyrazines, mainly 2-isopropyl-3-methoxypyrazine (IPMP) that, in very low concentrations, taints the wine (Pickering et al. 2005, 2006). It has been hypothesized that problems related to *H. axyridis* in vineyards were caused by the unusual abundance of soybean aphids in the Great Lakes region. Other examples of new pests threatening vineyards are discussed by Pfeiffer et al. (Chap. 19).

1.6 The Foundation of IPM: Sampling, Thresholds, and Modeling

Without an understanding of how many insects or mites are present in a vineyard, the manager cannot make informed decisions regarding the need (or not) to control pests. This issue is central to integrated pest management, and the development of sampling schemes that provide the required precision to determine pest abundance has occupied applied entomologists ever since the birth of IPM in the 1950s.

As discussed by Isaacs et al. (Chap. 2), the development of thresholds for use by vineyard managers based on knowledge of pest injury potential, crop value, and control tactic costs lags behind that of many other crops. This is in part a result of the challenges of setting thresholds for a crop that has a high capacity for compensatory growth and exhibits significant tolerance for injury from many pests, and where cosmetic damage to the berry may be irrelevant. Still, some key vineyard pests have sampling plans and action thresholds that are used to make control decisions, and these can support accurate decision-making in IPM programs. Development of these quantitative tools needs to take into account the inter-annual variability in the price of grapes, crop yield, and pest infestation. There are also diseases and other arthropods that may affect the vine tolerance to pests, but currently there is little information on how to adjust thresholds based on the presence of additional vine stresses, whether those are biotic or abiotic.

Issues related to modeling and prediction of pest development are addressed by Hardman (Chap. 3). Modeling can be a useful tool to enable accurate timing of insecticides for pest control, and can optimize resources in some situations. For example Bostanian et al. (2006) developed a model to optimize sampling efforts for cicadellids in cool-climate vineyards.

1.7 Alternatives to Chemical Control

1.7.1 Cultural Control

Historically, one of the most economically significant insect pests of vineyards has been the grape phylloxera. Resistant rootstocks have controlled grape phylloxera for more than 130 years (Granett et al. 2001). It is noteworthy that, in this case, insecticides have proved to be an inefficient means of control of this pest, and the use of resistant rootstocks has been highly effective (Powell, Chap. 10). Many other vineyard arthropods show varying levels of activity based on vine genetics and there is significant variation in vine susceptibility among grape cultivars and species. Knowledge of pest susceptibility is often not considered during the selection of cultivars for planting, but it can help vineyard managers understand where to focus pest management activities.

1.7.2 Biological Control

Flaherty and Wilson (1999) reviewed biological control of arthropods on grapes relevant to the taxa Homoptera (Cicadellidae, Pseudococcidae, Coccidae, Phylloxeridae), Lepidoptera (Tortricidae, Zygaenidae, Pyralidae, Sesiidae, Heliozelidae), Coleoptera (Curculionidae, Bostrichidae), Thysanoptera (Thripidae) and Acari (Tetranychidae, Tenuipalpidae, Eriopyidae). They concluded that most research efforts in the world have been directed to address the disruption of secondary pests by pesticides and that there are relatively few studies on biological control of primary pests. Hence, more efforts should be channelled to manage primary pests by native as well as imported biocontrol agents. Likewise, Mills and Daane (2005) stated that in 50 years of efforts to develop biological control in California vineyards, there have been few successes in classical biological control. It is possible that insecticide use will decrease as the demand for organic or sustainably-grown grapes increases and as there are further regulatory restrictions on pesticides. This would open new opportunities for alternative technologies, including biological control.

Different approaches to promote biological control in vineyards are addressed in this book, such as the use of natural enemies and pathogens (Walton et al., Chap. 5), ecosystem services (Tompkins et al., Chap. 7), and management of habitat diversity (Miles et al., Chap. 8). The last two authors discuss the concept of ecological engineering whereby the ecosystem across a farm is modified to enhance the delivery of biological control to vineyards.

Currently the most widespread and successful tactic for biological control is the conservation, augmentation, and dissemination of predatory mites to manage phytophagous mites. This technique is based on the premise that mite resurgence occurs only in response to the misuse of pesticides that kill predatory mites. Actually, phytophagous mites never attain pest status in wild grapes and are uncommon in vineyards practicing integrated pest management approaches because they are controlled by natural enemies. Therefore, by using pesticides that are effective against a particular pest(s) and at the same time innocuous to predatory mites, biological control of the phytophagous mite species can take place while the pesticide controls the targeted pest(s). Bostanian et al. (Chap. 4) discuss the fine tuning of the latest method to measure such effects with reduced-risk insecticides in a two tier evaluation program. Duso et al. (Chap. 9) discuss phytophagous mites and the International Organization for Biological Control recommendations currently used for managing mites in Europe.

1.7.3 Physical Control

Physical control methods encompass a number of technologies. Thermotherapy of vine plants is a recognized method to kill phytoplasmas in living vines before their exportation (Mannini 2007) (Olivier et al., Chap. 11).

Blua et al. (2005) investigated the effect of deploying 5-m-high screen barrier surrounding vineyards to exclude the sharpshooter *H. vitripennis*. When *H. vitripennis* were placed near or on the barrier, 6% flew over it. Such a tactic is compatible with insecticide applications, biological control, and is consistent with area-wide management strategies. However, in large commercial plantings, such physical barriers can be extremely costly.

The management of irrigation can control some phytophagous insects such as the variegated leafhopper, *Erythroneura variabilis* Beamer (Daane and Williams 2003). In California, there is a positive linear relationship between the volume of water applied to vines and the leafhopper density in August and September. Clearly, vines with lower vigor support smaller leafhopper populations, because of lower fecundity and less adult immigration. In Brazil, Chaves et al. (2007) demonstrated that, in regions affected by seasonal drought, grapevine irrigation can be an important practice to guarantee wine quality or even for vine survival. Water has to be managed to optimize source to sink balance and avoid excessive vine vigor. Therefore, in dry production regions irrigation management has to be used carefully to achieve an optimal balance between plant vigor and insect management objectives.

1.8 Semiochemicals

In vineyards, the most advanced uses of semiochemicals are for monitoring and mating disruption of pests such as grape berry moths. Use of synthetic sex pheromone dispensers is widespread in European vineyards for mating disruption of these pests, and is becoming more common in regions where these species have recently been introduced. This biologically-based approach to grape berry moth control is covered for European and North American species by Ioriatti et al. (Chap. 14) and Isaacs et al. (Chap. 15), respectively. Mating disruption also has potential for use against mealybugs (Daane et al., Chap. 12).

Host plant volatiles can also be employed in IPM programs, either for monitoring or for pest management. Recent investigations of the role of these semiochemicals in moth attraction to grape vines are revealing the complexities of chemical interactions among vines and their insect pests (Tasin et al. 2005; Cha et al. 2008a, b, 2011), and these studies may lead to refined tools for pest monitoring in the future. Attraction of natural enemies to vines using host plant volatiles, either deployed in dispensers or stimulated in the plant by application of an elicitor, is now being explored in combination with providing habitat for these insects in and around vineyards. Recent progress towards this 'attract and reward' strategy is reviewed by Simpson et al. (Chap. 6).

1.9 Chemical Control

Insecticides are widely used in viticulture and they remain important components of arthropod management programs. Bostanian et al. (Chap. 4) give an overview of conventional and reduced-risk insecticides and acaricides. In that chapter, these products are classified according to their mode of action, with comments on their effectiveness on the target arthropod as well as non-target predatory mites and other beneficials whenever the information is available. They also discuss such topics as polarity of insecticides, formulations, synergism, rain fastness, insecticide resistance and its management.

Bacillus thuringiensis Berliner is the main biopesticide used in vineyards. It is targeted primarily at lepidopteran larvae. Other biopesticides including living organisms such as fungi, nematodes, and viruses are used much less often (Walton et al., Chap. 5). Currently, few of these biopesticides are registered for use in viticulture, and they pose some challenges for effective use. Overall, they are unstable to heat, UV radiation, or desiccation. They are also slow in action when compared with conventional pesticides and have a short shelf life. Nevertheless, with the expansion of organic viticulture around the world, their use is expected to increase in the future.

1.10 Regulations

International movement of grapevine material and harvested grapes is increasingly common, and several countries and states enforce quarantine legislation to prevent the unwanted movement of grape pests. As an example, Olivier et al. (Chap. 11) describe legislation regarding *Flavescence Dorée*, a vine disease vectored by a cicadellid. The implications of having failed to prevent pest movements can be seen in the economic impact of grape phylloxera in vineyards (Powell, Chap. 10) and more recently in the arrival of *Lobesia botrana* (Denis & Schiffermüller) in the New World (Ioriatti et al., Chap. 14).

Many countries are also making pesticide regulations more restrictive, as they phase out the use of broad-spectrum insecticides and acaricides. While some regions of viticulture have been at the forefront of developing advanced IPM systems over the years, these regulatory changes are helping to force the transition of whole grape industries toward advanced IPM tactics relying less on chemical inputs and more on biological processes.

1.11 IPM in Organic Viticulture

The main characteristic of organic agriculture is the avoidance of synthetic inputs as described within organic certification programs. The details vary from one certification organization to another but they share the same philosophy of minimizing the impact of agriculture on the environment. As stated by Madge (2005), organic viticulture is a more holistic approach to management, not the standard approach based on selecting different inputs that respond to different problems. In terms of protection programs, organic certification implies that no synthetic pesticides are used. Such constraints open the door to alternative management methods. In theory, the pest community found in organic vineyards should be similar to conventional vineyards, but with a higher level of biological control. Little has been published on the implications of organic viticulture for IPM programs based on scientific experimentation, but there are some IPM components such as habitat manipulation that have been tested within organic vineyards (e.g. Nicholls et al. 2000) where chemical inputs were minimal. According to Madge (2005), quarantine and hygiene are the cornerstones of organic viticulture to prevent pests and diseases from accidentally entering the vineyard. These practices should also be important in conventional viticulture.

1.12 IPM Program Delivery

The translation of scientific knowledge into practical IPM programs is an everlasting and important challenge. Grape pest researchers from around the world are responding to this challenge, as shown in this book. Several countries or states have made considerable investments in employing people to aid in this dissemination of knowledge from the research community into the grower community. Such extension services can provide unbiased information to help growers make appropriate management decisions within their IPM programs (Flaherty et al. 1992), and this effort has been shown to pay off in terms of reduced grower costs, less use of pesticides, and a decrease in environmental contamination. However, with the future of these programs unclear in some regions where financial resources are limited and with greater access to technology among farmers or their advisors, information delivery is moving from a more personal approach to increasing use of electronic delivery. The internet allows the circulation of information at very low cost. For instance, much of the classic publication 'Grape Pest Management in California' (Flaherty et al. 1992) is now available free online at UC IPM Online (2011a). IPM guidelines are also available on the web, e.g. UC IPM Online (2011b), OMAFRA (2011). The Low Input Viticulture and Enology (LIVE 2011) is another example of resources available to persons interested in alternatives. Many regions of grape production also have locally-relevant and timely pest newsletters that are distributed during the growing season to provide growers with news on which pests are active, and with pest alerts and reminders on their IPM options available for pest management, e.g. Michigan State University Extension (2011). Finally, field guides for scouting are useful tools to provide rough identification of pests in a vineyard: an example is Isaacs et al. (2011), a pocket scouting guide covering vineyards of the north central and eastern United States, that is available in English and Spanish.

1.13 Challenges

Challenges posed by arthropods differ depending on localities. In dry areas such as California, Italy and southern France, phytophagous mites and mealybugs are driving the IPM systems. The major challenge of cool-climate viticulture is the climate limitations, as discussed by Lasserre (2001). Because cool-climate viticulture occurs at the edge of the agronomic possibilities, the choice of cultivars and the economics (including research funding) are limited. From a crop protection point of view, diseases are often much more important than entomological problems in these regions, as reflected in the number of fungicide sprays applied each year against downy mildew (*P. viticola*), powdery mildew (*E. necator*) and bunch rot (*B. cinerea*). As agricultural lands devoted to viticulture are relatively small in cool-climate regions, the build-up of entomological problems do not occur at the same pace and intensity compared to well established vine-growing regions.

As argued for the need for microbial insecticides in temperate orchards by Lacey and Shapiro-Ilan (2008), there are needs for biopesticides in vineyards. When compared with chemical insecticides there is currently much less use of microbial insecticides, and this area has great potential if the efficacy is high enough and the cost competitive. Likewise, relatively little has been published concerning the use of botanicals in vineyards. Gökçe et al. (2011) determined the ovicidal, larvicidal and anti-ovipositional activities of ethanol extracts that contained phenolics, terpenoids and alkaloids from four native plants from Turkey on grape berry moth, *Paralobesia viteana* (Clemens). Of the four plants, *Bifora radians* has the greatest potential for further development because it showed high ingestion and ovicidal activity, as well as anti-oviposition activity. However, as it is the case for all botanicals, it will take several years to be developed and registered because a registrant must document its field efficacy and its innocuity in order to comply with the regulations (Regnault-Roger et al. 2012).

1.14 Conclusion

In viticultural settings, the main challenge for integrated pest management remains the development and coordination of all information and technologies into a package that is optimally relevant to growers in a given area. Only dedicated research efforts such as those described in the pages of this book can provide the tools to achieve this goal.

References

Abbona EA, Sarandón SJ, Marasas ME, Astier M (2007) Ecological sustainability evaluation of traditional management in different vineyard systems in Berisso, Argentina. Agric Ecosyst Environ 119:335–345

Aluja M, Leskey TC, Vincent C (2009) Biorational tree-fruit pest management. CABI Publishing, Wallingford

Arroyo-García R, Ruiz-García L, Bolling L, Ocete R, Lopez MA, Arnold C, Ergul A et al (2006) Multiple origins of cultivated grapevine (*Vitis vinifera* L. ssp. *sativa*) based on chloroplast DNA polymorphisms. Mol Ecol 15:3707–3714

Backus EA, Serrano MS, Ranger CM (2005) Mechanisms of hopperburn: an overview of insect taxonomy, behavior, and physiology. Annu Rev Entomol 50:125–151

Barnard H, Dooley AN, Areshian G, Gasparyan B, Faull KF (2011) Chemical evidence for wine production around 4000 BCE in the Late Chalcolithic near Eastern Highlands. J Archaeol Sci 38:977–984

Bentley WJ, Varela L, Daane KM (2005) Grapes, insects ecology and control. In: Pimentel D (ed) Encyclopedia of pest management. Taylor & Francis, New York, pp 1–8. doi:10.1081/E-EPM-120041132

Bloesch B, Viret O (2008) Stades phénologiques repères de la vigne. Rev Suisse Vitic Arboric Hortic 40:1–4

Blua MJ, Campbell K, Morgan DJW, Redak RA (2005) Impact of a screen barrier on dispersion behaviour of *Homalodisca coagulata* (Hemiptera: Cicadellidae). J Econ Entomol 98:1664–1668

Bolduc E, Buddle CM, Bostanian NJ, Vincent C (2005) The ground-dwelling spiders (Aranae) of two vineyards in southern Quebec. Environ Entomol 34:635–645

Bostanian NJ, Vincent C, Goulet G, LeSage L, Lasnier J, Bellemare J, Mauffette Y (2003) The arthropod fauna of Quebec vineyards, with particular reference to phytophagous species. J Econ Entomol 96:1221–1229

Bostanian NJ, Bourgeois G, Vincent C, Plouffe D, Trudeau M, Lasnier J (2006) The population dynamics of leafhoppers in Quebec vineyards. Environ Entomol 35:1477–1482

Bouchard P, Lesage L, Goulet H, Bostanian NJ, Vincent C, Zmudzinska A, Lasnier J (2005) Weevil (Coleoptera: Curculionoidea) diversity and abundance in Quebec vineyards. Ann Entomol Soc Am 98:565–574

Bournier A (1976) Grape insects. Annu Rev Entomol 22:355–376

Brostrom G, Brostrom J (2009) The business of wine, an encyclopedia. Greenwood Press, Westport

Carisse O, Bacon R, Lasnier J, Lefebvre A, Levasseur A, Rolland D, Jobin T (2009) Grape disease management in Quebec. Agriculture and Agri-Food Canada Catalogue Number A52-146/2009E-PDF, Ottawa

Cha DH, Nojima S, Hesler SP, Zhang A, Linn CE Jr, Roelofs WL, Loeb GM (2008a) Identification and field evaluation of grape shoot volatiles attractive to female grape berry moth (*Paralobesia viteana*). J Chem Ecol 34:1180–1189

Cha DH, Hesler SP, Moser CL, Nojima S, Linn CE Jr, Roelofs WL, Loeb GM (2008b) Flight tunnel responses of female grape berry moth (*Paralobesia viteana*) to host plants. J Chem Ecol 34:622–627

Cha DH, Linn CE Jr, Teal PEA, Zhang A, Roelofs WL, Loeb GM (2011) Eavesdropping on plant volatiles by a specialist moth: significance of ratio and concentration. PLoS One 6:e17033. doi:10.1371/journal.pone.0017033

Chaves MM, Santos TP, Souza CR, Ortuño MF, Rodrigues ML, Lopes CM et al (2007) Deficit irrigation in grapevine improves water-use efficiency while controlling vigour and production quality. Ann Appl Biol 150:237–252

Daane KM, Williams LE (2003) Manipulating vineyards irrigation amounts to reduce insect pest damage. Ecol Appl 13:1650–1666

Dominé A (2010) Wine. h.f.ullmann, Potsdam

Esmenjaud D, Kreiter S, Martinez M, Sforza R, Thiéry D, Van Helden M, Yvon M (2008) Les ravageurs de la vigne, 2nd edn. Féret, Bordeaux

Flaherty DL, Wilson LT (1999) Biological control of insects and mites on grapes. In: Bellows TS, Fisher TW (eds) Handbook of biological control. Academic, San Diego, pp 853–869

Flaherty DL, Christensen LP, Lanini WT, Marois JJ, Philips PA, Wilson LT (1992) Grape pest management, 2nd edn. Publication 3343. University of California, Division of Agriculture and Natural Resource, Oakland

Fleury D, Paré J, Vincent C, Mauffette Y (2006) Feeding impact of *Lygus lineolaris* (Heteroptera: Miridae) on *Vitis vinifera*: a behavioural and histological study. Can J Bot 84:493–500

Gholami M, Coombe SG, Robinson SR (2004) Grapevine phloem sap analysis: 1-sucrose, amino acids, potassium concentrations, seasonal and diurnal patterns. Acta Hortic 640:143–153

Gökçe A, Isaacs R, Whalon ME (2011) Ovicidal, larvicidal and anti-ovipositional activities of *Bifora radians* and other plant extracts on the grape berry moth *Paralobesia viteana* (Clemens). J Pest Sci. doi:10.1007/s10340-011-0368-z

Goulet H, LeSage L, Bostanian NJ, Vincent C, Lasnier J (2004) Diversity and seasonal activity of ground beetles (Coleoptera: Carabidae) from two vineyards in southern Quebec. Ann Entomol Soc Am 97:1263–1272

Granett J, Walker MA, Kocsis L, Omer AD (2001) Biology and management of grape phylloxera. Annu Rev Entomol 46:387–412

Isaacs R, Schilder AMC, Zabadal T, Weigle T (2011) A pocket guide for grape IPM scouting in the north central and eastern U.S. Mich State Univ Ext Bull 2889, pp 110

Jackson RS (2008) Wine science, principles and applications, 3rd edn. Elsevier, Boston

Keller M (2010) The science of grapevines: anatomy and physiology. Elsevier, Boston

Kennedy J (2002) Understanding grape berry development. Practical Winery and Vineyards (July/August 2022). http://www.practicalwinery.com/julyaugust02/julaug02p14.htm

Kogan M, Hilton RJ (2009) Conceptual framework for integrated pest management (IPM) of tree-fruit pests. In: Aluja M, Leskey TC, Vincent C (eds) Biorational tree-fruit pest management. CABI, Wallingford, pp 1–31

Lacey LA, Shapiro-Ilan DI (2008) Microbial control of insect pests in temperate orchard systems: potential for incorporation into IPM. Annu Rev Entomol 53:121–144

Lasserre F (2001) L'essor du vignoble au Québec. Histoire de climats et de goûts. Cybergeo Eur J Geogr Politi Cult Représent Doc 190. http://www.cybergeo.eu/index3747.html

Lesage L, Bouchard P, Goulet H (2008) Leaf beetle diversity and abundance in two Quebec vineyards (Coleoptera, Chrysomelidae). Nouv Rev Entomol 25:3–16

LIVE (2011) Low Input Viticulture and Enology, inc. http://liveinc.org

Lucas E, Vincent C, Labrie G, Chouinard G, Fournier F, Pelletier F et al (2007a) The multicolored Asian ladybeetle *Harmonia axyridis* (Coleoptera: Coccinellidae) in Quebec agroecosystems ten years after its arrival. Eur J Entomol 104:737–743

Lucas E, Labrie G, Vincent C, Kovach J (2007b) The multicoloured Asian ladybeetle *Harmonia axyridis* – beneficial or nuisance organism? In: Vincent C, Goettel M, Lazarovits G (eds) Biological control: a global perspective. Case histories from around the world. CABI Publishing, Wallingford, pp 38–52

Madge D (2005) Organic viticulture: an Australian manual. Department of Primary Industries, Mildura, Victoria, Australia. http://www.dpi.vic.gov.au/agriculture/farming-management/organic-farming/organic-viticulture/organic-viticulture-manual

Mannini F (2007) Hot water treatment and field coverage of mother plant vineyards to prevent propagation material from phytoplasma infections. Bull Insectol 60:311–312

Michigan State University Extension (2011) AgBioResearch. http://www.grapes.msu.edu

Mills NJ, Daane KM (2005) Biological and cultural controls. Nonpesticide alternatives can suppress crop pests. Calif Agric 59:23–28

Nicholls CI, Parrella MP, Altieri MA (2000) Reducing the abundance of leafhoppers and thrips in a northern California organic vineyard through maintenance of full season floral diversity with summer cover crops. Agric For Entomol 2:107–113

OMAFRA (Ontario Ministry of Agriculture, Food and Rural Affairs) (2011) Ontario grape IPM. http://www.omafra.gov.on.ca/IPM/english/grapes/index.html

Pickering GJ, Lin Y, Reynolds A, Soleas G, Riesen R, Brindle I (2005) The influence of *Harmonia axyridis* on wine composition and aging. J Food Sci 70:128–135

Pickering GJ, Lin Y, Reynolds A, Soleas G, Riesen R, Brindle I (2006) The evaluation of remedial treatments for wine affected by *Harmonia axyridis*. Int J Food Sci Technol 41:77–86

Pinney T (2005) A history of wine in America: from prohibition to the present. University of California Press, Berkeley

Regnault-Roger C, Vincent C, Arnason JT (2012) Essential oils in insect control: low risk products in a high stakes world. Annu Rev Entomol 57:405–424

Reynier A (1997) Manuel de Viticulture, 7th edn. Lavoisier Tech et Doc, Paris

Tasin M, Anfora G, Ioriatti C, Carlin S, de Christofar A, Schmidt S et al (2005) Antennal and behavioural responses of grapevine moth *Lobesia botrana* females to volatiles from grapevine. J Chem Ecol 31:77–87

UC IPM Online (2011a) How to manage pest, grape, statewide IPM program, Agriculture and Natural Resources, University of California. http://www.ipm.ucdavis.edu/PMG/selectnewpest.grapes.html

UC IPM Online (2011b) UC IPM pest management guidelines: grape. Publication 3448. Agriculture and Natural Resources, University of California. http://www.ipm.ucdavis.edu/PDF/PMG/index.html

Vincent C, Bostanian NJ, Lasnier J (2009) Biodiversity and management of arthropods in cool-climate vineyards. In: Proceedings of the 2nd International Conference on northern Viticulture, Saint-Hyacinthe, QC, Canada, pp 189–199, 9–11 November 2009. http://www.eduportfolio.org/6644

Chapter 2
Pest Thresholds: Their Development and Use in Vineyards for Arthropod Management

Rufus Isaacs, Michael C. Saunders, and Noubar J. Bostanian

2.1 Introduction

Thresholds provide a quantitative basis upon which crop managers can decide whether arthropod pest populations are below, at, or exceeding a level that warrants the expense of activities to reduce the pest's density. These interventions may be cultural, biological, or chemical control practices that reduce the pest population below the economic threshold. Thresholds are an essential component of an IPM program, and their use can lead to significant reduction in pesticides applied to crops and lower costs of production for farmers (Pedigo et al. 1986).

The development, validation, and implementation of economic thresholds have been reported for a wide range of destructive arthropods that affect crop systems, as reviewed by Stern (1973) and Pedigo et al. (1986), although these reviews contained little discussion of vineyard systems. Economic thresholds have been developed for some key grape pests, but there is significant need for further development and refinement of thresholds for arthropod pests of vineyards, as well as a need for dissemination of the information and education about their use.

R. Isaacs (✉)
Department of Entomology, Michigan State University, East Lansing, MI 48824, USA
e-mail: isaacsr@msu.edu

M.C. Saunders
Department of Entomology, Pennsylvania State University,
University Park, PA 16802, USA
e-mail: mcs5@psu.edu

N.J. Bostanian
Horticulture Research and Development Center, Agriculture and Agri-Food Canada,
430 Gouin Blvd, Saint-Jean-sur-Richelieu, QC, Canada J3B 3E6
e-mail: noubar.bostanian@agr.gc.ca

N.J. Bostanian et al. (eds.), *Arthropod Management in Vineyards: Pests, Approaches, and Future Directions*, DOI 10.1007/978-94-007-4032-7_2,

Here, we discuss the theoretical basis of economic thresholds and the unique challenges inherent in developing and implementing threshold-based pest management in vineyards. A series of examples are provided to highlight how vine thresholds to arthropod pest injury have been developed, and we give examples of different approaches that can be employed for using thresholds in vineyard IPM programs.

2.1.1 Thresholds in Arthropod Pest Management

This field of science began when Stern et al. (1959) proposed the concept of integrated control wherein pesticides were to be used for management of pest densities only when natural mortality factors such as biological control or host plant resistance were insufficient for their control. One of their important insights was to formalize the concept that the expense of pest control was warranted only when the value of crop losses exceeded the pesticide and application costs. The Integrated Control Concept stresses integration of natural control strategies along with pesticides and it also gave birth to the idea of economic thresholds. Further investigations by Pedigo and colleagues (e.g. Pedigo et al. 1977, 1986; Poston et al. 1983) continued these ideas by formalizing definitions and by developing the mathematical framework for calculating economic thresholds.

Yield or quality loss assessment data are essential for developing economic thresholds, because they are the means by which an insect is judged a pest, they are the final criteria by which the efficacy of control measures are evaluated, and because they form the basis for decision-making in insect pest management programs. Adoption of threshold-based pest management can also allow time for natural controls such as parasitic wasps or predatory mites to feed on the pest, and maintain it below the threshold.

Despite their importance for full implementation of integrated pest management programs, relatively few studies have determined the quantitative relationships between pest density and yield loss or crop damage. There are many instances where locally-developed 'rules-of-thumb' are employed to provide working thresholds, but in the absence of thorough investigation and with a high per-hectare value to grape crops, these are often overly conservative. The development of research-based and validated pest thresholds is an area where further research effort should be directed. This is especially important for the newer regions of grape production where novel arthropod-vine interactions are developing in which the level of economic impact is not known, and where managers are likely to take a cautious approach in the absence of formal pest thresholds and associated decision tools. Still, established regions of grape production would also benefit from further refinement of thresholds or testing them under a broader range of conditions.

2.1.2 Relationships Between Pest Density and Crop Damage

Increasing density of a pest arthropod on a crop will eventually cause sufficient damage to result in yield loss and hence a decrease in income. The form of the

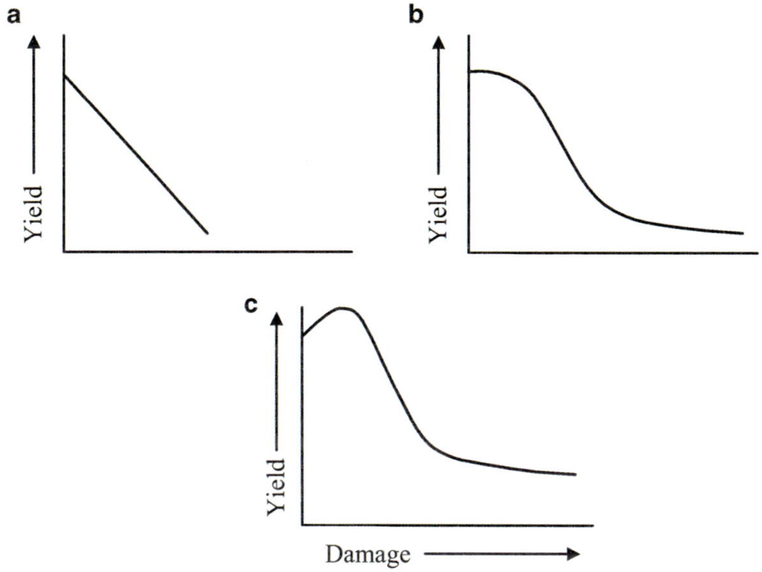

Fig. 2.1 A schematic representation of plant responses to insect damage: (**a**) Susceptive response, (**b**) Tolerant response, (**c**) Overcompensatory response (Adapted from Poston et al. 1983)

relationship between pest density and crop yield typically falls into one of the following three categories as described by Poston et al. (1983):

1. Susceptive response (Fig. 2.1a) whereby the pest causes direct damage and yield declines in direct proportion to the number of insects present. This is usually seen in insects that attack seeds. The total number of seeds damaged will be the product of the total number of seeds consumed during the lifespan of an insect and the total number of insects present per plant. In viticulture, many direct-feeding insect pests of grape clusters fall into this category, such as the various moth species, the larvae of which infest grape berries.
2. Tolerant response (Fig. 2.1b) whereby the plant can tolerate a certain pest density before yield is adversely affected. Once over this tolerance level, yield declines rapidly with increasing insect density, similar to the susceptive response. This is usually seen in insects that attack foliage or roots where a certain level of damage can be tolerated before yield is affected. The density of a pest that corresponds to this tolerance level has been designated 'tolerance limit' by Seinhorst (1965), 'threshold level' by Bardner and Fletcher (1974), 'damage boundary' (Pedigo et al. 1986) and 'carrying capacity' by Mailloux and Bostanian (1988). This type of response is common for foliage feeding insects that attack grapevines, wherein low levels of leaf feeding may have no effect on yield, but once the photosynthetic capacity of a vine is compromised, its performance declines rapidly.
3. Overcompensatory response (Fig. 2.1c) whereby the plant reacts to the presence of damage in such a manner that yield is actually increased above that which would have been achieved in the absence of the pest. This response is mostly

limited to early infestations and low levels of damage, such that damage greater than that causing overcompensation will eventually reduce yield. In addition to the level of injury, plant phenology and environmental conditions influence the ability of a plant to compensate (Bardner and Fletcher 1974). Examples of this type of effect on yield are most common in annual crops, whereas it is less likely for perennial crops in which fruit buds are set for the following year based on current year conditions.

2.1.3 The Relationship Between Injury, Yield Loss, and Revenue

Southwood and Norton (1973) generalized the relationship between pest density, crop damage, crop yield, crop price and revenue for insects attacking foliage, roots or crop product. In graphic form they showed that the amount of damage is linearly related to pest density (Fig. 2.2a). When insects damage foliage and roots, and these are not the crop product, then at low densities the price obtained for the crop would remain high and begin to decrease when pest density increases (Fig. 2.2b). The effect on price of damage to the crop product would be similar in form to the relationship between pest density and yield (Fig. 2.2c), whereas the effect on revenue would be a more extreme form of the pest-yield curve (Fig. 2.2d).

These response relationships are appropriate for pests where their density is proportional to damage of the harvested crop, such as moth larvae infesting fruit clusters. However, in some situations grape pest contamination can cause catastrophic loss of revenue where detection of the contaminant causes the harvested fruit to change from being accepted to rejected by a buyer/processor. For example, if larval infestations of grape berry moth in eastern US vineyards are high enough, then grape loads may fail the inspection at a juice processing plant and lead to complete loss of income after all the expenses for production have already been made (Hoffman et al. 1992). Another example is the presence of spiders in grape clusters destined for export. Despite their importance as biological control agents of key vineyard pests (e.g. Hanna et al. 2003), black widow spiders (*Latrodectus hesperus* Chamberlin & Ivie) are listed among the 14 major arthropod pests of California grapes because this arthropod may be present in grape clusters at harvest time (Bentley 2009). The threshold is essentially zero for this pest because fresh grapes exported to countries where this spider is not native is unacceptable by the buyers and detection of a single spider would cause rejection of the shipment and cancellation of the sales by the importer. Such extremely low thresholds for infestation may result in prophylactic spraying of vineyards in advance of harvest or use of post-harvest treatments. Sampling methods have been developed to allow quantification of spider populations in vineyards (Costello and Daane 1997), and analysis techniques are available to address pest populations with very low abundances (Venette et al. 2002), so progress towards threshold-based management may be possible even for pests with low abundance and high risk of economic loss. While this potential exists, the additional time required to sample for rare pests is often not considered

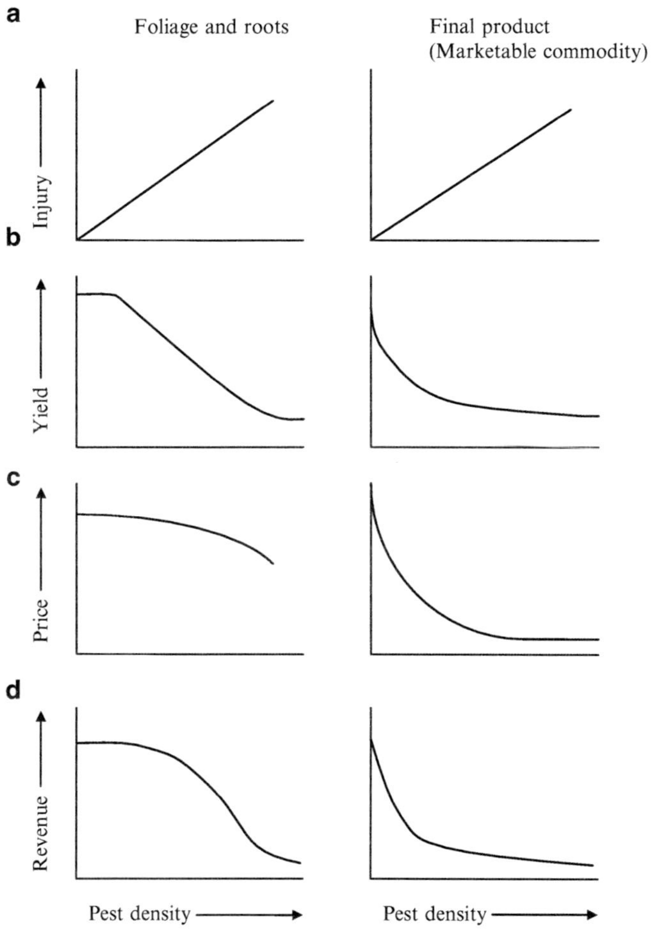

Fig. 2.2 Generalized patterns describing the relationships between pest density, damage, crop yield, price and revenue. (**a**) Pest on crop damage, (**b**) Pest damage on yield, (**c**) Pest damage on crop price, (**d**) Pest density on revenue (Adapted from Southwood and Norton 1973)

in the expense side of the economic threshold calculation, and this can add significant cost to the management of arthropod pests, particularly if they are small, cryptic, or at low abundance.

2.2 Economic Thresholds

The economic threshold concept was developed by Stern et al. (1959), who formalized the concept of the economic injury level (EIL) as the minimal population density that will cause economic damage and therefore justify the cost of artificial control measures. Injury is defined as the effect of the pest on host plant tissues, whereas damage is the measurable yield loss (Pedigo et al. 1986). Discrimination between these two is important because injury such as leaf feeding by a beetle does not necessarily lead to damage, i.e. yield loss.

The economic threshold (ET) is the level at which control measures should be made to stop an increasing pest population from reaching the EIL. The length of time between the ET and the EIL will be dependent on the speed with which the control measures show their impact on the pest and the rate of pest development. The EIL and ET are dynamic, varying from one cultivar to another and also with the cost of crop protection inputs and the crop value. For example, one would expect the same arthropod pest to have lower EIL and ET values in table grapes than in grapes destined for wine production and lower values still in vineyards with berries destined for juice production. The calculation of the EIL is performed using the following calculation: EIL = C/VYD, where EIL = number of insects per hectare (or other unit of production), all of which live to attain full injury potential and where C = cost of management activity per unit of production ($/ha), V = value of crop per unit of the produce ($/kg), Y = yield loss per insect per production unit (proportion defoliated/(insect/ha)), and D = damage per unit injury ((kg reduction/ha) or proportion defoliated). Using this simple equation, one can determine whether the pest is below threshold and therefore does not require control, or the point at which the value of the yield lost (damage) caused by the pest exceeds the cost.

The ET is technically connected to the EIL and it is based on experimental field studies (e.g. Bostanian and Mailloux 1990). In many systems there is also the less-defined action threshold (AT) which is based on local experience or less rigorous analyses of the population level warranting control measures. This is often used in situations where the ET and EIL have not been determined or published (Mitchell and Hutchison 2009). In such cases, an empirically determined AT serves as a functional threshold that when used to drive IPM decision-making, can result in lower pesticide use and more information-based management of pest populations. The use of ATs by growers and consultants is likely quite common in viticulture, but by definition these locally adopted thresholds may not be published in the scientific literature.

2.3 Challenges to Developing Pest Thresholds in Grapes

Before discussing grape arthropod pests for which thresholds have been developed, we consider it instructive to review the factors that can make thresholds challenging to develop for this crop. This is not meant to discourage future developments in this area, but rather to stimulate thorough consideration of the unique aspects of threshold development for grapes before studies are initiated. It is hoped that this section will stimulate collaboration between entomologists and viticulturalists so that thresholds are developed within the context of grapevine physiology and the realities of grape production for a specific region.

Gathering the information for calculation of the EIL and ET requires an understanding of how pest abundance affects yield. For direct pests that infest clusters and cause a loss of harvestable grapes, determining this relationship can be relatively straightforward. In this case, the number of insects detected within a sample of clusters (or the proportion of clusters infested) provides determination of whether a vineyard or part of a vineyard is below, at, or above the threshold. Such a system has been developed for *Lobesia botrana* (Denis & Schiffermuller) in which chemical control is applied for the first generation only when more than 50% of inflorescences are infested (Moschos 2006). In later generations the EIL is 5% or 15% of infested clusters for compact or loosely-bunched varieties, respectively, reflecting their relative susceptibility to rots (Ioriatti et al., Chap. 14). A similar sampling and threshold system was developed for *Paralobesia viteana* (Clemens) in *Vitis labrusca* vineyards (see below).

Understanding relationships between infestation and yield loss becomes more complicated when considering pests that feed on the shoots, leaves and roots of grapevines. The grapevine has a prodigious capacity to tolerate removal of vegetative tissues, as exemplified by the annual cycle of pruning used to maintain vine balance, and so the removal of leaf area by insects and mites may have little effect on long-term vine canopy growth. Of greater interest to vineyard managers are the effects on berry quality and long-term vine health. For the first of these, one needs to determine whether defoliation causes a significant reduction in berry quality. Berry quality is measured in its simplest form as percent soluble solids or degrees Brix that reflect the carbohydrate composition of the grape juice, but more detailed analysis will also include measurements of pH and titratable acidity as well as colorimetric evaluation or chemical analysis of the juice. These parameters are important for winemakers and juice processors, so studies that can investigate links between pest activity in the vineyard and fruit quality at harvest provide insights that simple yield measurements may miss. The second aspect of great interest to vineyard managers is whether defoliation this season will have a long term effect on vine growth. To answer that question requires multi-year investigations. To develop a thorough picture of the interaction between pests and vines these investigations should ideally be conducted in field-grown vines including sites that are of different cultivars, varying vine maturity and with different crop loads. This is because these factors can have a major influence on the result. In practice, studies of pest effects on crop quality and vine physiology are typically done across one or a few of these variables, such as the recent investigation of Japanese beetle feeding on different grape cultivars (Hammons et al. 2010a, b).

The perennial nature of grapevines is central to the difficulties of developing thresholds for this crop. Depending on the conditions, arthropod damage in the current season may have no significant effect this year but could reduce winter hardiness of buds leading to lower vine productivity in the following year. This type of effect has been seen with beetle defoliation of young grapevines, in which 'Norton', 'Chambourcin', and 'Cabernet Sauvignon' vines receiving high levels of defoliation (38–48% leaf area loss) had lower winter bud hardiness than vines that were protected from beetle feeding (Hammons et al. 2010a).

Grapevines also have a varying cycle of carbohydrate source and sink dynamics depending on the time of the growing season and on the relative size of the grape foliage canopy and berry load (Mullins et al. 1992). Knowledge of these dynamics can help managers balance the vines for optimal sustainable production, defined by Howell (2001) in a practical way as '*a collective methodology that produces highest yields of ripe fruit per unit land area with no reduction in vine vegetative growth and does so over a period of years at costs which return a net profit.*' While the viticultural science to achieve a balance between vine growth and production of ripe fruit is well advanced, there is much less understanding of how this balance is affected by pest injury. Some general patterns of variation in sensitivity to defoliation are clear, however. Pests that feed on buds such as noctuid cutworm larvae or the grape flea beetle, *Altica chalybea* (Illiger), can cause complete shoot loss, and so ATs are typically low with treatment being recommended at 1–2% bud damage. In contrast, once the vine canopy starts developing, 10–20% leaf area loss can often be tolerated by young and vigorous vines without significant reduction in growth during establishment. Once vines start producing grapes, the clusters are a large sink for carbohydrates after veraison, and so vines may then be expected to be more sensitive to reduction in leaf area at this time. Thresholds have been developed for leafhoppers and for mites, two groups of leaf-feeding arthropods that can build populations during the growing season, especially in dry growing regions, to levels that limit berry ripeness and reduce yield in the following season (Martinson et al. 1997).

Depending on the market that grapes are destined for, there are also varying sensitivities to crop infestation by vineyard managers. For example, infestation of clusters by grape berry moth, *P. viteana*, can cause yield loss but as mentioned above the presence of larvae of this species will also trigger a load rejection if levels exceed the threshold used by processors. In contrast, wineries in the same regions do not use these thresholds when harvested berries are being received at the winery, and tend to manage this insect more intensively because infestation of clusters by *P. viteana* can provide access for opportunistic rots that can affect the flavour profile of wine. Additionally, because of the different returns per hectare possible in wine vs. juice grapes, wine grape growers tend to have lower ATs for the presence of *P. viteana* infestation and are more likely to apply insecticides to prevent cluster infestation and associated diseases.

As explained by Pfeiffer et al. (Chap. 19), some insects that infest clusters at harvest time can also release secretions that can taint grape juice, leading to a risk of lower quality, off flavors in the finished product and potential loss of sales. These pests provide a special case for developing economic thresholds because a low density of infestation can have a large effect on juice quality. Sampling and management decisions have to be made immediately before harvest when growers are typically very busy with other activities on the farm. The level of economic effect is dependent on consumer perception of the taint chemicals. Despite these issues, economic thresholds for such pests have been developed or are currently in development, based on the principles developed originally in the 1950s and later formalized into the EIL and related parameters.

2.4 Manipulating Pest Injury to Determine Thresholds

A review of methods for measuring and statistically detecting relationships between arthropod feeding and yield loss is provided by Buntin (2001), with a focus on insects in field crops. If insects damage vineyards by infestation of berries or by feeding on clusters, the relationships between infestation level and harvestable or marketable yields are relatively simple to determine using the general methods described above and which have application to most common crop situations. Measurements across multiple vineyards with varying pest populations can be used to develop pest density-yield relationships, although variations in other factors among vineyards may then obscure the effect of the pest on yield. Exclusion techniques have been used to manipulate exposure of vines to pest injury, although this should be done with some caution due to the potential for cage effects. Inclusion of a sham cage treatment in experimental designs can help with interpretation of the results in terms of the relative effect of the cage and the insect injury treatment. A third approach to determine the relationship between infestation level and yield is to establish varying chemical treatments for cluster pests. Such an approach was used by Dennehy et al. (1990) who applied insecticides for *P. viteana* at different times of the season, showing the importance of late season control of this pest.

While developing economic thresholds for cluster feeding pests is relatively straightforward, pests that affect the grapevine canopy are more problematic. At the simplest level, manual removal of whole leaves can be applied to mimic different levels of canopy loss (e.g. May et al. 1969; Mansfield and Howell 1981; Howell et al. 1994). However, this approach may not cause the same damage response as actual feeding injury by arthropods on vines for the following reasons: (1) leaves may retain significant portions of photosynthetically-active tissue even after feeding injury, (2) vines may redistribute resources differently when whole leaves are removed compared with when parts are injured, or (3) the remaining leaf area may compensate for the lost area which could result in no functional damage, as seen in vines when whole leaf removal treatments have been applied (Candolfi-Vasconcelos and Koblet 1991).

The most common methods of manipulating levels of foliage-feeding arthropods are caging vines with varying numbers of insects or using chemical treatments to change the pest density. The former approach can allow greater control over the number of arthropods per vine, particularly for highly mobile insects. For some of the smallest arthropods such as mites and leafhoppers, manual reproduction of feeding injury is very challenging, and so levels of infestation are manipulated using either infestation of vines with varying numbers of individuals (e.g. Lenz et al. 2009) or by application of pesticides to reduce populations. In a comparison of vines receiving no treatments, weekly, or biweekly applications of carbaryl, Hammons et al. (2010a) compared sensitivity of different cultivars to feeding by Japanese beetle and determined how increased feeding level affected cluster quality and bud hardiness. Their study, while not used specifically to develop an economic threshold, demonstrates the utility of this approach for manipulating the level of defoliation.

Fig. 2.3 Discrete levels of grape leaf defoliation achieved using a manual paper hole-puncher to enable avoidance of major vascular tissues. Levels of defoliation are (from *left* to *right*) 10%, 20% and 30% of the leaf area, verified using a digital leaf scanner (From Mercader and Isaacs 2004)

For some experimental situations, greater control over defoliation may be needed than by allowing field-grown vines to be infested, and whole leaf removal may not be appropriate, as mentioned above. Manual hole punching has been used to apply discrete and defined levels of defoliation (Fig. 2.3) so that the effects of skeletonizing injury caused by scarab beetle feeding on growth of young vines could be determined (Mercader and Isaacs 2004). This approach avoids some of the issues related to whole leaf removal, plus it reflects more accurately the distribution of defoliation. Although this technique was used to apply a defined level of leaf area loss to all expanded leaves, it could also be used to investigate how distribution of injury across the canopy affects economic thresholds.

2.5 Measuring Effects on Grapevines

Members of the *Vitis* genus are perennial vines with indeterminate growth habit and a high capacity for compensatory growth in response to stress. This reflects their evolutionary history as plants adapted to living in shaded habitats, with the capacity to exploit light gaps and to efficiently intercept sunlight so that sufficient carbohydrates can be produced for ripening berries and for storing resources for the subsequent year's growth. Grapevines have been described as a system for turning sunlight into wine (Smart and Robinson 1991) and this is achieved most efficiently by balancing the carbon brought into the vine (through photosynthesis that converts CO_2 into carbohydrates) on the source side of the plant with that on the sink side (i.e. the ripening berries). There also needs to be sufficient carbon assimilated for maintaining the structures of the plant in terms of the starch reserves for overwintering survival and subsequent growth in the following spring. This balancing act that vineyard managers must pursue may be disrupted by the feeding activities of arthropods, leading to reduction in vine performance parameters. All of these parameters can be measured using standard viticultural techniques, that include but are not limited to shoot length, leaf area, leaf: berry ratio, root growth, berry mass and composition, bud hardiness, and the weight of wood pruned in the winter. An overview of the viticultural issues related to maintaining vine balance is provided by Howell (2001), and this is a good introduction to the types of variable responses vines may have to variable growing conditions.

Measurement of carbohydrate assimilation provides a window into the effects of canopy-feeding pests on vine physiology in response to stress such as arthropod feeding, and this approach has been used in concert with growth and berry measurements to provide a deeper understanding of arthropod-vine interactions. By measuring the concentration of CO_2 in the airstream entering a chamber that encloses part of a leaf, a whole leaf or shoot, or whole vine, the change in concentration can be used to calculate the photosynthetic activity of a vine plant, or any other plant. A review of this approach in the context of understanding fruit crop physiology is provided by Flore and Lakso (1989). Using such methods, the effect on photosynthesis of leaf injury has been investigated for scarab beetles (Mercader and Isaacs 2004), leafhoppers (Candolfi et al. 1993a; Lenz et al. 2009) and mites (Candolfi et al. 1993b, c). These studies generally demonstrate that vines have a significant tolerance for leaf injury before damage occurs, measured in terms of leaf area injured or pest-days accumulated. This information can be used to set thresholds, or to conduct further validation trials under vineyard conditions, that may then allow vineyard managers to accept a certain level of injury without the need for costly intervention.

2.6 Examples of Pest Thresholds Developed for Use in Vineyard Management

These examples are provided to demonstrate how arthropod thresholds have been determined for use in vineyard IPM programs. This is not an exhaustive treatment, but provides the reader with perspectives and references that can be used to adapt these approaches for other key pests in different viticultural regions.

2.6.1 Mites

Mites feed on the mesophyll tissues of grape leaves (Duso et al., Chap. 9), and therefore have the potential to compromise the photosynthetic capacity of vines. Whether this feeding is at a level sufficient to compromise yield, cluster quality or long term vine growth, has been the subject of numerous studies since mites are common pests in some of the primary regions of grape production. Threshold levels for some of the key mite species have been determined in studies using varying levels of mite infestation and sampling of net vine photosynthesis.

Measurement of photosynthesis has been used to evaluate the effects of mites on vine physiology, but these studies are often done on potted vines for logistical reasons. Although they provide valuable insights, extrapolation to vineyard situations with mature vines is still not advisable. Infestation of field grown vines by European red mite, *Panonychus ulmi* (Koch), significantly reduced whole vine photosynthesis only above 3,500 mite-days per leaf (peak of about 60 mites per leaf) (Candolfi et al. 1993b). In field trials, Kast (1989) found only a small decline in soluble sugars at the highest densities of 56 *P. ulmi* mites per leaf with no influence on berry yield. However, because of the high reproductive potential of this mite species, Girolami (1987) recommended an AT of 10 motile forms per leaf during midsummer, with higher thresholds of 20 mites per leaf after veraison and 30 mites per leaf in early spring during rapid shoot growth.

Candolfi et al. (1992), reported that the twospotted spider mite, *Tetranychus urticae* Koch, caused significant reduction of vine photosynthesis, transpiration, as well as stomatal and mesophyll conductance of potted vines. In a similar study, it was found that even a level of 60 *P. ulmi* mites per leaf at the peak infestation was not sufficient to reduce photosynthetic rate (Candolfi et al. 1993c). This led to recommendations that the AT of 2–5 mites per leaf used at that time was far too low. The same team showed that 6,000 mite-days per leaf caused 21–52% reduction in carbon assimilation and the highest sensitivity to mite feeding was at bloom (Candolfi et al. 1993a). Total plant dry weight was reduced by 12.6% when 7,000 mite-days per leaf had been accumulated during the growing season. Nevertheless, these effects did not translate into any reduction in berry yield or quality. By measuring the defoliation caused by different *T. urticae* densities, Arias and Nieto (1983) found a 0.05 Brix reduction in soluble sugars for each 10% leaf defoliation or from each week of defoliation, and 0.3 kg per vine reduction. This kind of information can be used in association with chemical input costs and price structures to determine the EIL for mites on grapes, but AT for *T. urticae* mites are generally high, based on their low individual damage potential and the potential of predatory mites to suppress populations.

2.6.2 Grape Berry Moth

Grape berry moth, *P. viteana*, is the key arthropod pest of grapes grown in the eastern US, requiring control in many regions to avoid significant economic injury. This insect has been the subject of much research since the turn of the twentieth century. Some of the earliest reports of crop loss and suggestions for control of GBM came from Delaware (Dozier et al. 1932), New York State (Hartzell 1910), Ohio (Gossard and Houser 1906), Missouri (Shepard and Rook 1952), and Michigan (Pettit 1933). In general, these researchers recommended a combination of cultural and chemical controls to manage GBM populations. Cultural controls included the pre or post season mounding of soil under trellises to suppress adult emergence from overwintering pupae, as well as collection and burning of leaf litter as a method of reducing pupal populations. Chemical controls in the early 1900s often involved insecticides such as calcium arsenate, nicotine, and then late in the century DDT was made available after WWII. According to Taschenberg (1948), application of insecticides should be made immediately after grape bloom, 10 days post bloom, and a third application in late July-early August. All the early researchers indicated that severity of infestations by *P. viteana* is hard to predict and that frequent scouting is needed to ascertain the need for chemical intervention. Most early approaches to the control of this pest followed these general recommendations.

The first significant revision to this schedule of insecticide applications was the Grape Berry Moth Risk Assessment Program (GBMRAP) (Martinson et al. 1991). In this approach, which is specific to grapes grown for juice processing, growers were asked to determine a GBM risk rating for their vineyard. The risk rating assessed three primary factors understood to predict damage severity: (1) presence of a wooded area immediately adjacent to the vineyard, (2) winter temperatures and snow cover, and (3) infestation history for the vineyard. According to this risk rating procedure, vineyards that either had a history of infestations in excess of 6% damaged clusters in July, or vineyards adjacent to wooded areas or hedgerows, or vineyards with prolonged winter snow cover or mild winter temperatures were considered to be at high risk. Vineyards with none of these characteristics were deemed low risk, and any vineyard not classified as high or low risk were classed as intermediate risk.

The risk rating of a vineyard allows a grower to more accurately determine the need for monitoring activities and insecticide applications against the different generations of this pest. Furthermore, this protocol permits vineyards to be subdivided according to within-vineyard risk (i.e. the six rows adjacent to a wooded area is considered at high risk while the vineyard interior is classed as low or intermediate risk). This brought about a more systematic approach to identifying risk to *P. viteana* damage and because vineyards are scouted at least once per year, the scouting results are used to re-evaluate the risk rating of the vineyard. For vineyards classified as high risk, insecticide treatments are scheduled for 10 days post bloom, early August, and a late August treatment that is based upon scouting done in the fourth week of August. For vineyards classified as intermediate risk, insecticide treatments are recommended for 10 days post bloom and an early August treatment based on scouting done in the third week in July. For low risk vineyards, only an early August treatment is recommended based upon scouting done in the third week of July.

Scouting procedures and thresholds for action are described in Martinson et al. (1991). Four areas in the vineyard are sampled. Two of these are from the vineyard center and two from the vineyard edge. Ten randomly-selected clusters from five vines are examined at each location (i.e. 50 clusters per location). The cluster counts from the two edge locations are combined (i.e. 100 clusters from the edge locations) and the interior cluster counts are also combined (also 100 total clusters). For the July sample date used for low and intermediate risk vineyards, the threshold for treatment in early August is 6% damaged clusters. For the late August sample date for high risk vineyards, the threshold for treatment is 15% damaged clusters.

The GBMRAP was widely adopted by juice grape growers, especially in New York State where it was developed. In addition to eliminating the pre-bloom application of insecticide that was endorsed by earlier researchers, this approach appeared to be an effective method for management of *P. viteana*. In recent years, however, control failures have become more common resulting in numerous loads of harvested grapes being rejected at processing plants. The reasons for this loss of performance of the GBMRAP are thought to include the loss of long-lasting broad-spectrum insecticides from this system during the late twentieth century. Temperature-based approaches to the timing of insecticide applications are being explored as a refinement to the timings of GBMRAP (Isaacs et al., Chap. 15).

2.6.3 Grape Leafhoppers

Several species of leafhopper feed on grape foliage in North America. In eastern vineyards, the eastern grape leafhopper, *Erythroneura comes* (Say), the threebanded leafhopper, *Erythroneura tricincta* Fitch, and the seasonally migrant potato leafhopper, *Empoasca fabae* Harris are the most common species affecting vines. Of the two *Erythroneura* spp., *E. comes* dominates in the northeast. In the west, the primary leafhopper species affecting grapes are the western grape leafhopper, *Erythroneura elegantula* Osborne, and the variegated leafhopper *Erythroneura variabilis* Beamer. These insects belong to the family Cicadellidae and both nymphs and adults feed on grape leaves using piercing-sucking mouthparts to extract the contents of leaf cells. This feeding activity leads to leaf stippling that can be so severe that the entire leaf may be pale yellow or white, and this can compromise photosynthesis.

Most grape varieties can withstand a large infestation of leafhoppers for a single season without apparent diminution in crop quality and quantity. However, repeated years of heavy infestations can lead to significant reduction in berry sugar content and crop size. Irrigated vineyards in warm regions have been shown to tolerate a 20% loss of functional leaf area without concomitant crop losses (Flaherty et al. 1992). In cooler locations such as northeastern North America, crop loss caused by these insects can occur at much lower levels of damage (Martinson et al. 1997). In studies done in grape growing regions throughout North America, it has been observed that well watered, vigorously growing vines are better able to tolerate leafhopper damage (Flaherty et al. 1992; Martinson et al. 1997). In most grape growing regions, natural and enhanced populations of egg parasitoids, especially *Anagrus* spp., are relied upon to help minimize leafhopper numbers and damage. Additionally, because most leafhopper pests of grapes have only one or two generations in a year, a single well-timed application of insecticide can often lead to effective control.

Because leafhoppers are indirect pests, many researchers have sought to establish thresholds of damage to grape leaves that can be measured in terms of crop quality and quantity. One of the earliest attempts to establish a threshold for grape leafhopper management (Jubb et al. 1978) resulted in the recommendation that sprays should be applied when >15% of the leaves showed stippling, or counts of leafhoppers on leaves exceeded an average of eight nymphs per leaf. These thresholds were refined by Martinson et al. (1991) in conjunction with the establishment of the GBMRAP. These efforts were linked because of concerns that reduced insecticide applications for grape berry moth control could lead to leafhopper outbreaks. In a multi-year study, these researchers found that even when untreated, only a small proportion of vineyards exceeded provisional treatment thresholds. According to the risk rating of a vineyard for grape berry moth, leafhopper scouting will occur in the fourth week of August for high risk sites, in the third week of July and the fourth week of August for intermediate risk sites, and 10 days post-bloom, third week of July, and the fourth week of August for low risk sites. Note that the frequency of grape berry moth insecticide applications diminishes with the risk rating for a vineyard, so the frequency of scouting for leafhoppers increases as the risk rating diminishes. Provisional ATs for each sampling date are as follows: fourth week of August, >10 nymphs per leaf; third week of July, >5 nymphs per leaf; and 10 days post-bloom, the presence of stippling and adult leafhoppers.

In western North America, ATs vary according to grape cultivar and crop use as well as leafhopper generation. For wine and raisin 'Thompson Seedless' grapes, treatment for first generation leafhoppers is indicated when numbers exceed 20 nymphs per leaf. For later generations the AT is reduced to 15–20 nymphs per leaf. For table grapes, the first generation threshold is 15 nymphs per leaf, whereas for later generations thresholds vary according to the maturation times of the grape cultivar. The AT for early maturing cultivars is 10 nymphs per leaf, for mid-season cultivars it is 5–10 nymphs per leaf and for late season it is 5–8 nymphs per leaf.

2.6.4 Multicolored Asian Lady Beetle

In the past 10 years the multicolored Asian lady beetle, *Harmonia axyridis* (Pallas), has become an established member of the predatory insect community in the main grape production regions of Europe and eastern North America (Koch 2003; Kenis et al. 2008). During the 2001 and 2003 seasons *H. axyridis* developed high populations and colonized ripening fruit in the Midwest and Great Lakes regions of the United States and Canada (Koch et al. 2004). In grapes this led to the 'lady beetle taint' of juice and wine with the defensive secretions of the beetles when the grapes were crushed (Pickering et al. 2004, 2005; Lucas et al. 2007). The taint can be partially masked by addition of flavor-imparting oak chips during vinification, but other tested treatments had little effect on the taint (Pickering et al. 2006). Winemakers and juice processors would like growers to have zero *H. axyridis* in harvested grapes, but knowledge of human sensory perception of the taint has been used to develop ETs that link numbers of beetles per cluster sample to the risk of having juice or wine with a detectable taint (Galvan et al. 2007a). The very low detection threshold that humans have for the taint chemicals, and the variability in their perception of taint in grape juice (Ross et al. 2007) or wine (Pickering et al. 2004) products has created a unique set of circumstances for developing thresholds for this pest. As mentioned by Pffeifer et al. (Chap. 19), similar approaches will be needed for other pre-harvest pests with the potential for contamination of harvested berries.

Galvan et al. (2007b) sampled multiple vineyards and compared eight sampling plans for their ability to detect one adult beetle per cluster. They demonstrated that *H. axyridis* beetles could be detected in grape clusters with a high chance of accurate decisions if approximately 25 clusters were sampled per vineyard using a binomial sampling plan. This study highlights the relative efficiency of binomial sampling plans for detecting rare insect pests with low thresholds. Using wine made from 'Frontenac' grapes that were spiked with varying levels of *H. axyridis,* Galvan et al. (2007a) used logistic regression methods and a tasting panel to determine the relationship between concentration of the taint in wine and human perception of the taint. This was used to calculate a level of infestation in clusters that would trigger taint detection with a certain frequency, i.e. the probability of wine being perceived as from infested clusters or not. At 1.9 beetles per kg of grapes, or 0.27 beetles per cluster, 10% of the panel conducting the discrimination tests were able to detect the taint, suggesting that growers should prevent MALB populations from reaching this level by using a lower value for the AT. This is similar to the detection threshold of 0.2 beetles per cluster for Riesling (Pickering et al. 2006), indicating that the ET is actually much lower than one beetle per cluster that had been used as a 'working AT threshold' when the beetle first became a pest of grapes. An IPM program based on the presence-absence sampling method for *H. axyridis* and using a practical AT of 10% of clusters infested with beetles was implemented in 2007, and this has been used by several growers in Minnesota and Wisconsin (Hutchison et al. 2010). There are various control options available for *H. axyridis* (Galvan et al. 2006; Kenis et al. 2008), and these can be applied as needed based on the sampling and thresholds now available for vineyard managers to follow.

2.6.5 Grape Cane Gallmaker

For some pests that infest vines, despite causing highly visible injury that may be of concern to vineyard managers, their injury to the vine has no measurable effects on berry quality or vine health so that high populations have no economic impact. This was found for infestation of *Labrusca* vines by the grape cane gallmaker, *Ampeloglypter sesostris* (LeConte), a curculionid whose oviposition causes highly-apparent red galls on shoots that are then weakened (Saunders and Tobin 2000). At the levels experienced in vineyards, up to five galls per vine, no effect on berry weights or sugar concentration was detected. This study highlights the value of studies relating insect infestation to vine productivity and quality to enable informed pest management decisions and minimize the dependence on chemical inputs.

2.7 Integrated Thresholds: The Future for Grape IPM?

For logistical, financial, and statistical power reasons, most studies of pest interactions with grapevines are conducted on one pest at a time, with vines that are growing under optimal or typical conditions. Research scientists also tend to study disease thresholds separately from arthropod thresholds, and often these studies do not consider variation in vine vigor or crop load, despite these factors all varying together within vineyards to potentially affect vine yield or quality. This lack of integration limits the applicability of the resulting thresholds for growers and other vineyard managers. It also makes it more challenging to implement thresholds within IPM programs if the answer to 'What level of infestation can I tolerate before there is an economic effect on my vineyard?' always starts with 'It depends'. Entomologists, plant pathologists and viticulturists have much to gain by collaborating in the study of pest thresholds for vineyards, and this will lead to new insights that will help those thresholds be dynamic and relevant components of vineyard management. The logistical and statistical difficulties inherent in conducting experiments to allow development of integrated threshold are not trivial, and would take significant resources to accomplish, but if multi-pest or multi-stress type studies can be accomplished the value to viticulture would be great. With modern portable digital technologies such as personal 'smart' phones, the data-rich inputs required to determine whether pest control is necessary can be added to decision tools in the vineyard, allowing more rapid evaluation of the need to control pest populations. But first, the science underpinning such decision-making must advance much further than its present state.

Acknowledgements The authors thank William Hutchison (University of Minnesota) and Craig Roubos (Michigan State University) for valuable comments on earlier drafts of this chapter. The lead author also acknowledges the financial support of the Michigan Grape and Wine Industry Council, National Grape Cooperative, and the USDA Viticulture East Program.

References

Arias GA, Nieto JC (1983) Estimación de las pérdidas producidas por la "araña amarilla común" (*Tetranychus urticae* Koch) en "Tierra de Barros" (Badajoz) y propuesta de un umbral de tolerancia económica. Bol Serv Def Plagas Insp Fitopat 9:227–252

Bardner RK, Fletcher E (1974) Insect infestations and their effects on the growth and yield of field crops: a review. Bull Entomol Res 64:142–160

Bentley WJ (2009) The integrated control concept and its relevance to current integrated pest management in California fresh market grapes. Pest Manag Sci 65:1298–1304

Bostanian NJ, Mailloux G (1990) Threshold levels and sequential sampling plans for tarnished plant bug in strawberries. In: Bostanian NJ, Wilson LT, Dennehy TJ (eds) Monitoring and integrated management of arthropod pests of small fruit crops. Intercept Ltd., Andover, pp 81–101

Buntin GD (2001) Techniques for evaluating yield loss from insects. In: Peterson RKD, Higley LG (eds) Biotic stress and yield loss. CRC Press, Boca Raton, pp 23–41

Candolfi MP, Boller EF, Wermelinger B (1992) Influence of the twospotted spider mite, *Tetranychus urticae*, on the gas exchange of Pinot noir grapevine leaves. Vitis 31:205–212

Candolfi MP, Jermini M, Carrera E, Candolfi-Vasconcelos MC (1993a) Grapevine leaf gas exchange, plant growth, yield, fruit quality and carbohydrate reserves influenced by the grape leafhopper, *Empoasca vitis*. Entomol Exp Appl 69:289–296

Candolfi MP, Wermelinger B, Boller EF (1993b) Influence of the red mite (*Panonychus ulmi* Koch) on yield, fruit quality and plant vigour of three *Vitis vinifera* varieties. Vitic Enol Sci 48:161–164

Candolfi MP, Wermelinger B, Boller EF (1993c) Photosynthesis and transpiration of "Riesling x Sylvaner" grapevine leaves as affected by the European red mite (*Panonychus ulmi* Koch) (Acari, Tetranychidae) feeding. J Appl Entomol 115:233–239

Candolfi-Vasconcelos MC, Koblet W (1991) Influences of partial defoliation on gas exchange parameters and chlorophyll content of field-grown grapevines: mechanisms and limitations of the compensation capacity. Vitis 30:129–141

Costello MJ, Daane KM (1997) A comparison of sampling methods used to estimate spider (Araneae) species abundance and composition in grape vineyards. Environ Entomol 26: 142–149

Dennehy TJ, Hoffman CJ, Nyrop JP, Saunders MC (1990) Development of low-spray, biological and pheromone approaches for control of grape berry moth, *Endopiza viteana* Clemens, in the eastern United States. In: Bostanian NJ, Wilson L, Dennehy TJ (eds) Monitoring and integrated management of arthropod pests of small fruit crops. Intercept Ltd., Andover, pp 261–282

Dozier HL, Williams LL, Butler HG (1932) Life history of the grape-berry moth in Delaware. Univ Del Agric Exp Stn Bull, 176:1–47

Flaherty, DL, Christensen LP, Lanini WT, Marois JJ, Phillips PA, Wilson LT (1992) Grape pest management, 2nd edn. Publication 3343. University of California, Division of Agriculture and Natural Resources, Oakland

Flore JA, Lakso AN (1989) Environmental and physiological regulation of photosynthesis in fruit crops. Hortic Rev 11:111–157

Galvan TL, Burkness EC, Hutchison WD (2006) Efficacy of selected insecticides for management of the multicolored Asian ladybeetle on wine grapes near harvest. Plant Health Prog. doi:10.1094/PHP-2006-1003-01-RS

Galvan TL, Burkness EC, Vickers Z, Stenburg P, Mansfield AK, Hutchison WD (2007a) Sensory-based threshold for multicolored Asian lady beetle-related taint in winegrapes. Am J Enol Vitic 58:518–522

Galvan TL, Burkness EC, Hutchison WD (2007b) Enumerative and binomial sequential sampling plans for multicolored Asian lady beetle (Coleoptera: Coccinellidae) in wine grapes. J Econ Entomol 100:1000–1010

Girolami V (1987) Mites of vineyards and control strategies. In: Balkema AA (ed) Integrated pest control in viticulture. Commission of the European Communities, Rotterdam, pp 185–194

Gossard HA, Houser JS (1906) The grapeberry worm. Ohio Agric Exp Stn Circ 63:1–16

Hammons DL, Kurtural SK, Potter DA (2010a) Impact of insecticide-manipulated defoliation by Japanese beetle (*Popillia japonica*) on grapevines from vineyard establishment through production. Pest Manag Sci 66:565–571

Hammons DL, Kurtural SK, Potter DA (2010b) Japanese beetle defoliation reduced primary bud cold hardiness during vineyard establishment. Am J Enol Vitic 61:130–134

Hanna R, Zalom FG, Roltsch WJ (2003) Relative impact of spider predation and cover crop on population dynamics of *Erythroneura variabilis* in a raisin grape vineyard. Entomol Exp Appl 107:177–191

Hartzell FZ (1910) A preliminary report on grape insects. N Y Agric Exp Stn Bull 331:489–581

Hoffman CJ, Dennehy TJ, Nyrop JP (1992) Phenology, monitoring, and control decision components of the grape berry moth (Lepidoptera: Tortricidae) risk assessment program in New York. J Econ Entomol 85:2218–2227

Howell GS (2001) Sustainable grape productivity and the growth-yield relationship: a review. Am J Enol Vitic 52:165–174

Howell GS, Candolfi-Vasconcelos MC, Koblet W (1994) Response of Pinot noir grapevine growth, yield, and fruit composition to defoliation the previous growing season. Am J Enol Vitic 45:188–191

Hutchison WD, Galvan TL, Burkness EC, Koch RL (2010) *Harmonia axyridis* as an economic pest of wine grapes in the U.S.: progress in developing an IPM program and potential impact in Europe. IOBC/WPRS Bull 58:47–52

Jubb G, Obourn T, Petersen D (1978) Pilot pest management program for grapes in Erie County, Pennsylvania. J Econ Entomol 71:913–916

Kast WK (1989) Untersuchungen zur Befall-Verlust-Relation und Bekämpfungsschwelle bei der Obstbaumspinnmilbe (*Panonychus ulmi* Koch) an Reben. Dtsch Weinbau Jahr 40:199–209

Kenis M, Roy HE, Zindel R, Majerus MEN (2008) Current and potential management strategies against *Harmonia axyridis*. BioControl 53:235–252

Koch RL (2003) The multicolored Asian lady beetle, *Harmonia axyridis*: a review of its biology, uses in biological control, and non-target impacts. J Insect Sci 3:32

Koch RL, Burkness EC, Wold Burkness SJ, Hutchison WD (2004) Phytophagous preferences of the multicolored Asian ladybeetle (Coleoptea: Coccinellidae) for autumn-ripening fruit. J Econ Entomol 97:539–544

Lenz MS, Isaacs R, Howell GS, Flore J (2009) Vegetative growth responses of Pinot gris (*Vitis vinifera* L.) grapevines to infestation by potato leafhoppers (*Empoasca fabae* Harris). Am J Enol Vitic 60:130–137

Lucas E, Labrie G, Vincent C, Kovach J (2007) The multicoloured Asian ladybeetle *Harmonia axyridis* – beneficial or nuisance organism? In: Vincent C, Goettel M, Lazarovits G (eds) Biological control: a global perspective. Case histories from around the world. CABI Publishing, Wallingford, pp 38–52

Mailloux G, Bostanian NJ (1988) Economic injury level model for tarnished plant bug, *Lygus lineolaris* (Palisot de Beauvois) (Hemiptera: Miridae) in strawberry fields. Environ Entomol 17:581–586

Mansfield TK, Howell GS (1981) Response of soluble solids accumulation, fruitfulness, cold resistance, and onset of bud growth to differential defoliation stress at véraison in Concord grapevines. Am J Enol Vitic 32:200–205

Martinson TE, Hoffman CJ, Dennehy TJ, Kamas JS, Weigle T (1991) Risk assessment of grape berry moth and guidelines for management of the eastern grape leafhopper. N Y Food Life Sci Bull 138:1–10

Martinson TE, Dunst R, Lakso A, English-Loeb G (1997) Impact of feeding injury by eastern grape leafhopper (Homoptera: Cicadellidae) on yield and juice quality of Concord grapes. Am J Enol Vitic 48:291–302

May P, Shaulis NJ, Antcliff AJ (1969) The effect of controlled defoliation in the Sultana vine. Am J Enol Vitic 20:237–250

Mercader RJ, Isaacs R (2004) Phenophase-dependent growth responses to foliar injury in *Vitis labruscana* Bailey var. Niagara during vineyard establishment. Am J Enol Vitic 55:1–6

Mitchell PD, Hutchison WD (2009) Economic risk and decision making in IPM. In: Radcliffe EB, Hutchison WD, Cancelado RE (eds) IPM: concepts, tactics, strategies, and case studies. Cambridge University Press, Cambridge, pp 35–50

Moschos T (2006) Yield loss quantification and economic injury level estimation for the carpophagous generations of the European grapevine moth *Lobesia botrana* Den. et Schiff. (Lepidoptera: Tortricidae). Int J Pest Manag 52:141–147

Mullins MG, Bouquet A, Williams LE (1992) Biology of the grapevine. Cambridge University Press, Cambridge

Pedigo LP, Hammond RB, Poston FL (1977) Effects of green cloverworm larval intensity on consumption of soybean leaf tissue. J Econ Entomol 70:159–162

Pedigo LP, Hutchins SH, Higley LG (1986) Economic injury levels theory and practice. Annu Rev Entomol 31:341–368

Pettit RH (1933) The principal grape insects of Michigan. Mich State Coll Agric Exp Stn Bull 239:1–7

Pickering GJ, Lin JY, Riesen R, Reynolds A, Brindle I, Soleas G (2004) Influence of *Harmonia axyridis* on the sensory properties of white and red wine. Am J Enol Vitic 55:153–159

Pickering G, Lin J, Reynolds A, Soleas G, Riesen R, Berindle I (2005) The influence of *Harmonia axyridis* on wine composition and aging. J Food Sci 70:5128–5135

Pickering G, Lin J, Reynolds A, Soleas G, Riesen R (2006) The evaluation of remedial treatments for wine affected by *Harmonia axyridis*. Int J Food Sci Technol 41:77–86

Poston FL, Pedigo LP, Welch SM (1983) Economic injury levels: reality and practicality. Bull Entomol Soc Am 29:49–53

Ross C, Ferguson H, Keller M, Walsh D, Weller K, Spayd S (2007) Determination of ortho-nasal aroma threshold for multicoloured Asian ladybeetle in a Concord grape juice. J Food Qual 30:855–863

Saunders MC, Tobin PC (2000) Grape cane gallmaker (Coleoptera: Curculionidae) and its impact on cultivated grapes. J Econ Entomol 93:795–799

Seinhorst JW (1965) The relationship between nematode density and damage to plants. Nematologica 11:137–154

Shepard PH, Rook G (1952) Grape spray schedule and description of grape pests. Miss State Fruit Exp Stn Circ 33:1–9

Smart R, Robinson M (1991) Sunlight into wine; a handbook for wine grape canopy management. Australian Industrial Publishers Pty Ltd., Underdale

Southwood TRE, Norton GA (1973) Economic aspects of pest management strategies and decisions. Meml Ecol Soc Aust 1:168–184

Stern VM (1973) Economic thresholds. Annu Rev Entomol 18:259–280

Stern VM, Smith RF, van den Bosch R, Hagen KS (1959) The integrated control concept. Hilgardia 29:81–101

Taschenberg EF (1948) Evaluation of spray programs for the control of the grape berry moth, *Polychrosis viteana* Clemens. N Y State Agric Exp Stn Geneva Tech Bull 283:3–70

Venette RC, Moon RD, Hutchison WD (2002) Strategies and statistics of sampling for rare individuals. Annu Rev Entomol 47:143–174

Chapter 3
Modeling Arthropods to Support IPM in Vineyards

John Michael Hardman

3.1 Introduction

This chapter describes the use of mathematical modeling in support of vineyard integrated pest management (IPM) programs. In IPM, models are used to represent aspects of the agroecosystem that include the crop, insect and mite pests and their natural enemies and external factors (driving variables) such as weather, pesticide applications and horticultural practices (Getz and Gutierrez 1982; Baumgärtner et al. 1988). We are all familiar with verbal, descriptive models which appear in scientific papers or technical articles and give a picture of aspects of the system by means of graphs, tables and verbal descriptions. If they are based on a solid foundation of knowledge, descriptive verbal models will not only clearly portray our understanding of arthropod dynamics in vineyards, but can also suggest solutions, even for complex pest problems. As an example, Mizell et al. (2008) present a detailed overview of the ecology of the glassy-winged sharpshooter, *Homalodisca vitripennis* (Germar). This leafhopper feeds on a broad array of wild and cultivated plants, including grapevines, and is a vector for the bacterium *Xylella fastidiosa* Wells et al., the causative agent of Pierce's disease of grapevines. Using a flow diagram, a series of tables, and detailed verbal description based on extensive research, the authors describe a conceptual model that integrates insect behavior, life history strategies and their associated risks, with the nutritional requirements of each life stage. The model not only describes the insect-host system, but also shows how appropriate manipulation of plant communities could effectively suppress *H. vitripennis* and *X. fastidiosa,* thus protecting crops such as citrus and grapes. In contrast to verbal models, mathematical models can be used to predict aspects of system behavior

J.M. Hardman (✉)
Atlantic Food and Horticulture Research Centre, Agriculture and Agri-Food Canada,
32 Main Street, Kentville, NS, Canada B4N 1J5
e-mail: Michael.Hardman@agr.gc.ca

N.J. Bostanian et al. (eds.), *Arthropod Management in Vineyards: Pests, Approaches, and Future Directions*, DOI 10.1007/978-94-007-4032-7_3,

given an understanding of initial conditions and driving variables. The advantages of mathematical models compared with verbal descriptive models can include greater clarity of structure, clearer exposure of underlying assumptions, and generation of the correct dynamic consequences of the functions contained in the model (Forester 1968). The model is useful in its developmental stages if it clarifies thought, captures and records what we know, and allows us to see the consequences of our assumptions, whether those assumptions are later found to be right or wrong. Later on, the mathematical model succeeds if it opens the road to improving the accuracy with which we can represent reality.

How have these advantages of mathematical models benefitted vineyard IPM? We shall approach this question by considering the types of mathematical models that have been used in vineyard IPM and then explore the usefulness of each.

3.2 Phenology Models Used in Vineyard IPM

The primary objective of phenological models is to predict time of appearance of specific developmental stages of an insect pest to help select appropriate sampling dates or to time control operations. But when there are complicating factors, the very process of developing and using phenological models can also advance understanding of insect population dynamics.

Phenological models typically comprise one to several regression models. In the simplest case cumulative counts or cumulative proportions of seasonal counts are computed from cumulative degree-days. Simpler phenology models focus on the effects of temperature on rates of development and do not involve other aspects of population dynamics such as effects of biotic and abiotic factors on age-specific rates of survival and fecundity. Often the functions are distributed delays which can emulate the variability observed in recruitment curves. Typically, predicted distributions are validated with observed data to ensure accuracy in the predicted time of appearance of the life stages.

The seven examples listed in Table 3.1 progress from this straightforward pattern (Bostanian et al. 2006) to more complex cases where authors had to address additional factors. These include severe drought (Gallardo et al. 2009); strong year-to-year contrasts in climate, making it impossible to use one function for all years (Lopez et al. 2003; Moravie et al. 2006); site-specific effects of microclimate (Moravie et al. 2006); and differing temperature responses of insect populations either on a local scale in rugged terrain (Moravie et al. 2006) or on a much larger regional scale (Gallardo et al. 2009). In other cases, complexity was a consequence of developing more comprehensive models which include acquisition of a grapevine pathogen and its transmission by an insect vector (Bressan et al. 2006) or predicting the effects of temperature on the phenology of all life stages from egg to adult and including the effect of day length on diapause, which permits prediction of the number of generations that can occur in a particular locality (Tobin et al. 2003, 2008). Examples are detailed in the following paragraphs.

Table 3.1 Phenology models used in grape pest management

Pest species	Location	Use in IPM	Complicating factors	References
Potato leafhopper, *Empoasca fabae*, five species of *Erythroneura* leafhoppers	Quebec, Canada	Time sampling activities for control decisions	None	Bostanian et al. (2006)
Grape vine moth, *Lobesia botrana*	Ribera del Guadiana, southwestern Spain	Time application of biorational control measures	Rainfall, geographic limits of model	Gallardo et al. (2009)
Termite, *Kalotermes flavicollis*	Andalusia, Spain	Timing insecticide application to control dispersing adults	Yearly climatic effects	Lopez et al. (2003)
Grape vine moth, *Lobesia botrana*, grape berry moth, *Eupoecilia ambiguella*	La Côte region, Switzerland	Decision for mating disruption and timing of insecticide applications	Microclimatic effects, genetic isolation of moths, yearly climatic effects	Moravie et al. (2006)
Leafhopper, *Scaphoideus titanus*; Phytoplasma Flavescence dorée (FDP)	Languedoc region, France	Predict cumulative nymph hatch, proportion of FDP-infective leafhoppers, time control measures	Infection of vector, latent period, infection of grapevines, effect of leafhopper density on FDP spread	Bressan et al. (2006)
Grape berry moth, *Paralobesia* (=*Endopiza*) *viteana*	Eastern US, Ontario, Canada	Determine effects of temperature and day length on number of generations	DD[a] for egg to adult development, effect of photoperiod on diapause induction	Tobin et al. (2003)
Grape berry moth, *Paralobesia* (=*Endopiza*) *viteana*	New York, Pennsylvania, Michigan, US	Explore potential for an extra generation given various levels of climatic warming	Time of oviposition in relation to diapause-inducing photoperiods	Tobin et al. (2008)

[a]Degree-days

Bostanian et al. (2006) developed a phenological model for a complex of five species of leafhoppers in the genus *Erythroneura* and the potato leafhopper, *Empoasca fabae* (Harris), found in Quebec vineyards, Canada. Their model, a four parameter Weibull function, uses degree-days (DD) after March 1 accumulated above a threshold of 8°C to predict cumulative abundance of leafhoppers. They found that surveys should be initiated at 630 DD (5% cumulative abundance), maximum abundance occurs between 850 and 860 DD (50% cumulative abundance) and sampling can be terminated at 1,140 DD (95% cumulative abundance). With this knowledge, sampling effort can be timed so that field scouts can optimally determine if cumulative leafhopper abundance exceeds the economic action threshold.

Gallardo et al. (2009) developed a phenological model to forecast activity of the second and third flight periods of the European grapevine moth, *Lobesia botrana* (Denis & Schiffermüller), in the southern part of the Ribera del Guadiana in southwestern Spain. They did not attempt to predict the first flight period in detail because of low male captures and low damage caused by the first generation. However, data related to the first flight were used to compute DD for the emergence of the second and third generations. A sine wave function was used to compute cumulative DD above a base of 7°C with an upper threshold of 30°C. Because of non-linearities in the effect of low and high temperatures on rate of development, they used a probit scale for cumulative moth-days and a common log scale for cumulative DD in their regressions. A complicating factor in their study was the effect of severe drought in 1996 which led to extremely low male captures. For this reason the sparse capture data for 1996 were not included in single year or multi-year analyses. Regressions for second and third generation flights for each of the other 12 years gave good fits to the data. Moreover, when they conducted regressions based on data pooled over the 12 years, the models derived from pooled data also gave good fits with correlation coefficients of 0.836 and 0.873 for second and third flights, respectively. They concluded that the second and third generation models would be useful for prediction of flight periods of *L. botrana* in the future and hence could assist IPM. However, they also found that the model developed in the Ribera del Guadiana was not accurate for *L. botrana* populations in Andalusia, a grape growing region about 300 km away.

Lopez et al. (2003), working in sherry vineyards in Andalusia, southwestern Spain, used sticky cards to monitor the flight of adults of the termite *Kalotermes flavicollis* (F.). Adults are the only termite life stage that can be conveniently controlled with insecticides. Using trap capture data from a sherry vineyard from 1993 to 1996 they developed a logistic model to predict cumulative percentage of flying adults caught in traps as a function of cumulative DD above a base temperature of 0°C. The logistic model gave good fits to cumulative captures in individual years with R^2 values >95%. However, because of changing climatic conditions from year to year, their logistic model fitted to data from all 4 years gave an R^2 of only 61.1%. Nonetheless, based on the 4-year model, the authors concluded that for any given year it should be possible to destroy over 90% of emerged adult termites if an insecticide is applied at 730 DD.

Moravie et al. (2006) developed models for emergence times of the grape vine moth, *L. botrana,* and the grape berry moth, *Eupoecilia ambiguella* Hübner, in the La Côte region in western Switzerland, to correctly time control measures. One such measure, mating disruption, is normally begun before 5% of total flight occurs. Insecticide applications, however, may be initiated at 5%, 50% or 80% of cumulative flight. Data on moth captures were collected from four sites over 12 years (1992–2003). They used the sine method to compute DD above 10°C after 1 January. For each site and each year they fitted a three parameter Weibull distribution to compute the cumulative proportion that emerged as a function of cumulative DD (t):

$$F(t) = 1 - \exp\left[-\left\{1 + \kappa(\mathrm{t} - \mu)/\tau\right\}^{1/\kappa}\right]$$

where $1 + \kappa(\mathrm{t} - \mu)/\tau > 0$ and κ, $\tau > 0$. The shape parameter κ controls the initial slope of the curve depending on whether the flight starts slowly or quickly. The scale parameter τ is proportional to the duration of the flight period. The location parameter μ indicates when flight starts.

In multi-year analyses with each moth species, they found that all three parameters in the Weibull distribution were affected by site and year. They suggested two causes for the influence of site, which was treated as a fixed effect. First, local microclimate at each site was affected by altitude plus proximity of dwellings or adjacent vegetation such as vineyards and woodlands. Second, local moth populations may respond differently to the same temperature regimes due to genetic isolation. Both species have a limited dispersal range of 1–3 km. The authors suggested two ways to take account of the effects of microclimate: either record weather variables at each locality or use data from appropriate reference sites representing the range of climatic variability in the region, taking pains to include sites with very early or very dispersed flight periods.

In their analyses they found the factor 'year' was best treated as having a random effect on the Weibull parameters. With *E. ambiguella* they could predict flight using Weibull parameters adjusted for the local site effect, plus DD averaged over several years because the flight period was very concentrated. However, with *L. botrana*, which has a more dispersed flight period, they could only improve predictions by including counts in the current year that cover the first 15–20% of the flight period. In practice, this is too late for estimating the need for mating disruption, but it would be useful for timing insecticide applications, which correspond to 50% or 80% of cumulative moth emergence.

In the Languedoc region of France, Bressan et al. (2006) collected nymphs and adults of cicadellid leafhoppers, *Scaphoideus titanus* Ball, a vector of Flavescence Dorée phytoplasma (FDP). Sampling was done in four vineyards with a high incidence of the grape pathogen. They also used the ELISA technique to detect incidence of the pathogen in young nymphs, older nymphs and adults. Failure of young nymphs, but not fifth instars and adults, to infect caged grapevines indoors indicated a 30 day latent period before leafhoppers became infective. Using all of these data they were able to develop logistic models to predict the proportion of nymphs hatching, the proportion of leafhoppers that are infected but in a latent state for transmission, and the proportion of leafhoppers that can infect grapevines as a function of cumulative DD. The logistic models gave a good fit to 2 vineyards with a 4-year data set for the proportion of leafhopper nymphs hatching. However, the 4 vineyards × 4-year logistic models for infected (latent period) and infective leafhoppers overestimated the respective proportions with FDP in the second half of the season. There is no mention of difficulties with year or site effects. The three logistic functions predicted three successive, well separated events: (1) the rise in hatched nymphs; (2) the rise in infected nymphs; and (3) the rise in nymphs and adults able to infect the grapevines. Their three-function model also suggested that the higher the density of leafhoppers in a vineyard the earlier a given density of infective insects will occur. The authors cautioned that their model was based on data from highly FDP-infective vineyards. Therefore, it may not be advisable to use the same model in vineyards with lower levels of infection. Nonetheless, their model could be used to correctly time control measures to reduce potentially FDP-infective *S. titanus* adults in highly infested vineyards.

Tobin et al. (2003) developed a comprehensive phenology model to predict the temporal dynamics of the grape berry moth, *Paralobesia* (=*Endopiza*) *viteana* Clemens in various grape growing regions in the eastern US and the Niagara peninsula in Ontario, Canada. The first component of their model was a Gompertz function to compute the cumulative proportion of total flight of first generation male moths emerging from overwintered pupae as a function of cumulative DD. Parameter estimates were based on data for male captures in pheromone traps over a 3-year period at a research station in Pennsylvania. The second component was a logistic function to compute the proportion of individuals completing egg to adult development as a function of cumulative DD. This function was derived from data from laboratory rearing of *P. viteana* (Tobin et al. 2001). For the last component of the phenology model they used published data on diapause induction (Nagarkatti et al. 2001) to compute the proportion of eggs that develop into diapausing pupae as a function of decreasing day length shortly after the summer solstice. In the final phase of the study they explored the behavior of the phenology model for different geographic locations in the eastern US. For input data they used daily photoperiod and daily maximum and minimum temperatures from six representative grape growing regions over the interval 1991–2000. Using maximum and minimum daily temperatures, daily DD above 8.4°C were computed for each year and then 10-year averages were calculated for the DD. Using DD and photoperiod as inputs, they employed the full phenology model to predict time of appearance of life stages and to determine how many *P. viteana* generations there would be in each year at each locality. The model predicted as few as two and a partial third generation in cooler, northern locations and as many as three and a partial fourth generation in warmer, southern locations. Predictions agreed with past observations of *P. viteana* phenology so that the authors concluded that the model provides a plausible explanation for the differences in voltinism seen in different grape growing regions. Because this phenology model was used in a geographic perspective, it is included both in Table 3.1 for phenology models and Table 3.2 for geographic models.

Table 3.2 Geographic models used in grape pest management

Pest species	Location	Uses for the model	References
Grape berry moth, *Paralobesia* (=*Endopiza*) *viteana*	Eastern US, Ontario Canada	Determine number of generations in different grape growing regions in eastern US and Ontario, Canada	Tobin et al. (2003)
Vine mealybug, *Planococcus ficus*	California, US	Potential efficacy of three natural enemies in different regions of the state; regions likely to have higher *P. ficus* populations	Gutierrez et al. (2008)
Glassy-winged sharp-shooter, *Homalodisca coagulata*; grape pathogenic bacterium, *Xylella fastidiosa*	California, US	Invasion risk for California and other grape growing regions world wide	Hoddle (2004)

In a later study, Tobin et al. (2008) used the 2003 phenology model to explore historical fluctuations in the estimated annual number of generations of the grape berry moth at single locations in Pennsylvania, New York and Michigan. For each location, they estimated the potential number of generations of grape berry moth from the early 1900s to 2005. Next they examined the relationship between predicted number of generations and DD accumulated over varying periods in the summer to see which intervals could most accurately predict the number of generations. The best predictor was the number of DD accumulated before August 3, which was the date at which day length at the study sites would induce 90% of oviposited eggs to develop into diapausing pupae. Lastly they explored what would happen at each site if daily temperatures averaged over the period 1996–2005 had been 1–5°C warmer. Little change would occur with increases in the range 0–2°C. Greater increases, however, would be sufficient to shift the phenology so that oviposition of the second generation could occur before diapause-inducing photoperiods. This shift would greatly increase the risk of an economically damaging late summer generation.

3.3 Population Models Used in Grape IPM

While phenology models predict time of appearance of one or more life stages of an insect, models simulating population dynamics predict daily densities for all life stages (Getz and Gutierrez 1982; Baumgärtner et al. 1988). In these models the insect population consists of cohorts of individuals of the same physiological age. Cohorts pass through age classes. Sets of age classes comprise the major life stages of eggs, larvae, pupae and adults. Changes in age structure and density are affected by age-specific rates of development, survival and fecundity, and these in turn are affected by environmental factors such as weather, natural enemies and food supply. Because these models are mechanistic rather than statistical they are typically used

Table 3.3 Population models used in grape pest management

Pest species	Location	Uses for the model	References
Grape vine moth, *Lobesia botrana*	Emilia-Romagna, Italy	Forecasting flight period and analysis of population dynamics	Baumgärtner and Baronio (1989)
Grape leafhopper, *Empoasca vitis*	Tessin, southern Switzerland	Analysis of population dynamics including impact of egg parasitoids	Cerutti et al. (1992)
Grape vine moth, *Lobesia botrana*, grape berry moth, *Eupoecilia ambiguella*	Regions adjoining Rhine and Main rivers, Germany	Understanding effects of weather on population dynamics, timing control operations	Schmidt et al. (2001)
Grape vine moth, *Lobesia botrana*, grape berry moth, *Eupoecilia ambiguella*	Regions adjoining Rhine and Main rivers, Germany	Forecasting population dynamics, optimizing spray frequency and dosage	Schmidt et al. (2003)
European red mite *Panonychus ulmi*	Region of Zurich, northern Switzerland	Understanding of plant-mite interactions, impact of site specific factors, temperature, rainfall	Wermelinger et al. (1992)
Vine mealybug, *Planococcus ficus*	California, US	Realistic simulation of growth of grapevines and population dynamics of vine mealybug and three natural enemies under specified climatic conditions	Gutierrez et al. (2008)

not just for improved IPM tactics but also to gain better understanding of population dynamics. Examples of population models are listed in Table 3.3 and described below.

3.3.1 Single Species Population Models

In the Rhineland region of Germany, the initial purpose for developing models for the grape vine moth, *L. botrana*, and the grape berry moth, *E. ambiguella*, was to forecast the time of appearance and density of first instar larvae, the only life stage susceptible to softer plant protection measures such as *Bacillus thuringiensis* Berliner. Armed with this information, growers could optimize timing of applications and adjust insecticide rates to larval density. However difficulties arose in making consistently reliable predictions. Thus necessity dictated a progression from a simpler statistical model relating phenology to temperature to two increasingly detailed population models to explore the possible effects of several weather variables on rates of development, survival and fecundity of age-structured populations. Moreover, the population models were not only used for practical purposes but also to gain better understanding of population dynamics.

The initial phenology model for *L. botrana* and *E. ambiguella* was based on 7 years of weather data, male captures in pheromone traps, and densities of eggs and larvae in vineyards (Hoppmann and Holst 1993). The model was used to predict the period of flight activity, the start and end of oviposition, and the periods of egg hatch and development of the different larval instars. When the model was tested in three grapevine-growing areas in Germany they found that calculated dates for pesticide application only coincided with first appearance of larvae in 80% of the cases. Inaccurate predictions were associated with weather extremes or unexplained, sudden changes in population dynamics. One underlying problem was reliance on pheromone trapping data which provided no information on ovipositing females (Schmidt et al. 2001).

To remedy this situation Schmidt et al. (2001) developed an age structured Leslie process model to simulate effects of weather variables on daily densities of the major life stages of the two moth species. Required biological inputs included grapevine phenology, counts of overwintered pupae in the vineyard, and male captures in pheromone traps. A major aspect of the first Leslie model was inclusion of Weibull functions to simulate the effects of temperature on age-specific fecundity of females and the longevity of adults of both sexes. This was based on data for moths reared in controlled climate chambers with different temperature regimes. Effects of temperature on rates of development and survival of eggs and larval instars were based on iterative least squares parameter estimation using the 7-year vineyard sampling data. When tested against vineyard time series data for *L. botrana* and *E. ambiguella,* the model proved reliable in simulating the start and duration of the life stages, but simulated densities of eggs, larvae and adults often exceeded observed values in early season and declined more slowly in late season. Preliminary simulations suggested that additional weather variables such as rain, relative humidity or wind that were not included in the model may have reduced egg and larval survival in vineyards.

Development of the second Leslie process model involved exploring and incorporating the impact of these other weather variables in the model (Schmidt et al. 2003). This final step was restricted to *L. botrana,* the species with more complete sampling data from vineyards. Comparisons of observed and simulated densities of eggs, larvae, and adults indicated that wind and rain did not affect survival. Relative humidity (RH), however, was expected to have an impact because it was already known that egg and larval survival rates are reduced both by drought and high humidity. Ultimately, good matches with observed densities were obtained with inclusion of a four parameter Weibull function where temperature and relative humidity both affected mortality. The function assumed that mortality increased when RH exceeded 90% (only affecting eggs and first instar larvae), was near zero in the favorable intermediate RH range, and increased again when RH dropped below 40% (affecting eggs and all larval instars). Parameters for the Weibull function were estimated iteratively by selecting values that minimized sums of squares of differences between observed and simulated densities in vineyards for each year from 1992 to 1998. With this version of the model, both time of appearance and densities were accurately simulated. Thus, the second Leslie process model could potentially permit growers to optimize not only timing of application but also ensure that concentrations of *B. thuringiensis* or conventional insecticides would not be excessive or insufficient to prevent economic loss.

Baumgärtner and Baronio (1989) described their model as one that simulated the phenology of *L. botrana* in the Emilia-Romagna region of northern Italy. However, because their model was mechanistic rather than statistical and could simulate densities of life stages it could be considered a population model. Their model included non-linear functions to determine temperature dependent developmental rates of eggs, larvae and pupae. With adults, however, rates of development of females and their reproductive patterns were driven by DD, i.e. rates of development were linear functions of temperature. These developmental functions were used to compute instantaneous values for the duration of each life stage in a time-varying distributed delay model. The model was validated by comparing observed and simulated numbers of males caught in pheromone traps for 4 separate years.

3.3.2 Multi-species Population Models

Tritrophic models, which simulate interactions among the host plant, plant-feeding arthropods, and their natural enemies, are recommended for yielding fuller understanding and providing more reliable conclusions in crop-pest systems (Baumgärtner et al. 1988; Gutierrez et al. 1988). However, two of the vineyard models shown in Table 3.3 are bitrophic and only one is tritrophic. The bitrophic models simulate the dynamics of an insect pest and a parasitoid, and the interactions between grapevines and a mite pest, respectively. The tritrophic model simulates interactions between grapevines, an insect pest and three natural enemies.

Cerutti et al. (1992) developed a model simulating the dynamics of the grape leafhopper *Empoasca vitis* Goethe in the Canton of Ticino (Tessin) in southern Switzerland. Their demographic model, based on life table data for the leafhopper, employed distributed delays (Manetsch 1976; Vansickle 1977) to handle variations in rate of passage of cohorts through the life stages. Simulated densities matched observed densities quite well in a vineyard where egg parasitoids strongly affected leafhopper dynamics, but the fit was less accurate in another vineyard where unexplained early emigration of adult leafhoppers had a greater impact on the population.

Wermelinger et al. (1992) used the metabolic pool approach to model both the growth of grapevines and the dynamics of the European red mite, *Panonychus ulmi* (Koch) in northern Switzerland. Linkages between the comprehensive grape model and that of the mites were facilitated because both models shared the same basic structure. Effects of feeding by the various stages of motile mites were simulated by transfer of dry matter from grape leaves to mites, and by reduction in photosynthetic surface and leaf nitrogen. In return, these changes impacted the mites because simulated rates of development, survival and reproduction decreased with reduced ingestion and lower leaf nitrogen. Higher temperatures were particularly important in promoting mite population growth, whereas rainfall tended to reduce populations by decreasing survival of juveniles and adults and inhibiting oviposition. The authors compared simulated densities of *P. ulmi* with those observed in three vineyards over a 3-year period. For the model to work they had to assume different levels of mite fecundity in each vineyard, probably due to a combination of site specific factors (grape cultivar, soil fertility, fertilizer use). The authors concluded that the model correctly predicted that *P. ulmi* populations start at low subeconomic levels during the damage-sensitive early phases of vine development, not reaching injurious levels until late season when damage would be much less likely to affect quality and yield. The model also correctly simulated the impact of temperature, rainfall and leaf nitrogen on the plant-herbivore interaction.

In a two-phase project, Gutierrez et al. (2008) employed mathematical modeling to determine why biological control of the vine mealybug *Planococcus ficus* (Signoret) was unsuccessful in California. In the first phase they developed a tritrophic model that simulated interactions among grapevines, the mealybug and its natural enemies: two parasitoids, *Anagyrus pseudococci* (Girault) and *Leptomastidea abnormis* (Girault), and a coccinellid predator, *Cryptolaemus montrouzieri* Mulsant. Weather-driven, age-biomass structured demographic models of the mealybug and its natural enemies were parameterized using laboratory data and field observations. Temperature was used to define the thermal limits and developmental rates of each species, while daily rates of growth, survival and fecundity were scaled to resource supply/demand ratios. Once the model was developed and verified they moved on to phase two where population dynamics of the mealybug and its natural enemies were simulated at 108 locations in California over a 10-year period using local weather data. At this point their tritrophic model became a geographic model.

3.4 Geographic Models Used in Grape IPM

All previous models were single site or single region models because the focus was on phenology or population dynamics at one site or region. Here we present differing types of models each of which has a geographic application. The first is derived from a phenology model that predicts the number of generations of a multivoltine species in different geographic regions. The second is based on a tritrophic population model which simulates dynamics of an insect pest and its natural enemies under different climatic conditions. The third uses a climatic window approach where biological and climatic data are fed into a generic, commercially available model that predicts the risk of colonization of different regions by a grape pathogen and its insect vector. The three geographic models are listed in Table 3.2 and are described below.

The phenology model of Tobin et al. (2003) can be used to predict how many generations of grape berry moth, *P. viteana,* could occur in different grape growing regions in the eastern US and how this potential could change if there were temperature increases due to climate change (Tobin et al. 2008). Details on phenology models are given in Sect. 3.2. The relevance to IPM is that the extent of economic damage is highly dependent on the size of the third generation.

As mentioned above, Gutierrez et al. (2008) used their tritrophic model to simulate dynamics of grape mealybugs and their natural enemies at 108 locations in California over a 10-year period using local weather data. Next, using GIS software, they mapped the 10-year, 108 site output and analyzed the data using multiple linear regression and marginal analysis. Both the site-specific simulation model and the geographic model yielded important insights on different aspects of the tritrophic system. The geographic model predicted mealybug populations would be higher in the cooler wine growing regions of California (in the north and in the Sierra Nevada foothills) with lower mealybug densities in the hotter south. This is because the mealybug cannot develop well at temperatures >35°C and seeks cooler sites under the bark or in the root zone of grapevines during hotter periods, where it is less likely to be attacked by natural enemies (a spatial refuge). Also the model indicates the lady beetle *C. montrouzieri*, would be at higher numbers in warmer regions and at lower numbers in cooler regions because it does not readily survive cold winters. Model predictions also indicated that *Anagyrus pseudococci* (Girault) has a larger impact on the mealybug than the other parasitoid, *Leptomastidea abnormis* Macchiati or the coccinellid, *C. montrouzieri*. This is because the widely established *A. pseudococci* has the best climatic match to the vine mealybug and its predicted average geographical distribution and patterns of abundance are similar to those of its prey. However, simulation output also suggests that *A. pseudococci* has a density-dependent response to vine mealybug that is insufficient for economic control. The second parasitoid, *L. abnormis*, is both less widely distributed and has a lower density-dependent response to the mealybug because of a low per capita effective search rate. Another weakness of *L. abnormis*, is its greater vulnerability to interference by ants, which in effect provide a temporal refuge for the vine mealybug from attack by the three natural enemies. Given the deficiencies of the three natural enemies, the analysis of simulation output suggests that reducing the temporal refuge of the mealybug by controlling ants or limiting the movement of mealybugs to spatial refuges are key elements for economic control of this pest.

The authors also used simulation modeling to explore the likely consequences of average temperature increases of 2–4°C in California as predicted in state of the art climate models. With these increases, the model predicts mealybug would increase generally throughout the state. The level of biological control would decrease despite increases in the density of *A. pseudococci* and increases in the favorable range and density of *L. abnormis* and *C. montrouzieri*.

Hoddle (2004) used a commercially available climate modeling program, CLIMEX, to determine the potential geographic range of *X. fastidiosa*, and its vector, the polyphagous xylem-feeding cicadellid leafhopper, *Homalodisca coagulata* (Say). Using weather data, CLIMEX first employs a hydrological model to calculate weekly soil moisture from rainfall data and from estimated rates of evaporation. Next it computes weekly growth indices which summarize the response of the species to prevailing temperatures, day length and moisture. Because periods of unfavorably cold, wet, hot or dry weather can impede or reverse population growth and persistence, stress indices are computed to quantify these negative effects. Growth and stress indices are then combined into eco-climate indices, scaled from 1 to 100, which indicate favorability of a given site for population growth and persistence. Input parameters required by CLIMEX to predict the potential geographic range of the pathogen and insect vector were based partly on knowledge of climatic conditions in the known home range of the species and partly on the known effects of temperature on both species. Using these parameters, the author first verified the model by confirming that the known home ranges in various countries actually had suitable climatic conditions for population growth and persistence. He then proceeded to predict the potential geographic ranges in countries and regions where the pathogen and cicadellid vector had not yet been detected. He concluded that many regions with tropical, subtropical, mild-temperate and moderate Mediterranean climates are suitable for colonization by *H. coagulata* and *X. fastidiosa*. Needless to say, these regions at risk included a large portion of grape growing regions around the world.

3.5 Conclusion

As we have seen from this survey, phenology models, population models and geographic models each can assist vineyard IPM in particular ways. Phenology models improve IPM tactics by predicting the best times for sampling or control operations. Sometimes a single equation with a specific set of parameters will suffice for an extensive region over a number of years. But in other cases separate sets of parameters may have to be computed for each region or each locality within a region to make accurate predictions. Sometimes climatic changes from year to year may necessitate a few early season samples to assist in tuning predictions. In some cases, invalidations may lead to further research and result in better understanding of pest phenology. Further research and better understanding are also needed when more complex phenology models are developed, whether to predict time of acquisition and transfer of pathogens by an insect vector or when a model is used to predict phenology of multiple life stages and to predict the number of generations of multivoltine species.

Population models simulate the effects of the host plant, weather and natural enemies on the density and age structure of insect and mite pests in vineyards. These models, by simulating the impact of these factors on pest dynamics and plant growth, can enhance understanding of plant/pest/natural enemy interactions in the agroecosystem and indicate ways to improve IPM both at the tactical level and the strategic level. Thus, at the tactical level there is emphasis on timing pest control products and adjusting concentrations to prevent economic injury. At the strategic level there can be examination of plant/pest interactions to determine the threat of economic loss at different times in the season under different weather conditions, or examination of the effectiveness of different natural enemies under different climatic conditions. As with phenology models, invalidations of population models can stimulate further research that leads to better understanding and more accurate predictions.

Geographic models give predictions and insights on a geographic scale. A simpler phenological model can predict the potential number of pest generations in different geographic locations, with and without climate change. A more elaborate analysis using output from a tritrophic population model can explore the risk of pest increase in different regions and indicate where biological control is more effective and which biotic and abiotic factors can affect the degree of control. Geographic envelope models can indicate the potential for insect vectors and grape diseases to invade new regions and indicate which climatic factors can serve as barriers.

This survey indicates a number of precautions are needed in the development and use of vineyard models. One complication is that of genetic variation in insect responses to temperature and possibly other weather variables. In some cases responses of a pest species to temperature are uniform enough that a model can be valid over a geographic region or even a number of regions. In other cases, however, pest populations in different regions, or even in localities within a region, respond differently to weather variables. Thus parameters based on rearing insects in controlled climatic conditions or determined iteratively from vineyard samples may have to be determined separately for populations from each region. Where microclimate is an issue, it may be necessary to have separate weather stations in different localities or at least have a representative network of stations. If pest species can respond differently to temperature, it is only reasonable to assume that a similar response can also occur in natural enemy populations. Clearly it is important to be aware of these issues and if necessary, conduct further research.

Another cautionary note is given by Gutierrez et al. (2008) in the context of predicting the likely effects of climate change. It is risky to consider only the effects of higher temperatures on the pest species in isolation, as some have done, because warmer conditions can also affect the host plant and natural enemies of the pest, as well as interactions among these species. Hence, the use of a mechanistic tritrophic model was regarded as a more reliable approach for more complex systems (Gutierrez et al. 2008).

Lastly, what steps should be followed to enhance the contribution of mathematical models to vineyard IPM? In this review we find that vineyard researchers developed models for at most two pest species but vineyards are beset by a complex of pests. Moreover, it may not be ideal to simply develop separate pest models that are used in isolation. A multi-species decision support system involving insect pests in fruit orchards may serve as a paradigm (Samietz et al. 2007). Over a period of several years Swiss researchers developed SOPRA, a web based forecasting tool to optimize timing of monitoring and use of control measures for eight major insect pests in fruit orchards. For each species they determined relationships between temperature and stage-specific developmental rates by rearing insects at constant temperatures. They also developed biophysical models to estimate temperatures experienced by hibernating larvae or pupae in the soil, or within or on stem surfaces. Once biophysical and phenology models for each insect were extensively validated, they were incorporated into SOPRA. Through SOPRA the simulation results are made available to consultants and growers and the predicted phenologies are used as a decision support system in the Alpine valleys and regions north of the Alps in Switzerland. Moreover, using SOPRA growers can select from a variety of strategies to keep pests below damaging levels while minimizing environmental impact. The authors concluded that after several years of use, with proper timing of monitoring and pest control measures, decision support by SOPRA increased the efficacy of pest management and reduced adverse side effects. In the longer term such decision support systems could also assist IPM practice in vineyards.

Acknowledgments The author thanks Paul Hildebrand and Debbie Moreau for useful advice in preparing this chapter.

References

Baumgärtner J, Baronio P (1989) Phenological model of *Lobesia botrana* Den. and Schiff. (Lepidoptera, Tortricidae) as related to the environment of Italy's Emilia-Romagna region. Boll Ist Entomol Univ Studi Bologna 43:157–170

Baumgärtner J, Gutierrez AP, Klay A (1988) Elements for modelling the dynamics of tritrophic population interactions. Exp Appl Acarol 5:243–263

Bostanian NJ, Bougeois G, Vincent C, Plouffe D, Trudeau M, Lasnier J (2006) Modeling leafhopper nymphs in temperate vineyards for optimal sampling. Environ Entomol 35:1477–1482

Bressan A, Larrue J, Padieu EB (2006) Patterns of phytoplasma-infected and infective *Scaphoideus titanus* leafhoppers in vineyards with high incidence of Flavescence dorée. Entomol Exp Appl 119:61–69

Cerutti F, Baumgärtner J, Delucchi V (1992) Research on the grapevine ecosystem in Tessin: IV. Modelling the population dynamics of *Empoasca vitis* Goethe (Hom., Cicadellidae, Typhlocybinae). Boll Ist Entomol Guido Grande Stud Bologna 46:179–200

Forester JW (1968) Principles of systems. Wright-Allen Press, Cambridge, MA

Gallardo A, Ocete R, Lopez MA, Maistrello L, Ortega F, Semedo A, Soria FJ (2009) Forecasting the flight activity of *Lobesia botrana* (Denis & Schiffermüller) (Lepidoptera, Tortricidae) in southwestern Spain. J Appl Entomol 133:626–632

Getz WM, Gutierrez AP (1982) A perspective on systems analysis in crop production and insect pest management. Annu Rev Entomol 27:447–466

Gutierrez AP, Yaninek JS, Wermelinger B, Herren HR, Ellis CK (1988) Analysis of biological control of cassava pests in Africa: III. Cassava green mite *Mononychellus tanajoa*. J Appl Ecol 25:941–950

Gutierrez AP, Daane KM, Ponti L, Walton VM, Ellis CK (2008) Prospective evaluation of the biological control of vine mealybug: refuge effects and climate. J Appl Ecol 45:524–536

Hoddle MS (2004) The potential adventive geographic range of glassy-winged sharpshooter, *Homalodisca coagulata* and the grape pathogen *Xylella fastidiosa*: implications for California and other grape growing regions of the world. Crop Prot 23:691–699

Hoppmann D, Holst H (1993) Forecasting of cycles in the developmental cycles of the grape moths (*Eupocellia ambiguella* and *Lobesia botrana*) and their relationships on weather conditions. Vitic Enol Sci 48:172–175

Lopez MA, Ocete R, González-Andujar JL (2003) Logistic model for describing the pattern of flight of *Kalotermes flavicollis* in sherry vineyards. EPPO Bull 33:331–333

Manetsch TJ (1976) Time-varying distributed delays and their use in aggregative models of large systems. IEEE Trans Syst Manag Cybern 6:547–553

Mizell RF, Tipping C, Andersen PC, Brodbeck BV, Hunter WB, Northfield T (2008) Behavioral model for *Hormalodisca vitripennis* (Hemiptera: Cicadellidae): optimization of host plant utilization and management implications. Environ Entomol 37:1049–1062

Moravie M-A, Davison AC, Pasquier D, Charmillot P-J (2006) Bayesian forecasting of grape moth emergence. Ecol Model 197:478–489

Nagarkatti S, Tobin PC, Saunders MC (2001) Diapause induction in the grape berry moth, *Endopiza viteana* (Clemens) (Lepidoptera: Tortricidae). Environ Entomol 30:540–544

Samietz J, Graf B, Höhn H, Schaub L, Höpli HU (2007) Phenology modelling of major insect pests in fruit orchards from biological basics to decision support: the forecasting tool SOPRA. EPPO Bull 37:255–260

Schmidt K, Hoppmann D, Holst H, Berkelmann-Löhnertz B (2001) Prediction of grape moths dynamics using age structured models. IOBC/WPRS Bull 24(7):127–134

Schmidt K, Hoppmann D, Holst H, Berkelmann-Löhnertz B (2003) Identifying weather-related covariates controlling grape berry moth dynamics. EPPO Bull 33:517–524

Tobin PC, Nagarkatti S, Saunders MC (2001) Modelling development in grape berry moth (Lepidoptera: Tortricidae). Environ Entomol 30:692–699

Tobin PC, Nagarkatti S, Saunders MC (2003) Phenology of grape berry moth (Lepidoptera: Tortricidae) on cultivated grape at selected geographic locations. Environ Entomol 32:340–346

Tobin PC, Nagarkatti S, Loeb G, Saunders MC (2008) Historical and projected interactions between climate change and voltinism in a multivoltine species. Global Change Biol 14:951–957

Vansickle J (1977) Attrition in distributed delay models. IEEE Trans Syst Manag Cybern 7:635–638

Wermelinger B, Candolfi MP, Baumgärtner J (1992) A model of the European red mite (Acari, Tetranychidae) population dynamics and its linkage to grapevine growth and development. J Appl Entomol 114:155–166

Chapter 4
Pesticides for Arthropod Control in Vineyards

Noubar J. Bostanian, John C. Wise, and Rufus Isaacs

4.1 Introduction

Humans have long relied on tactics to protect themselves and their food and fiber from damaging insects. Chemical approaches to arthropod management are relatively recent additions to the multiple tactics available for preventing or reducing damage. Despite the availability of modern synthetic insecticides, about one-third of the world's food crops are systematically destroyed by arthropods during growth, harvest and storage (Ware and Whiticare 2004). This estimate is even higher in developing countries. Because of constraints in space and of the vast scientific information currently available, this chapter, without claiming completeness, provides the reader with basic information on pesticides currently available to control arthropods in vineyards. For more in depth information we refer readers to earlier reviews of insecticides and their modes of action by O'Brien (1960), Corbett et al. (1984), Casida and Quistad (1998), Ishaaya and Degheele (1998), Ware and Whiticare (2004), Stenersen (2004) and Yu (2008). We also guide readers towards Isman's review of botanical insecticides (2006) and Dekseyer's review of acaricides (2005).

N.J. Bostanian (✉)
Horticulture Research and Development Centre, Agriculture and Agri-Food Canada, 430 Gouin Blvd, Saint-Jean-sur-Richelieu, QC, Canada J3B 3E6
e-mail: noubar.bostanian@agr.gc.ca

J.C. Wise
Applied Insecticide Toxicology Laboratory, Michigan State University, East Lansing, MI 48824, USA
e-mail: wisejohn@cns.msu.edu

R. Isaacs
Department of Entomology, Michigan State University, East Lansing, MI 48824, USA
e-mail: isaacsr@msu.edu

N.J. Bostanian et al. (eds.), *Arthropod Management in Vineyards: Pests, Approaches, and Future Directions*, DOI 10.1007/978-94-007-4032-7_4,

Modern grape producers are subject to some powerful forces that are shaping their pest management programs. First, in many parts of the world there is concern over the negative side-effects of pesticides, and this has led to increasing restrictions on the availability of certain chemicals for crop protection. These restrictions have also created an opportunity for more selective insecticides to penetrate the marketplace, and there have been a number of new chemical classes with unique modes of action registered for use in vineyards in recent years. Considerable efforts are also being channelled to produce grapes in organic vineyards. Nevertheless, currently 790,000 ha of vineyards planted globally are managed using conventional management tactics, including insecticide and acaricide applications to prevent yield loss and to ensure high quality grapes. Within each geographic region, integrated pest management programs have been developed that fit the local needs for insect control with the available insect control tools. With the recent rapid changes in availability of some insecticides that were the foundation of these programs, development of resistance to pesticides and the expansion of grape plantings into new regions, there remains a need for testing insecticides and acaricides against key insect and mite pests to meet viticultural goals. As mentioned elsewhere in this book, pest communities are also constantly changing, particularly with the current levels of international trade that bring them into new regions. The movement of grape phylloxera from the New World to France is a classic example, and this pest is now distributed to most major grape production regions (Powell, Chap. 10).

4.2 Insecticide/Acaricide Formulations

Most pesticides cannot be used in their pure form and they have to be formulated as mixtures before application. As a rule, the active ingredient is mixed with inert ingredients to make a combination that is effective and safe to use. Several attributes of the active ingredient and the inert ingredients have to be taken into consideration, including the melting or boiling point, rate of hydrolysis, specific gravity, solubility, vapor pressure, UV degradation, and the inherent biological activity of the active ingredient. The compatibility of the inert ingredient(s) with the active ingredient, compatibility with the container, and the physical properties of the final mixture must also be known. The formulation is then evaluated for homogeneity, particle size, storage stability, retention by the target, wetting, penetration and translocation in plants, residual nature on the target, in the soil, efficacy on the target, effect on non-target beneficials, and hazards to applicators.

4.2.1 Formulation Types

There are numerous formulations of pesticides, each with their properties with relevance to viticultural pest management, and appropriateness for use under different conditions. The earliest formulated pesticides were typically dusts. Dusts contain two ingredients, the toxicant (about 1–10% of the mixture) and an inert carrier. They are the easiest to manufacture and apply in the field. Dusts are, however, the least effective and least economical of the pesticide formulations. They tend to drift during application, resulting in poor deposition on the target. Another shortcoming of dusts is their inhalation hazard. As far as dermal toxicity to humans is concerned, however, dusts are safer than liquid formulations.

Wettable powder (WP) has been a popular formulation for use in fruit and small fruit crops. These are comprised of the toxicant, a wetting agent, and an inert diluent which is usually an adsorptive clay. The wetting agent can be a blend of two or more surfactants. The toxicant comprises 25–75% of the mixture. As the name suggests, these formulations are mixed with water and applied as a spray. They are relatively safe on foliage, but the spray mix must be agitated continuously to avoid settling. Emulsifiable concentrates (EC) are also designed to be applied with water. When mixed with water a stable opaque emulsion is formed, requiring minimum agitation. It is comprised of the toxicant, a solvent for the toxicant and an emulsifier. The non-toxic components may be a mixture of two or more substances. EC formulations penetrate the skin more readily than dusts and WPs. The toxicant content of ECs is indicated in terms of weight/volume instead of weight/weight as with the WPs. Because it is about 25–50% by weight, they are more dangerous if spilled on an applicator. The solvent may also increase the penetration into plant tissue and thus cause phytotoxicity. EC formulations are not popular in viticulture because of the possibility of phytotoxic effects. Suspendable concentrates (SC) or flowables (F) are very fine WP formulations, where the particle size of the active ingredient is 1–5 μm. The particles are suspended in water with surfactants and various additives. Typically these formulations contain 50–90% of toxicant and are therefore applied using smaller volumes of formulated product than other types. Oils may be added if penetration of plants is needed. Water soluble powders (WSP or SP) contain the technical-grade toxicant as a finely ground solid. When added to water these dissolve quickly and can be applied as an invisible solution, with obvious advantages for producers of fresh or dried grapes. Solution (S) formulations have the technical grade toxicant dissolved in a solvent that is highly water soluble. When added to water it dissolves completely and can be applied with a sprayer. Granule (G) formulation pesticides are coarse dusts with particle sizes ranging from 149 to 841 μm. They are manufactured by either impregnating the otherwise inert granule with the toxicant that will be released when the granule breaks up or by surface coating using a volatile solvent. The inert ingredient may be clay or other materials such as corn cobs, pecan and walnut shells or tobacco stems. Granular formulations are mostly used for soil insecticides. Water-dispersible granules (WG or WDG), also known as dry flowable (DF), typically contain 50–95% toxicant, along with the dispersant, binder and diluents. Once in the spray tank, the granules disintegrate and disperse in water and are easily applied by conventional sprayers. These formulations produce less dust than WP and therefore they are safer to use.

In ultra low volume (ULV) formulations, the undiluted technical material (liquids) or the solid material are dissolved in a solvent and applied in an extremely fine spray at 0.6–3.5 l/ha. Such applications can be very effective, possibly because of the absence of inert ingredients (Terriere 1982). They are typically applied by airplane or other specialized equipment.

Some insecticides are applied as aerosols, in which the toxicant is dissolved in volatile petroleum, and the resulting solution is atomized through a jet by means of a propellant gas under pressure (Ware and Whiticare 2004). This creates a fogging action that can reach insects in tight spaces, such as in grape clusters.

Finally, there are various controlled release (CR) devises and formulations that protect the active ingredient and extend the duration of activity. The toxicant may be wrapped up in a polymer carrier (Scher 1999), or in reservoir devices where the toxicant is enclosed in a thin polymeric material to become a microcapsule of 1–100 μm in diameter. In monolithic devices, the toxicant is uniformly dissolved within a polymer matrix to become microcapsules of 1–100 μm in diameter. The toxicant is released by: (a) diffusion through the membrane, (b) digestion of the membrane by an enzyme, microorganism, or chemical process (c) destruction of the membrane by temperature, or moisture. These formulations can reduce worker exposure and can minimize pesticide impact on the environment through lower evaporation and/or leaching. The main disadvantage is cost and longer lasting residues.

4.2.2 Polarity of Insecticides

The polarity of insecticides is an important factor for cuticular penetration. The insect cuticle may be considered as a two-phase system: the epicuticle (outer layer) with hydrophobic properties and the procuticle (inner layer) with hydrophilic properties. Therefore, whether an insecticide is lipid soluble or water soluble, its ability to move across the whole cuticle depends on whether it can pass through the hydrophobic and hydrophilic barriers. The efficiency of such movement will depend on the oil–water partition coefficient of the insecticide, the nature of the surfactant, the solvent, and the nature of the cuticle (Terriere 1982). The partition coefficient and water solubility also dictate the ability of the pesticide to penetrate into leaf and fruit cuticles and move in the vascular system, respectively. This will also influence whether the residue will remain on the plant after rain, as well as where they will be distributed in the grapevine where insects are feeding.

4.2.3 Synergism and Antagonism

Synergism is a process where the toxicity of two compounds together exceeds the expected toxicity from the sum of their effects when applied separately and the non-toxic compound is the synergist. Antagonism is said to occur when precisely the opposite effect to synergism is noted. In other words the toxicity of two compounds applied together is less than that expected from the sum of their effects when applied separately. The synergistic ratio is defined as the increase in toxicity caused by the non-toxic compound, and is obtained by estimating the (LD_{50} of toxicant alone)/(LD_{50} of the mixture). If this value is greater than one, synergism has occurred and the non-toxic compound is a synergist. If the value is less than one, antagonism has occurred and the non-toxic compound is an antagonist.

Many synergists such as sesamin, sesamolin, piperonyl butoxide and sesamex contain the active methylenedioxyphenyl moiety. Currently piperonyl butoxide is used with pyrethrin to manage drosophilids in California, US. These synergists were originally developed to be used with pyrethrins. However, they have been noted to synergize some but not all carbamates, oranophosphates, pyrethroids and chlorinated hydrocarbons.

Synergists increase the toxicity of pyrethroids as much as 50-fold. However, synergist/insecticide ratios are often high. For example a 5–10:1 ratio is needed for OP synergists used against houseflies (O'Brien 1960). It has now been shown that these synergists act by inhibiting cytochrome P450 monooxygenases and result in the accumulation of the toxicant.

4.3 Background Neurobiology and Transmission of an Impulse

Currently most insecticides in use act upon nerve impulse transmission. Hence, a brief review explaining the transmission of impulses will be necessary to explain the significance of the interference caused by these insecticides. For an in depth treatment the reader is referred to physiology textbooks such as Nation (2008).

The nervous system is composed of nerve cells termed neurons. At rest, the nerve cell membrane, which is made of a double layer of lipids, is relatively impermeable to sodium, permeable to chloride ions and has a controlled permeability to potassium ions. There are pores or channels in the membrane that allow various ions to pass when open. These channels are gated, and there are two types of gates: ligand-ion gated and voltage-ion gated channels. In ligand-ion gated channels the gates open or close by various signal molecules (chemical messengers such as a neurotransmitter). The binding sites are normally located on a different portion of the protein (allosteric binding site) relative to where the ion conduction pore is situated. An example of such a channel is the nicotinic acetylcholine receptor. It consists of a pentamer of protein subunits, with two binding sites for acetylcholine, which when bound change the configuration of the receptor and result in the opening of an internal pore. This pore allows Na^{+} ions to flow down their gradient into the cell. In voltage-gated ion channels, the gates are activated by changes in electrical potential difference near the channel. Voltage-gated sodium and calcium channels consist of a single polypeptide with four homologous domains. Voltage-gated potassium channels are composed of four separate polypeptide chains each with one domain. Voltage-gated ion channels open when the voltage falls below a threshold. The arrival of an impulse at a point along the axon results in a drop of a voltage difference of −70 to +30 mV at that point. When the voltage-gated sodium channels open, sodium rushes into the cell and increases the voltage drop. Then the voltage-gated potassium channels open, and the potassium rushes out. In insects, voltage-gated sodium channels are located in neurons only. A slight voltage drop is also experienced further down the axon causing the entry and exit of sodium and potassium ions, and in this way, the impulse travels down the axon until it reaches the synapse where it declines. Following the passage of an impulse at any point the sodium channels close, followed slowly by the closing of the potassium channels. The efflux of potassium ions compensates for the influx of sodium ions and re-establishes the resting potential. The sodium ions are continually pumped out and potassium ions pumped in, with the expenditure of energy by the ion pump, and this activity maintains the resting potential with a concentration difference of ions between the inside and outside of a nerve cell.

When the impulse arrives at the presynaptic membrane, the drop in potential allows calcium ions to flow into the terminal via voltage-gated calcium channels. These channels are present in nerve terminals and muscles. They are usually closed, but open in response to a drop in voltage. The increase in free calcium ions inside the cell is transient as the synaptic knob can remove calcium from the cytoplasm by pumping it out of the cell or taking it up into intracellular bodies. Calcium ions at a concentration of 1–10 μmol reduce the energy barrier between the membranes of the cell and the membranes of vesicles within. The vesicles fuse and discharge their content (acetylcholine) into the synaptic cleft. In less than a millisecond the acetylcholine diffuses across the cleft and binds with specific receptor proteins on the postsynaptic membrane. The receptors are normally closed, but open in response to acetylcholine binding, and allow sodium to flow in and potassium out. Each channel molecule needs two acetylcholine molecules to open. The electrical voltage at the postsynaptic membrane drops because of the influx of sodium. The magnitude of the drop depends on the number of gates that have been opened and for how long. If sufficient number of gates are opened long enough, the voltage difference across the postsynaptic membrane decreases sufficiently to open the voltage gated sodium channels, so that voltage difference is further decreased, and an action potential is achieved (Fig. 4.1).

There are also synapses that deliver substances that do not decrease the membrane potential at the postsynaptic membrane. On the contrary they increase it by binding to specific receptor sites. These synapses are inhibitory because when activated they inhibit the transfer of signals from the excitatory synapses. Most chloride channels are of this nature. Although chloride ions cannot freely move across the membrane, the outside and inside concentrations are such that they could do this with ease. Because of voltage differences, the outside-inside concentration difference may be extensive (e.g. 575 μmol outside and 40 μmol inside). Opening the chloride channels allows the chloride ions to enter the cell because the concentration is much higher on the outside than the inside. This chloride influx reduces the effect of sodium influx caused by the opening of the sodium channels. The most important inhibitory neurotransmitter is gamma-aminobutyric acid (GABA).

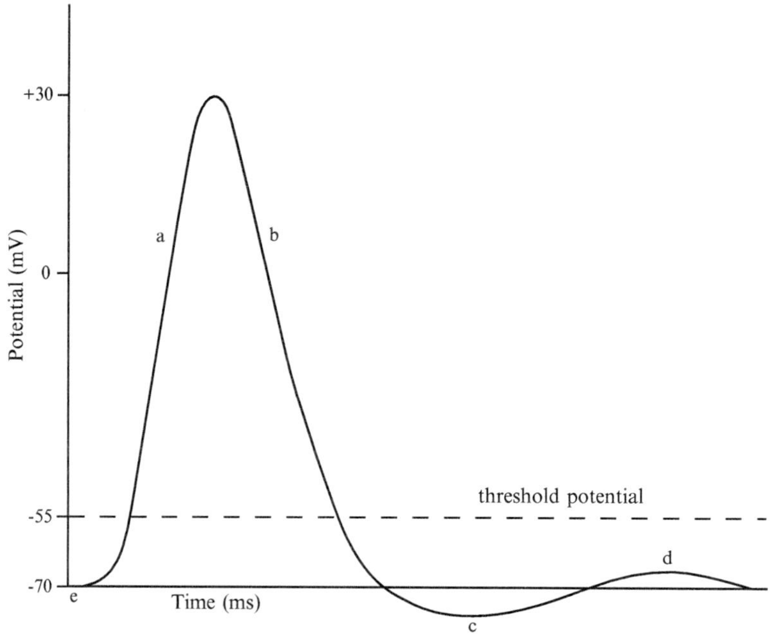

Fig. 4.1 A model action potential. (**a**) The rising phase: Influx of Na^+ ions into the axon from outside. (**b**) The falling phase: Na^+ permeability has decreased while K^+ permeability has increased causing the exflux of K^+ ions out of the axon. (**c**) The positive phase: This caused by enhanced permeability to K^+ ions even when the membrane is repolarised to the resting potential. (**d**) Negative after potential: This is caused by high local concentration of K^+ ions outside the axon resulting in increased net K^+ ion influx that delays equilibration. (**e**) Resting potential: The axon contains relatively a low concentration of Na^+ and relatively high levels of K^+ (Adapted from Corbett et al. 1984)

4.4 Classification of Insecticides According to Their Mode of Action

Each pesticide has at least three names, including a trade name that most growers, extension educators, and crop consultants would recognize, a common chemical name proposed by the manufacturer and endorsed by government agencies and professional societies, and a chemical name that describes the chemical composition and structure of the active ingredient. The chemical name is based on the principles of chemical nomenclature set by the International Union of Pure and Applied Chemistry (IUPAC). Pesticides may be classified according to their chemical structure as well as by their mode of action, as described in this chapter.

As an aid to resistance management (Sect. 4.11.4), the Insecticide Resistance Action Committee (IRAC) arranges insecticides and acaricides into groups according to their modes of action. As a general rule, rotation of compounds between different IRAC groups to minimize selection pressure on a given pest will help avoid or delay the development of resistant pest populations. For convenience, we provide the IRAC groups for each major group of insecticides and acaricides described in this chapter.

4.4.1 Acetylcholinesterase (AChE) Inhibitors (IRAC Group 1A and 1B)

4.4.1.1 Organophosphate and Carbamate Insecticides

Organophosphates (OPs) cover a very large group of toxic compounds containing phosphorus. The inclusion of different groups around the phosphorus center produces toxicants with a wide range of physical and biological attributes, chemical stability, and selectivity. Organophosphates can be synthesized with exact lipophilic-hydrophilic balance required for movement within plants (e.g. dimethoate). They have a very specific mode of action, and their biological activity is easily destroyed by the simplest chemical or biochemical modification. They are also relatively unstable in biological systems.

Organophosphates can be considered as esters of alcohols with a phosphorus acid or as anhydrides of phosphorus acid with another acid. They are divided into three groups of derivatives: aliphatic, phenyl and heterocyclic. The aliphatic derivatives such as dimethoate are the simplest in structure and have a wide range of toxicities. The heterocyclic derivatives are the most complex (e.g. azinphosmethyl) and one or more of the carbon atoms in the ring are displaced by oxygen, nitrogen or sulphur. The residual toxicity is the longest and they often have several metabolites. In between these two groups in complexity and persistence lie the phenyl derivatives (e.g. ethyl parathion). An essential characteristic of OPs is the electrophilic phosphorus atom, which is brought about partly by polarization of the P=O bond and partly by the electron withdrawal properties of the other groups in the molecule. The two most important non-enzymatic features of OPs are hydrolysis and isomerization. The speed of hydrolysis is directly related to the concentration of the alkaline medium, because it is the OH^- ion that causes the hydrolysis, and it attacks the partially-charged P moiety. The conversion of P=S to P=O is isomerization and it is termed 'desulfuration'. All compounds containing =S are 'latent inhibitors' and become direct inhibitors after desulfuration. Most OPs would be poor toxicants if they were not activated in living systems. The reactions are desulfuration, hydroxylation, thioether oxidation and cyclization, all four of which take place in vertebrates and insects.

Carbamates are another important class of anticholinesterases. They are derivatives of carbamic acid ($HOOCNH_2$). Three groups describe most of the carbamate insecticides: (1) aryl N-methylcarbamate such as carbaryl, (2) dimethylcarbamyl ester of a heterocyclic hydroxyl compound such as pirimicarb, and (3) N-methylcarbamyl ester of an oxime, also called oxime carbamates such as methomyl and thiocarb. As a class they tend to be fast acting and degrade rapidly. When compared to OPs, most carbamates are more systemically active in plants indicating that they have high water solubility and they are not easily metabolized by plants.

They are direct inhibitors and bioactivation is not important. They are moderately broad-spectrum in effectiveness (not as much as organophosphates), being used as insecticides, acaricides and even molluscicides. Carbaryl and methomyl are examples of carbamates currently used in North American vineyards (Anonymous 2010, 2011). Carbaryl is effective against grape berry moth, hoplia beetle, cutworms, omnivorous leafroller, orange tortrix, grape leaffolder and leafhoppers. Methomyl is recommended against grape leaffolder, cutworms, mealybugs, omnivorous leafroller, orange tortrix, thrips, and western grapeleaf skeletonizer. Unless predatory mites have been shown to be resistant to carbamates (Hassan et al. 1987; Hardman et al. 2000), this family of insecticides may provoke spider mite outbreaks in vineyards. Hence, careful planning in timing their application, predator mite prey ratios, and other ecological concerns should be taken into consideration when applying them in vineyards.

Organophosphates and carbamates exert their toxic action by inhibition of cholinesterase. Following the transmission of an impulse and the release of acetylcholine, a return to normality is brought about by hydrolysis of acetylcholine by acetylcholinesterase to acetic acid and choline. In the presence of OPs and carbamates, this hydrolysis is inhibited, resulting in the accumulation of acetylcholine. The inhibition can be explained in three steps. The first step is the formation of a complex (E-OH.AZ). The second step involves phosphorylation, acetylation, or carbamylation (E-OA) of the enzyme at the esteratic site of the amino acid serine in the acetylolinesterase, while the remainder of the molecule is attached to the anionic site. The third and last step involves hydrolysis (i.e., dephosphorylation, deacetylation, or decarbamylation) to yield the original enzyme (E-OH). These events are summarized by Corbett et al. (1984) as follows:

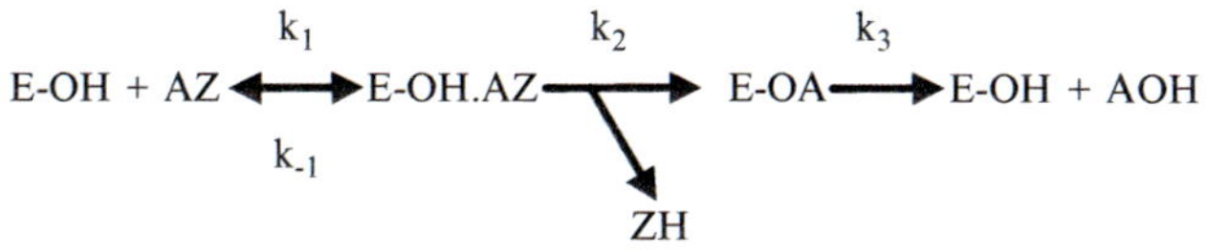

where E-OH is the enzyme and AZ is the inhibitor (A is the phosphorylating or carbamylating group (acetyl group, the dialkyl phosphoryl group, or the methylcarbamyl group)) and Z represents the leaving group (i.e. choline in uninhibited acetylcholine, p-nitrophenol in paraoxon, or 1-naphthol in carbaryl). Kinetic studies have shown that with organophosphates, k_2 is moderately fast and k_3 is extremely slow. Therefore, E-OA accumulates. The values of k_1, k_{-1} and k_2 are such that under normal conditions no E-OH.AZ is ever present. With carbamates, k_2 is much slower and k_3 even more slower, so that under normal conditions low concentrations of the complex E-OH.AZ and high concentrations of the carbamylated enzyme E-OA (half life is a few minutes) have been observed. With organophosphates, then the inhibitory action is reversed very slowly; with carbamates, decarbamylation takes place within a few hours, and the inhibition is readily reversed. The accumulation of acetylcholine in the synapses of insects and of vertebrates leads to repetitive firing followed by blockage of nerve transmission. In mammals, this inhibition takes place at the neuromuscular junction, and death is brought about by paralysis of the intercostal muscles, which results in asphyxiation. Organophosphates can also cause neuropathology of the visual system or effects on cognitive functions (learning and memory). These attributes have caused their uses to be restricted since 1996 in several countries. Nevertheless, in North America, there are still few OPs that are effective and recommended in vineyards. The list includes: chlorpyrifos to control spiders, climbing cutworms, and vine mealybug; diazinon against cutworms, false chinch bug, chinch bug, grape berry moth, omnivorous leafroller and leafhoppers; dimethoate against grape bud beetle, leafhoppers, thrips, mealybugs and sharpshooter; phosmet against grape bud beetle, omnivorous leafroller, Japanese beetle, grape berry moth, and light brown apple moth; malathion against false chinch bug and the multicolored Asian lady beetle (Anonymous 2010, 2011).

4.4.2 Voltage-Gated Sodium Channel Modulators (IRAC Group 3)

4.4.2.1 Pyrethroids

Pyrethrin, the pyrethroid group (permethrin, cypermethrin, bifenthrin, lambda-cyhalothrin, cyfluthrin, deltamethrin, esfenvalerate, etofenprox and tefluthrin) and the botanical insecticide sabadilla are major examples of insecticides that bind to sodium channels and cause a delay in sodium channel closing. The lipophilic attributes of these compounds are central to their chemical reactivity. Another compound that is thought to interfere in this manner is DDT and its analogues (e.g. dicofol). Nerve preparations from squid (Lund and Narahashi 1981) and frogs (Vijverberg et al. 1982) have indicated that pyrethroids delay the closing of a small percentage of the sodium channels, which open on depolarization (Na^+ gushes into the axon), while the majority of the channels behave normally. The delay of the closure of sodium channels causes the membrane potential to remain above the threshold for a longer period. Furthermore these insecticides show a negative after-potential which means that the axon has not recovered to its resting stage. This results in a greater release of transmitter triggering a number of spike potentials, a process known as repetitive spiking. These repeated action potentials lead to postsynaptic hyperstimulation, resulting in hyperactivity, tremors and rigid paralysis (Corbett et al. 1984; Matsumura 1985).

With pyrethroids, the time needed to close the modified channels depends on the pyrethroid in question (Vijverberg et al. 1982). It is short with alpha-cyano pyrethroids including permethrin. Gammon et al. (1981) classified pyrethroids causing tremors as Type I and those causing convulsions with salivation as Type II. Another attribute of Type I pyrethroids is that their toxicity is inversely related to temperature. In contrast, the toxicity of Type II pyrethroids is directly related to temperature. Though pyrethroids are highly lipophilic, they are not stored to a significant extent in fatty tissues or other tissues in mammals. This is because of their rapid metabolism with the production of metabolites that are highly soluble in water and can be conjugated and excreted. They are viscous liquids with a high boiling point and a low vapour pressure. These attributes determine their fast action on insects, slow penetration into leaves and low systemic movement in plants. Thus, they are very effective contact insecticides against Lepidoptera larvae and eggs, as well as against larvae and adults of several Coleoptera, Diptera and Heteroptera pests. Though pyrethroids are persistent in nature they are relatively harmless to mammals and birds compared with other broad-spectrum insecticides and do not have any phytotoxic attributes (Elliot 1977; Elliot et al. 1978). In contrast pyrethrin, the natural product obtained from the ground flowers of chrysanthemum, has little residual activity and decomposes quickly once it is applied. Pyrethroids are toxic to certain predatory mites (Bostanian et al. 1985) and non-toxic to others (Hardman et al. 2000; Bostanian and Laroque 2001). Hence, correct identification of the predator, careful planning in timing the application, and the residual toxicity of the pyrethroid to the predatory mite(s) should be taken into consideration when applying them in vineyards. Permethrin is effective and recommended against climbing cutworms, leafhoppers and grape berry moth and cypermethrin against multicolored Asian lady beetle in Ontario, Canada (Anonymous 2010). In California, fenpropathrin is recommended against spiders, false chinch bug and sharpshooters, whereas the natural pyrethrin-piperonyl butoxide mixture is recommended against vinegar flies and leafhoppers (Anonymous 2011).

4.4.3 Voltage Dependent Sodium Channel Blockers (IRAC Group 22)

4.4.3.1 Oxadiazines

Indoxacarb is an oxadiazine. It is a latent inhibitor that becomes quickly a direct inhibitor as soon as it is metabolized by an esterase/amidase to its corresponding N-decarbomethoxylated metabolite (DCJW). This metabolite interferes with the normal functioning of Na channels but in a different manner. DCJW suppresses action potentials (Wing et al. 1998, 2005), leading to flaccid paralysis and death. Insects exposed to this compound stop feeding within a few hours, become less mobile and can show slight tremors and convulsions. It has contact and stomach activity against lepidopteran larvae and some ovicidal action. It is moderately translaminar and innocuous to some beneficials and toxic to others (Bostanian et al. 2004).

4.4.4 Voltage-Gated Calcium Channel Modulators (IRAC Group 28)

4.4.4.1 Diamides

Calcium channels are present in nerve and muscle terminals and have an important role in neurotransmitter release in presynaptic nerve endings. Following depolarization caused by an action potential in the nerve terminal, calcium channels are activated and result in an influx of Ca^{2+} ions. In insects these ions stimulate the release of amino glutamate which diffuses across the synaptic cleft and binds to a receptor-operated ion channel resulting in the influx of Na^{+} and Ca^{2+} ions. This influx then activates the sarcoplasmic reticulum calcium channels in muscles which then release Ca^{2+} ions in the muscle filaments resulting in muscle contraction. Flubendiamide, a benzenedicarboxamide insecticide, induces intracellular Ca^{2+} release mediated by a calcium channel ryanodine receptor causing the contraction of insect muscle and body (Tohnishi et al. 2005). In insects these receptors are 500-fold more sensitive (Cordova et al. 2006). Hence, insecticides with this mode of action are very safe from a user's point of view. It is effective against the grape berry moth and climbing cutworms. It is totally innocuous to *Galendromus occidentalis* (Nesbitt) (Lefebvre et al. 2011) and *Neoseiulus fallacis* (Garman) (Lefebvre et al. 2012), two predatory mites, dominant in eastern and western North American vineyards. Among insecticides of botanical origin, ryanodine the active component of ryania, activates calcium channels which release an excess of calcium ions into the protein fibers causing tremendous increases in oxygen consumption followed by flaccid paralysis and death of insects and vertebrates (Bloomquist 1999).

Chlorantraniliprole and cyantraniliprole are selective ryanodine receptor activators, stimulating the release of Ca^{2+} ions from the sarcoplasmic reticulum. They interfere with normal muscle contraction (Cordova et al. 2006). Lepidoptera exposed by ingestion to these insecticides are immobilized and cease to feed. Chlorantraniliprole is currently increasing in use in vineyards for management of lepidopteran pests. It is effective against the grape berry moth and the climbing cutworms (Anonymous 2010). Laboratory studies have shown it to be marginally toxic only to the larvae of the predatory mite *G. occidentalis* (Lefebvre et al. 2011). In contrast it is totally innocuous to *N. fallacis* (Lefebvre et al. 2012).

4.4.5 Glutamate-Gated Chloride Channel Agonists (IRAC Group 6) and Antagonists (IRAC Group 2)

4.4.5.1 Chloride Channel Agonists and Chloride Channel Antagonists

Gamma-aminobutyric acid (GABA) is located in the central nervous system and at peripheral neuromuscular junctions. It incites inhibitory actions by its ability to increase Cl^{-} ion permeability of the nerve membrane at these locations. It has been described as a revolving tap mechanism. Gamma-aminobutyric acid activation (agonists) rotates the tap in the open position, whereas receptor antagonists oppose this activation. In insects they are found in the central nervous system and also at peripheral neuromuscular junctions. In GABA receptors, the arrival of an action potential triggers the release of GABA from the presynaptic terminal, and GABA binds to a postsynaptic receptor protein containing a chloride channel. As a result, the chloride channel is opened and Cl^- ions flow into the post synaptic neuron. The increased Cl^- ions cause hyperpolarization of the membrane and produce an inhibitory postsynaptic potential (Buckingham and Sattelle 2005). Avermectins, which are a mixture of macrocyclic lactone antibiotics isolated from *Streptomyces avermitilis*, open the chloride channel and act as partial agonists (Albrecht and Sherman 1987). In susceptible arthropods they cause loss of motor function and paralysis. Abamectin is very toxic to mites and less toxic to lepidopteran and homopteran insects (Lasota and Dybas 1991). In the laboratory it was very toxic to *Orius insidiosus* (Say) (Anthocoridae) adults and *Aphidius colemani* Viereck (Braconidae) adults after 9 days of exposure (Bostanian and Akalach 2004). In California it is recommended against the western grapeleaf skeletonizer (Anonymous 2011) and in Ontario (Canada) against phytophagous mites (Anonymous 2010). Among acaricides, milbemectin is a GABA agonist that degrades rapidly when exposed to light. However, because of its translaminar attributes, it provides residual contact activity against pest mites.

In contrast, GABA antagonists (IRAC Group 2) such as cyclodienes, lindane and phenylpyrazole insecticides, bind to the chloride channel and block its activation by GABA. The absence of inhibition results in hyperexcitation of the CNS. Currently endosulfan is the only cyclodiene organochlorine still recommended in North America. It is effective against leafhoppers in California and phylloxera in Ontario.

4.4.6 Nicotinic Acetylcholine Receptor (nAChR) Agonists (IRAC Group 4)

4.4.6.1 Neonicotinoids

The nicotinic acetylcholine receptors (nAChR) are present on both post- and presynaptic nerve terminals as well as on motor and sensory neurons and cell bodies of interneurons (Jeschke and Nauen 2005). Nicotine and the neonicotinoids (imidacloprid, acetamiprid, thiamethoxam, nitenpyram, thiacloprid, dinotefuran and clothianidin) mimic acetylcholine by acting as agonists to bind at the nAChR. This attribute results in an influx of sodium ions and the generation of action potentials. In normal circumstances, the action of acetylcholine is stopped by acetylcholinesterase which quickly hydrolyzes the neurotransmitter. However, these agonists are not degraded by acetylcholinesterase and continuous activation leads to over stimulation of cholinergic synapses, resulting in hyperexcitation, convulsion, paralysis, and eventual death of the insect.

The neonicotinoids are active against key homopteran and some coleopteran and lepidopteran pests. Imidacloprid is recommended against leafhoppers, grape phylloxera in eastern North America, and against mealybugs, sharpshooters, grape phylloxera, western grapeleaf skeletonizer, and thrips, in western North America. Acetamiprid is recommended against banded grape bug, leafhoppers and grape berry moth in New York State and against western grapeleaf skeletonizer and leafhoppers in California. They act as contact and stomach poisons and can be used as foliar, drench or seed dressing treatments. They exhibit systemic and translaminar properties and high residual activity (Elbert et al. 1998). The effects of neonicotinoids on predatory mites range from no effect (Laurin and Bostanian 2007) to unacceptable levels of mortality (Bostanian et al. 2010). Consequently, each product should be evaluated in depth for its toxicity to the particular beneficial(s) present in grape production regions.

4.4.7 Nicotinic Acetylcholine Receptor (nAChR) Allosteric Modulators (IRAC Group 5)

4.4.7.1 Spinosyns

The symptoms of intoxication are similar to nicotinic acetylcholine receptor (nAChR) agonists described above. However, instead of binding directly to the active site, compounds in this group (spinosad and spinetoram) bind to a specific allosteric site remote from the active site, and in the process they contribute to the activation of the active site (Dunbar et al. 1998). This causes spontaneous muscle contractions and tremors. They consist of macrocyclic lactones obtained from the fermentation of the soil actinomycete bacterium *Saccharopolyspora spinosa*. Unlike the neonicotinoids, the toxicity of spinosyns to homopterans such as aphids is low but they are particularly active against Lepidoptera, Thysanoptera and Diptera. Spinosad is recommended against grape leaffolder, cutworms, omnivorous leafroller, and western grapeleaf skeletonizer in California and against grape berry moth in eastern North America. Hence, these compounds can be used with neonicotinoids in pest management programs with lower environmental impact. They have relatively low mammalian toxicity but, like neonicotinoids, their toxicity to non-target beneficials is variable (Laurin and Bostanian 2007; Bostanian et al. 2010; Lefebvre et al. 2011). They should be evaluated for their effects before they are considered for inclusion in IPM programs for vineyards. Some formulations of spinosyn are also registered for use in organic viticulture.

4.4.8 Oxidative Phosphorylation via Disruption of Proton Gradient (IRAC Group 13)

4.4.8.1 Pyrroles

Chlorfenapyr is a pyrrole pro-insecticide-acaricide. Its activity depends upon the oxidative removal of N-ethoxymethyl group by mixed function oxidases to form

CL303268. This compound acts as an uncoupler of oxidative phosphorylation when ingested. Once in a cell, it breaks the close coupling between the respiratory chain and phosphorylation resulting in the loss of respiratory control. While electron transport along the chain occurs at full speed, ATP production ceases. The affected cells are starved of energy causing the eventual death of the organism (Hunt and Tracy 1998).

4.4.9 Microbial Disruptors of Insect Midgut Membranes (IRAC Group 11)

4.4.9.1 Endotoxins

The surface of *Bacillus thuringiensis* var. *kurstaki* (Btk) spores form crystalline inclusions that contain the δ-endotoxin. When the spores are ingested by insects, they germinate and the alkaline media in the lepidopteran midgut dissolve the inclusions and release protoxins. These are then activated by proteases in the midgut into smaller toxin molecules. Once activated these toxins bind to receptor sites on the microvillar membrane of the midgut epithelial cells. The toxins then disturb the osmotic balance of the cells by forming pores in the epithelium. The cells swell and lyse, with eventual destruction of the midgut. Fluids from the highly alkaline intestine pass into the hemolymph, and raise its pH from 6.8 to 8. This leads to a generalized paralysis and death of the poisoned insect. Graf (2011) categorizes poisoned insects into three types. Type I includes insects that exhibit general paralysis with leakage in the midgut followed by rapid death. Type II includes insects that exhibit gut paralysis without leakage of the gut contents resulting in slow death. Type III includes insects that die from septicaemia and not the toxin, following the germination of the Bt spores. The septicaemia is caused by micro-organisms such as *Enterococcus faecalis* found in the midgut. Compared to pesticides that affect the nervous system, the δ-endotoxins are slower in action but they are specific to Lepidoptera with no side effects on humans or the environment, including natural enemies. Currently Btk is recommended against grape leaffolder, omnivorous leafroller, and the western grapeleaf skeletonizer in California, and it has shown activity against grape berry moth in eastern North America (R. Isaacs and J. C. Wise, unpubl. data). *Bacillus thuringiensis* formulations are also approved for use in organic viticulture.

4.4.10 Insecticides that Affect Insect Development

Insect growth regulators (IGRs) alter growth and development of insects. These compounds disrupt insect growth and development as juvenile hormones or as chitin synthesis inhibitors.

4.4.10.1 Juvenile Hormone Mimics (IRAC Group 7)

Juvenoids

The juvenile hormone analogues interfere with the development and emergence of insects as adults. Upon ingestion, larvae usually stop feeding and undergo incomplete or premature molts which eventually result in their death. Methoprene, hydroprene, fenoxycarb, and pyriproxyfen are juvenile hormone mimics and show their maximum effect when applied at the beginning of metamorphosis (Ishaaya 2001). They have activity against egg and larval stage insects, and there may be sublethal effects on fecundity.

4.4.10.2 Ecdysone Receptor Agonists (IRAC Group 18)

Diacylhydrazines

This group is comprised of diacylhydrazines (tebufenozide, methoxyfenozide, halofenozide and chromafenozide) which act as nonsteroidal ecdysone agonists. The group binds to the ecdysteroid receptor binding proteins and interferes with the normal processes of development. They induce premature incomplete molts that result in larval mortality (Smagghe et al. 2004), and are considered molting accelerating compounds. Tebufenozide and methoxyfenozide are very effective against lepidopteran pests, and they are highly selective with no adverse effects on natural enemies (Dhadialla et al. 1998). Methoxyfenozide is recommended against omnivorous leafroller and western grapeleaf skeletonizer in California. It is used increasingly in eastern North America for control of grape berry moth (Isaacs et al. 2005). It is totally innocuous to *G. occidentalis* (Bostanian et al. 2009a) and *N. fallacis* (Bostanian et al. 2010).

4.4.10.3 Inhibitors of Chitin Biosynthesis (IRAC Groups 15 and 16)

Benzoylphenylureas

Chitin inhibitors inhibit the production of new exoskeletons when insects are molting. Consequently, the cuticle is unable to support the insect and withstand the rigors of molting, leading to its death. Benzoylphenylureas (diflubenzuron, teflubenzuron, flufenenoxuron, lufenuron, and novaluron (IRAC Group 15 type 0)) and buprofezin (IRAC Group 16 type 1) are examples of chitin inhibitors in insects, whereas etoxazole and flufenoxuron are chitin inhibitors in mites. Benzoylphenylureas are very effective against Lepidoptera. They show their effect mainly by ingestion, but in some species they suppress fecundity and show ovicidal and contact toxicity (Ishaaya and Horowitz 1998). Novaluron (currently not registered in grapes in North America) has more contact and translaminar activity when compared with other benzoylphenylureas. It is also effective against coleopteran larvae and leafminers

(Ishaaya et al. 2002). Novaluron was found to be slightly more toxic to larvae of *N. fallacis* than *G. occidentalis*. It had no direct effect on other growth stages of these two predatory mites (Lefebvre et al. 2011, 2012). Currently, buprofezin is recommended against leafhoppers and vine mealybug in California.

4.4.10.4 Inhibitors of Acetyl CoA Carboxylase (Lipid Metabolism Enzyme) (IRAC Group 23)

Tetronic and Tetramic Acid Derivatives

Spirotetramat is a keto-enol derivative of tetronic acid. It acts mostly by ingestion where it inhibits lipogenesis that leads to a diminution of growth regulators and fertility. It has systemic and translaminar attributes. Furthermore, it is characterized by ambimobility within plants. In other words, it moves into new leaves formed after treatment, and it also can move from foliage applications to roots. Its spectrum of activity is restricted to Homoptera, but it provides a high level of control of phylloxera both on foliage and on the roots (Powell, Chap. 10). In laboratory studies, it is toxic to *G. occidentalis* adults (Lefebvre et al. 2011), and marginally toxic to *N. fallacis* adults (Lefebvre et al. 2012). Other pesticides in this group include spirodiclofen which is strictly an acaricide and spiromesifen which is an insecticide-acaricide.

4.5 Classification of Acaricides According to Their Mode of Action

Phytophagous mites are generally considered induced pests in vineyards. Usually they become troublesome when pesticides used to manage other arthropods and diseases are toxic to their natural enemies that would otherwise keep them below their economic injury level. Once mites become a problem, acaricides are used to shift the balance in favor of the predators. Management of mites with acaricides can be a continual struggle for vineyard managers, especially in hot arid regions of grape production. They rapidly develop resistance to acaricides due to their high reproductive capacity and very short life cycles. A highly desired feature of acaricides is their innocuity to key predatory mites so that they can be utilized to re-establish natural control (biological control) of phytophagous mites, if possible within the same growing season.

4.5.1 Neurotoxins

Historically compounds that interfere in one way or another with nerve activity have been the largest group of acaricides. Currently dicofol, an analogue of DDT, is the only chlorinated hydrocarbon acaricide still recommended in certain areas. It is not compatible with IPM programs as it is toxic to predatory mites and lady beetles (Anonymous 2011). Several of the early organophosphates (such as parathion), carbamates and pyrethroids have acaricidal attributes. Recently newer pyrethroids have been added to the arsenal of acaricides. Acrinathrin, halfenprox and lubrocythrinate are all fluorinated pyrethroids with activity on mites. Bifenazate, a novel carbazate compound, and milbemectin which is a mixture of two macrolides both provide mite control. For additional details on these novel acaricides the reader is referred to Dekeyser (2005).

4.5.2 *Acaricides and Insecticides Affecting Respiration (IRAC Groups 20 and 21)*

The group fenazaquin, fenpyroximate, pyrimidifen, pyridaben, and tebufenpyrad all have acaricide/insecticide activity, and they are all inhibitors of mitochondrial complex I transport system (Group 21). Acequinocyl and fluacrypyrim are strictly acaricidal and inhibitors of mitochondrial complex III transport system (Group 20). These structurally diverse acaricides interfere with the normal process of mitochondrial respiration, and have varying levels of selectivity against predatory mites.

4.5.3 *Inhibitors of Growth (IRAC Groups 15 Type 0 and Group 10)*

The group includes compounds that interfere with the normal utilization of lipids leading to death. This includes etoxazole which is an oxazoline, and the following benzoylphenylureas: flucycloxuron, flufenoxuron. All three inhibit chitin biosynthesis in insects and mites (IRAC Group 15 type 0). The two tetronic acid derivatives, spirodiclofen and spiromesifen (insecticide/acaricide) also interfere with growth. Spirodiclofen inhibits lipid biosynthesis. Spiromesifen is suspected to have the same mode of action. Clofentezine (IRAC Group 10) inhibits embryo development, and is characterized by long residual toxic activity to eggs and larvae. Its exact mode of action is not known and it has no effect on adult mites. All these acaricides are generally slow to act, but they provide long-lasting control.

4.6 Insecticide Mode of Activity: General Considerations

Describing the mode of action is the critical first step in understanding the fundamental mechanism by which an insecticide or an acaricide controls a pest. It also serves a key role in compound classification for resistance management. We consider, however, that documenting the mode of activity provides additional field-relevant information about insecticides that allows pest management practitioners to more accurately predict and evaluate the performance of insecticides for vineyard pest management programs.

The mode of activity is reflected in field-assessable symptoms on an organism caused by the action of a pesticide (Wise and Whalon 2009). It captures the influence of key ecological elements of the Plant-Insect-Chemical Triad (Wise and Whalon 2009) on the ultimate performance of an insecticide or an acaricide under vineyard conditions. Many of the pesticides discovered in the twentieth century, such as organophosphates, pyrethroids and carbamates, rely on a singular lethal mode of activity to kill, and to control the target pest. These same compounds are typically lethal to three or more insect life stages (Fig. 4.2), such that their efficacy reflects breadth not only in terms of pest spectrum but also of life-stage activity. Adulticides are directly toxic to the adult life-stage of the insect by contact or ingestion. Compounds that kill larvae by contact or ingestion are called larvicides, but the term ovi-larvicidal is used when the larva ingests the toxin while emerging from the egg. Ovicides are toxic to insect eggs either when the egg is laid on top of plant-applied residues or when the insecticide is sprayed over top of the eggs. Many of the recently discovered insecticides that offer elements of lethal activity do so with narrower life-stage selectivity, with implications for control of vineyard pests (Isaacs et al. 2005). For example, the IGR methoxyfenozide is highly lethal to grape berry moth eggs and larvae, but is benign in terms of any direct lethal effects on the adult stage (Isaacs et al. 2005). For practical purposes of integrating insecticides and acaricides into a grape IPM program, describing the life-stage selectivity is a key step in determining the optimal timing of a vineyard spray against the target pest.

Insecticides with sublethal activity are not directly toxic to the insect life stage that they come into contact with, but the subsequent generation of the pest is affected in terms of reduced fecundity or viability of eggs (Fig. 4.3). Thus, sublethal activity can provide crop protection by pre-empting the insect larvae that would otherwise infest the crop, which also suppresses the pest population over time. Many IGR insecticides express this mode of activity to some degree, but the ecdysone agonists tebufenozide and methoxyfenozide are well known for fecundity and fertility effects on Lepidoptera insects (Xiaoping and Barrett 1999; Pineda et al. 2006). The benzoylphenylurea novaluron is exceptional in terms of the breadth of insect orders on which it is known to reduce egg viability via adult exposure, including Lepidoptera, Coleoptera, and Diptera (Hoffmann et al. 2008; Gökçe et al. 2009).

An example of a non-lethal mode of activity is that of repellency. An insecticide that is a repellent causes the pest to actively avoid the treated substrate, and the behavior modification serves to protect the crop from injury. The best examples of this mode of activity include pyrethrin and azadirachtin compounds. It is important to distinguish between repellent and antifeedant activity. Antifeedants and oviposition deterrents reduce the desirability of the crop plant as a food source or oviposition substrate for the pest, but in many cases the pest will remain in the plant canopy without feeding or further infesting the crop (Fig. 4.3). Interestingly, the oviposition deterrence first described for neonicotinoids on the plum curculio (Wise et al. 2006),

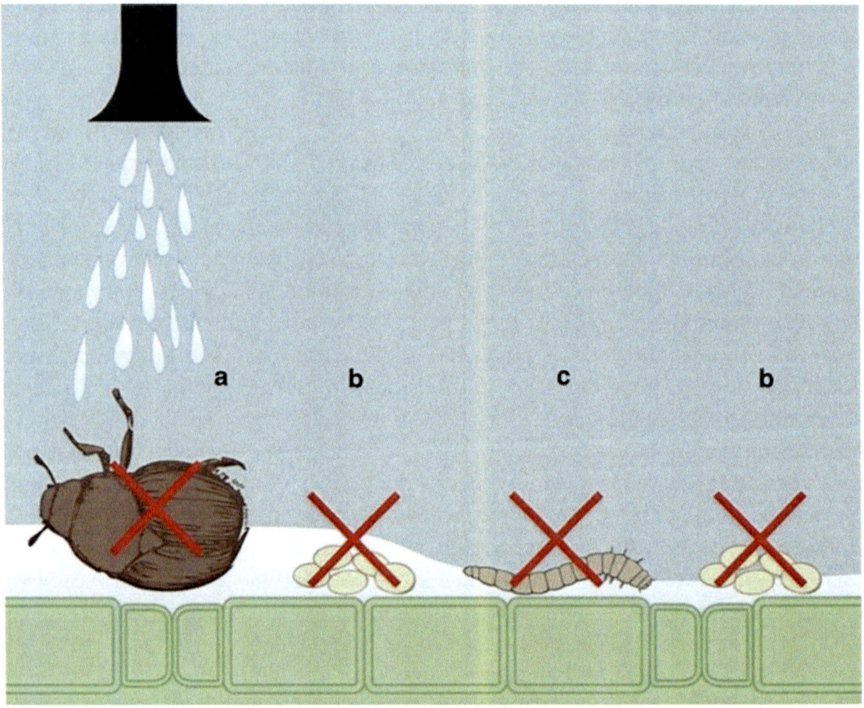

Fig. 4.2 Insecticides with lethal activity cause direct mortality to the pest, and are toxic to one or more insect life-stages, respectively called (**a**) adulticidal, (**b**) ovi-larvicidal, (**c**) larvicidal, and (**d**) ovicidal (Image by Marlene Cameron and John Wise)

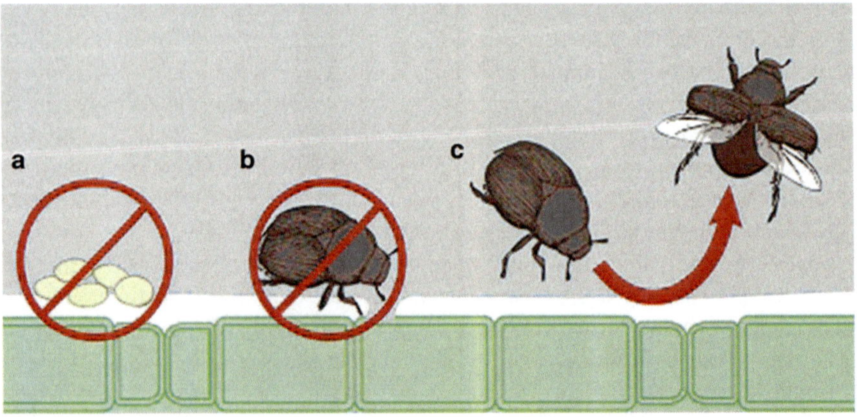

Fig. 4.3 Some insecticides have non-lethal (not directly lethal) modes of activity that suppress the pest population or protect the crop from infestation or injury, respectively called (**a**) sublethal activity, (**b**) antifeedant and oviposition deterrence, and (**c**) repellency (Image by Marlene Cameron and John Wise)

is now better understood as a secondary precipitate of an antifeedant effect on an organism, whose oviposition behavior is dependent on a feeding sequence to prepare the host site for egg deposition (Hoffmann et al. 2010). Thus in some cases, oviposition deterrence and antifeedant activity are inherently linked.

Curative activity is the lethal action of an insecticide on a pest post-infestation, resulting from the transitory penetration of the insecticide into plant tissue (Wise and Whalon 2009). The neonicotinoids are the best known for having curative activity on the egg and larval stages of insect pests in fruit (Hoffmann et al. 2009; Wise et al. 2009). This mode of activity is not generally considered as a first line of defense in insect pest management, because it does not prevent all injury and infestation of the crop. However, the impetus of using insecticides with this activity lies primarily with situations where there is value in eliminating the probability of a living life-stage being found in the crop at harvest. During post-harvest inspections, the detection of fruit infested with live insects can trigger a rejection by a buyer, processor or winemaker, whereas the presence of blemished fruit alone may reduce yield, but is unlikely to cause a load rejection.

4.7 Insecticide Movement in Vines: Enhancing Plant Protection

In contrast to conventional insecticides, many newer insecticides act through a suite of mechanisms against insects to achieve crop protection. The expression of the various modes of activity for a given compound are often linked to the residue profile of the chemical on the plant (Wise et al. 2007, 2009). The spatial and temporal dimensions of the residue profile in effect regulate the mechanisms at work on the pest. For example, when first applied as a foliar spray, neonicotinoids provide a surface residue that for most lepidopteran, coleopteran and dipteran pests will be highly lethal as a contact poison. As the surface residues decline and the compound penetrates the plant tissue, the antifeedant and oviposition deterrent modes of activity become the dominant means of plant protection. The neonicotinoids also tend to have plant penetrative characteristics, sufficient to express curative activity on a pest post infestation. The systemic characteristics of neonicotinoids vary among the individual compounds, and the extent of penetration also depends on the crop, type of plant tissue (leaves versus fruit), and maturity of plant tissue (Tomizawa and Casida 2005) (Table 4.1). The different penetrative and translocative capabilities of these compounds are largely associated with their water solubility and octanol/water partitioning coefficients (Log K_{ow}) (Chowdhury et al. 2001; Buchholz and Nauen 2002).

The spinosyns, avermectins, IGRs and diamides show limited penetrative capabilities, predominantly in leaf tissue rather than in fruit (Chowdhury et al. 2001; Wise et al. 2009) (Table 4.1). There are several forms of locally systemic movement in leaves that are important to distinguish because of their implications to vineyard-level IPM. Translaminar movement represents the penetration of a foliar applied insecticide from the cuticular surface of the leaf, through the epidermis layer and

Table 4.1 Summary of performance characteristics for each of the major insecticide classes, including the primary mode of activity responsible for insect pest control, the life-stage activity expected from direct lethal action, and the plant systemic capabilities in grape leaves. These provide a general view of insecticide characteristics, although there will be variation within classes

Chemical class	Modes of activity	Life-stage activity	Plant systemic capabilities
Organophosphates	Lethal, curative	Adult, larva/nymph, egg	Cuticle penetration
Carbamates	Lethal	Adult, larva/nymph, egg	Translaminar
Pyrethroids	Lethal, repellent	Adult, larva/nymph, egg	Cuticle penetration
Oxadiazines	Lethal	Adult, larva/nymph, egg	Cuticle penetration
Neonicotinoids	Lethal, antifeedant Oviposition deterrent, curative	Adult, larva/nymph, egg	Translaminar, acropetal (xylem mobile)
Insect growth regulators	Lethal, sublethal	Larva/nymph, egg	Translaminar
Spinosyns	Lethal	Larva/nymph, egg	Translaminar
Diamides	Lethal	Larva/nymph, egg	Translaminar
Avermectins	Lethal	Larva/nymph	Translaminar
Tetronic acid derivatives	Lethal	Adult, larva/nymph	Translaminar, acropetal, basipetal

distributing into the mesophyll on the abaxial side. This movement results in a reservoir of active ingredient within the plant tissue that is protected from UV degradation and is active on pests that feed on the plant. Acropetal movement represents a horizontal mobility in the plant xylem from treated area of the leaf tissue to the marginal ends. This can be particularly valuable when the target pests prefer the youngest leaf tissue on an actively growing shoot, such as potato leafhopper, *Empoasca fabae* Harris. Basipetal translocation is movement of the insecticide within the phloem from the site of application in the downward direction. The term fully systemic is often used to represent insecticides that can be translocated throughout the plant by xylem or phloem elements of the plant's vascular system. For most fully systemic insecticides the plant xylem is the primary route of delivery, which is typically initiated from root targeted application techniques. A foliar application of spirotetramat, a keto-enol derivative of tetronic acid, can be translocated acropetally and basipetally in addition to being ambimobile (movement through the xylem and phloem), thus providing the opportunity of systemic delivery both up and down the plant, with great potential for control of grape phylloxera (Powell, Chap. 10).

The residue profile of an insecticide in the vine is influenced not only by its penetrative and translocative properties, but also the growth pattern of the vine. After a foliar spray there is a finite amount of insecticide distributed across the surface. If the vine is actively growing, a dilution effect on the insecticide residues will occur. In a study of potato leafhopper control on grape leaves (Timmeren et al. 2011), when imidacloprid was applied as a foliar spray the residues in mature leaves declined 65% over 21 days. This decline represents the environmental degradation of this compound for the given time duration, since the size of the leaves was constant.

The decline of imidacloprid residues on immature leaves collected over 21 days after the spray was 96%, representing a combination of compound degradation and growth dilution from the rapid expansion of leaf tissue over time. Foliar and soil applied imidacloprid treatments on mature grape leaves for potato leafhopper control were equally effective for 27 days, whereas application on young leaves showed diminished activity on leafhopper nymphs as the leaf tissue expanded over time. In contrast, when imidacloprid was applied systemically through a soil treatment, the systemic movement within the vine and leaf overcame the dilution effect of the growing plant, and potato leafhopper control remained uniform over the 28 day period.

4.8 Pesticide Delivery and Deposition

Delivery of the insecticide or acaricide to the crop and to the pest is an important, but often under appreciated component of chemical control. Twentieth century grape pest management included the introduction of the air-assisted ground sprayer and organophosphate insecticides which, when combined, made pest control relatively simple and highly effective. Application technology development continued through the latter half of the century with improved versions of the airblast sprayer, including high speed air-assisted rotary atomizer (AARA), electrostatic, and various types of tower sprayers with electronic sensors. The objective of many of these newer models was to improve performance while using less water (diluent), and reducing pesticide waste and drift. This was achieved in many respects, but with the assumption that the toxic attributes of the product to be delivered did not change, i.e. a broad-spectrum contact nerve poison. Discovery of new insecticides since the 1980s have resulted in the introduction of a range of pesticides that are selective, slow-acting, ingestion-dependant compounds. As a result, there has been concern regarding the effectiveness of airblast sprayers developed to deliver neurotoxin insecticides. A recent study (Wise et al. 2010a) demonstrated that insecticide performance can be optimized by considering three key variables: sprayer type, water volume, and type of insecticide used in vineyard IPM. Water volume was shown to influence the quality of pesticide deposition by the airblast sprayer and by the AARA sprayer. Coverage by the airblast sprayer is generally improved as water volume is increased, to the point of run-off (grape canopy water holding capacity), after which there is a diminishing return as portions of the insecticide are lost to the ground (Fig. 4.4). The AARA is designed for highest performance at a lower water volume than the airblast, and indeed we found deposition on clusters of *Vitis labrusca* L. juice grapes with heavy leaf canopy to be best with this sprayer at the lower volumes of water. In contrast with wine grapes, where thinning is done to expose fruit clusters to light, deposition was best as water volume increased. The coverage parameters for the AARA were significantly different in some respects than the conventional airblast sprayer. For example, the diameter size of deposits and percent area covered by insecticides delivered by the AARA were less than for those delivered by the airblast sprayer. When the insecticide being delivered was a contact nerve poison there

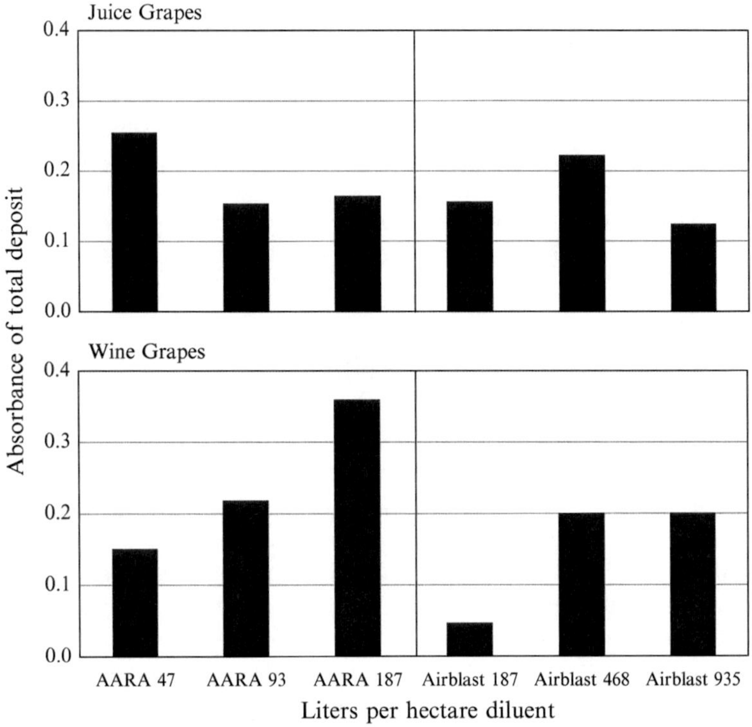

Fig. 4.4 Deposition of kaolin on juice and wine grape clusters when applied using two types of sprayers, each tested at three water volumes. Relative amounts of kaolin were quantified using spectrometry to measure absorbance at 400 nm, and values were corrected for the absorbance measured on untreated clusters

was no difference between sprayers in insect control, but when an ingestion-dependent IGR (tebufenozide) was used, the airblast sprayer demonstrated better control of grape berry moth with its superior coverage at higher water volumes. Similar findings emphasize the importance of cluster coverage for pest control in wine grapes in Europe (Viret et al. 2003).

An alternative approach to foliar sprays for pesticide delivery to control vine pests is through the vascular system of the plant. Systems for addition of insecticide to the irrigation system (chemigation) are available and can offer a low energy approach to delivery to the root zone for systemic insecticides that then move into the foliage (Giddings 2004). This provides the added benefit of long-term delivery of insecticide to protect the foliar canopy (Byrne and Toscano 2006). Recent research indicates how effective soil application can be for control of insects that feed at the tip of growing vine shoots, such as potato leafhopper (*E. fabae*), that would otherwise be feeding in a residue-free area if non-systemic insecticides were used (Timmeren et al. 2011, 2012).

4.9 Rainfastness

The chemo dynamic properties of crop protection products also influence their susceptibility to wash-off from precipitation. This issue is more important in some grape-growing regions of the world than others, but it requires attention since it can be a cause of control failure. Variation among insecticides in their rainfastness has been well documented in other crops (Mashaya 1993), but many regions of grape production have significant rainfall that require an understanding of rain-residue dynamics.

In a recent study using a rainfall simulation chamber, the organophosphate phosmet was shown to lose nearly 50% of its residues on the fruit from a 0.5 in. (12.7 mm) of rainfall, whereas residues from the neonicotinoid thiamethoxam remained relatively constant (Fig. 4.5) (Hulbert et al. 2011). The wash-off patterns

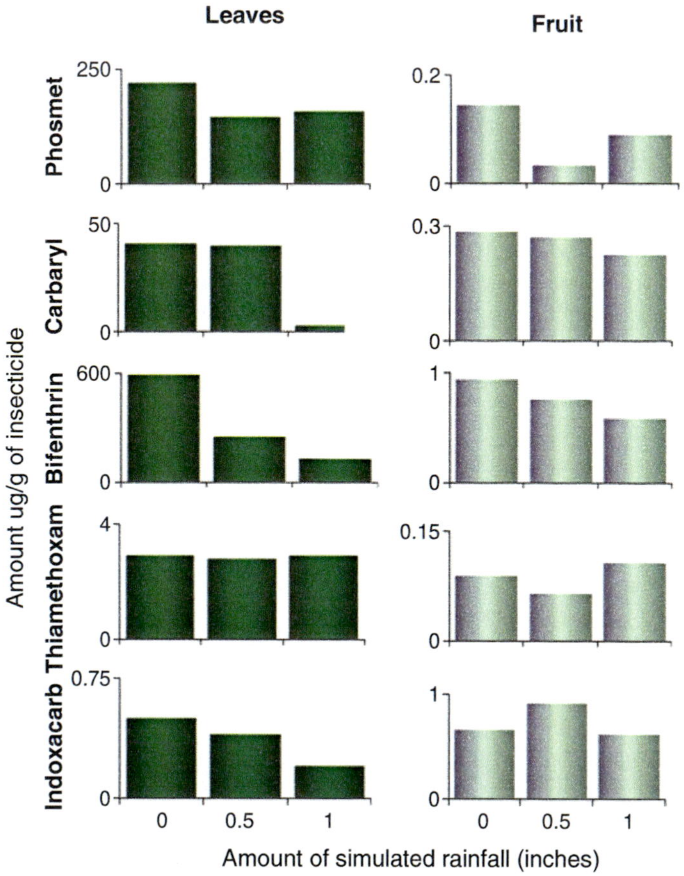

Fig. 4.5 Residue wash-off profiles of grape leaves and fruit from three levels of simulated rainfall, 24 h after field application with airblast sprayer

for the carbamate carbaryl and the pyrethroid bifenthrin were different, with a reduction of residues as the amount of rainfall increased. There are several key factors that influence the impact of precipitation on the performance of a pesticide on the target pest. First is the actual loss of insecticide residue from a precipitation event. The residue wash-off potential is influenced by the water solubility of the compound, its binding to the plant cuticle, the proportion of residues on the plant surface versus within the plant tissue, and the nature of the rainfall in terms of volume, duration and intensity. Our research suggests that the duration of a precipitation event is relatively unimportant, but the amount of rainfall will significantly impact the insecticide residues remaining on the fruit and leaves of the plant. Note that the wash-off potential for a given compound can be different for fruit than leaves (Hulbert et al. 2011). For an indirect pest that feeds primarily on leaves, the rainfastness of a compound on foliage is the most relevant. The second factor is the inherent toxicity of the insecticide on the target pest. A given compound may be highly susceptible to wash-off, but if the target pest is very sensitive to the compound there may be sufficient residues remaining to protect the crop.

In general, organophosphate insecticides have the highest susceptibility to wash-off from precipitation, although their toxicity level to most insect pests can often overcome the necessity for an immediate re-application. Neonicotinoid insecticides are moderately susceptible to wash-off, although residues that have penetrated into plant tissue are highly rainfast, while surface residues less so. Pyrethroid, carbamate, oxadiazine and IGR insecticides are moderately susceptible to wash-off, and vary in their toxicity to the range of relevant grape pests. Diamide and spinosyn insecticides have proven to be highly rainfast. Thus for the decision making process, whether to re-apply or not after a rainfall event, a vineyard manager must consider the nature of the precipitation event, the rainfastness attributes of the compound, knowledge of the pest and relative toxicity of the insecticide (Wise et al. 2010b).

4.10 Effects of Pesticides to Non-target Beneficials Other than Bees

The implementation of IPM programs in vineyards has made the need for an appreciation of the effects of pesticides to non-target arthropods, especially natural predators and parasitoids, essential. To that end the reader is referred to Candolfi et al. (2000), who discuss techniques developed for testing insecticide and acaricide effects on beneficials. For predatory mites several methods have been developed each with its pros and cons. The most popular one is the 'excised leaf method' which is extensively used by members of the Western Paleartic region of IOBC/WPRS (Hassan 1985; Oomen 1988). The method is simple for it measures the toxicity of a limited dry residue picked up by tarsal contact and some diet contamination. Unfortunately, the findings are prone to statistical Type II errors. This is because toxicity caused by overspray of the test arthropod, its food and water is not recorded, whereas

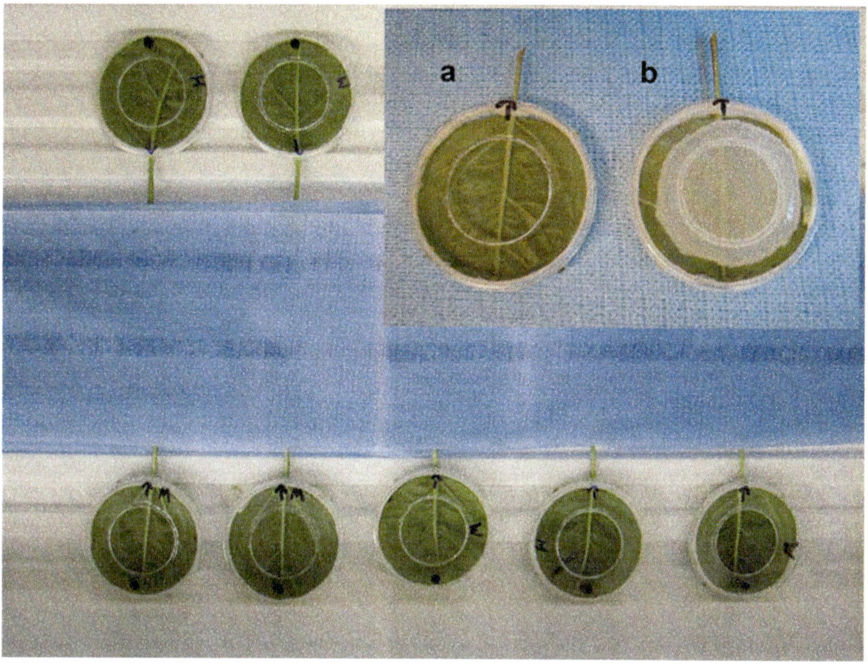

Fig. 4.6 The modified excised leaf disc method (**a**) for predatory mites, the window has no cover and is surrounded with a very thin coat of Tangle-Trap®, (**b**) for tetranychid mites, the window is covered with a 40-µm Pecap® polyester screen because tetranychid mites get trapped in the Tangle-Trap® (Modified from Bostanian et al. 2009b)

in the vineyard such exposure takes place repeatedly with an unknown fraction of the pest-predator population. This technique was recently refined to the 'modified excised leaf disc method' where the target predator is exposed for a longer period to contact and residual toxicity. In the 'worst case laboratory' exposure (Fig. 4.6), toxicity to adults, nymphs, eggs, and effects on fecundity are measured over several days with minimal mortality in the control (Bostanian et al. 2009b).

Understanding the time-course of toxic action of pesticides is essential for development of bioassay procedures, yet standardization of bioassay procedures overlooks this problem. This is currently very critical, because while earlier chemistries (OPs, carbamates and pyrethroids) expressed their full toxicity at the latest within 48 h, reduced-risk insecticides such as spirotetramat, spinetoram, novaluron, chlorantraniliprole and flubendiamide express their toxic effects several days after treatment (Lefebvre et al. 2011). Finally, estimating LC_{50} values for adults and calculating the toxicity quotient of field rate to LC_{50} for adults may provide a better appreciation of the toxicological data. The effect of the pesticide on the different growth stages along with its impact on fecundity and the toxicity quotient for adults are used to (1) accept the compound for inclusion without further second tier field testing, (2) recommend the compound for field testing at different scenarios (early, mid- and late season), the outcome of which would dictate the suitability of the compound for inclusion, sometimes with restrictions, in an IPM program, or (3) reject the compound outright unless alternatives are unavailable, at which time, the compound would be field evaluated and possibly under certain circumstances, it would be appropriate for inclusion in an IPM program (Lefebvre et al. 2011).

It is strongly urged that investigators that extrapolate data from one species to another, or on the same species from one region to another, exercise great caution as diametrically opposite effects are not uncommon. Bostanian and Laroque (2001) reported imidacloprid to be non-toxic to the adults of the stigmaeid mite *Agistemus fleschneri* Summers and to the anystid *Anystis baccarum* (L.) (Laurin and Bostanian 2007). In contrast, it was toxic to the phytoseiids *G. occidentalis* and *N. fallacis* (Bostanian et al. 2009a, 2010). Thiamethoxam was non-toxic to adult *G. occidentalis* (Bostanian et al. 2009a) but moderately toxic to adult *N. fallacis* (Bostanian et al. 2010). The effects of pesticides on beneficial insects in vineyards are also being studied. For example, Walton and Pringle (1999) reported on the effects of chlorpyrifos, endosulfan, cypermethrin, penconazole, and mancozeb to *Coccidoxenoides peregrinus* (Timberlake). Thomson et al. (2000) reported on the effects of sulfur to *Trichogramma carverae* (Oatman & Pinto). James (2004) reported on the effects of buprofezin to *Harmonia axyridis* (Pallas), *Stethorus punctum picipes* Casey, *Orius tristicolor* (White), *Geocoris pallens* Stål, and *Geocoris punctipes* (Say). In summary, in order to obtain laboratory data that would be useful for implementing pesticide usage strategies, it is essential that investigators understand the attributes of the pesticide in question, the target organism (pest or beneficial) and the agroecosystem in which these organisms are present.

4.11 Insecticide Resistance

Genetic selection of pests for resistance to insecticides and acaricides is one of the most serious obstacles to effective pest management. Insecticide resistance by insects and mites exemplifies the selection principle of evolution. True resistance occurs only when there is a structural genetic change that can be inherited. In contrast, tolerance is the natural ability of an arthropod population to withstand the toxic effects of a specific insecticide. It may come about by physiological adaptation within a single generation, but it is lost as soon as the arthropods are not re-exposed to the toxicant. Currently, resistance to insecticides and acaricides is not a major concern for grape pests in North America, thanks to resistance management programs. Nevertheless, wide variation among vineyards in the susceptibility to carbaryl in populations of grape berry moth was detected in New York and Pennsylvania (Nagarkatti et al. 2002). Similarly, populations of *Drosophila melanogaster* Meigen, have exhibited varying tolerance to a range of insecticides (Bridae et al. 1997).

4.11.1 Attributes and Factors Affecting Resistance

The following 10 issues should be considered when addressing potential tolerance or resistance of pest populations to chemical control methods:

1. Preadaptation. In insects and mites, genes controlling the resistance mechanism are already in place at very low frequencies prior to the selection by pesticides. These genes have been selected by other toxicants in the history of the arthropod and have been retained in low level heterozygous phenotypes until selected a second time by a pesticide.
2. Gene frequency. The frequency of resistance (R) alleles in natural populations is usually very low and ranges from 0.0001 to 0.01. Nevertheless, exceptions do exist and these often promote the rapid development of resistance.
3. Dominance of R alleles. Resistant populations would evolve faster if the resistance is dominant and slower if it is recessive.
4. Gene numbers and dominance. Often a single gene may be responsible for resistance and in this case the level of resistance can be high. An example is organophosphate resistance among spider mites. However, multiple genes can also be involved in the development of resistance as reported for carbaryl in houseflies (Georghiou 1972). Resistance to OPs and carbamates is most of the time dominant or incompletely dominant. DDT, Bt and spinosyn resistance is most of the time recessive. Pyrethroid resistance is frequently incompletely recessive.
5. Generation turnover. Typically, 10–15 generations are needed for the development of resistance in insects and mites. Therefore, arthropods with several generations per year such as spider mites tend to develop resistance much faster than arthropods with one or two generations per year such as lepidopteran pests.
6. Population mobility. The frequency of resistance is diluted with the influx of susceptible strains into a population. Furthermore, the selection of resistant strains is considerably slower in mobile arthropods when compared to sedentary arthropods. This is because with mobile arthropods considerable dilution of resistance by hybridization with susceptible individuals occurs in the field.
7. Persistence of pesticides. Persistent pesticides promote faster the development of resistance than pesticides with short residual activity. The dose-response curves of chlorinated hydrocarbon insecticides (DDT) are much flatter than the dose response curves for organophosphate and carbamate insecticides. This suggests that there may be several mechanisms for selection for the chlorinated hydrocarbons and also a wider range of concentrations for resistance selection. In this respect Brown and Pal (1971) reported that houseflies developed DDT resistance in 2 years whereas malathion resistance took 5 years.
8. Timing of pesticide applications. An early season application against immature forms may sometimes be effective even against a highly resistant strain of the arthropod. Edge and James (1986) in Australia and Flexner et al. (1987) in the Pacific Northwest of the United States have reported that small early season populations consisting primarily of larvae and overwintering adults of organotin

resistant strains of *Tetranychus urticae* Koch can be more easily controlled than the same population later in the season. The reason for this difference may be the greater susceptibility in larval and early nymphal stages of the mite.

9. Reversion to the wild type. Resistance is the result of an artificial selection caused by applying a pesticide. Consequently, there is always the possibility of reversion to the wild type once the applications are discontinued (Edge and James 1986). Nevertheless, resistance rebounds slowly when selection is resumed (Flexner et al. 1987).
10. Biotic fitness. Resistance to a xenobiotic may lower the biotic fitness of the resistant strain compared with the susceptible wild strain. Reduced fecundity in houseflies with metabolic resistance to insecticides had been reported as a physiological cost by Roush and Plapp (1982). A similar observation was noted in *Panonychus ulmi* (Koch) to cyhexatin (Flexner et al. 1987). In such situations, the R allele frequency declines during the interval between treatments of the same compound and the R strain may eventually be wiped-out (Georghiou 1980).

4.11.2 Mechanisms of Resistance

Two types of resistance are recognized: behavioral and physiological. Behavioral resistance is defined as the ability of an arthropod to avoid a dose of toxicant that would otherwise be lethal. It is mainly stimulus-dependent and is a matter of hypersensitivity or hyperirritability of the arthropod exposed to the compound. Arthropods with behavioral resistance respond to lower concentrations of insecticides, suggesting that their receptors are more sensitive to detect the presence of xenobiotics than the wild strains. Consequently, upon contact with xenobiotics, these arthropods will quickly seek refuge away from a treated surface. Physiological resistance is dependent on three factors: (1) reduced penetration, (2) enhanced detoxification, and (3) target site insensitivity. These three parameters do not occur alone and are known to interact with each other.

4.11.2.1 Reduced Penetration

This attribute appears to be widespread and along with the other mechanisms confers considerable resistance to insecticides, but by itself it is of slight importance (Plapp 1986). In this respect Ahmad et al. (2006) reported deltamethrin resistance in strains of *Helicoverpa armigera* (Hübner) caused by a slowing of cuticular penetration.

4.11.2.2 Enhanced Detoxification

Cytochrome P450 monooxygenases. Resistance may be caused by increased oxidative metabolism caused by cytochrome P450 monooxygenases. It results in the production of less toxic metabolites. The metabolites, even if they are more toxic, are very unstable and do not reach the site of action due to changes in polarity or are neutralized by other factors. The P450 enzyme system is non-specific, and it plays an important role in the development of cross-resistance. It has been shown to be a major mechanism of resistance for OPs, carbamates, pyrethroids, neonicotinoids, abamectin, juvenoids and chlorinated hydrocarbons other than cyclodienes. Resistance is caused by an overexpression of P450 genes (due to mutations) resulting in increased enzyme production.

Glutathione S-transferases (GSTs). Currently seven GST genes have been identified with insecticide resistance by gene amplification or overexpression (Li et al. 2007). Gene amplification means there are multiple copies of the structure genes that direct the synthesis of detoxifying enzymes. Gene overexpression implies that the genes are out of control because of mutations and produce more detoxifying enzymes leading to sequestration of GST with the xenobiotic and detoxifying it (Kostaropoulos et al. 2001). GSTs are important in the development of organophosphate resistance in insects and predacious mites. For an in-depth understanding of the molecular mechanisms of resistance, the reader is referred to Li et al. (2007).

Hydrolases. Carboxylesterases play an important role in resistance to ester containing insecticides such as OPs, carbamates and pyrethroids. At the molecular level this is brought about by gene amplification (Hemingway 2000) and esterase mutation (Claudianos et al. 1999).

4.11.2.3 Target Site Insensitivity

There are three types of target site insensitivity: (1) nerve insensitivity, (2) altered acetylcholinesterase, (3) reduction in midgut binding sites.

1. Nerve insensitivity. This is brought about by one or more point mutations that modify the affinity of the insecticide for its receptor site on the sodium channel. For example knockdown resistance (kdr) to pyrethroids in *D. melanogaster* is caused by a point mutation in the sodium channel gene which modifies the affinity of the insecticide to its receptor site on the sodium channel. Resistance to pyrethroids in the Arizona strain of *Bemisia tabaci* (Gennadius) is caused by three point mutations (Morin et al. 2002). This type of resistance has been reported in organochlorine, pyrethroid, neonicotinoid, and phenylpyrazole insecticides.
2. Altered acetylcholinesterase. This type of resistance is the result of one or more point mutations in the AChE genes. This mutation renders several insect and mite species insensitive to OPs and carbamate insecticides. In the Colorado potato beetle a point mutation of AChE replaced serine by glycine and rendered this beetle resistant to azinphosmethyl (Zhu et al. 1996).
3. Reduction in midgut binding sites. In this type of resistance, Bt toxin (Cry proteins) is unable to bind to the intestinal lining. For example, a change of the

Bt R-4 gene in Bt resistant pink bollworms reduces the binding target site of Bt toxin Cry1Ac (Morin et al. 2003). Another site are glycolipids as receptors for Bt toxin. In resistant strains the sugar structure of the glycolipid molecule is altered, and the Bt crystal toxin is unable to attach itself and express its toxicity (Griffitts et al. 2005).

4.11.3 Cross and Multiple Resistance

Cross-resistance implies that a strain that has become resistant to an insecticide automatically becomes resistant to another insecticide even though it has never been exposed to it. Hassall (1990) showed that while selecting for permethrin resistance in houseflies, the flies also became resistant to methomyl, DDT, dichlorvos and naled. It may be brought about by: (a) nonspecific enzymes such as cytochrome P450 monooxygenases, (b) mutation at an insecticide target site, (c) delayed cuticular penetration. There have also been reports of negative cross-resistance whereby resistance to an insecticide renders the population susceptible to another. A classical example is increased azinphosmethyl resistance in the twospotted spider mite which increased the susceptibility of the mite to fenvalerate (Chapman and Penman 1979). Multiple resistance implies that a strain comes into contact with two or more different insecticides, and it is caused by sequential selection of populations with other replacement insecticides. In cross-resistance a single defense mechanism confers resistance to several insecticides. In multiple resistance we have resistance to several insecticides brought about by different mechanisms.

4.11.4 Resistance Management

Since the 1970s several tactics based largely on computer models have been forwarded for combating resistance to pesticides. These tactics are essentially based on operational procedures whereby the rate, timing and frequency of treatment along with the ecology of the arthropod pest are taken into consideration to anticipate the selection pressure imposed by the treatments. Several authors have emphasized that there is no single universal prescription for combating resistance under all situations (Sawicki 1981; Georghiou 1983; Roush 1989; Denholm and Rowland 1992; Stenersen 2004). Tactics must be tailored to individual or pest complexes based on the genetics of the pest, availability of chemicals with different modes of action and the precision with which these factors can be applied. In the context of this chapter, we have found the proposals made by Georghiou (1983) to be useful and they are summarized below.

Management by moderation (application of minimal low concentrations). This tactic aims to reduce selection pressure by preserving susceptible insects in the population through the use of low treatment rates, less frequent treatments, short lived residues, and if possible untreated refuges.

Management by saturation (application of very high concentrations). This tactic aims to kill even the resistant insects. It is aimed against the heterozygotes. The idea is to make resistance functionally recessive. Whenever possible, synergists are applied to suppress detoxifying enzymes. It is applicable when gene frequencies for resistance are low, and when immigration of susceptible individuals is high. Otherwise there is always the risk that high concentrations would accelerate selection (Tabashnik and Croft 1982).

Managing by treating only the most susceptible life stage. Most of the time, treatments are applied at the most damaging life stage. However, other life stages (often younger instars) may be more susceptible targets and reduce the development of resistance. This is because in early life stages, genes for metabolic mechanisms of resistance are not expressed or poorly expressed.

Managing by multiple treatments. This tactic is based on applying more than one unrelated insecticide to the target. The compounds may be applied together as mixtures, alternatively in rotation or in spatial patterns known as mosaics. All tactics in this section depend on the absence of cross-resistance.

Since the 1970s, combinations of the tactics described above have been implemented by vineyard managers and IPM practitioners.

4.12 Conclusion

Pesticides vary widely in their spectrum of activity, mode of action, and persistence of activity against insects and mites. Cost is also an important factor that drives adoption by growers. The range of different pesticide types available to grape producers has increased in the last decade. This has brought a broader spectrum of options and greater opportunity for applications that are pest group-specific and least disruptive to natural enemies, although these are often more expensive and their adoption is sometimes driven by resistance to other older pesticides. Understanding the mode of activity at the interface of the berry and leaf surface, the insect, and the pesticide has brought new insights into how these more selective, sometimes systemic, new insecticides achieve control of insects and mites. Researchers and pest management advisors will be increasingly challenged to understand the unique properties of new pesticides, and to make this information accessible and available to vineyard decision-makers.

Grape pest management is an information-intensive endeavour that integrates cutting-edge technology to maximize profit while respecting the natural resources on which farms depend. In the coming years, development of technologies that allow real-time access to combined GPS mapping, scouting information, weather data, and pest development models will provide greater ability to target pesticides and other pest control tactics to the time and place where they are most effective. If pest action thresholds are met, grape growers will need a thorough understanding of how best to integrate insecticides into grape management programs. This will require continued evaluation of new insecticides for their suitability within grape IPM programs coupled with ongoing development of associated tools to ensure their appropriate deployment. These include degree day models, sampling plans, thresholds, and improved knowledge of insecticide persistence and stability under varying weather conditions as well as any toxic effects on biocontrol agents. Pesticides are likely to remain a part of vineyard IPM programs, with their level of adoption depending on pest pressure, quality of educational programs, and the degree of regulation imposed on their use. In summary, applied entomologists must understand thoroughly the agroecosystem in which pests, diseases, and beneficials are present, as well as the attributes of the different pesticides available, in order to develop chemical usage strategies within IPM programs that can be economically and environmentally acceptable to growers and consumers.

Acknowledgements We thank the staff of the Trevor Nichols Research Center, Dan Hulbert, Keith Mason, and Steve Van Timmeren for technical assistance. Support for some of the research presented here was provided by the Michigan Grape and Wine Industry Council, National Grape Cooperative, and the Michigan Agricultural Experiment Station.

References

Ahmad M, Denholm I, Bromilow RH (2006) Delayed cuticular penetration and enhanced metabolism of deltamethrin in pyrethrin-resistant strains of *Helicoverpa armigera* from China and Pakistan. Pest Manag Sci 62:805–810

Albrecht CP, Sherman M (1987) Lethal and sublethal effects of avermectin B1 on three fruit fly species (Diptera: Tephritidae). J Econ Entomol 80:344–347

Anonymous (2010) Fruit production recommendations 2010–2011, Publication 360. Ontario Ministry of Agriculture, Food and Rural Affairs, Toronto

Anonymous (2011) UC IPM online, Statewide IPM program. http://www.ipm.ucdavis.edu/PMG/r302900611.html

Bloomquist JR (1999) Insecticides: chemistries and characteristics. National IPM Network, University of Minnesota. http://www.ipmworld.umn.edu/chapters/bloomq.htm

Bostanian NJ, Akalach M (2004) The contact toxicity of indoxacarb and five other insecticides to *Orius insidiosus* (Hemiptera: Anthocoridae) and *Aphidius colemani* (Hymenoptera: Braconidae) beneficials used in the greenhouse industry. Pest Manag Sci 60:1231–1236

Bostanian NJ, Laroque N (2001) Laboratory tests to determine the intrinsic toxicity of four fungicides and two insecticides to the predacious mite *Agistemus fleschneri*. Phytoparasitica 29:215–222

Bostanian NJ, Bélanger A, Rivard I (1985) Residues of four synthetic pyrethroids and azinphos-methyl on apple foliage and their toxicity to *Amblyseius fallacis* (Acari: Phytoseiidae). Can Entomol 117:143–152

Bostanian NJ, Vincent C, Hardman JM, Larocque N (2004) Toxicity of indoxacarb to two species of predacious mites and a predacious mirid. Pest Manag Sci 60:483–486

Bostanian NJ, Thistlewood HA, Hardman JM, Laurin M-C, Racette G (2009a) Effect of seven new orchard pesticides on *Galendromus occidentalis* in laboratory studies. Pest Manag Sci 65:635–639

Bostanian NJ, Beudjekian S, McGregor E, Racette G (2009b) A modified excised leaf disc method to estimate the toxicity of slow acting reduced-risk acaricides to mites. J Econ Entomol 102:2084–2089

Bostanian NJ, Hardman JM, Thistlewood HA, Racette G (2010) Effects of six selected orchard insecticides on *Neoseiulus fallacis* (Acari: Phytoseiidae) in the laboratory. Pest Manag Sci 66:1263–1267

Bridae JM, Cuany A, Amichot M, Brun A, Babault M, Le Mouel T et al (1997) Cytochrome P-450 field insecticide tolerance and development of laboratory resistance in grape vine populations of *Drosophila melanogaster* (Diptera: Drosophilidae). J Econ Entomol 90:1514–1520

Brown AWA, Pal R (1971) Insecticide resistance in arthropods, 2nd edn. World Health Organization, Geneva

Buchholz A, Nauen R (2002) Translocation and translaminar bioavailability of two neonicotinoid insecticides after foliar application to cabbage and cotton. Pest Manag Sci 58:10–16

Buckingham SD, Sattelle DB (2005) GABA receptors of insects. In: Gilbert LI, Iatrou K, Gill SS (eds) Comprehensive molecular insect science, vol 5. Elsevier, London, pp 107–142

Byrne FJ, Toscano NC (2006) Uptake and persistence of imidacloprid in grapevines treated by chemigation. Crop Prot 25:831–834

Candolfi MP, Blümel S, Forster R, Bakker FM, Grimm C, Hassan SA et al (2000) Guidelines to evaluate side-effects of plant protection products to non-target arthropods, IOBC, BART and EPPO Joint Initiative. IOBC/WPRS, Darmstadt

Casida JE, Quistad GB (1998) Golden age of insecticide research: past, present, or future? Annu Rev Entomol 43:1–16

Chapman RB, Penman DR (1979) Negatively correlated cross-resistance to synthetic pyrethroid in organophosphorus-resistant *Tetranychus urticae*. Nature 281:298–299

Chowdhury A, Jepson PC, Howse PE, Ford MG (2001) Leaf surfaces and the bioavailability of pesticide residues. Pest Manag Sci 57:403–412

Claudianos C, Russel RJ, Oakeshort JG (1999) The same amino acid substitution in orthologous esterases confers organophosphate resistance on the house fly and a blowfly. Insect Biochem Mol Biol 29:675–686

Corbett JR, Wright K, Baillie AC (1984) The biochemical mode of action of pesticides, 2nd edn. Academic, London

Cordova D, Benner EA, Sacher MD, Rauh JJ, Sopa JS, Lahm GP et al (2006) Anthranilic diamides: a new class of insecticides with a novel mode of action, ryanodine receptor activation. Pestic Biochem Physiol 84:196–214

Dekeyser MA (2005) Review, Acaricide mode of action. Pest Manag Sci 61:103–110

Denholm I, Rowland MW (1992) Tactics for managing pesticide resistance in arthropods; theory and practice. Annu Rev Entomol 37:91–112

Dhadialla TS, Carson GR, Le DP (1998) New insecticides with ecdysteroidal and juvenile hormone activity. Annu Rev Entomol 43:545–569

Dunbar SJ, Goodchild JA, Cutler PM (1998) Actions of natural products on insect nicotinic receptors. In: Proceedings of the 9th International Congress of Pesticide Chemistry, IUPAC Book of Abstracts 1: 4B-040, London

Edge VE, James DG (1986) Organotin resistance in *Tetranychus urticae* Koch (Acari: Tetranychidae) in Australia. J Econ Entomol 79:1477–1483

Elbert A, Nauen R, Leicht W (1998) Imidacloprid, a novel chloronicotinyl insecticide: biological activity and agricultural importance. In: Ishaaya I, Degheele D (eds) Insecticides with novel modes of action: mechanism and application. Springer, Berlin/Heidelberg, pp 50–73

Elliot M (1977) Synthetic pyrethroids. In: Elliot M (ed) Synthetic pyrethroids, ACS symposium series 42. American Chemical Society, Washington, DC, pp 11–28

Elliot M, Janes NF, Potter CV (1978) The future of pyrethroids in insect control. Annu Rev Entomol 23:443–469

Flexner JL, Croft BA, Westigard PH (1987) Effect of organotin formulations on organotin resistance of *Tetranychus urticae* Koch (Acarina: Tetranychidae). J Econ Entomol 81: 766–769

Gammon DW, Brown MA, Casida JE (1981) Two classes of pyrethroid action in the cockroach. Pestic Biochem Physiol 15:181–191

Georghiou GP (1972) The evolution of resistance to pesticides. Annu Rev Ecol Syst 3:133–168

Georghiou GP (1980) Insecticide resistance and prospects for its management. Residue Rev 76:131–135

Georghiou GP (1983) Management of resistance in arthropods. In: Georghiou GP, Saito T (eds) Pest resistance to pesticides. Plenum Press, New York, pp 769–792

Giddings J (2004) Drip irrigation – a grape grower's guide. NSW Agriculture, Orange

Gökçe A, Kim S-HS, Wise JC, Whalon ME (2009) Reduced egg viability in codling moth *Cydia pomonella* (L.) (Lepidoptera: Tortricidae) following adult exposure to novaluron. Pest Manag Sci 65:283–287

Graf J (2011) Shifting paradigm on *Bacillus thuringiensis* toxin and a natural model for *Enterococcus faecalis* septecemia. http://www.biomedsearch.com/attachments/00/21/84/68/21846827/mBio.00161-11.pdf

Griffitts JS, Haslam SM, Yang T, Garczynski SF, Mulloy B, Morris H et al (2005) Glycolipids as receptors for Bacillus thuringiensis crystal toxin. Science 307:922–925

Hardman JM, Moreau DL, Snyder M, Gaul SO, Bent ED (2000) Performance of a pyrethroid resistant strain of the predator mite *Typhlodromus pyri* (Acari: Phytoseiidae) under different regimes. J Econ Entomol 93:509–604

Hassall KA (1990) The biochemistry and uses of pesticides, 2nd edn. VCH Publishers, New York

Hassan SA (1985) Standard methods to test side-effects of pesticides on natural enemies of insects and mites developed by the IOBC/WPRS working group 'Pesticides and beneficial organisms'. OEPP/EPPO Bull 15:214–255

Hassan SA, Albert R, Bigler F, Blaisinger P, Bogenschütz H, Boller E et al (1987) Results of the 3rd joint pesticide testing programme by the IOBC/WPRS-working group 'Pesticides and beneficial organisms'. Z Angew Entomol 103:92–107

Hemingway J (2000) The molecular basis of two contrasting metabolic mechanisms of insecticide resistance. Insect Biochem Mol Biol 30:1009–1015

Hoffmann EJ, Middleton SM, Wise JC (2008) Ovicidal activity of organophosphate, oxadiazine, neonicotinoid and insect growth regulator chemistries on northern strain plum curculio, *Conotrachelus nenuphar*. J Insect Sci 8:1–6

Hoffmann EJ, Vandervoort C, Wise JC (2009) Curative activity of insecticides against plum curculio (Coleoptera: Curculionidae) in tart cherries. J Econ Entomol 102:1864–1873

Hoffmann EJ, Vandervoort C, Wise JC (2010) Plum curculio (Coleoptera: Curculionidae) adult mortality and associated fruit injury after exposure to field-aged insecticides on tart cherry branches. J Econ Entomol 103:1196–1205

Hulbert D, Isaacs R, Vandervoort C, Wise JC (2011) Rainfastness and residual activity of insecticides to control Japanese beetle (Coleoptera: Scarabaeidae) in grapes. J Econ Entomol 104:1656–1664

Hunt DA, Tracy MF (1998) Pyrrole insecticides: a new class of agriculturally important insecticides functioning as uncouplers of oxidative phosphorylation. In: Ishaaya I, Degheele D (eds) Insecticides with novel modes of action: mechanisms and application. Springer, Berlin, pp 138–151

Isaacs R, Mason KS, Maxwell E (2005) Stage-specific control of grape berry moth, *Endopiza viteana* (Clemens) (Lepidoptera: Tortricidae), by selective and broad-spectrum insecticides. J Econ Entomol 98:415–422

Ishaaya I (2001) Biochemical sites of insecticide action and resistance. Springer, Berlin

Ishaaya I, Degheele D (1998) Insecticides with novel modes of action: mechanisms and application. Springer, Berlin

Ishaaya I, Horowitz AR (1998) Insecticides with novel modes of action: an overview. In: Ishaaya I, Degheele D (eds) Insecticides with novel modes of action: mechanisms and application. Springer, Berlin, pp 1–24

Ishaaya I, Horowitz AR, Tirry L, Barazani A (2002) Novaluron (Rimon) a novel IGR-mechanism, selectivity and importance in IPM programs. Meded Rijksuniv Gent Fac Landbouwkd Toegep Biol Wet 67:617–626

Isman MB (2006) Botanical insecticides, deterrents, and repellents in modern agriculture and an increasingly regulated world. Annu Rev Entomol 51:45–66

James DJ (2004) Effect of buprofezin on survival of immature stages of *Harmonia axyridis*, *Stethorus punctum picipes* (Coleoptera: Coccinellidae), *Orius tristicolor* (Hemiptera: Anthocoridae), and *Geocoris* spp. (Hemiptera: Geocoridae). J Econ Entomol 97:900–904
Jeschke P, Nauen R (2005) Neonicotinoid insecticides. In: Gilbert LI, Iatrou K, Gill SS (eds) Comprehensive molecular insect science, vol 5. Elsevier, London, pp 53–105
Kostaropoulos I, Papadopoulos AI, Metaxakis A, Boukouvala E, Papadopoulou-Mourkidou E (2001) Glutathione S-transferase in the defence against pyrethroids in insects. Insect Biochem Mol Biol 31:313–319
Lasota JA, Dybas RA (1991) Avermectin, a novel class of compounds: implications for use in arthropod pest control. Annu Rev Entomol 36:96–117
Laurin M-C, Bostanian NJ (2007) Laboratory studies to elucidate the residual toxicity of eight insecticides to *Anystis baccarum* (Acari: Anystidae). J Econ Entomol 100:1210–1214
Lefebvre M, Bostanian NJ, Thistlewood HMA, Mauffette Y, Racette G (2011) A laboratory assessment of toxic attributes of six 'reduced risk insecticides' on *Galendromus occidentalis* (Acari: Phytoseiidae). Chemosphere 84:25–30
Lefebvre M, Bostanian NJ, Mauffette Y, Racette G, Thistlewood HMA, Hardman JM (2012) Laboratory assessment on the toxicological attributes of new insecticides on mortality and fecundity of *Neoseiulus fallacis* (Acari: Phytoseiidae). J Econ Entomol (in press)
Li X, Schuler NA, Berenbaum MR (2007) Molecular mechanisms of metabolic resistance to synthetic and natural xenobiotics. Annu Rev Entomol 52:231–253
Lund AE, Narahashi T (1981) Kinetics of sodium channel modification by the insecticide tetramethrin, in squid axon membranes. J Pharmacol Exp Ther 219:464–473
Mashaya N (1993) Effect of simulated rain on efficacy of insecticide deposits on tobacco. Crop Prot 12:55–58
Matsumura F (1985) Toxicology of insecticides, 2nd edn. Plenum Press, New York
Morin S, Williamson MS, Goodson SJ, Brown JK, Tabashnik BE, Dennehy TJ (2002) Mutations in the *Bemisia tabaci* para sodium channel gene associated with resistance to a pyrethroid plus organophosphate mixture. Insect Biochem Mol Biol 32:1781–1791
Morin S, Biggs RW, Sisterson MS, Shriver L, Ellers-Kirk C, Higginson D et al (2003) Three cadherin alleles associated with resistance to *Bacillus thuringiensis* in pink bollworm. Proc Natl Acad Sci U S A 100:5004–5009
Nagarkatti S, Tobin PC, Munza AJ, Saunders MC (2002) Carbaryl resistance in populations of grape berry moth (Lepidoptera: Tortricidae) in New York and Pennsylvania. J Econ Entomol 95:1027–1032
Nation JL (2008) Insect physiology and biochemistry, 2nd edn. CRC Press, Boca Raton
O'Brien RD (1960) Toxic phosphorus esters. Academic, New York
Oomen PA (1988) Guideline for the evaluation of side-effects of pesticides *Phytoseiulus persimilis* A.-H. IOBC/WPRS Bull 11:51–63
Pineda S, Smagghe G, Scheider M, Del Estal P, Vinuela E, Mabel Martinez A, Budia F (2006) Toxicity and pharmacokinetics of spinosad and methoxyfenozide to *Spodoptera littoralis* (Lepidoptera: Noctuidae). Environ Entomol 35:856–864
Plapp FW Jr (1986) Genetics and biochemistry of insecticide resistance in arthropods: prospects for the future. In: Pesticide resistance; strategies and tactics for management. National Academy Press, Washington, DC, pp 74–86
Roush RT (1989) Designing resistance management programs: how can you choose? Pestic Sci 26:423–441
Roush RT, Plapp FW Jr (1982) Effects of insecticide resistance on biotic potential of the house fly (Diptera: Muscidae). J Econ Entomol 75:708–713
Sawicki RM (1981) Problems in countering resistance. Philos Trans R Soc Lond B 295:143–151
Scher HB (1999) Controlled release delivery system for pesticides. Marcel Dekker, New York
Smagghe G, Bylemans D, Medina P, Budia F, Avilla J, Vinuela E (2004) Tebufenozide distorted codling moth larval growth and reproduction, and controlled field populations. Ann Appl Biol 145:291–298
Stenersen J (2004) Chemical pesticides: mode of action and toxicology. CRC Press, Boca Raton

Tabashnik BE, Croft BA (1982) Managing pesticide resistance in crop-arthropod complexes: interactions between biological and operational factors. Environ Entomol 11:1137–1144

Terriere LC (1982) The biochemistry and toxicology of insecticides. Oregon State University, Corvalis

Thomson LJ, Glenn DC, Hoffmann AA (2000) Effects of sulfur on *Trichogramma* egg parasitoids in vineyards: measuring toxic effects and establishing release windows. Aust J Agric 40:1165–1171

Timmeren S van, Wise JC, Isaacs R (2012) Soil application of neonicotinoid insecticides for control of insect pests in wine grape vineyards. Pest Manag Sci. doi: 10.1002/ps.2285

Timmeren S van, Wise JC, Vandervoort C, Isaacs R (2011) Comparison of foliar and soil formulations of neonicotinoid insecticides for control of potato leafhopper, *Empoasca fabae* (Homoptera: Cicadellidae), in wine grapes. Pest Manag Sci 67:560–567

Tohnishi M, Nakao H, Furuya T, Seo A, Kodama H, Tsubata K et al (2005) Flubendiamide, a novel insecticide highly active against lepidopterous insect pests. J Pestic Sci 30:354–360

Tomizawa M, Casida JE (2005) Neonicotinoid insecticide toxicology: mechanisms of selective action. Annu Rev Pharmacol Toxicol 45:247–268

Vijverberg HPM, Van der Zalm JM, Van der Bercken J (1982) Similar mode of action of pyrethroids and DDT on sodium channel gating in myelinated nerves. Nature 295:601–603

Viret O, Siegfried W, Holliger E, Raisigl U (2003) Comparison of spray deposits and efficacy against powdery mildew of aerial and ground-based spraying equipment in viticulture. Crop Prot 22:1023–1032

Walton VM, Pringle KL (1999) Effects of pesticides used on table grapes on the mealybug parasitoid *Coccidoxenoides peregrinus* (Timberlake) (Hymenoptera: Encyrtidae). S Afr J Enol Vitic 20:31–34

Ware GW, Whiticare DW (2004) The pesticide book, 6th edn. Meister Pro Information Resources, Willoughby

Wing KD, Andaloro JT, McCann SF, Saldago VL (1998) A novel oxadiazine insecticide is bioactivated in lepidopteran larvae. Arch Insect Biochem Physiol 37:91–103

Wing KD, Andaloro JT, McCann SF, Saldago VL (2005) Indoxacarb and the sodium channel blocker insecticides: chemistry, physiology and biology in insects. In: Gilbert LI, Iatrou K, Gill SS (eds) Comprehensive molecular insect science, vol 6. Elsevier, London, pp 32–53

Wise JC, Whalon ME (2009) A systems approach to IPM integration, ecological assessment and resistance management in tree fruit orchards. In: Ishaaya I, Horowitz R (eds) Biorational control of arthropod pests. Springer, Dordrecht, pp 325–345

Wise JC, Coombs AB, Vandervoort C, Gut LJ, Hoffmann EJ, Whalon ME (2006) Use of residue profile analysis to identify modes of insecticide activity contributing to control of plum curculio in apples. J Econ Entomol 99:2055–2064

Wise JC, Kim K, Hoffmann EJ, Vandervoort C, Gökçe A, Whalon ME (2007) Novel life stage targets against plum curculio, *Conotrachelus nenuphar* (Herbst), in apple integrated pest management. Pest Manag Sci 63:737–742

Wise JC, Vanderpoppen R, Vandervoort C (2009) Curative activity of insecticides on *Rhagoletis pomonella* (Diptera: Tephritidae) in apples. J Econ Entomol 102:1884–1890

Wise JC, Jenkins P, Schilder A, Vandervoort C, Isaacs R (2010a) Sprayer type and water volume influence pesticide deposition and control of insect pests and diseases in juice grapes. Crop Prot 29:378–385

Wise JC, Schilder A, Zandstra B, Hanson E, Gut JL, Isaacs R, Sundin G (2010b) Michigan fruit management guide, MSU Extension Bulletin E-154. Michigan State University, East Lansing

Xiaoping S, Barrett B (1999) Fecundity and fertility changes in adult codling moth (Lepidoptera: Tortricidae) exposed to surfaces treated with tebufenozide and methoxyfenozide. J Econ Entomol 92:1039–1044

Yu SJ (2008) The toxicology and biochemistry of insecticides. Taylor & Francis Group/CRC Press, Boca Raton

Zhu KY, Lee SH, Clark JM (1996) A point mutation of acetylcholinesterase associated with azinphosmethyl resistance and reduced fitness in Colorado potato beetle. Pestic Biochem Physiol 55:100–108

Chapter 5
Biological Control of Arthropods and Its Application in Vineyards

Vaughn M. Walton, Kent M. Daane, and Pia Addison

5.1 Introduction

Biological control has been defined as: '*The action of parasites, predators or pathogens in maintaining another organism's population density at a lower average than would occur in their absence*' (De Bach 1964). Therefore, successful biological control of arthropod pests relies on the presence and viability of effective predators, parasitoids and/or entomopathogens in sufficient numbers and at critical seasonal periods to provide population regulation.

Predators, such as entomophagous mites, lady beetles and lacewings are free-living organisms and consume a large number of prey (Huffaker et al. 1976), while parasitoids have immature developmental stages that are found on or within a single host and their feeding results in host death (Reuter 1913). Pathogens are disease-causing organisms that can kill or debilitate the host and include nematodes, protozoa, bacteria, fungi, and viruses (Federici 1999). In this chapter, biological control agents are grouped into three key categories: predaceous arthropods, parasitic arthropods and pathogens. The brief review of the predaceous vineyard arthropods discussed herein will include predatory spiders and mites (Arachnidae, Phytoseiidae,

V.M. Walton (✉)
Department of Horticulture, Oregon State University, Corvallis, OR 97331-7304, USA
e-mail: waltonv@hort.oregonstate.edu

K.M. Daane
Department of Environmental Science, Policy and Management, University of California, 130 Mulford Hall, Berkeley, CA 94720-3114, USA
e-mail: daane@uckac.edu

P. Addison
Department of Conservation Ecology and Entomology, Faculty of AgriSciences, University of Stellenbosch, Private Bag X1, Matieland 7602, South Africa
e-mail: pia@sun.ac.za

N.J. Bostanian et al. (eds.), *Arthropod Management in Vineyards: Pests, Approaches, and Future Directions*, DOI 10.1007/978-94-007-4032-7_5,

Anystidae and Stigmaeidae) and predatory insects (Chrysopidae, Carabidae, Coccinellidae, Syrphidae and predaceous Heteroptera). Parasitic vineyard arthropods discussed herein include Diptera (e.g. Tachinidae) and Hymenoptera (e.g. Ichneumonidae, Braconidae and Chalcidoidea). Pathogens include entomopathogenic nematodes, protozoa, fungi, bacteria and viruses.

Three basic types of biological control strategies are practiced: conservation, classical, and augmentation (De Bach 1964). Conservation of natural enemies is the most important practice available when comparing the three types of biological control (De Bach 1964; Landis et al. 2000). Conservation biological control attempts to optimize the impact of biocontrol agents by enhancing life-sustaining resources of natural enemies in the environment and limiting negative factors (Price 1997; Landis et al. 2000). Many factors can interfere with the effectiveness of natural enemies. Certain cultural practices can kill natural enemies or make the crop habitat unsuitable. To be effective, many natural enemy species need access to several resources besides the targeted pest species. These may include alternate prey/hosts, adult food resources (e.g. nectar), shelter (e.g. overwintering habitats), a food supply throughout the season, and appropriate microclimates to support development (Rabb and Guthrie 1970). For example, floral resources may be an important alternative food source for parasitoids (Altieri et al. 2005). Enhanced availability of these resources should in theory increase the longevity and fecundity of resident parasitoids (Price 1997). As another example, pesticide applications may kill natural enemies or have indirect effects through reduction in the numbers or availability of hosts (Elzen and King 1999; Landis et al. 2000; Mills and Daane 2005).

Classical biological control is the introduction of natural enemies to a new location where they did not originate or do not occur naturally (De Bach 1964; Fisher and Andres 1999; Legner and Bellows 1999). In many instances, the resident complex of natural enemies to manage an invasive arthropod pest may be inadequate. To obtain the needed natural enemies, candidate species are often imported from the native ranges of the exotic pest. Today, the importation of any natural enemies is guided by strict protocols, with all imported organisms passed through a rigorous quarantine process to determine if there may be any non-target impacts (Fisher and Andres 1999). For these reasons, rarely are generalist predators or entomopathogens imported and released. Once screened in quarantine and approved by government agencies for field release, these beneficial arthropods are then mass-reared to be released and field evaluated (Etzel and Legner 1999). Follow-up studies are conducted to determine if the natural enemy becomes successfully established at the site of release, and to assess the long-term benefit of its presence (Gutierrez et al. 1999).

The third biological control technique involves the supplemental release of natural enemies (De Bach and Hagen 1964; Elzen and King 1999). A relatively small number of natural enemies may be released at the most optimal time of the season (inoculative release) or many may be released (inundative release) several times per season. Inundative release of natural enemies may become more promising with the use of less toxic pesticides and development of simple rearing techniques, which will allow abundant production of beneficials (Elzen and King 1999). Limiting factors of this technology include cost, widespread availability of quality products, and lack of immediate visible impact.

5.2 Predaceous Arthropods

Hagen et al. (1999) provide a detailed list of terrestrial arthropod predators of insect and mite pests used in biological control systems.

5.2.1 Arachnidae

Spiders can comprise up to 95% of predators in vineyards and can have an important impact on pest populations (Costello and Daane 1999). Spiders rarely show specificity toward prey (Riechert and Lockley 1984). Generalist feeding is one of several traits spiders possess to ensure individuals survive over periods of food shortage; it is nevertheless possible for such generalist predators to exhibit density-dependent responses to changes in prey numbers or to provide a constant level of prey kill in the vineyard (inverse density-dependent response). As such, spiders can have a strong stabilizing impact on prey populations (Clark and Grant 1968) even though they do not fit the mold of the specialist predator or parasitoid (Riechert and Lockley 1984).

In California, Hogg and Daane (2010) studied the impact of surrounding oak-woodlands on spider populations in vineyards. This research indicated most spiders overwintered outside of the crop habitat, with most species ballooning into the vineyard during the season (Hogg et al. 2010). Spider migration from more diverse surrounding habitats did not occur in spring, but started in midsummer. Their work also looked at the success of native and invasive spider species (Hogg and Daane 2011a), and showed a strong impact of landscape heterogeneity on spider presence and the success of invasive species (Hogg and Daane 2011b). These findings are supported by several studies that indicated that increased landscape heterogeneity resulted in increased spider community heterogeneity (Bolduc et al. 2005; Prischmann et al. 2005b; Isaia et al. 2006).

Straw and compost may play an important role in managing weeds and reducing chemical inputs in vineyards and may also provide suitable habitat for beneficial arthropods. Thomson and Hoffmann (2007) found that the abundance of parasitic Hymenoptera, spiders and ground beetles increased with the addition of such mulches. In the canopy, predatory and parasitic Diptera and predatory Heteroptera also increased due to mulching. However, no impact on pests was recorded (Thomson and Hoffmann 2007). Costello and Daane (1998) found that cover cropping resulted in lower leafhopper densities but these densities could not be explained when looking at spider population levels. The lower density of leafhoppers in the cover crop treatment resulted from poorer host plant quality and was not necessarily because of increased biological control levels (Daane and Costello 1998; Costello and Daane 2003).

Hanna et al. (2003) found increased population levels of spiders in vineyards that had cover crops, but did not find that these increased spider populations necessarily resulted in lower pest densities.

Cultural practices, such as partial root zone drying, and its impact on natural enemies of key pests, were studied in southeastern Australia using pitfall traps and sticky traps (Thomson 2006). No impact was found on two generalist predator orders: the Coleoptera and Araneae. Black widow spiders, despite playing a role in biological control, are often found as contaminants of table grapes in California vineyards (Hernandez et al. 2005) and are seen as a pest.

Nash et al. (2010) investigated the impacts of often-used pesticides on many beneficial arthropods and found that repeated pesticide applications negatively impacted spider populations. In contrast, Booth et al. (2003) found that often-used pesticides such as tebufenozide and chlorpyrifos had no measurable impact on ground-dwelling wolf spiders, indicating that application of different pesticides should not be expected to produce an equivalent effect on all beneficial species.

5.2.2 Phytoseiidae

Phytoseiid mites are important biological control agents and essential elements of some pest management systems (McMurtry 1982). Phytoseiid mite species are characterized as specialized predators of *Tetranychus* species, selective predators of tetranychid mites, generalist predators and specialist pollen feeders/generalist predators (McMurtry and Croft 1997). Phytoseiid mites kill both spider mites and eriophyid mites in vineyards, and there is increasing evidence that the different phytoseiid life stages all contribute to keep pest mite numbers below economic thresholds. It is important that predatory mite populations persist at low population levels, even when starved. Some mites have the ability to use alternate food sources, while others use cannibalism and predation on other phytoseiid species in order to survive these conditions. Predaceous mite populations are also dependent on vineyard conditions. For example, just as water stress has been shown to impact phytophagous mites, vine condition has also been shown to impact the effectiveness (English-Loeb 1990) and densities (Stavrinides et al. 2010) of predatory mite populations. Nevertheless, the following phytoseiid species contribute to phytophagous mite control in unique ways.

Typhlodromus pyri Scheuten (Acari: Phytoseiidae) is dependent on alternative resources when prey is in low supply. Leaf trichomes play such a role and feeding on these tissues will result in higher populations of *T. pyri* in vineyards. Engel and Ohnesorge (1994) found that a supply of additional pollen during field experiments contributed to sustaining the majority of *T. pyri* predatory mites during the initial period of the season when few non-plant derived food sources are available (Engel and Ohnesorge 1994), thereby allowing more rapid colonization when pest populations increase. As an example, *T. pyri* is a generalist predatory mite often found in vineyards that can feed on eriophyid mites as well as spider mites, pollen and leaf trichomes (Loughner et al. 2008). Mass releases of the pesticide-resistant spider mite predator *Galendromus* (=*Typhlodromus*) *occidentalis* (Nesbitt) showed that large releases of phytoseiid mites could result in wind dispersal after release, especially after foliage quality declined (Hoy et al. 1982).

In western US vineyards, sulfur is a key fungicide for control of powdery mildew (*Uncinula necator* [Schweinitz] Burrill). Investigations on the impact of sulfur on pest mites such as *Tetranychus pacificus* McGregor and predatory mites such as *G. occidentalis* were done in the central California vineyards of Fresno County (Costello 2007). No differences were found between treatments that received different rates of sulfur and those without sulfur. Stavrinides and Mills (2009) found that the fungicides trifloxystrobin and tebuconazole had no influence on *T. pacificus* or *G. occidentalis* population growth rates. This study demonstrated the value of simultaneous testing of demographic effects of pesticides on pests and natural enemies in order to more fully understand the impacts of applied pesticides on biological control. Posenato et al. (2003) stated that high doses of sulfur during hot periods resulted in temporary reductions in several predatory mite species. These findings are confirmed by laboratory studies on *T. pyri* (Gadino et al. 2011). However, more *T. pacificus* were found in sulfur treatments during the period of pre-bloom. Sulfur use in field-applied trials in South Africa also caused a reduction in numbers of *Amblyseius addoensis* Van der Merwe & Ryke (Schwartz 1993).

Field and semi-field trials have been conducted on insecticide treatment effects on beneficials. Research on the impact of the anthranilic diamide insecticide, chlorantraniliprole, did not show any long or short-term impact on *Kampimodromus aberrans* (Oudemans), a beneficial mite species (Marchesini et al. 2008). The insecticide imidacloprid severely impacted population growth of *G. occidentalis*, but resulted in virtually no impact of the pest mite *T. pacificus* (Stavrinides and Mills 2009). Flubendiamide was totally innocuous while spinetoram and spirotetramat were toxic to *G. occidentalis* in laboratory studies. Novaluron, clothianidin and chlorantraniliprole were found to have some toxicity and are recommended for field evaluation (Lefebvre et al. 2011). Parexan N® (pyrethrin + sesame oil) is currently used in organic vineyards to control *Scaphoideus titanus* Ball, the insect vector of Flavescence Dorée (Gusberti et al. 2008). Shibao et al. (2006) reported that Parexan N® was highly toxic to *Amblyseius andersoni* (Chant), whereas imidacloprid had little effect on *Euseius sojaensis* (Ehara).

5.2.3 *Anystidae*

Anystidae play an important role in phytophagous mite management in orchards and vineyards (Laurin and Bostanian 2007). Anystidae are voracious generalists, feeding on any prey that they can overpower. These species are relatively large, fast-moving, orange-red mites that reproduce parthenogenetically. Anystidae are found on agricultural crops grown from temperate to subtropical regions. In Canada, *Anystis* spp. was reported feeding on *Panonychus ulmi* (Koch) on peach (*Prunus* spp.)

trees in southern Ontario and have even been reported to have preyed on Lepidoptera eggs on artichoke. Anystidae have been found on grapes (*Vitis* spp.) feeding on phytophagous mites (James and Whitney 1991), but knowledge is currently limited. The predatory mite *Anystis baccarum* (L.) is commonly found in vineyards (Laurin and Bostanian 2007). Near Moscow, Russia, *A. baccarum* has been the most common predaceous mite feeding on phytophagous mites on black currants, *Ribes nigrum* L. Laurin and Bostanian (2007) found five insecticides that were not toxic to *A. baccarum*: methoxyfenozide, acetamiprid, thiamethoxam, imidacloprid, and spinosad.

5.2.4 Stigmaeidae

Stigmaeidae are commonly found in the predatory mite complex but usually are not dominant in terms of the abundance (Johann et al. 2009). Stigmaeidae are less dominant generalist feeders (Slone and Croft 2000) that often occur together with other predators and are seen as vulnerable species when there is a scarcity of food due to the possibility of intraguild predation. Laboratory tests showed that the following pesticides: trifloxystrobin, myclobutanil, flusilazole, kresoxim-methyl, imidacloprid and lambda-cyhalothrin were neither toxic nor affected adversely the fecundity of the stigmaeid *Agistemus fleschneri* Summers (Bostanian and Larocque 2001).

5.3 Predaceous Insects

5.3.1 Chrysopidae

A thorough overview of Chrysopidae in grapes including seasonality, impacts of habitat structure and biocontrol can be found in Szentkiralyi (2001). A total of 36 chrysopid and hemerobiid species combined have been recorded in European, North American and Indian vineyards (Szentkiralyi 2001). Chrysopidae feed on pollen, nectar and honeydew supplemented with mites, aphids and other small arthropods (Szentkiralyi 2001). Lacewings are often in contact with pesticides used in cropping systems. Research assessing the impacts of often-used pesticides on many beneficial arthropods including Neuroptera found that repeated pesticide applications negatively impacted these populations (Booth et al. 2003; Nash et al. 2010).

Chrysopidae are commercially reared and for sale as biological control agents of insect and mite pests in agriculture and gardens. They are often distributed as eggs when released inoculatively. They are highly aggressive and cannibalistic in confined quarters and care needs to be taken to provide adequate food before distribution. Several members of *Chrysoperla* (Chrysopidae), as well as *Mallada signatus* (Schneider) (Chrysopidae), have hitherto attracted wider study and are readily available as captive-bred eggs to deploy for hatching in pest-infested plant cultures. Biological control experiments with mass releases of green lacewings (*Chrysoperla* spp.) in vineyards produced some reduction in pest insect numbers (Daane et al. 1996). However, Daane and Yokota (1997) studied mass releases of green lacewings and, looking at delivery, release rate, timing and developmental stage, they showed that release rates of 19,768 green lacewings per ha resulted in significant but not economically viable reductions of leafhopper (*Erythroneura variabilis* Beamer) densities. The primary issue was an inadequate delivery system that could place large numbers of viable lacewings on the vines (Daane and Yokota 1997). The successful augmentative release of green lacewings must consider not only the appropriateness of the targeted prey, but also the suitability of the lacewing species released with the environmental conditions of the crop system, and biology, and ecology of the target prey (Tauber et al. 2000).

5.3.2 Carabidae

An overview of Carabidae (Hagen et al. 1999) indicates limited knowledge in terms of commercial biological control. Carabid survival strategies range from ectoparasitism on insect hosts to obligatory carnivore feeding. Carabidae can also be herbivores or omnivores. Comprehensive studies show that carabids generally appear more omnivorous. Carabidae have been shown to have an impact on pest insects in perennial crops and can reduce up to 60% of tethered codling moth larvae per night (Riddick and Mills 1994).

The fact that beetles are omnivorous indicates that supply of seeds as an alternative food source may play a role in increased carabid beetle populations and therefore may aid in suppressing pest populations. One way of increasing beetle populations is by creating beetle banks, thus providing seeds and refuge habitats for beetles during periods when prey is in shortage (Frank et al. 2011). Findings of a recent study (Prasad and Snyder 2006) suggest that beetle banks may actually skew carabid guild composition in favor of omnivores when seed density increases. Also, an increase in carabid beetle populations will not necessarily lead to improved pest control (Prasad and Snyder 2006). Overall, the results from this study suggested that both intraguild predation and the presence of alternative prey could limit conservation biological control that is dependent on generalist predators. Still, there have been few studies that document the species composition and abundance, or the contribution of carabids to biological control in vineyards.

5.3.3 Coccinellidae

The historical importance, biology and case studies of successful biocontrol strategies using coccinellids are discussed in Hagen et al. (1999). In Coccinellidae there

Fig. 5.1 *Nephus bineavatus*, a specialist Coccinellid beetle to control mealybugs in California and South Africa

is a contrast between coccidophagous species that have had more success in biological control compared to aphidophagous species. Two factors are suggested to play a role in the first group. Coccidophagous species tend to be small and feed continuously compared to aphidophagous species that have a long period of inactivity due to satiation (Mills 1982). Aphidophagous species lay their eggs in batches on the host plant leaf surfaces, making them prone to density-dependent cannibalism. Coccidophagous species lay their eggs singly or on small groups concealed beneath the prey where they can escape cannibalism. Coccidophagous species were also considered to have a shorter generation time than their host species, a generally advantageous characteristic. Coccinellidae are important natural enemies of pests including important mealybug, scale, and phytophagous mites found in vineyards (Obrycki and Kring 1998; Hagen et al. 1999). The coccinellid *Cryptolaemus montrouzieri* Mulsant has been used in biological control through mass releases in citrus in California (Fisher 1963) and *Nephus bineavatus* Mulsant (Coccinellidae) has been used for control of grape mealybug, *Pseudococcus maritimus* (Ehrhorn) (Heteroptera: Pseudococcidae) (Smith 1923) (Fig. 5.1).

The most recent studies on the impacts of pesticides on Coccinellids were conducted by Mani and Thorntakarya (1988), Walton and Pringle (2001) and Nash et al. (2010). Fungicides had a lesser impact on coccinellids than did insecticides.

Biocontrol agents may be seen as pests, if they contaminate the grape crop close to harvest. One such example is the multicolored Asian ladybird beetle, *Harmonia axyridis* (Pallas). This insect is one of the most voracious polyphagous predators in the world (Lucas et al. 2007) and is not only a nuisance pest in houses in fall, but also may taint wines due to their large numbers in berries during the harvest period

(Lucas et al. 2007). The major peak in *H. axyridis* numbers in Minnesota and Wisconsin grapes were found to occur between veraison and harvest (Galvan et al. 2009). These beetles were closely associated with vineyards neighboring soybean fields in the Niagara Peninsula in Canada (Bahlai and Sears 2009). Management of these beetles in soybean fields did not impact numbers of pest beetles in vineyards however, but recent declines in soybean aphid populations in the Midwest have been associated with lower incidence of problems with *H. axyridis* infestation in grape clusters.

5.3.4 *Syrphidae*

Syrphidae are important predators of aphids and other Heteroptera (Chambers 1988) that may occasionally occur in vineyards. The impact of syrphids has seldom been assessed and their role may be underestimated due to their nocturnal feeding behavior. The majority of predaceous syrphids are multivoltine and the range of prey that can be consumed by syrphid larvae can be extensive. Syrphids have been used in classical biological control but have failed to establish in target regions (Waage et al. 1984). Syrphids play an important role, however, in perennial biological control systems and often are the first to colonize and prey on pest aphid populations (Dib et al. 2010). Stutz and Entling (2011) studied the black cherry aphid, *Myzus cerasi* F. They found that generalist predators including syrphids in wooded habitats controlled it. Stutz and Entling (2011) found that aphid colonies where ant populations were excluded showed a dramatic decrease in pest aphid populations compared to aphid colonies where ants were allowed to tend. Significantly lower densities of earwigs and syrphids were also noted on trees isolated from woody habitats than on trees adjacent to forest.

5.3.5 *Heteroptera*

The majority of Heteroptera are phytophagous but thousands of species are predatory. Biological control literature details the role of insects in the families Anthocoridae, Berytidae, Lygaeidae, Phymatidae, and Reduviidae (Hagen et al. 1999) as biocontrol agents. Most predaceous heteropteran species also feed on plant tissues or secretions. These predators have been found in a range of agricultural crops. *Orius insidiosus* (Say) is well known and economically important, and it plays a very important role in biocontrol because of its attraction to volatiles in corn silks (Hagen et al. 1999; Crowder et al. 2010) that bring *O. insidiosus* to corn silks where eggs of *Helicoverpa zea* (Boddie) and the European corn borer *Ostrinia nubilalis* (Hübner) occur (Reid and Lampman 1989). It was found that surrounding vegetation plays an important role to promote Heteropteran generalist predators in perennial cropping systems such as apples (Miliczky and Horton 2007). Some of these species were particularly abundant on the plants when they were flowering and included *Orius tristicolor* (White), *Deraeocoris brevis* (Uhler) and *Nabis alternatus* Parsh. Crowder et al. (2010) argue that species diversity is enhanced by mitigating ecosystem disruption by organic farming and this may lead to increased abundance of generalists predatory bugs such as *N. alternatus* and *Geocoris bullatus* (Say). An increase of these generalists is believed to enhance ecosystem stability and exert stronger pest insect control.

5.4 Parasitic Insects

5.4.1 Tachinidae

Tachinidae are important biological control agents of many pests, but primarily lepidopterans, and various species have been introduced and successfully established in classical biological control programs (Feener and Brown 1997). All species are parasitic in the larval stage, and tachinid parasitoids are generally polyphagous (Belshaw 1994). Tachinid parasitoids are able to attack key lepidopteran vineyard pests such as *Lobesia botrana* (Denis & Schiffermüller) and *Eupoecilia ambiguella* (Hübner) but the rates of parasitism are low, probably because they are generalist parasitoids (Martinez et al. 2006).

5.4.2 Ichneumonidae

All ichneumonids are primary parasitoids, and many are important parasitoids of vineyard pests, primarily lepidopterans. For example, in European vineyards the ichneumonid *Campoplex capitator* Aubert was the most common species collected from *L. botrana* larvae. Its incidence was higher during the spring compared to summer. The overall parasitism rate found on one experimental vineyard varied from 23% in 2000 to 53% in 2001, and the increased rate of parasitism was mainly due to *C. capitator* (Xuéreb and Thiéry 2006). Ichneumonids can be manipulated via augmentative release, with larvae and pupae generally the preferred stages for biological control releases. However, such programs are rare. More commonly, ichneumonid populations are manipulated by providing a better habitat for adult survival, by changes in ground cover in the vineyard (see below).

5.4.3 Braconidae

Most braconids are primary parasitoids (both external and internal) on other insects, especially upon the larval stages of Coleoptera, Diptera, and Lepidoptera, but also some aphids. Endoparasitoid species often display elaborate physiological adaptations to enhance larval survival within the host, such as the co-option of endosymbiotic viruses for compromising host immune defenses. These viruses suppress the immune system and allow the parasitoid to grow inside the host undetected. Because of this highly modified system of host immunosuppression it is not surprising that there is a high level of parasitoid-host specificity. It is this specificity that makes Braconidae powerful and important biological control agents. Examples in vineyards include biocontrol of tortricids (Thiéry et al. 2001; Jenkins and Isaacs 2007a).

Braconidae are dependent on alternative food sources. Floral resources are often seen as a particularly important habitat resource for parasitoids and other beneficials (Begum et al. 2006; Berndt et al. 2006; Campos et al. 2006; Scarratt et al. 2008). For example, Scarratt and Wratten (2004) used biological markers to show that the addition of flowering buckwheat *Fagopyrum esculentum* Moench resulted in flights of the braconid *Dolichogenidea tasmanica* (Cameron) up to 30 m from flowering plants within a 7-day sampling period. This information may also give growers an idea of how far the impact of surrounding vegetation may reach into a vineyard. Bell et al. (2006) found, however, that increased floral resources (*Alyssum* spp.) did not result in increased parasitism rates of the leafroller *Epiphyas postvittana* (Walker) (Lepidoptera: Tortricidae) by *Dolichogenidea* spp. The parasitoid source population was believed to come from a nearby orchard, which was thought to have a much greater impact in providing parasitoids with resources than the added *Alyssum* plants. It was suggested that much higher levels of resource provision were needed to positively influence parasitism rates (Bell et al. 2006).

Grape berry moth, *Paralobesia* (=*Endopiza*) *viteana* (Clemens) (Tortricidae), can find host plant material in both vineyards and adjacent wild patches of vines (Jenkins and Isaacs 2007a). These authors cut the wild grapevine as a cultural control strategy, but did not find any differences in pest infestation in vineyards nor were there any differences in percent parasitism of Braconid and Ichneumonid parasitoids or natural enemy densities between the sanitized and non-sanitized vineyard treatments.

5.4.4 Chalcidoidea

Chalcidoidea is a superfamily that includes important species in applied biological control. Some of the key families are the Aphelinidae which are parasitoids of Hemiptera; the Trichogrammatidae which include key egg parasitoids, primarily of lepidopteran pest species; the Mymaridae which include the hemipteran egg parasitoids; the Chalcididae which comprise numerous species that attack lepidopteran vineyard pests and to a lesser extent dipteran and coleopteran pest species; and perhaps most importantly in vineyards the Encyrtidae which contain many important mealybug parasitoid species.

Thomson and Hoffmann (2010) studied the abundance of natural enemies in surrounding habitats and found an increase of chalcidoid parasitoids in vineyards when surrounding woody vegetation was present. Wilson et al. (1991) reported that smaller natural enemies such as *Anagrus* spp. (Mymaridae) were influenced by local surrounding vegetation, whereas larger, more mobile natural enemies were less influenced. Williams and Martinson (2000) found that *Anagrus* sp. emerges from alternate hosts during the winter, completes a summer generation on the alternate hosts, and then moves from that host to the nearest adjacent edges of the vines and finally into the vineyard. The same trend of higher numbers of natural enemies with nearby surrounding woodland vegetation was found in studies by Altieri et al. (2005), Nicholls et al. (2001, 2008), and Thomson and Hoffmann (2009). Murphy et al. (1998) found that surrounding woody habitats played a more important role than adjacent non-crop pastures. In studies concerning host species suitability, prune trees were good alternate hosts that allowed overwintering of *Anagrus epos* Girault (Murphy et al. 1998). Recent work has provided better identification of these parasitoids which were formerly called *Anagrus epos* Girault, but are actually a complex of closely related species (Triapitsyn 1998). Currently, four different *Anagrus* species are known to attack grape leafhopper eggs in North America. The two most common in California are *Anagrus erythroneurae* S. Triapitsyn & Chiappini and *Anagrus daanei* S. Triapitsyn. A third species is *Anagrus tretiakovae* F. Triapitsyn, which is found in warmer regions (e.g., Arizona, New Mexico, Mexico). *Anagrus epos* does attack grape leafhopper eggs, but it is more typically found in colder regions (e.g., Colorado, Illinois, US, and Canada).

To manipulate these small wasps and other natural enemies, in California 'vegetational corridors', riparian forests connecting to vineyards were investigated to channel biological control agents into vineyards (Altieri et al. 2005; Nicholls et al. 2001, 2008). Planting of dog roses (*Rosa canina* L.) along the perimeters of vineyards was investigated in order to see if these flowers could serve to substitute natural surrounding habitat (Böll et al. 2006) for populations of *Anagrus atomus* (L.), a natural enemy of the leafhopper *Empoasca vitis* (Göthe) (Cicadellidae). In the study, young shoots of *R. canina* were used as egg laying sites of *E. vitis* and mean parasitism rate of the host eggs was 59%. Once established, dog roses supported as many parasitoids as wild surrounding dog rose in adjacent hedges (Böll et al. 2006), suggesting that additional plantings could serve as a substitute. However, the details of pest populations were not discussed in this study.

The importance of botanical diversity for the presence of egg parasitoids of grape leafhopper *S. titanus* and cicadellids (*E. vitis*) was studied in northern Italy (Rigamonti 2006) and Switzerland (Remund and Boller 1996), respectively. In the first study, an increase in botanical diversity did not result in an increase in parasitism. Surrounding brambles also did not result in higher parasitoid numbers. In the second study (Remund and Boller 1996), hedgerows provided important overwintering habitats for the egg parasitoids *A. atomus* and *Stethynium triclavatum* Enock. *Prunus* spp. trees can encourage overwintering populations of *A. epos* and control of *Erythroneura elegantula* Osborn. Murphy et al. (1996) showed that *A. epos* was mainly found downwind from these trees. Corbett and Rosenheim (1996) also discussed the importance of prune trees as an overwintering habitat for beneficials. Ponti et al. (2005) showed that surrounding hedges including *Rubus ulmifolius*

Schott and *Ulmus minor* Miller were hosts of the non-pest leafhoppers *Ribautiana tenerrima* (Herrich-Schäffer) (Cicadellidae), *Arboridia parvula* (Boheman) (Cicadellidae), and a species of the genus *Zygina* in Italian vineyards. *Ribautiana tenerrima* served as the main food source of the *Anagrus* spp. during the early part of the season, after which these parasitoids moved from the surrounding hedges to the adjacent vineyards where pest leafhopper populations served as a host during the latter part of the season.

English-Loeb et al. (2003) showed that sentinel leafhopper eggs had higher parasitism rates in vineyards planted with buckwheat compared to those without buckwheat. Parasitism rates of leafhopper eggs were higher when *Anagrus* parasitoids had access to flowering buckwheat rather than buckwheat without flowers in this study. Studies by Berndt et al. (2000) on leafroller parasitoids showed similar trends. In a conservation biocontrol study, Sharley et al. (2008) studied the impact of tilling of inter-rows and found that populations of spiders, millipedes, centipedes, earwigs and Trichogrammatidae were decreased by tillage on the soil surface. In Berndt et al. (2006), the impact of several flowering shoots on the egg parasitoid *Trichogramma carverae* Oatman & Pinto (Trichogrammatidae), a parasitoid of the leafroller pest *E. postvittana* was studied under greenhouse conditions. It was found that *Lobularia maritima* (L.) Desvaux and *F. esculentum* increased fecundity of this parasitoid. In this study, it was concluded that *L. maritima* might offer an additional advantage as a food source for these parasitoids. Vine architecture can provide physical refuge to protect mealybug pest populations from koinobiont encyrtid parasitoids such as *Anagyrus pseudococci* (Girault) (Encyrtidae) (Daane et al. 2008a, b). It was found that *A. pseudococci* are a less successful parasite toward mealybugs in hidden locations of the vine, including tightly packed vine clusters, under bark and under the soil surface. These findings can be used as tools to more efficiently manage pests by removal or addition of refuges.

In classical biological control the impacts of climate and the effectiveness of biological control agents on specific pests are closely related. California has a varied climate ranging from hot and dry desert regions with large diurnal temperature differences to cool moist coastal climates with much lower temperature variations. The invasive vine mealybug *Planococcus ficus* (Signoret) (Pseudococcidae) has colonized vineyards in these different climate regimes and provides an excellent opportunity to model the impact of climate on efficacy of key biological control agents (Gutierrez et al. 2008). Two parasitoids of the vine mealybug, *A. pseudococci* and *Leptomastidea abnormis* (Girault) (Encyrtidae), as well as the predator *C. montrouzieri*, were examined across ecologically varied regions of California. Temperature was used to define developmental parameters of each species, and resource supply and demand ratios to scale daily population increase rates. Generally the model predicted lower densities of *P. ficus* in hot desert climates and higher densities in cooler northern coastal California climates. These models coincide well with field observations of vine mealybug (Gutierrez et al. 2008).

Encyrtids are important in vineyards, as they are parasitoids of mealybugs and scale insects (Walton and Pringle 1999; Daane et al. 2008a). Ants such as *Formica perpilosa* Wheeler (Formicidae), *Anoplolepis steingroeveri* (Forel), *Linepithema*

Fig. 5.2 Ants tending grape mealybug, preventing parasitism and predation (Photo by Daniel Dalton)

humile (Mayr) and *Crematogaster peringueyi* Emery, are indirect vineyard pests as they provide biological refuges to heteropteran pests like *P. ficus*, the grape mealybug, *P. maritimus*, and the obscure mealybug, *Pseudococcus viburni* (Signoret) (Daane et al. 2007; Tollerup et al. 2007; Mgocheki and Addison 2009b) (Fig. 5.2). Ants tend heteropteran pests, and disrupt biological control by predators and parasitoids including the above-mentioned encyrtids, leading to increased parasitoid mortality (Cooper et al. 2008). Management of ant colonies has led to marked increases of parasitism and, ultimately, biological control of these pests (Daane et al. 2007; Mgocheki and Addison 2009b). In pest management systems the control of ants is seen as a prerequisite for biological control. Nelson and Daane (2007) suggested that the ant-mealybug symbiotic relationships should be disrupted in order to enhance biological control. Physical refuges can be found inside the host insect. Parasitized mealybugs were less often predated on by *C. montrouzieri* 4 days after parasitism due to the onset of mummification (Mustu et al. 2008).

Mass rearing (Fig. 5.3) and inundative releases of the encyrtid *Coccidoxenoides perminutus* (Timberlake) (Encyrtidae) against *P. ficus* (Fig. 5.4) in South African vineyards indicated that repeated early releases were successful if conducted for consecutive years in table grape blocks (Walton 2003). Trichogrammatidae have been studied in classical, conservation and inundative control efforts to manage holometabolous orders as well as Heteroptera, Thysanoptera and others (Glenn and

Fig. 5.3 Mass rearing of *Coccidoxenoides perminutus* on butternut squash for mass release on vine mealybug, *Planococcus ficus*

Fig. 5.4 *Coccidoxenoides perminutus* parasitizing vine mealybug, *Planococcus ficus*

Hoffmann 1997). Commercial mass releases of *Trichogramma* spp. (Trichogrammatidae) for control of light brown apple moth, *E. postvittana,* were investigated in Victoria, Australia (Glenn and Hoffmann 1997). In this study, parasitism rates were higher in release sites within the first 2 days after release, and this method is considered as an economically viable alternative to control light brown apple moth. The mass releases of the egg parasitoid *Trichogramma minutum* Riley (Trichogrammatidae) against the grape berry moth *P. viteana* showed significant decreases of this pest insect to below 3% infestation in low risk vineyards compared to high risk vineyards where damage levels were maintained below 15% (Nagarkatti et al. 2003).

Regarding conservation biological control, pesticide use in the vineyard is one of the more important practices that impacts natural enemies, including the Chalcidoidea. Sulfur, which is used in most vineyards, did not have any impact on the reproductive success of *A. erythroneurae* and *A. daanei*, both egg parasitoids of the grape leafhopper, *E. elegantula* (Jepsen et al. 2007a, b). However, another research group showed that sulfur affected the mortality of immature *T. carverae* adults, as well as the fitness in both laboratory and field studies (Thomson et al. 2000). Jepsen et al. (2007a, b) showed some effect of sulfur on *Anagrus* spp. in the laboratory, but did not find any impacts on these parasitoids when studies were repeated in the field. Mgocheki and Addison (2009a) found that fipronil and α-cypermethrin were highly toxic to *Anagyrus* species near *pseudococci* (Encyrtidae) (Fig. 5.5) and *C. perminutus*. Buprofezin, mancozeb and an insecticidal soap were less toxic. In this study the pupal stage was found to be generally less susceptible to pesticides. Walton and Pringle (1999) found that herbicides were less toxic to *C. perminutus* than some of the commonly used insecticides. *Scymnus coccivora* Aiyar (Coccinellidae) and *Leptomastix dactylopii* Howard are key beneficials against scale insects in Indian vineyards (Mani and Thorntakarya 1988), and mancozeb, sulfur, carbendazim, Bordeaux mixture and dicofol were determined to be safe to both these natural enemies. Field observations of managed and unmanaged vineyards in south central Washington State have shown that unmanaged vineyards generally have higher densities of pest *Erythroneura* spp. leafhoppers (Cicadellidae), but also had marginally higher parasitism levels of leafhopper eggs, and higher numbers of wasps, spiders, and big-eyed bug densities (Prischmann et al. 2005a). Similarly, work with the grape berry moth *P. viteana* was undertaken by Jenkins and Isaacs (2007b) with the goal of finding reduced-risk spray programs that had a positive impact on biological control of this pest, but no differences in parasitism were found between the reduced-risk spray programs and the conventional programs in a 3-year study. These authors, however, did find that reduced-risk spray programs led to similar or greater control of *P. viteana*. Insecticide persistence trials showed that *Trichogramma* were less affected when sulfur was applied 6 days prior to their release. Field experiments with sulfur in Germany (Ibrahim et al. 2004) resulted in no *Trichogramma* species found for two consecutive seasons where sulfur was intensively used. The appearance of new compounds on the pesticide market warrants continued non-target pesticide work. This field of study is essential for informed pest management decision-making by growers.

Parasitoids in the Scelionidae family are endoparasitoids of arthropod eggs, and have shown a clear pattern of host specificity possibly matched only by the

Fig. 5.5 *Anagyrus* sp. near *pseudococci*, a parasitoid of *Planococcus ficus*

Braconidae (Austin et al. 2005). Nevertheless, Scelionidae utilize a wide spectrum of hosts, possibly identifying suitable hosts by airborne and surface kairomones on eggs, or sex pheromones from adult hosts or host plant volatiles. For example, Scelionidae were attracted by methyl salicylate in wine grape vineyards that also had *Alyssum* planted in the vineyard as a reward for the attracted parasitoids (Austin et al. 2005). The scelonid egg parasitoid *Telenomus euproctidis* Wilcox was attracted to egg masses laid by wingless immobile female *Orgyia postica* (Walker) (Lymantriidae). Virgin females, a solvent extract of pheromone glands, and a synthetic sex pheromone, (6Z 9Z 11S 12S)-11, 12-epoxyhenicosa-6, 9-diene (posticlure), also attracted this parasitoid in the field, demonstrating that *T. euproctidis* uses the sex pheromone of female *O. postica* as a kairomone to locate host eggs (Arakaki et al. 2011). Scelionidae are a diverse group and their application in biological control has been a basis for their use as a model for kairomone research (Austin et al. 2005).

5.5 Pathogens and Entomopathogenic Nematodes

Pathogens that can cause insect disease fall into four groups: viruses, bacteria, fungi and protozoa. A good overview is provided in Federici (1999). The most common strategy to use pathogenic agents is by using them as microbial insecticides. Sometimes fewer applications are needed than pesticides because of high host specificity and less negative impacts on non-target beneficials.

5.5.1 *Viruses*

A good example of a virus used as an insect control agent is the nuclear polyhedrosis virus of the European spruce sawfly, *Gilpinia hercyniae* (Hartig) (Diprionidae). One of the reasons for the use of these types of viruses is that they are easily isolated and produced from pest populations, but there are no viruses currently used for arthropod control in vineyards. The major problem with these viruses is that their host range is narrow. Additional limitations to the use of viruses are the slow speed of kill, little residual activity and lack of effective in-vitro mass production. The slow speed of kill can be overcome by using recombinant DNA technology.

5.5.2 *Bacteria*

Bacteria are most widely and successfully used as pathogen for insect control as they are easy to mass-produce, formulate and use in large-scale operational systems, they kill the target insect quickly, can kill a wide range of economically important pests and are safer than synthetic chemical pesticides. The most widely used bacteria is *Bacillus thuringiensis* Berliner that produces four major endotoxin proteins and is used against lepidopteran pests in vineyards, fields, vegetable crops and forests.

5.5.3 *Fungi*

Fungi infect insects through the cuticle and therefore make them attractive as biological control agents. Currently there are few commercial fungal insecticides. Two prominent commercial biological control agents include the terrestrial fungi *Metarhizium anisopliae* (Metchnikoff) Sorokin (Fig. 5.6) and *Beauveria bassiana* (Balsamo) Vuillemin (Federici 1999), which have very broad host range capable of infecting insects of most orders. These fungi are generally used in environments that are cooler and moist such as beetle larvae in soil. For effective management, large quantities of the biocontrol agents are required because of the limited time that the conidia are viable. A review of biological control of grape phylloxera can be found in Kirchmair et al. (2009). In this paper several biocontrol efforts are highlighted including use of the entomopathogenic fungi *B. bassiana, M. anisopliae* and *Paecilomyces farinosus* (Holmskjold) A.H.S. Brown & G. Smith. The commercial *M. anisopliae*-based product was determined to be promising as an alternative control product for grapevine phylloxera. The black vine weevil, *Otiorhynchus sulcatus* (F.) (Curculionidae) is an important pest of horticultural crops, including grapes in the United States, Canada, Australia and New Zealand, and damage to grapes due to adult feeding on clusters and larval feeding on root systems in Europe and central

Fig. 5.6 Healthy black vine weevil larvae compared to *Metarhizium anisopliae*-infected black vine weevil larvae (Photo by Betsey Miller)

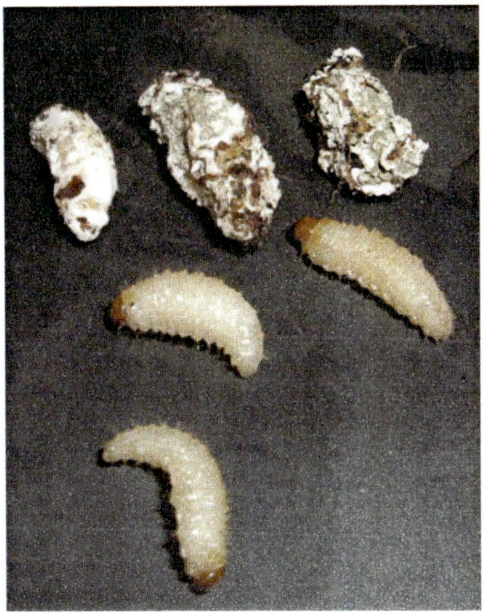

Washington (Bedding and Miller 1981). Several new products such as Met 52® Granular (Novozymes, Bagsvaerd, Denmark) are currently commercially available, making this field of study an option, especially in ground-dwelling pests.

5.5.4 *Protozoa*

Protozoa are slow acting but potentially useful as classical biological control agents due to the fact that they can result in longer-term population management. Microsporidia may play the most important role in biocontrol in the future. Currently microsporidia have resulted in inconsistent control (Federici 1999).

5.5.5 *Nematodes*

Steinernema and *Heterorhabditis* nematodes have been used most commonly for commercial insect pest control. These nematodes infect host insects via the anus, mouth or spiracles. They begin to feed on the hemolymph, and upon defecation, release symbiotic bacteria which quickly colonize the insect, killing it within 1–3 days. The best results compare favorably to synthetic insecticides in commercial settings and have been found to manage cryptic insects such as beetle grubs including black vine weevil (Fisher et al. 2009).

5.6 Pheromones, Confusion Techniques and Sterile Male Technique

The impact of pheromones and herbivore induced plant volatiles on biological control will be briefly discussed, as there are often direct interactions with beneficial arthropods. Several examples of kairomonal activity have been recorded in vineyards. Plants produce herbivore-induced plant volatiles (HIPVs) in response to attack by herbivores, which act to repel pests and attract natural enemies (James and Price 2004; Khan et al. 2008). In addition some damaged plants may produce volatile signals that warn other plants of impending attack. James and Grasswitz (2005) showed that sticky cards baited with methyl salicylate, methyl jasmonate, and 3-hexenyl acetate trapped significantly higher numbers of *Metaphycus* spp. (Encyrtidae) than unbaited traps. They also discussed the possibility of these compounds being produced by plants themselves. In Khan et al. (2008) the possible uses of synthetic HIPVs to recruit natural enemies are discussed. James (2006) found increased numbers of the green lacewing *Chrysoperla oculata* Say (Chrysopidae) attracted to sticky cards baited with undiluted methyl salicylate in a Washington vineyard. The practical use of these compounds needs continued investigation in order to better understand the optimal use of these tools.

Mansour et al. (2010) saw significantly increased parasitism rates of *P. ficus* by *A.* sp. near *pseudococci* in blocks treated with the vine mealybug sex pheromone (S)-(+)-lavandulyl senecioate indicating kairomonal impact by this pheromone on the parasitoids. *Lobesia botrana* has also been monitored and managed using pheromones (Gabel and Renczés 1985). Koclu et al. (2005) showed increased predators of *L. botrana* in vineyards that were treated with mating disruption compared to those that were not treated with mating disruption. The populations of chrysopid and coccinellid insects increased in the mating disruption vineyards compared to the control check vineyard.

5.7 Conclusion

Management of key vineyard pests is dependent on using a combination of biological control, cultural controls, mating disruption and pesticide sprays. The integrative use of these techniques will necessitate a continued increase of understanding of biological control agent behavior and non-target impacts, especially with the eventual loss of often-used pesticides, whether through legislative restrictions, non-target impacts, or development of resistance. Future work is needed on the interactions between herbivore-induced plant volatiles, optimizing mating disruption and determination of non-target impacts of new pesticide chemistries. Currently, very little information is available on the use of entomopathogenic nematodes and fungi. It is anticipated that several studies in this field will be forthcoming.

Acknowledgements The authors would like to thank Daniel Todd Dalton for reviewing this document.

References

Altieri MA, Ponti L, Nicholls CI (2005) Manipulating vineyard biodiversity for improved insect pest management: case studies from northern California. Int J Biodivers Sci Manag 1:191–203

Arakaki N, Yamazawa H, Wakamura S (2011) The egg parasitoid *Telenomus euproctidis* (Hymenoptera: Scelionidae) uses sex pheromone released by immobile female tussock moth *Orgyia postica* (Lepidoptera: Lymantriidae) as kairomone. Appl Entomol Zool 46:195–200

Austin AD, Johnson NF, Dowton M (2005) Systematics, evolution, and biology of scelionid and platygastrid wasps. Annu Rev Entomol 50:553–582

Bahlai CA, Sears MK (2009) Population dynamics of *Harmonia axyridis* and *Aphis glycines* in Niagara Peninsula soybean fields and vineyards. J Entomol Soc Ont 140:27–39

Bedding RA, Miller LA (1981) Use of a nematode, *Heterorhabditis heliothidis*, to control black vine weevil, *Otiorhynchus sulcatus*, in potted plants. Ann Appl Biol 99:211–216

Begum M, Gurr GM, Wratten SD, Hedberg PR, Nicol HI (2006) Using selective food plants to maximize biological control of vineyard pests. J Appl Ecol 43:547–554

Bell VA, Brightwell RJ, Lester PJ (2006) Increasing vineyard floral resources may not enhance localised biological control of the leafroller *Epiphyas postvittana* (Lepidoptera: Tortricidae) by *Dolichogenidea* spp. (Hymenoptera: Braconidae) parasitoids. Biocontrol Sci Technol 16:1031–1042

Belshaw R (1994) Life history characteristics of Tachinidae (Diptera) and their effect on polyphagy. In: Hawkins BA, Sheehan W (eds) Parasitoid community ecology. Oxford University Press, Oxford, pp 145–162

Berndt LA, Wratten SD, Frampton C (2000) The use of buckwheat flowers to enhance efficiency of leafroller parasitoids in a New Zealand vineyard. In: Hoddle MS (ed) Proceedings of the California conference on biological control II, Riverside, CA, pp 121–123, 11–12 July 2000

Berndt LA, Wratten SD, Scarratt SL (2006) The influence of floral resource subsidies on parasitism rates of leafrollers (Lepidoptera: Tortricidae) in New Zealand vineyards. Biol Control 37:50–55

Bolduc E, Buddle CM, Bostanian NJ, Vincent C (2005) Ground-dwelling spider fauna (Araneae) of two vineyards in southern Quebec. Environ Entomol 34:635–645

Böll S, Schwappach P, Herrmann JV (2006) Planting dog roses – an efficient method to promote mymarid populations in vineyards? IOBC/WPRS Bull 29(11):175–181

Booth LH, Bithell SL, Wratten SD, Heppelthwaite VJ (2003) Vineyard pesticides and their effects on invertebrate biomarkers and bioindicator species in New Zealand. Bull Environ Contam Toxicol 71:1131–1138

Bostanian NJ, Larocque N (2001) Laboratory tests to determine the intrinsic toxicity of four fungicides and two insecticides to the predacious mite *Agistemus fleschneri*. Phytoparasitica 29:215–222

Campos L, Franco JC, Monteiro A, Lopes C (2006) Influence of cover cropping on arthropods associated to a vineyard in Estremadura. Cien Tec Vitivinic 21:33–46

Chambers RJ (1988) Syrphidae. In: Minks AK, Harrewijn P (eds) Aphids, their biology, natural enemies and control, vol 2B. Elsevier, Amsterdam, pp 259–270

Clark RD, Grant PR (1968) An experimental study of the role of spiders as predators in a forest litter community. Ecology 49:1152–1154

Cooper ML, Daane KM, Nelson EH, Varela LG, Battany MC, Tsutsui ND, Rust MK (2008) Liquid baits control Argentine ants sustainably in coastal vineyards. Calif Agric 62:177–183

Corbett A, Rosenheim JA (1996) Impact of a natural enemy overwintering refuge and its interaction with the surrounding landscape. Ecol Entomol 21:155–164

Costello MJ (2007) Impact of sulfur on density of *Tetranychus pacificus* (Acari: Tetranychidae) and *Galendromus occidentalis* (Acari: Phytoseiidae) in a central California vineyard. Exp Appl Acarol 42:197–208

Costello MJ, Daane KM (1998) Influence of ground covers on spider (Araneae) populations in a table grape vineyard. Ecol Entomol 23:33–40

Costello MJ, Daane KM (1999) Abundance of spiders and insect predators on grapes in central California. J Arachnol 27:531–538

Costello MJ, Daane KM (2003) Spider and leafhopper (*Erythroneura* spp.) response to vineyard ground cover. Environ Entomol 32:1085–1098

Crowder DW, Northfield TD, Strand MR, Snyder WE (2010) Organic agriculture promotes evenness and natural pest control. Nature 466:109–112

Daane KM, Costello MJ (1998) Can cover crops reduce leafhopper abundance in vineyards? Calif Agric 52:27–33

Daane KM, Yokota GY (1997) Release strategies affect survival and distribution of green lacewings (Neuroptera: Chrysopidae) in augmentation programs. Environ Entomol 26:455–464

Daane KM, Yokota GY, Zheng Y, Hagen KS (1996) Inundative release of common green lacewings (Neuroptera: Chrysopidae) to suppress *Erythroneura variabilis* and *E. elegantula* (Homoptera: Cicadellidae) in vineyards. Environ Entomol 25:1224–1234

Daane KM, Sime KR, Fallon J, Cooper ML (2007) Impacts of Argentine ants on mealybugs and their natural enemies in California's coastal vineyards. Ecol Entomol 32:583–596

Daane KM, Cooper ML, Triapitsyn SV, Walton VM, Yokota GY, Haviland DR et al (2008a) Vineyard managers and researchers seek sustainable solutions for mealybugs, a changing pest complex. Calif Agric 62:167–176

Daane KM, Bentley WJ, Millar JG, Walton VM, Cooper ML, Biscay P, Yokota GY (2008b) Integrated management of mealybugs in California vineyards. Acta Hortic 785:235–252

De Bach P (1964) The scope of biological control. In: De Bach P (ed) Biological control of insect pests and weeds. Chapman and Hall, London, pp 3–20

De Bach P, Hagen KS (1964) Manipulation of entomophagous species. In: De Bach P (ed) Biological control of insect pests and weeds. Chapman and Hall, London, pp 429–455

Dib H, Simon S, Sauphanor B, Capowiez Y (2010) The role of natural enemies on the population dynamics of the rosy apple aphid, *Dysaphis plantaginea* Passerini (Hemiptera: Aphididae) in organic apple orchards in southeastern France. Biol Control 55:97–109

Elzen GW, King EG (1999) Periodic release and manipulation of natural enemies. In: Bellows TS, Fisher TW (eds) Handbook of biological control. Academic, San Diego, pp 253–270

Engel VR, Ohnesorge B (1994) The role of alternative food and microclimate in the system *Typhlodromus pyri* (Acari, Phytoseiidae) – *Panonychus ulmi* (Acari, Tetranychidae) on grape vines. II. Field experiments. J Appl Entomol 118:224–238

English-Loeb GM (1990) Plant drought stress and outbreaks of spider mites: a field test. Ecology 71:1401–1411

English-Loeb G, Rhainds M, Martinson T, Ugine T (2003) Influence of flowering cover crops on *Anagrus* parasitoids (Hymenoptera: Mymaridae) and *Erythroneura* leafhoppers (Homoptera: Cicadellidae) in New York vineyards. Agric For Entomol 5:173–181

Etzel LK, Legner EF (1999) Culture and colonization. In: Bellows TS, Fisher TW (eds) Handbook of biological control. Academic, San Diego, pp 125–198

Federici BA (1999) A perspective on pathogens as biological control agents for insect pests. In: Bellows TS, Fisher TW (eds) Handbook of biological control. Academic, San Diego, pp 517–548

Feener DH, Brown BV (1997) Diptera as parasitoids. Annu Rev Entomol 42:73–97

Fisher TW (1963) Mass culture of Cryptolaemus and Leptomastix, natural enemies of citrus mealybugs. University of California, Berkeley, CA. Agric Exp Stn Bull 797:1–38

Fisher TW, Andres LA (1999) Quarantine. In: Bellows TS, Fisher TW (eds) Handbook of biological control. Academic, San Diego, pp 103–124

Fisher JR, Bruck DJ, Bañados P, Dale A (2009) Biology and control of root weevils on berry and nursery crops in Oregon. Acta Hortic 777:345–351

Frank SD, Shrewsbury PM, Denno RF (2011) Plant versus prey resources: influence on omnivore behavior and herbivore suppression. Biol Control 57:229–235
Gabel B, Renczés V (1985) Factors affecting the monitoring of flight activity of *Lobesia botrana* and *Eupoecilia ambiguella* (Lepidoptera, Tortricidae) by pheromone traps. Acta Entomol Bohemoslov 82:269–277
Gadino AN, Walton VM, Dreves AJ (2011) Impact of vineyard pesticides on a beneficial arthropod, *Typhlodromus pyri* (Acari: Phytoseiidae), in laboratory bioassays. J Econ Entomol 104: 970–977
Galvan TL, Burkness EC, Koch RL, Hutchison WD (2009) Multicolored Asian lady beetle (Coleoptera: Coccinellidae) activity and wine grape phenology: implications for pest management. Environ Entomol 38:1563–1574
Glenn DC, Hoffmann AA (1997) Developing a commercially viable system for biological control of light brown apple moth (Lepidoptera: Tortricidae) in grapes using endemic *Trichogramma* (Hymenoptera: Trichogrammatidae). J Econ Entomol 90:370–382
Gusberti MM, Jermini E, Wyss E, Linder C (2008) Efficacy of insecticides against *Scaphoideus titanus* in organic vineyards and their side effects. Rev Suisse Vitic Arboric Hortic 40: 173–177
Gutierrez AP, Caltagirone LE, Meikle W (1999) Evaluation of results. In: Bellows TS, Fisher TW (eds) Handbook of biological control. Academic, San Diego, pp 243–252
Gutierrez AP, Daane KM, Ponti L, Walton VM, Ellis CK (2008) Prospective evaluation of the biological control of vine mealybug: refuge effects and climate. J Appl Ecol 45:524–536
Hagen KS, Mills NJ, Gordh G, McMurtry JA (1999) Terrestrial arthropod predators of insect and mite pests. In: Bellows TS, Fisher TW (eds) Handbook of biological control. Academic, San Diego, pp 383–503
Hanna R, Zalom FG, Roltsch WJ (2003) Relative impact of spider predation and cover crop on population dynamics of *Erythroneura variabilis* in a raisin grape vineyard. Entomol Exp Appl 107:177–191
Hernandez P, Daane KM, Lawson A, Yokota G (2005) Efficacy of several pesticides of table grape to control the black widow spider *Latrodectus hesperus* (Araneae: Theridiidae) collected from California table grape vineyards. The 2005 ESA annual meeting and exhibition, Fort Lauderdale, FL, 15–18 Dec 2005. http://www.esa.confex.com/esa/2005/techprogram/paper_21957.htm
Hogg BN, Daane KM (2010) The role of dispersal from natural habitat in determining spider abundance and diversity in California vineyards. Agric Ecosyst Environ 135:260–267
Hogg BN, Daane KM (2011a) Diversity and invasion within a predator community: impacts on herbivore suppression. J Appl Ecol 48:453–461
Hogg BN, Daane KM (2011b) Ecosystem services in the face of invasion: the persistence of native and nonnative spiders in an agricultural landscape. Ecol Appl 21:565–576
Hogg BN, Gillespie RG, Daane KM (2010) Regional patterns in the invasion success of *Cheiracanthium* spiders (Miturgidae) in vineyard ecosystems. Biol Invasions 12: 2499–2508
Hoy MA, Barnett WW, Reil WO, Castro D, Cahn D, Hendricks LC et al (1982) Large-scale releases of pesticide-resistant spider mite predators. Calif Agric 36:8–10
Huffaker CB, Simmonds FJ, Liang JE (1976) The theoretical and empirical basis of biological control. In: Huffaker CB, Messenger PS (eds) Theory and practice of biological control. Academic, New York, pp 42–78
Ibrahim R, Holst H, Basedow T (2004) Natural occurrence and distribution of *Trichogramma* spp. in vineyards of Rheingau (Hessia, Germany). Mitt Dtscher Ges Allg Entomol 14:213–216
Isaia M, Bona F, Badino G (2006) Influence of landscape diversity and agricultural practices on spider assemblage in Italian vineyards of Langa Astigiana (northwest Italy). Environ Entomol 35:297–307
James DG (2006) Methyl salicylate is a field attractant for the goldeneyed lacewing, *Chrysopa oculata*. Biocontrol Sci Technol 16:107–110
James DG, Grasswitz TR (2005) Synthetic herbivore-induced plant volatiles increase field captures of parasitic wasps. BioControl 50:871–880

James DG, Price TS (2004) Field-testing of methyl salicylate for recruitment and retention of beneficial insects in grapes and hops. J Chem Ecol 30:1613–1628

James DG, Whitney J (1991) Biological control of grapevine mites in inland south-eastern Australia. Aust N Z Wine Ind J 6:3

Jenkins PE, Isaacs R (2007a) Cutting wild grapevines as a cultural control strategy for grape berry moth (Lepidoptera: Tortricidae). Environ Entomol 36:187–194

Jenkins PE, Isaacs R (2007b) Reduced-risk insecticides for control of grape berry moth (Lepidoptera: Tortricidae) and conservation of natural enemies. J Econ Entomol 100:855–865

Jepsen SJ, Rosenheim JA, Matthews CE (2007a) The impact of sulfur on the reproductive success of *Anagrus* spp. parasitoids in the field. BioControl 52:599–612

Jepsen SJ, Rosenheim JA, Bench ME (2007b) The effect of sulfur on biological control of the grape leafhopper, *Erythroneura elegantula*, by the egg parasitoid *Anagrus erythroneurae*. BioControl 52:721–732

Johann L, Klock CL, Ferla NJ, Botton M (2009) Mites (Acari) associated with grapevine (*Vitis vinifera* L.) in Rio Grande do Sul. Pontifícia Universidade Católica do Rio Grande do Sul, Faculdade de Biociências, Porto Alegre, Brazil. Biociencias 17(1):1–19

Khan ZR, James DG, Midega CAO, Pickett JA (2008) Chemical ecology and conservation biological control. Biol Control 45:210–224

Kirchmair M, Neuhauser S, Strasser H, Voloshchuk N, Hoffmann M, Huber L (2009) Biological control of grape phylloxera – a historical review and future prospects. Acta Hortic 816:13–17

Koclu T, Altindisli FO, Ozsemerci F (2005) The parasitoids of the European grapevine moth (*Lobesia botrana* Den.-Schiff.) and predators in the mating disruption-treated vineyards in Turkey. OILB/SROP Bull 28(7):293–297

Landis DA, Wratten SD, Gurr GM (2000) Habitat management to conserve natural enemies of arthropod pests in agriculture. Annu Rev Entomol 45:175–201

Laurin M-C, Bostanian NJ (2007) Laboratory studies to elucidate the residual toxicity of eight insecticides to *Anystis baccarum* (Acari: Anystidae). J Econ Entomol 100:1210–1214

Lefebvre M, Bostanian NJ, Thistlewood HMA, Mauffette Y, Racette G (2011) A laboratory assessment of toxic attributes of six 'reduced risk insecticides' on *Galendromus occidentalis* (Acari: Phytoseiidae). Chemosphere 84:25–30

Legner EF, Bellows TS (1999) Exploration for natural enemies. In: Bellows TS, Fisher TW (eds) Handbook of biological control. Academic, San Diego, pp 87–102

Loughner R, Goldman K, Loeb G, Nyrop J (2008) Influence of leaf trichomes on predatory mite (*Typhlodromus pyri*) abundance in grape varieties. Exp Appl Acarol 45:111–122

Lucas É, Labrie G, Vincent C, Kovach J (2007) The multicolored Asian ladybird beetle: beneficial or nuisance organism? In: Vincent C, Goettel MS, Lazarovits G (eds) Biological control: a global perspective. CABI, Wallingford, pp 38–52

Mani M, Thorntakarya TS (1988) Studies on the safety of different pesticides to the grape mealybug natural enemies, *Anagyrus dactylopii* (How.) and *Scymnus coccivora* Ayyar. Indian J Plant Prot 16:205–210

Mansour R, Suma P, Mazzeo G, Buonocore E, Lebdi KG, Russo A (2010) Using a kairomone-based attracting system to enhance biological control of mealybugs (Hemiptera: Pseudococcidae) by *Anagyrus* sp. near *pseudococci* (Hymenoptera: Encyrtidae) in Sicilian vineyards. J Entomol Acarol Res 42:161–170

Marchesini E, Mori N, Pasini M, Bassi A (2008) Selectivity of Rynaxypyr® towards beneficial arthropods in different agroecosystems. In: Brunelli A (ed) Giornate fitopatologiche, vol 1. Cervia (RA), 12–14 Marzo, Bologna, pp 71–76

Martinez M, Coutinot D, Hoelmer K, Denis J (2006) Suitability of European Diptera tachinid parasitoids of *Lobesia botrana* (Denis & Schiffermüller) and *Eupoecilia ambiguella* (Hübner) (Lepidoptera Tortricidae) for introduction against grape berry moth, *Paralobesia viteana* (Clemens) (Lepidoptera Tortricidae), in North America. Istituto Sperimentale per la Zoologia Agraria, Firenze, Italy, Redia 89:87–97

McMurtry JA (1982) The use of phytoseiids for biological control: progress and future prospects. In: Hoy MA (ed) Recent advances in knowledge of the phytoseiidae, vol 3284. University of California Press, Berkeley, pp 23–48

McMurtry JA, Croft BA (1997) Life-styles of phytoseiid mites and their roles in biological control. Annu Rev Entomol 42:291–321

Mgocheki N, Addison P (2009a) Effect of contact pesticides on vine mealybug parasitoids, *Anagyrus* sp. near *pseudococci* (Girault) and *Coccidoxenoides perminutus* (Timberlake) (Hymenoptera: Encyrtidae). S Afr J Enol Vitic 30:110–116

Mgocheki N, Addison P (2009b) Interference of ants (Hymenoptera: Formicidae) with biological control of the vine mealybug *Planococcus ficus* (Signoret) (Hemiptera: Pseudococcidae). Biol Control 49:180–185

Miliczky E, Horton DR (2007) Natural enemy fauna (Insecta, Araneae) found on native sagebrush steppe plants in eastern Washington with reference to species also found in adjacent apple and pear orchards. Pan-Pac Entomol 83:50–65

Mills NJ (1982) Voracity, cannibalism and coccinellid predation. Ann Appl Biol 101:144–148

Mills NJ, Daane KM (2005) Non-pesticide alternatives (biological and cultural controls) can suppress crop pests. Calif Agric 59:23–28

Murphy BC, Rosenheim JA, Grannett J (1996) Habitat diversification for improving biological control: abundance of *Anagrus epos* (Hymenoptera: Mymaridae) in grape vineyards. Environ Entomol 25:495–504

Murphy BC, Rosenheim JA, Dowell RV, Grannett J (1998) Habitat diversification tactic for improving biological control: parasitism of the western grape leafhopper. Entomol Exp Appl 87:225–235

Mustu M, Kilincer N, Ulgenturk S, Kaydan MB (2008) Feeding behavior of *Cryptolaemus montrouzieri* on mealybugs parasitized by *Anagyrus pseudococci*. Phytoparasitica 36:360–367

Nagarkatti S, Tobin C, Saunders MC, Muza AJ (2003) Release of native *Trichogramma minutum* to control grape berry moth. Can Entomol 135:589–598

Nash MA, Hoffmann AA, Thomson LJ (2010) Identifying signature of chemical applications on indigenous and invasive non-target arthropod communities in vineyards. Ecol Appl 20:1693–1703

Nelson EH, Daane KM (2007) Improving liquid bait programs for Argentine ant control: bait station density. Environ Entomol 36:1475–1484

Nicholls CI, Parrella M, Altieri MA (2001) The effects of a vegetational corridor on the abundance and dispersal of insect biodiversity within a northern California organic vineyard. Landsc Ecol 16:133–146

Nicholls CI, Altieri MA, Ponti L (2008) Enhancing plant diversity for improved insect pest management in northern California organic vineyards. Acta Hortic 785:263–278

Obrycki JJ, Kring TJ (1998) Predaceous Coccinellidae in biological control. Annu Rev Entomol 43:295–321

Ponti L, Ricci C, Veronesi F, Torricelli R (2005) Natural hedges as an element of functional biodiversity in agroecosystems: the case of a central Italy vineyard. Bull Insectol 58:19–23

Posenato G, Marchesini E, Graziani N, Vandini G, Ferrari D, Parrilla MM et al (2003) Use of dinocap in control strategy against powdery mildew and side effects on predatory mites. Informatore Agrario 59:73–78

Prasad RP, Snyder WE (2006) Polyphagy complicates conservation biological control that targets generalist predators. J Appl Ecol 43:343–352

Price PW (1997) Insect ecology. Wiley, New York

Prischmann DA, James DG, Wright LC, Teneyck RD, Snyder WE (2005a) Effects of chlorpyrifos and sulfur on spider mites (Acari: Tetranychidae) and their natural enemies. Biol Control 33:324–334

Prischmann DA, James DG, Gingras SN, Snyder WE (2005b) Diversity and abundance of insects and spiders on managed and unmanaged grapevines in southcentral Washington State. Pan-Pac Entomol 81:131–144

Rabb RL, Guthrie FE (1970) Concepts of pest management. North Carolina State University Press, Raleigh

Reid CD, Lampman RL (1989) Olfactory responses of *Orius insidiosus* (Hemiptera: Anthocoridae) to volatiles of corn silks. J Chem Ecol 15:1109–1115

Remund U, Boller E (1996) Importance of hedgerow plants for the egg parasitoids of the green grapevine leafhopper in eastern Switzerland. Obst-und Weinbau 132:238–241

Reuter OM (1913) Lebensgewohnheiten und Instinkte der Insekten. Friedlander, Berlin

Riddick EW, Mills NJ (1994) Potential of adult carabids (Coleoptera: Carabidae) as predators of fifth-instar codling moth (Lepidoptera: Tortricide) in apple orchards in California. Environ Entomol 23:1338–1345

Riechert SE, Lockley T (1984) Spiders as biological control agents. Annu Rev Entomol 29:299–320

Rigamonti IE (2006) Preliminary observations on the role of botanical diversity on the presence of egg parasitoids of grape leafhoppers in northern Italy. IOBC/WPRS Bull 29(11):187–192

Scarratt SL, Wratten SD (2004) Using a rubidium marker to study the dispersal of a parasitoid from floral resources. In: Hoddle MS (ed) Proceedings of the California conference on biological control IV, Berkeley, CA, pp 137–140, 13–15 July 2004

Scarratt SL, Wratten SD, Shishehbor P (2008) Measuring parasitoid movement from floral resources in a vineyard. Biol Control 46:107–113

Schwartz A (1993) Occurrence of natural enemies of phytophagous mites on grapevine leaves following application of fungicides for disease control. S Afr J Enol Vitic 14:16–17

Sharley DJ, Hoffmann AA, Thomson LJ (2008) The effects of soil tillage on beneficial invertebrates within the vineyard. Agric For Entomol 10:233–243

Shibao M, Ehara S, Hosomi A, Tanaka H (2006) Effect of insecticide application on the population density of yellow tea thrips, *Scirtothrips dorsalis* Hood (Thysanoptera: Thripidae) and *Euseius sojaensis* (Ehara) (Acari: Phytoseiidae) on grapes. Jpn J Appl Entomol Zool 50:247–252

Slone DH, Croft BA (2000) Changes in intraspecific aggregation and the coexistence of predaceous apple mites. Oikos 91:153–161

Smith HS (1923) Successful introduction and establishment of the ladybird, *Scymnus bineavatus* Mulsant, in California. J Econ Entomol 16:516–518

Stavrinides MC, Mills NJ (2009) Demographic effects of pesticides on biological control of Pacific spider mite (*Tetranychus pacificus*) by the western predatory mite (*Galendromus occidentalis*). Biol Control 48:267–273

Stavrinides M, Daane KM, Lampien B, Mills NJ (2010) Plant water stress, leaf temperature and spider mite (Acari: Tetranychidae) outbreaks in California vineyards. Environ Entomol 39:1232–1241

Stutz S, Entling MH (2011) Effects of the landscape context on aphid-ant-predator interactions on cherry trees. Biol Control 57:37–43

Szentkiralyi F (2001) Lacewings in fruit and nut crops. In: McEwen P, New TR, Whittington AE (eds) Lacewings in the crop environment. Cambridge University Press, Cambridge, pp 172–238

Tauber MJ, Tauber CA, Daane KM, Hagen KS (2000) New tricks for old predators: implementing biological control with *Chrysoperla*. Am Entomol 46:26–38

Thiéry D, Xuéreb A, Villemant C, Sentenac G, Delbac L, Kuntzman P, Lozzia C (2001) The parasites of grape tortricids: noticed on several species present in 3 French vine regions. OILB/SROP Bull 24:135–141

Thomson LJ (2006) Influence of reduced irrigation on beneficial invertebrates in vineyards. Aust J Exp Agric 46:1389–1395

Thomson LJ, Hoffmann AA (2007) Effects of ground cover (straw and compost) on the abundance of natural enemies and soil macro invertebrates in vineyards. Agric For Entomol 9:173–179

Thomson LJ, Hoffmann AA (2009) Vegetation increases the abundance of natural enemies in vineyards. Biol Control 49:259–269

Thomson LJ, Hoffmann AA (2010) Natural enemy responses and pest control: importance of local vegetation. Biol Control 52:160–166

Thomson LJ, Glenn DC, Hoffmann AA (2000) Effects of sulfur on *Trichogramma* egg parasitoids in vineyards: measuring toxic effects and establishing release windows. Aust J Exp Agric 40:1165–1171

Tollerup K, Rust MK, Klotz JH (2007) *Formica perpilosa*, an emerging pest in vineyards. J Agric Urban Entomol 24:147–158

Triapitsyn SV (1998) *Anagrus* (Hymenoptera: Mymaridae) egg parasitoids of *Erythroneura* spp. and other leafhoppers (Homoptera: Cicadellidae) in North American vineyards and orchards: a taxonomic review. Trans Am Entomol Soc 124:96–97

Waage JK, Carl KP, Mills NJ, Greathead DJ (1984) Rearing entomophagous insects. In: Singh P, Moore RF (eds) Handbook of insect rearing, vol 1. Elsevier, Amsterdam, pp 45–66

Walton VM (2003) Development of an integrated pest management system for vine mealybug, Planococcus ficus (Signoret), in vineyards in the Western Cape Province, South Africa. PhD dissertation, University of Stellenbosch, South Africa

Walton VM, Pringle KL (1999) Effects of pesticides used on table grapes on the mealybug parasitoid *Coccidoxenoides peregrinus*. S Afr J Enol Vitic 20:31–34

Walton VM, Pringle KL (2001) Effects of pesticides and fungicides used on grapevines on the mealybug predatory beetle *Nephus 'boschianus'* (Coccinellidae, Scymnini). S Afr J Enol Vitic 22:107–110

Williams LI, Martinson TE (2000) Colonization of New York vineyards by *Anagrus* spp. (Hymenoptera: Mymaridae): overwintering biology, within-vineyard distribution of wasps, and parasitism of grape leafhopper, *Erythroneura* spp. (Homoptera: Cicadellidae), eggs. Biol Control 18:136–146

Wilson LT, Carmean I, Flaherty DL (1991) *Aphelopus albopictus* Ashmead (Hymenoptera: Dryinidae): abundance, parasitism, and distribution in relation to leafhopper hosts in grapes. Hilgardia 59:16

Xuéreb A, Thiéry D (2006) Does natural larval parasitism of *Lobesia botrana* (Lepidoptera: Tortricidae) vary between years, generation, density of the host and vine cultivar? Bull Entomol Res 96:105–110

Chapter 6
Chemical Ecology Providing Novel Strategies Against Vineyard Pests in Australia

Marja Simpson, Vanessa J. Connick, Yann Guisard, Olivia L. Reynolds (née Kvedaras), Anthony Saliba, and Geoff M. Gurr

6.1 Introduction

Chemical ecology has been recognized as an important and distinct research area for over three decades and it deals with the chemical mechanisms which help control intra- and interspecific interactions amongst forms of life. All organisms use chemical signals to transmit information as a form of communication (Dicke 2009). Research in the field of chemical ecology involves the identification and synthesis of the chemical substances as well as the measurement of the ecological consequences of signal transfer (Dicke and Takken 2006).

M. Simpson • G.M. Gurr (✉)
EH Graham Centre for Agricultural Innovation (NSW Department of Primary Industries and Charles Sturt University), Charles Sturt University, Leeds Parade, 883, 2800, Orange, NSW, Australia
e-mail: marjasimpson@bigpond.com; ggurr@csu.edu.au

V.J. Connick • Y. Guisard
National Wine & Grape Industry Centre, Faculty of Science, School of Agricultural & Wine Sciences, Charles Sturt University, Locked Bag 588, Wagga Wagga, NSW 2678, Australia
e-mail: vconnick@csu.edu.au; yguisard@csu.edu.au

A. Saliba
National Wine & Grape Industry Centre, Faculty of Science, School of Agricultural & Wine Sciences, Charles Sturt University, Locked Bag 588, Wagga Wagga, NSW 2678, Australia
e-mail: asaliba@csu.edu.au

O.L. Reynolds (née Kvedaras)
EH Graham Centre for Agricultural Innovation (NSW Department of Primary Industries and Charles Sturt University), Elizabeth Macarthur Agricultural Institute, Woodbridge Road, Menangle, NSW 2568, Australia
e-mail: olivia.reynolds@industry.nsw.gov.au

N.J. Bostanian et al. (eds.), *Arthropod Management in Vineyards: Pests, Approaches, and Future Directions*, DOI 10.1007/978-94-007-4032-7_6,

Plants have evolved a wide range of constitutive and induced defense mechanisms against herbivore feeding or oviposition, including the production of herbivore-induced plant volatiles (HIPVs) (Karban and Baldwin 1997). Natural enemies use these volatile signals to locate their hosts or prey (Thaler 1999; Bernasconi Ockroy et al. 2001; Dicke 2009; Dicke et al. 1990). Herbivore induced plant volatile blends differ depending on the attacking herbivore and plant species involved. They can be extremely complex and vary qualitatively and quantitatively, with several compounds released more commonly than others (Dicke et al. 1998; Van den Boom et al. 2004). There is also evidence of plant-plant and within-plant communication through volatiles (Frost et al. 2008). For instance, specific volatiles such as methyl salicylate (MeSA), methyl jasmonate (MeJA), ethylene and the green-leaf volatiles (GLVs) can activate jasmonic acid-dependent defense reactions in neighbouring plants, or other parts of the same plant, boosting the production of endogenous aromatic and terpenoid volatile compounds that enhance the induced defense response of plants (Yan and Wang 2006; Ton et al. 2007; Tamogami et al. 2008).

This chapter provides a concise review of the ways in which chemical ecology research is generating new avenues for pest management and considers the utility of these novel technologies to major vineyard pest problems. Although our emphasis is on vineyard protection in Australia, as wine and table grapes are a substantial and valuable industry, many of the issues apply generally to other agricultural crops.

6.2 The Conservation Biological Control Context

In agriculture, integrated pest management (IPM) is a pest-control strategy which uses a variety of complementary practices. One of these, conservation biological control (CBC), involves cultural practices that preserve and enhance the efficacy of natural enemy populations through the modification of the biotic environment and pesticide usage (Eilenberg et al. 2001; Gurr et al. 2004; Tompkins et al., Chap. 7). Interest in biological control by growers, researchers and policy makers, as an alternative to pesticides, has intensified in recent years because of environmental and human health concerns, pest resistance and associated expenses. Adoption of biological control by growers has been limited by a range of factors, including the cost of mass produced agents (in inundative biological control) and the risks associated with introducing exotic agents (in classical biological control). Both of these issues are avoided through the use of CBC. However, this method is a comparatively new approach and therefore is constrained by a relative paucity of information.

Research on CBC in various crop systems is aimed principally at increasing the efficacy and reliability of this pest control method. Many studies have identified that habitat manipulation in the form of floral plant species distributed appropriately within or around the crop leads to increased abundance, residency and diversity of natural enemies, reduced pest numbers and increased parasitism rates (Baggen and Gurr 1998; Landis et al. 2000; English-Loeb et al. 2003; Bostanian et al. 2004; Lee and Heimpel 2005; Gurr et al. 2005; Berndt et al. 2006). A key challenge is to establish sufficient numbers of natural enemies exactly when and where they are required; an objective similar to biological insecticide use. Natural enemy populations may otherwise be too slow to establish and have limited ability to keep pests below economic thresholds.

6.3 Vineyard Pests Occurring in Australia and Their Natural Enemies

6.3.1 Important Vineyard Pest Species in Australia

The light brown apple moth (LBAM) *Epiphyas postvittana* (Walker) (Torticidae) is a polyphagous leafroller indigenous to Australia that feeds on native and introduced plant species. It is also widely distributed in New Zealand, Great Britain, and several other countries (Danthanarayana 1975; Buchanan 1977). It is regarded as the most serious pest in Australian vineyards as well as a major pest of other horticultural crops including pome, stone and citrus fruits.

Several common mite species, such as grapeleaf blister mite and grapeleaf bud mite *Eriophyes vitis* (Pagenstecher) (=*Colomerus vitis* (Pagenstecher)) (Eriophyidae), grape rust mite *Calepitrimerus vitis* Nalepa (Eriophyidae), and bunch mite *Brevipalpus* spp. (Tenuipalpidae) attack grapevines and are considered minor pest species. However, under favourable conditions they can cause economic damage (Nicholas et al. 1994; James et al. 1995).

Less frequently occurring pests are the twospotted spider mite *Tetranychus urticae* Koch (Tetranychidae), mealybugs *Pseudococcus* spp. (Pseudococcidae), and grapevine scale *Parthenolecanium persicae* (F.) (Coccidae). Regional pests include weevils (Curculionidae), Rutherglen bug *Nysius vinitor* Bergroth (Coccidae), fig longicorn *Acalolepta vastator* (Newman) (Cerambycidae), larvae of pink cutworm *Agrotis munda* Walker (Noctuidae), vine moth *Phalaenoides glycinae* Lewin (Noctuidae), thrips (Thysanoptera), wingless grasshoppers *Phaulacridium vittatum* (Sjöstedt) (Acrididae), and katydids *Caedicia* spp. (Tettigoniidae) (Nicholas et al. 1994; Thomson et al. 2007). Grapevine phylloxera *Daktulosphaira vitifoliae* (Fitch) (Phylloxeridae) is restricted to a small number of quarantined areas in the states of Victoria and New South Wales, with the majority of the Australian winegrowing regions remaining free of this pest. Phylloxera causes grapevine roots to develop fleshy, yellow galls resulting in weak shoot growth, reduced cropping and premature yellowing of foliage in autumn (Buchanan et al. 1994; Herbert et al. 2008; Forneck and Huber 2009: Powell, Chap. 10). A summary of the major grapevine pests occurring in Australia is presented in Table 6.1.

Table 6.1 Major grapevine pests, crop damage caused and their natural enemies in Australia

		Major grapevine pests		
		Light brown apple moth (*Epiphyas postvittana*) (Lepidoptera: Tortricidae)	Mites (*Brevipalpus* spp., *Colomerus vitis, Calepitrimerus vitis Tetranychus urticae*)	Mealybugs (*Pseudococcus* spp.)
Crop damage		Larvae eat shoots, flowers and fruit. The wounds to berries make the bunches further susceptible to mould entry by *Botrytis* spp. and *Aspergillus* spp.	Blister mites cause galls on leaves. Bud mites feed on bud scales, leaf, bunch primordia. Grape rust mite mites distort leaves, shorten growing roots. Bunch mites and *T. urticae* cause leaf necrosis	Mealybugs excrete sticky honeydew causing mould covering on grape bunches and leaves
Natural enemies Guild/order/family	**Species**	**Light brown apple moth – reference number**	**Mites**	**Mealybugs**
Parasitoids				
Hymenoptera				
Braconidae	*Dolichogenidea tasmanica*	Larvae- 1, 2, 3, 4		
	Bassus unimaculata	Larvae 1		
	Bracon spp.	Larvae 1		
Trichogrammatidae	*Trichogramma* spp.	Eggs 1, 2, 3		
Ichneumonidae	*Australoglypta latrobei*	Larvae 1, 2		
	Exochus spp.	Larvae 1, 2		
	Eriborus epiphyas	Larvae 1		
	Temelucha minuta	Larvae 1		
	Oedemopsis hobartensis	Larvae 1		
	Phytodietus celsissimus	Larvae 1		
	Glabridorsum stokesii	Pupae 1		
	Xanthopimpla rhopaloceros	Pupae 1, 4		
	Phytodietus spp.	Larvae 2		

Table 6.1 (continued)

Natural enemies Guild/order/family	Species	Light brown apple moth – reference number	Mites	Mealybugs
Bethylidae	*Eupsenella* spp.	Larvae 1		
	Goniozus jacintae	Larvae 1, 2, 4		
Chalcididae	*Brachymeria* spp.	Pupae 1, 2, 4		
Encyrtidae	*Anagyrus fusciventris*			3, 5
	Tetracnemoidea brevicornis			3, 5
Pteromalidae	*Ophelosia* spp.			3, 5
Diptera	*Voriella uniseta*	Larvae, pupae 1, 2, 4		
Tachinidae	*Trigonospila brevifacies*	Larvae 1, 2		
Predators			3	3, 5
Neuoptera				
Hemerobiidae	*Micromus* spp.	Eggs, larvae 3, 4	3	
Chysopidae	*Chrysopa* spp.	Eggs, larvae 1, 4		3, 5
	Mallada signata	Eggs, larvae 6		
Coleoptera				3, 5
Coccinellidae	*Rhizobius ruficollis*			
	Cryptolaemus montrouzieri			3, 5
	Stethorus spp.		3	
	Diomus notescens	Larvae 4		
Melyridae	*Dicranolaius bellulus*	Larvae 4		
Hemiptera				
Pentatomidae	*Oechalia schellenbergii*	Larvae 3		
Miridae	*Melanotrichus australianus*	1, 4		
Araneae				
Theridiidae	*Achaearanea veruculata*	Larvae 1, 4, 6		
Thomisidae	*Diaea* spp.	Larvae 1, 4, 6		

(continued)

Table 6.1 (continued)

Natural enemies Guild/order/family	Species	Light brown apple moth – reference number	Mites	Mealybugs
Dermaptera				
Forficulidae	*Forficula auricularia*	1, 2, 3, 4		
Acari			3	
Phytoseiidae	*Amblyseius victoriensis*			
	Typhlodromus doreenae		3	
Diptera	*Diadiplosis koebelei*			5
Cecidomyiidae				

[1]Paull and Austin (2006)
[2]Danthanarayana (1983)
[3]Geier and Briese (1980)
[4]MacLellan (1973)
[5]Furness (1976)
[6]Buchanan and Amos (1992)

6.3.2 Important Beneficial Parasitoids and Predators in Australian Vineyards

The pest-predator interactions in grapevines are complex and multidimensional. A summary of the important beneficial parasitoids and predators of major grapevine pests is presented in Table 6.1.

6.4 Plant Interactions with the Environment and Their Chemical Defense Mechanisms

Plants have evolved various direct and indirect responses to arthropod pests. Karban and Baldwin (1997) classified plant defenses into constitutive or induced. Constitutive defenses exist independently of plant damage and are activated prior to contact with the attacker, whereas induced responses are changes in the plant as a result of damage by an attacker (Levin 1976). A decrease in negative consequences from induced attacks on the plant is termed induced defense (Karban and Baldwin 1997). Plant defense can involve mechanical or chemical mechanisms. However, only direct and indirect chemical defense mechanisms will be discussed in this chapter. Chemical volatile compounds can be released from plant leaves, flowers and fruits into the atmosphere, and from roots into the soil. From more than 90 plant families 1,700 volatile compounds have been described (Dudareva et al. 2006). Typically, volatile compounds are lipophilic liquids of sufficiently low molecular weight and high vapor pressure to allow transport across membranes; thus volatile compounds can be released into the atmosphere or soil (Dudareva et al. 2006). These compounds can be classified into three major groups: terpenoids, phenylpropanoids/benzenoids, and fatty acid derivatives (Dudareva et al. 2006).

The primary function of such volatile compounds is to defend plants against herbivores and pathogens, or to provide a reproductive advantage by attracting pollinators and seed dispersers to the plant. Volatiles emitted by plants in response to herbivore damage are also referred to as HIPVs.

6.5 Induced Plant Defenses

The relationship between HIPV blends and tri-trophic insect-plant interactions has been the subject of research for the past 20 years. The majority of studies have been conducted in laboratory environments. Research has established that two types of volatile blends can be distinguished depending on the type of plant damage: herbivore-induced or mechanical. These blends can differ quantitatively and qualitatively with some compounds in common (Whitman and Eller 1990; Van den Boom et al. 2004; Dudareva et al. 2006). The first type includes the production of novel compounds

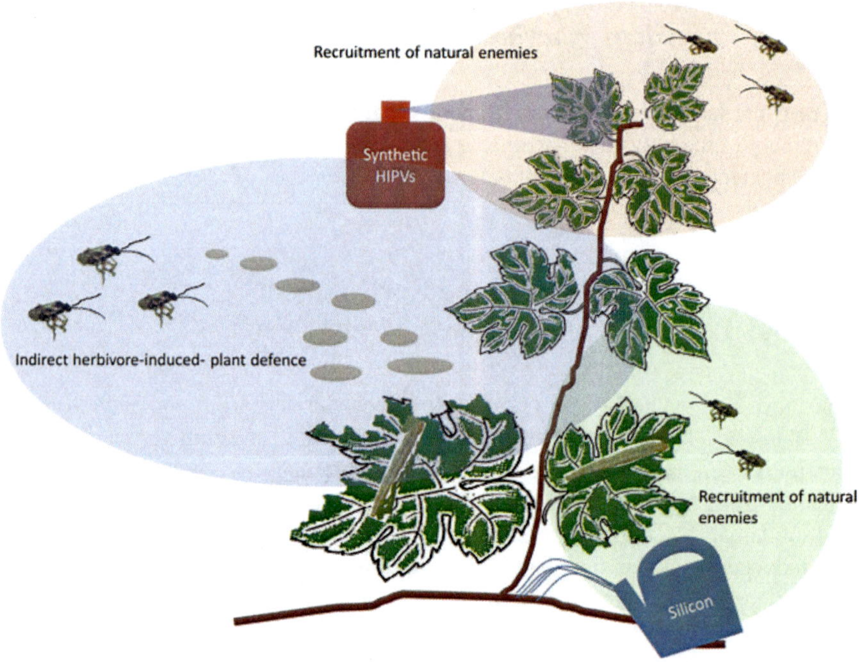

Fig. 6.1 Emission of HIPVs and recruitment of natural enemies in response to herbivore damage, application of synthetic HIPVs and silicon

which are the major components of volatile blends. The second type includes the production of the same components produced in lesser quantities after mechanical damage. Among the induced volatile substances released from plants are green-leaf volatiles (GLVs) which are important signalling cues and are released immediately after the plant tissue has been damaged mechanically or by a herbivore (Yan and Wang 2006). Green-leaf volatiles are six carbon alcohols, aldehydes and derivative esters (Whitman and Eller 1992) and are responsible for the odor of damaged leaves, for example, fresh mowed grass, and therefore are also defined as typical wound signals (Whitman and Eller 1990; Dudareva et al. 2006).

Herbivorous arthropods can be directly affected by emitted HIPVs due to their toxic, repelling and deterring properties which can result in their death or retard development (Dicke 1999). Herbivore-induced plant volatiles can also affect herbivores indirectly by attracting natural enemies of the attacking herbivore, which can protect the plant from further damage (Dicke 1999; Dicke et al. 1999; Turlings and Ton 2006; Dudareva et al. 2006). A schematic representation of an increase in volatile compounds released by plants in response to the application synthetic HIPVs and of silicon or subsequent herbivore feeding is shown in Fig. 6.1.

Emissions of HIPVs occur not only in response to herbivore feeding but also from the deposition of insect eggs on plant parts (oviposition) or from insect feeding on plant roots, thus attracting the natural enemies that use these eggs as hosts or root feeders as prey respectively (Hilker and Meiners 2006; Turlings and Ton 2006). In addition, HIPV compounds also act as plant to plant signals (Ruther and Kleier 2005; Dudareva et al. 2006; Yan and Wang 2006), triggering responses in neighbouring undamaged plants, effectively warning of impending attack. Plants may then produce their own direct and indirect defenses to respond faster to impending or future herbivore attack (Engelberth et al. 2004; Turlings and Ton 2006; Baldwin et al. 2006).

However, HIPVs not only attract beneficial insects, but can also attract herbivores that may cause increased damage to the plant (Dudareva et al. 2006). Bolter et al. (1997) demonstrated in a laboratory study the attraction of the herbivorous Colorado potato beetle (CPB) *Leptinotarsa decemlineata* (Say) (Coleoptera: Chrysomelidae) to herbivore-damaged plants during and after feeding. Colorado potato beetle adults were attracted to small potato plants infested with CPB and beet armyworm *Spodoptera exigua* (Hübner) (Lepidoptera: Noctuidae) larvae. However, Whitman and Eller (1990) argued that once a plant is under herbivore attack, attracting beneficials may outweigh the disadvantage of attracting additional herbivores.

Variations in composition of HIPV blends occur within an individual plant species as well as between different cultivars of the same species. Takabayashi et al. (1994) demonstrated that predators were preferentially attracted to volatiles of young cucumber leaves infested with *T. urticae* compared to older ones also infested with *T. urticae* as a response of the plant to directing predators to its growing parts. The attacking herbivore and abiotic factors such as light intensity, humidity, water stress, availability of nutrients and wind may also have an effect on HIPV composition (Takabayashi et al. 1994; Dudareva et al. 2006; Kessler et al. 2006). In an olfactometer, Takabayashi et al. (1994) assessed that uninfested lima bean leaves placed under high light intensity were more attractive to predatory mites than under lower light intensity. Gouinguene and Turlings (2002) described higher induced volatile emissions by corn plants when the soil was relatively dry, relative humidity was between 45% and 65% and air temperature was between 22°C and 27°C, with high light intensity and continuous fertilisation of the soil.

Little is known about the volatile profile induced by feeding or ovipositing arthropod herbivores on grapevines. Van den Boom et al. (2004) identified the volatile profiles induced by feeding of the twospotted spider mite on several plant species including grapevines. They also compared the volatile blend induced by spider mite feeding with blends emitted from mechanically damaged and healthy grapevine leaves. Results revealed that spider mite-infested grapevine leaves emitted several dominating novel compounds including MeSA, (3*E*)-4,8-dimethyl-1,3,7-nonatriene, β-caryophyllene and α- humulene. Additionally, the amount of (3*E*)-4,8-dimethyl-1,3,7-nonatriene was four times higher in spider mite-infested leaves compared with blends released from uninfested leaves. Loughrin et al. (1997) identified 19 compounds emitted from grapevine (*Vitis labrusca* L.) leaves after damage by Japanese beetle *Popillia japonica* Newman (Coleoptera: Scarabaeidae), the majority comprising aliphatic aldehydes, alcohols, esters, and terpene hydrocarbons, which were produced about 50 times higher in beetle-damaged vines compared to undamaged vines. Liu et al. (2006) in recreating Loughrin's et al. (1997) experiment, found peak emission time of volatiles produced from *V. labrusca* damaged by *P. japonica* occurred at 2.3–2.8 h after feeding began.

6.6 The Role of Silicon in Plant Defense Against Pests

The role of silicon (Si) in plant defense has been recognised since the early 1900s. Although Si is involved in inducing plant defenses against arthropod pests, its involvement is a relatively new field of study (Reynolds et al. 2009). Silicon is the second most abundant element in the Earth's crust (Epstein 1994) and is present in plants in amounts equivalent to, and sometimes in excess of, those of Ca, Mg, S, and P (Epstein 1999), although the extent to which it is an essential plant nutrient is not known. However, the importance of Si as an element that is especially beneficial for plants exposed to abiotic (e.g., drought, salinity, and heavy metal toxicity) and biotic (e.g., insects and pathogens) stresses is now beyond doubt. More recently, it has been found that Si also directly enhances induced resistance in plants attacked by pests by acting as a signal in inducing systemic chemical defenses in plants (Gomes et al. 2005; Kvedaras and Keeping 2007; Kvedaras et al. 2010).

6.6.1 Silicon Uptake by Vitis vinifera

The content of Si in plant tissue varies depending on the plant species, with dicotyledons generally containing lower concentrations (≥1% dry weight) compared to grasses (1–5% dry weight) and wetland grasses (10–15% dry weight) (Jones and Handreck 1967; Epstein 1994; Mitani and Ma 2005; Matichenkov and Bocharnikova 2007). Silicon accumulators are defined as plants which contain >1% Si dry weight and show a Si:Ca molecular ratio >1; plants which contain 0.5–1% Si (Si:Ca molecular ratio <1) are defined as intermediate accumulators; and plants which contain <0.5% Si are termed non-accumulators (Takahashi and Miyake 1977; Ma 2007). Silicon accumulators generally contain 8–20 times as much Si in their leaves as non-accumulators (Takahashi and Miyake 1977; Adatia and Besford 1986).

There is little published research on the role of Si in *Vitis vinifera* L., a dicotyledon, and a Si non-accumulator. However, it is known that Si uptake by *V. vinifera* via the transpiration stream is a passive process; a view reinforced by studies finding constant amounts of soluble SiO_2 in the 'transport regions' of the plant (including stem and petiole), and an accumulation of SiO_2 in older leaves (Blaich and Wind 1989; Blaich and Grundhöfer 1997). *Vitis vinifera* is able to restrict Si levels in the xylem and the concentration of silica in the roots is correlated with, but is considerably higher than the SiO_2 content in the soil solution, indicating that $Si(OH)_4$ might be concentrated in the root symplasts around the xylem from where it diffuses into the transpiration stream according to the speed of water flow (Lafos 1995; Blaich and Grundhöfer 1997).

Only one study reports the uptake of varying concentrations of soil-applied Si by *V. vinifera*. This study, by Blaich and Grundhöfer (1997), applied K_2SiO_3 to the soil growing *V. vinifera* at 10 and 112 mg/kg SiO_2, as well as conducting supplementary experiments using 200 and 400 mg/kg SiO_2, forming an oversaturated solution. The results of that study indicated that: (1) Si solubility was dependent on soil temperature – solubility, for at 5°C it was 50% less than at 20°C, (2) Si solubility in soil was reflected in Si content in *V. vinifera* leaves, (3) Si content in different *V. vinifera* cultivars did not vary significantly, (4) uptake of Si by *V. vinifera* depended on transpiration rate of the plant and decreased in dry conditions, (5) older leaves of plants grown on 112 ppm SiO_2 contained higher Si content (2% SiO_2) than younger leaves (0.5% SiO_2) with greater Si content in the leaf periphery, (6) Si content in shoots and petioles was less than 10% Si content in leaves, (7) soluble silica in *V. vinifera* leaves grown on 112 ppm SiO_2 made up 30% total silica content in young leaves, 15% in medium age leaves, and 7% in older leaves, and (8) Si content in grapevines could be further enhanced when applied to the soil at 200 and 400 ppm SiO_2, although $Si(OH)_4$ in *V. vinifera* tissue saturated at 150 ppm at 20°C. These findings are consistent with what is known about the uptake of Si in plants in general (Jones and Handreck 1967; Takahashi and Miyake 1977). Matichenkov and Bocharnikova (2007) consider soluble Si in the soil solution at 20–40 ppm to be a 'low level of deficiency' and >40 ppm Si as 'without deficiency'. However, additional research is required to determine Si requirements of grapevines, and the minimum and maximum threshold levels of Si required in the soil solution to provide plant protection against arthropod attack. Also unknown is the time lag between Si application and the resulting effects on enhanced pest-resistance of grapevines.

6.6.2 Role of Silicon in Induced Plant Chemical Defenses

There is recent evidence to suggest that soluble Si is involved in induced chemical defense to insect herbivore attack, through the enhanced production of defensive enzymes (Gomes et al. 2005; Kvedaras and Keeping 2007; Ranger et al. 2009) or possibly the enhanced release of plant volatiles for attraction of biological control agents (Kvedaras et al. 2010). Silicon, either alone or together with *Schizaphis graminum* (Rondani) (Aphididae) pre-infestation, elicited a significant increase in the defensive enzymes, peroxidase, polyphenoloxidase, and phenylalanine ammonia-lyase activity in wheat (Gomes et al. 2005). Similarly, Ranger et al. (2009) identified and quantified phenolic acids and flavonols in leaf tissue of *Zinnia elegans* Jacquin treated with potassium silicate and infested with the green peach aphid *Myzus persicae* (Sulzer) (Aphididae). This study found significant elevations in the defensive enzymes 5-caffeoylquinic acid, *p*-coumaroylquinic acid, rutin, and a slight elevation in guaiacol peroxidase activity. Further, the total cumulative fecundity and the intrinsic rate of increase of *M. persicae* were reduced on Si-treated plants, which were believed to be caused in part by the defense-related compounds.

More recently, Kvedaras et al. (2010) hypothesized that Si may also enhance induced chemical defenses against arthropod attack by altering and enhancing the volatile compounds emitted by an attacked plant. The authors were able to demonstrate that Si-treated plants with a pest-infestation were more attractive to natural enemies than Si-untreated plants with a pest infestation. Further, this effect was reflected in elevated biological control in the field. Additional studies to measure and identify the compounds produced by pest-infested plants, particularly those which may be strongly affected by treatment of the plants with Si are now being pursued. Research aimed at understanding the role of Si in HIPV production is gaining momentum although this area remains novel in horticultural and indeed agricultural research.

6.7 Novel Vineyard Pest Management Strategies in Australia

6.7.1 The Use of Synthetic HIPVs for Enhancing Recruitment and Residency of Beneficial Arthropods

Laboratory and field trials have assessed the potential and practicality of synthetic HIPVs for application in IPM programs, particularly the increased searching activity of beneficial arthropods and direct recruitment into the crop at certain times of the year (Khan et al. 2008). Several synthetic HIPVs, such as MeSA, MeJA, methyl anthranilate (MeA), cis-3-hexen-1-ol (He), cis-3-hexenyl acetate (HA) and benzaldehyde (Be), have been tested for their efficacy to attract beneficial insects in field studies in North American grapevine- and hop-yards. James (2003a, b) provided direct field evidence that the abundance of the green lacewing *Chrysopa nigricornis* Burmeister (Neuroptera: Chrysopidae), *Geocoris pallens* Stål (Hemiptera: Geocoridae), hoverflies (Diptera: Syrphidae), and the lady beetle *Stethorus punctum picipes* (Casey) (Coleoptera: Coccinellidae) were significantly increased on sticky traps baited with controlled released dispensers (CRD) containing MeSA in a hop-yard over several months. The large population of predatory insects in the MeSA-baited hop yard led to a reduction of spider mites below the economic threshold for the rest of the season. Similarly, a subsequent study conducted in vine- and hop-yards by James and Price (2004) demonstrated that brown lacewings *Hemerobius* spp. (Neuroptera: Hemerobiidae), *Orius tristicolor* (White) (Hemiptera: Anthocoridae), braconid wasps (Hymenoptera) and flies from the families Empididae and Sarcophagidae (Diptera) were attracted to CRD containing MeSA. James (2005) also showed attraction in the field of braconids to MeJA and MeA, the predaceous fruit fly *Thaumatomyia glabra* (Meigen) (Diptera: Chloropidae) to MeJA-baited traps and attraction of *O. tristicolor, S. punctum picipes, Anagrus daanei* Triapitsyn (Hymenoptra: Mymaridae), braconid wasps, and micro-Hymenoptera to

He. The GLV, and HA increased abundance of *O. tristicolor*, the predatory mirid *Deraeocoris brevis* (Uhler) (Hemiptera: Miridae) and *S. punctum picipes* (James 2003a). Numbers of *O. tristicolor*, *S. punctum picipes* and flies from the families Tachinidae and Sarcophagidae (Diptera) were also increased by Be (James 2005).

James and Grasswitz (2005) provided field evidence for attraction of MeSA, MeJA and HA to two parasitic wasp families, *Anagrus* spp. (Hymenoptera: Mymaridae) and *Metaphycus* spp. (Hymenoptera: Encyrtidae) in vineyards. However, previous field studies by James (2003b, 2005), which deployed the same HIPVs, did not show any response by these parasitic wasps. The authors suggested the concentrations of HIPVs used could have caused the variance in attraction, with the higher HIPV rates possibly acting as a repellent to the hymenopteran species. Another explanation could have been that the dispersion of HIPVs from an earlier study in the same vineyard (James 2005) led to signalling of the plants to produce their own HIPV blends which then led to attraction of *Anagrus* spp. and *Metaphycus* spp. in the latter study (James and Grasswitz 2005). The potential for seasonal differences and concentration effects highlights the need for further research on the role of HIPVs as pest management tools.

James (2011) provided field evidence of plant signalling functions of MeSA and HA in inducing indirect defense responses in hop and grapevine plants when sprayed with botanical oil pesticides containing small amounts of these HIPVs. In that study, the abundance of some carnivorous and parasitic insects was greater near HIPV-treated grape and hop plants. James (2011) explained that the higher numbers of natural enemies was unlikely to have been caused by direct attraction to botanical oils containing HIPVs due to their rapid evaporation after spraying and the small amounts of HIPVs involved. The author hypothesized that the treated plants altered their physiological response by emitting endogenous HIPV blends that lead to the attraction of natural enemies. Similar field studies which use synthetic spray-applied HIPVs are currently being conducted by the authors in Australian vineyards (Fig. 6.2). This work investigates several different synthetic HIPVs, spray-applied to *V. vinifera* at three different release rates, and monitored for their attraction by beneficial arthropods over extended time periods. The work is also testing these spray-applied HIPVs in combination with floral resources (Fig. 6.3) to explore whether this combined approach termed 'attract and reward' (Khan et al. 2008) could further boost conservation biological control.

6.7.2 *The Use of Silicon to Enhance Plant Defense Mechanisms*

Recent studies have started to scrutinise the role that Si plays in enhancing natural enemy attraction to the plant, thought to be through the qualitative and quantitative changes of the volatile profile emitted by plants under attack by arthropod pests (Fig. 6.1, Kvedaras et al. 2010). To date, Kvedaras et al. (2010) is the only published study that demonstrates increased natural enemy attraction to pest-infested plants and subsequent enhancement of field biological control through the plausible role of

Fig. 6.2 Spray application of synthetic HIPVs

Fig. 6.3 Buckwheat integrated in an Australian vineyard to enhance CBC

Fig. 6.4 Olfactometer bioassay with potted grapevines treated with silicon enclosed in air-tight bell-jars to determine their attractiveness to beneficial arthropods

Si on HIPV production. There are therefore numerous opportunities for research on the role of Si on HIPV production, including in *V. vinifera.*

Laboratory and field studies in Australia looking at the function of Si in HIPV production in *V. vinifera* have commenced. Grapevines have been treated with varying concentrations of Si, and the volatile compounds emitted as a result of arthropod attack absorbed onto solid-phase-microextraction (SPME) fibres, before being identified and quantified using gas chromatography–mass spectrometry (GC-MS). Arthropods may respond to HIPVs at low concentrations, below the level of analytical detection (D'Alessandro and Turlings 2005; Gouinguene et al. 2005; Dicke 2009). Therefore, a Y-tube olfactometer was used in conjunction with the SPME/GC-MS analysis to determine the attractiveness of the plant to beneficial arthropods (Fig. 6.4).

6.7.3 The Role of Functional Structural Plant Models in Vineyard Chemical Ecology Research

Simulation models are an important component of agricultural and ecological research, including experimental work using sensitivity analysis. Functional structural plant models (FSPM) are computer-based mechanistic models describing the structure and functions of plants and plant organs, usually using Lindenmayer Systems, a botanical formalism (Prusinkiewicz 2004). Botanical and physiological information drawn from the published literature are used to construct photorealistic three dimensional virtual plants (Prusinkiewicz 2004).

A range of commercially important crops and weeds have already been modelled, including grapevines (*V. vinifera*), peach (*Prunus persica* (L.) Batsch), wheat (*Triticum eastivum* L.), rice (*Oryza sativa* L.), maize (*Zea mays* L.), barley (*Hordeum vulgare* L.), faba bean (*Vicia faba* L.), yellow starthistle (*Centaurea solstitialis* L.), rye grass (*Lolium perenne* L.), and many others. Functional structural plant models are capable of simulating functions such as carbon capture and partitioning, water movements, nitrogen cycling, hormone synthesis as well as acropetal or basipetal flows. In addition, FSPM are able to describe interactions between the plant and their surroundings, such as light interception, air flow movements and exogenous spray depositions (Prusinkiewicz et al. 2007; Dorr et al. 2008). Finally, the behaviour of virtual pests and predators, as well as disease epidemiology can be simulated and new hypotheses tested without the associated cost of complex field experiments (Hanan et al. 2002; Skirvin 2007). Functional structural plant models are therefore a multi-variate integration tool rarely available in field studies and positively contribute to pre-experimental studies of complex ecosystems such as vineyards. Herbivore-induced plant volatile research using FSPM has not yet been reported in the literature. This is not due to a computational challenge but to our limited knowledge of the synthesis and emission of these compounds and hence represents an opportunity for investigative ecological research.

6.8 Conclusion

Plants emit HIPVs as part of their defense mechanisms, which can act as either a direct attractant or repellent to surrounding arthropods, as well as warning nearby plants of impending attack. Novel strategies using chemical ecology include the use of exogenously applied HIPV compounds or Si, which may trigger the plant to produce its own semiochemicals, or enhance and/or alter the existing volatile profile, thus boosting the defenses of the plant against arthropod pests. These inducing agents ‘switch on’ or ‘prime’ the plant for imminent pest attack, and are the focus of current research in the field of pest herbivore management in vineyards and provide opportunities to enhance CBC.

In order to apply these novel strategies in vineyards, research needs to identify whether insects from the second and fourth trophic level are attracted or repelled by these semiochemicals which may result in the attraction of organisms other than the target pest species. Research should also focus on the duration of effects of the inducing agents, and blends of HIPVs should be tested in addition to single compounds. Further, studies should address the practicality of field application and whether these strategies are compatible with other grapevine management practices, including the application of pesticide sprays.

References

Adatia MH, Besford RT (1986) Effects of silicon on cucumber plants grown in recirculating nutrient solution. Ann Bot 58:43–351

Baggen LR, Gurr GM (1998) The influence of food on *Copidosoma koehleri* (Hymenoptera: Encyrtidae), and the use of flowering plants as a habitat management tool to enhance biological control of potato moth, *Phthorimaea operculella* (Lepidoptera: Gelechiidae). Biol Control 11:9–17

Baldwin IT, Halitschke R, Paschold A, von Dahl CC, Preston CA (2006) Volatile signaling in plant-plant interactions: "Talking trees" in the genomics era. Science 311:812–815

Bernasconi Ockroy ML, Turlings TCJ, Edwards PJ, Fritzsche-Hoballah ME, Ambrosetti L, Bassetti P, Dorn S (2001) Response of natural populations of predators and parasitoids to artificially induced volatile emissions in maize plants (*Zea mays* L.). Agric For Entomol 3: 201–209

Berndt LA, Wratten SD, Scarratt SL (2006) The influence of floral resource subsidies on parasitism rates of leafrollers (Lepidoptera: Tortricidae) in New Zealand vineyards. Biol Control 37:50–55

Blaich R, Grundhöfer H (1997) Uptake of silica by grapevines from soil and recirculating nutrient solutions. Vitis 36:161–166

Blaich R, Wind R (1989) Inducible silica incrusts in cell walls of *Vitis* leaves. Vitis 28:73–80

Bolter CJ, Dicke M, vanLoon JJA, Visser JH, Posthumus MA (1997) Attraction of Colorado potato beetle to herbivore-damaged plants during herbivory and after its termination. J Chem Ecol 23:1003–1023

Bostanian NJ, Goulet H, O'Hara J, Masner L, Racette G (2004) Towards insecticide free apple orchards: flowering plants to attract beneficial arthropods. Biocontrol Sci Technol 14: 25–37

Buchanan GA (1977) The seasonal abundance and control of light brown apple moth, *Epiphyas postvittana* (Walker) (Lepidoptera: Tortricidae) on grapevines in Victoria. Aust J Agric Res 28:125–132

Buchanan GA, Amos TG (1992) Grape pests. In: Coombe BG, Dry PR (eds) Viticulture: practices, vol 2. Winetitles, Adelaide, pp 209–231

Buchanan GA, Furness GO, Charles JG (1994) Grape phylloxera. In: Nicholas P, Magarey P, Wachtel M (eds) Diseases and pests, Grape production series, Number 1. Winetitles, Adelaide, pp 71–73

D'Alessandro M, Turlings TCJ (2005) *In situ* modification of herbivore-induced plant odours: a novel approach to study the attractiveness of volatile organic compounds to parasitic wasps. Chem Senses 30:739–753

Danthanarayana W (1975) The bionomics, distribution and host range of light brown apple moth, *Epiphyas postvittana* (Walk.) (Tortricidae). Aust J Zool 23:419–437

Danthanarayana W (1983) Population ecology of the light brown apple moth, *Epiphyas postvittana* (Lepidoptera: Tortricidae). J Anim Ecol 52:1–33

Dicke M (1999) Evolution of induced indirect defense of plants. In: Tollrian R, Harvell CD (eds) The ecology and evolution of inducible defenses. Princeton University Press, Princeton, pp 62–88

Dicke M (2009) Behavioural and community ecology of plants that cry for help. Plant Cell Environ 32:654–665

Dicke M, Takken W (2006) Chemical ecology: from gene to ecosystem. Springer, Dordrecht

Dicke M, Vanbeek TA, Posthumus MA, Bendom N, Vanbokhoven H, Degroot AE (1990) Isolation and identification of volatile kairomone that affects acarine predator–prey interactions – involvement of host plant in its production. J Chem Ecol 16:381–396

Dicke M, Takabayashi J, Posthumus MA, Schutte C, Krips OE (1998) Plant-phytoseiid interactions mediated by herbivore-induced plant volatiles: variation in production of cues and in responses of predatory mites. Exp Appl Acarol 22:311–333

Dicke M, Gols R, Ludeking D, Posthumus MA (1999) Jasmonic acid and herbivory differentially induce carnivore-attracting plant volatiles in lima bean plants. J Chem Ecol 25:1907–1922

Dorr G, Hanan J, Adtkins S, Hewitt A, O'Donnell C, Noller B (2008) Spray deposition on plant surfaces: a modelling approach. Funct Plant Biol 35:988–996

Dudareva N, Negre F, Nagegowda DA, Orlova I (2006) Plant volatiles: recent advances and future perspectives. Crit Rev Plant Sci 25:417–440

Eilenberg J, Hajek A, Lomer C (2001) Suggestions for unifying the terminology in biological control. BioControl 46:387–400

Engelberth J, Alborn HT, Schmelz EA, Tumlinson JH (2004) Airborne signals prime plants against insect herbivore attack. Proc Natl Acad Sci USA 101:1781–1785

English-Loeb G, Rhainds M, Martinson T, Ugine T (2003) Influence of flowering cover crops on *Anagrus* parasitoids (Hymenoptera: Mymaridae) and *Erythroneura* leafhoppers (Homoptera: Cicadellidae) in New York vineyards. Agric For Entomol 5:173–181

Epstein E (1994) The anomaly of silicon in plant biology. Proc Natl Acad Sci USA 91:11–17

Epstein E (1999) Silicon. Annu Rev Plant Physiol Plant Mol Biol 50:641–664

Forneck A, Huber L (2009) (A)sexual reproduction – a review of life cycles of grape phylloxera, *Daktulosphaira vitifoliae*. Entomol Exp Appl 131:1–10

Frost CJ, Mescher MC, Carlson JE, De Moraes CM (2008) Plant defense priming against herbivores: getting ready for a different battle. Plant Physiol 146:818–824

Furness GO (1976) The dispersal, age structure and natural enemies of the long-tailed mealybug *Pseudococcus longispinus* (Targioni-Tozzetti) in relation to sampling methods and control. Aust J Zool 24:237–247

Geier PW, Briese T (1980) The light brown apple moth, *Epiphyas postvittana* (Walker): 4. Studies on population dynamics and injuriousness to apples in the Australian Capital Territory. Aust J Ecol 5:63–93

Gomes FB, de Moraes JC, dos Santos CD, Goussain MM (2005) Resistance induction in wheat plants by silicon and aphids. Sci Agric (Piracicaba, Braz) 62:547–555

Gouinguene SP, Turlings TCJ (2002) The effects of abiotic factors on induced volatile emissions in corn plants. Plant Physiol 129:1296–1307

Gouinguene SP, Pickett JA, Wadhams LJ, Birkett MA, Turlings TCJ (2005) Antennal electophysiological responses of three parasitic wasps to caterpillar-induced volatiles from maize (*Zea mays mays*), cotton (*Gossypium herbaceum*) and cowpea (*Vigna unguiculata*). J Chem Ecol 31:1023–1038

Gurr GM, Scarratt SL, Wratten SD, Berndt L, Irvin NA (2004) Ecological engineering, habitat manipulation and pest management. In: Gurr GM, Wratten SD, Altieri MA (eds) Ecological engineering for pest management: advances in habitat manipulation for arthropods. CSIRO Publishing, Melbourne, pp 1–12

Gurr GM, Wratten SD, Tylianakis J, Kean J, Keller M (2005) Providing plant foods for natural enemies in farming systems: balancing practicalities and theory. In: Wäckers FL, van Rijn PCJ, Bruin J (eds) Plant-provided food for carnivorous insects: a protective mutualism and its applications. Cambridge University Press, New York, pp 326–347

Hanan J, Prusinkiewicz P, Zalucki M, Skirvin D (2002) Simulation of insect movement with respect to plant architecture and morphogenesis. Comput Electron Agric 35:255–269

Herbert KS, Hoffmann AA, Powell KS (2008) Assaying the potential benefits of thiamethoxam and imidacloprid for phylloxera suppression and improvements to grapevine vigour. Crop Prot 27:1229–1236

Hilker M, Meiners T (2006) Early herbivore alert: insect eggs induce plant defense. J Chem Ecol 32:1379–1397

James DG (2003a) Field evaluation of herbivore-induced plant volatiles as attractants for beneficial insects: methyl salicylate and the green lacewing, *Chrysopa nigricornis*. J Chem Ecol 29:1601–1609

James DG (2003b) Synthetic herbivore-induced plant volatiles as field attractants for beneficial insects. Environ Entomol 32:977–982

James DG (2005) Further field evaluation of synthetic herbivore-induced plant volatiles as attractants for beneficial insects. J Chem Ecol 31:481–495
James DG (2011) Grape and hop plants sprayed with botanical oil pesticides containing herbivore-induced plant volatiles attract insect predators and parasitoids. Environ Entomol (in press)
James DG, Grasswitz TR (2005) Synthetic herbivore-induced plant volatiles increase field captures of parasitic wasps. BioControl 50:871–880
James DG, Price TS (2004) Field-testing of methyl salicylate for recruitment and retention of beneficial insects in grapes and hops. J Chem Ecol 30:1613–1628
James DG, Whitney J, Rayner M (1995) Phytoseiids (Acari, Phytoseiidae) dominate the mite fauna on grapevines in Canberra district vineyards. J Aust Entomol Soc 34:79–82
Jones LHP, Handreck KA (1967) Silica in soils, plants and animals. Adv Agron 19:107–149
Karban R, Baldwin IT (1997) Induced responses to herbivory. The University of Chicago, Chicago
Kessler A, Halitschke R, Diezel C, Baldwin IT (2006) Priming of plant defense responses in nature by airborne signaling between *Artemisia tridentata* and *Nicotiana attenuata*. Oecologia 148:280–292
Khan ZR, James DG, Midega CAO, Pickett JA (2008) Chemical ecology and conservation biological control. Biol Control 45:210–224
Kvedaras OL, Keeping MG (2007) Silicon impedes stalk penetration by the borer *Eldana saccharina* in sugarcane. Entomol Exp Appl 125:103–110
Kvedaras OL, An M, Choi YS, Gurr GM (2010) Silicon enhances natural enemy attraction and biological control through induced plant defenses. Bull Entomol Res 100:367–371
Lafos K (1995) Die Aufnahme und Verteilung von Silicium in Reben (*Vitis* spp.). Geisenheimer Berichte, Veröffentlichungen der Forschungsanstalt Geisenheim, Geisenheim, Germany
Landis DA, Wratten SD, Gurr GM (2000) Habitat management to conserve natural enemies of arthropod pests in agriculture. Annu Rev Entomol 45:175–201
Lee JC, Heimpel GE (2005) Impact of flowering buckwheat on Lepidopteran cabbage pests and their parasitoids at two spatial scales. Biol Control 34:290–301
Levin DA (1976) The chemical defenses of plants to pathogens and herbivores. Annu Rev Ecol Syst 7:121–159
Liu YH, Zeng RS, Liu DL, Luo SM, Wu HW, An M (2006) Modelling dynamics of plant defense volatiles using the An-Liu-Johnson-Lovett model. Allelopathy J 18:215–224
Loughrin JH, Potter DA, Hamilton-Kemp TR, Byers ME (1997) Diurnal emission of volatile compounds by Japanese beetle-damaged grape leaves. Phytochemistry 45:919–923
Ma JF (2007) Uptake of silicon by different plant species. In: Bäeuerlein E (ed) Handbook of biomineralization. Wiley-VCH Verlag GmbH & Co, Weinheim, pp 113–124
MacLellan CR (1973) Natural enemies of the light brown apple moth, *Epiphyas postvittana*, in the Australian Capital Territory. Can Entomol 105:681–700
Matichenkov V, Bocharnikova E (2007) Si in horticultural industry. In: Dris R, Jain SM (eds) Production practices and quality assessment of food crops, plant mineral nutrition and pesticide management, vol 2. Springer, Dordrecht, pp 217–228
Mitani N, Ma JF (2005) Uptake system of silicon in different plant species. J Exp Bot 56:1255–1261
Nicholas P, Magarey P, Wachtel M (1994) Diseases and pests. Winetitles, Adelaide
Paull C, Austin AD (2006) The hymenopteran parasitoids of light brown apple moth, *Epiphyas postvittana* (Walker) (Lepidoptera: Tortricidae) in Australia. J Aust Entomol Soc 45:142–156
Prusinkiewicz P (2004) Art and science for life: designing and growing virtual plants with L-systems. In: Davidson C, Fernandez T (eds) Nursery crops: development, evaluation, production and use, Proceedings of the XXVI International Horticultural Congress, Toronto, Canada, 11–17 August 2002, pp 15–28
Prusinkiewicz P, Lane B, Mech R (2007) Manipulating virtual plants. In: Prusinkiewicz P, Hanan J, Lane B (eds) Proceedings, 5th international workshop on functional-structural plant models, Napier, New Zealand, 4–9 Nov 2007, pp 24–1 to 24–4
Ranger CM, Singh AP, Frantz JM, Canas L, Locke JC, Reding ME, Vorsa N (2009) Influence of silicon on resistance of *Zinnia elegans* to *Myzus persicae* (Hemiptera: Aphididae). Environ Entomol 38:129–136

Reynolds OL, Keeping MG, Meyer JH (2009) Silicon-augmented resistance of plants to herbivorous insects: a review. Ann Appl Biol 155:171–186

Ruther J, Kleier S (2005) Plant-plant signaling: ethylene synergizes volatile emission in Zea mays induced by exposure to (Z)-3-Hexen-1-ol. J Chem Ecol 31:2217–2222

Skirvin D (2007) Modelling plant canopies for biocontrol and biodiversity. In: Vos J, Marcellis LFM, de Visser PHB, Struik PC, Evers JB (eds) Functional structural plant modelling in crop production. Springer, Wageningen, pp 253–264

Takabayashi J, Dicke M, Posthumus MA (1994) Volatile herbivore-induced terpenoids in plant mite interactions-variation caused by biotic and abiotic factors. J Chem Ecol 20:1329–1354

Takahashi E, Miyake Y (1977) Silica and plant growth. In: Proceedings, the international seminar of soil and environment, fertiliser management intensive agriculture. Society of the Science of Soil and Manure, Tokyo, Japan, pp 603–611

Tamogami S, Ralkwal R, Agrawal GK (2008) Interplant communication: airborne methyl jasmonate is essentially converted into JA and JA-Ile activating jasmonate signaling pathway and VOCs emission. Biochem Biophys Res Commun 376:723–727

Thaler JS (1999) Jasmonate-inducible plant defenses cause increased parasitism of herbivores. Nature 399:686

Thomson LJ, Sharley DJ, Hoffmann AA (2007) Beneficial organisms as bioindicators for environmental sustainability in the grape industry in Australia. Aust J Exp Agric 47:404–411

Ton J, D'Alessandro M, Jourdie V, Jakab G, Karlen D, Held M et al (2007) Priming by airborne signals boosts direct and indirect resistance in maize. Plant J 49:16–26

Turlings TCJ, Ton J (2006) Exploiting scents of distress: the prospect of manipulating herbivore-induced plant odours to enhance the control of agricultural pests. Curr Opin Plant Biol 9:421–427

Van den Boom CEM, van Beek TA, Posthumus MA, de Groot A, Dicke M (2004) Qualitative and quantitative variation among volatile profiles induced by *Tetranychus urticae* feeding on plants from various families. J Chem Ecol 30:69–89

Whitman DW, Eller FJ (1990) Parasitic wasps orient to green leaf volatiles. Chemoecology 1:69–76

Whitman DW, Eller FJ (1992) Orientation of *Microplitis croceipes* (Hymenoptera: Braconidae) to green leaf volatiles: dose–response curves. J Chem Ecol 18:1743–1753

Yan ZG, Wang CZ (2006) Wound-induced green leaf volatiles cause the release of acetylated derivatives and a terpenoid in maize. Phytochemistry 67:34–42

Chapter 7
Enhancing Ecosystem Services in Australasian Vineyards for Sustainability and Profit

Jean-Marie Tompkins, Steve D. Wratten, and Marja Simpson

7.1 Introduction

Worldwide, vineyards occupy over eight million ha with most being intensively managed as monocultures. In Australasia, conventionally managed vineyards typically consist of row upon row of *Vitis vinifera* L. with a bare earth or mown rye grass (*Lolium perenne* L.) floor (Tompkins 2008). Though such management may maximize profit in the short term, there is an increasing awareness that this may not be so in the future as social, environmental and economic pressures for sustainable wine production develop (Boller 1992; Nicholls et al. 2001; Forbes et al. 2009). Commentators on the conventional agricultural model have predicted that intensive agricultural management is not sustainable in the long-term, not only because it relies on finite fossil fuel resources (Hubbert 1981), but also because some of its practices degrade the natural capital and its functions (Tilman et al. 2002; MEA 2005; Kassam et al. 2009).

For vineyards, these functions, which may be termed ecosystem services (ES) (Daily 1997), include water supply and wastewater filtration, erosion control, nutrient cycling and the biological control of grape pests (Fiedler et al. 2008). Intensification of viticulture over the last few decades has however substituted out these ES for conventional inputs of synthetic fertilizers and pesticides, irrigated water, new vine

J.-M. Tompkins (✉) • S.D. Wratten
Department of Ecology, Bio-Protection Research Centre, Lincoln University,
PO Box 84, Lincoln 7647, New Zealand
e-mail: jean.tompkins@lincoln.ac.nz; steve.wratten@lincoln.ac.nz

M. Simpson
EH Graham Centre for Agricultural Innovation, Faculty of Science, School of Agricultural & Wine Sciences, Industry & Investment NSW and Charles Sturt University,
Leeds Parade, PO Box 883, Orange, NSW 2800, Australia
e-mail: mrsimpson@csu.edu.au

N.J. Bostanian et al. (eds.), *Arthropod Management in Vineyards: Pests, Approaches, and Future Directions*, DOI 10.1007/978-94-007-4032-7_7,

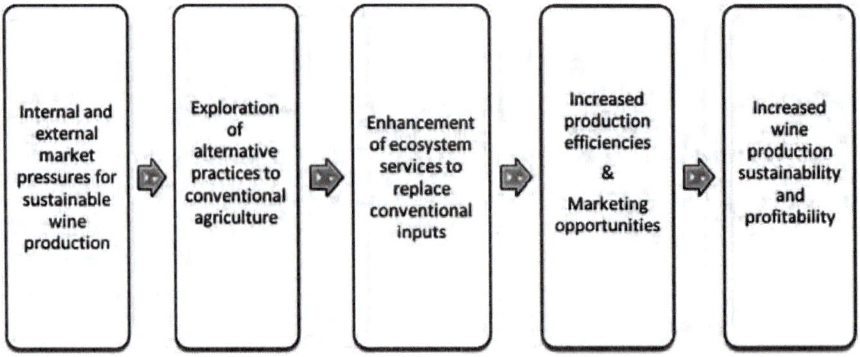

Fig. 7.1 The restoration of ecosystem services (ES) within vineyards has the potential to improve the sustainability and profitability of wine production

cultivars and mechanized weed and vine management. While such substitution in agriculture has driven huge production gains and profits, it has simultaneously degraded many of those ES its practices replaced. In New Zealand a report prepared by the Parliamentary Commissioner for the Environment (PCE) (2004) demonstrated that intensive agricultural practices being undertaken in this country were polluting the environment and damaging ES. Recent research undertaken by Sandhu et al. (2008) further confirmed this by demonstrating that conventional arable farming in the region of Canterbury was reducing ES which were of significant financial value. The Parliamentary report concluded by recommending that the natural capital of New Zealand and the ES which agriculture relies upon be protected (PCE 2004). Australia has also published reports with similar conclusions (DEWHA 2009; Australia21 2011).

The growing awareness of ES degradation within Australasian farms, coupled with an acknowledgment of the long term unsustainability of many conventional practices, has resulted in mounting internal and external market pressures for the return of ES within vineyards. Such restoration has the potential to improve the sustainability of wine production (Fig. 7.1) and indeed many see the enhancement of ES within all global crop production systems as a key step towards agriculture's long term sustainability and profitability (Pinstrup-Andersen and Pandya-Lorch 1998; Pretty and Hine 2001; Tilman et al. 2002; Kremen 2005; Reid et al. 2005; Robertson and Swinton 2005; FAO 2007; Kassam et al. 2009; Sachs et al. 2009; Charles et al. 2010).

Australasian wine growers utilizing conventional practices are facing rising costs, largely due to ever increasing oil prices driving up fertilizer and pesticide expenditures, not to mention transport costs, which are considerable due to distant export markets. These internal cost pressures have been accompanied by external consumer pressures for wine produced using environmentally responsible practices.

Recent work in New Zealand has shown that wine consumers want to be informed about which wines have been produced using environmentally sustainable practices, and that there exists a significant demand for sustainably produced wine in this country (Forbes et al. 2009). Similarly, a growing awareness of environmental issues in overseas markets, including New Zealand and Australia's primary wine export destinations Great Britain and the United States, is leading to increased demand for verifiably 'green' products (Campbell 1999). Wine consumers are known to place value on intangible dimensions of wine production such as sustainable vineyard practices (Hall and Mitchell 2008), practices which in turn may be marketed to the increasingly environmentally aware consumer (Bisson et al. 2002). For New Zealand wine exports, although supplying less than 2% of the world's wine, its share of the high-end price-point market is 18% (P. Manson, NZWG, personal communication). This makes the 'green' demand of consumers an especially pertinent consideration for New Zealand wine exports as it is this high-end target market where consumer demands for sustainable production is greatest.

These changing consumer tastes have not gone unnoticed by the Australasian wine industry players who have initiated schemes to ensure the sustainability of their wines to their customers. The establishment in 1995 of Sustainable Winegrowing New Zealand (SWNZ 2010) demonstrated the commitment of the industry to protect the environmental integrity of New Zealand's wine production, and justify the established 'clean, green' image brand. Similarly the chairman of Australia's Grape and Wine Research and Development Corporation, Dennis Mutton, has stressed within their 5-year research and development plan that sustainable wine production will be an important focus as the market becomes ever more competitive (GWRDC 2007). Individual wine operations are also aligning their production to meet the demands of these 'green' customers. Banrock Station (2011), a well known Australian wine label, informs consumers that it directs part of its profits into the protection and restoration of wetlands in the region where the wine is purchased. Another wine maker, Grove Mill (2011) of Marlborough in New Zealand, undertakes similar restoration initiatives and markets this upon their wine bottles with a stated philosophy to produce 'premium quality wines with minimal environmental impact'.

Clearly, adopting practices which are environmentally responsible as well as potentially more cost efficient than conventional practices is the directive for Australasian wine makers. Therefore enhancing ES within their vineyards appears to be a logical action to take. Several ES which could relieve those internal cost pressures mentioned above and also meet external consumer demands have been identified. These are discussed in the following section.

7.2 Ecosystem Services in Vineyards

Ecosystem services may be grouped into four categories comprising provisioning, cultural, regulating, and supporting services (MEA 2005). The latter two categories of these include ES which vineyard operators could enhance to improve the delivery of other ES. These may include soil formation and structure, nutrient management, biodiversity, biological pest control, winery wastewater filtration, weed suppression and greenhouse gas sequestration. Ecosytem services which the vineyard provides to

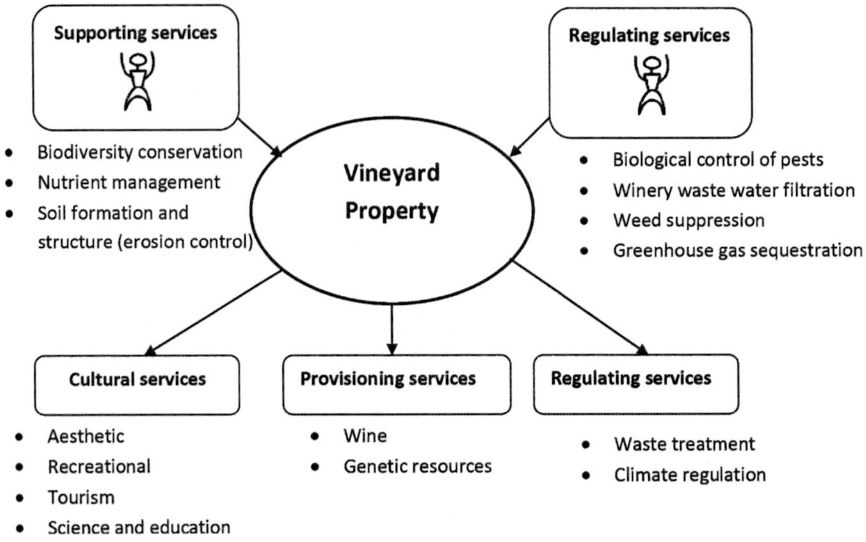

Fig. 7.2 The dynamics of ES in relation to vineyards. The human figure represents the potential for on-farm manipulations to enhance ES vineyards provide

humans could then include not only wine but also aesthetic benefits, waste treatment, climate regulation, genetic resources and opportunities for recreation, education and tourism. This consumption and production of ES in relation to a vineyard property is illustrated in Fig. 7.2. The illustration acknowledges that management techniques may be used to protect or enhance those services upon which vineyard systems rely (or consume), and thereby maximise the services they then provide to humans.

As with the production of other largely conventionally produced crops, wine grapes are grown in simplified agroecosystems or monocultures (Fig. 7.3) (Nicholls et al. 2008) that are likely to be lacking in these ES which would support the generation of other ES directly enjoyed by people. Therefore, enhancing these supportive ES would optimize services of value to people. Potential ES for enhancement in Australasian vineyards are now discussed. Specific actions that could be taken to achieve ES are summarized in Table 7.1. It also collates studies which have explored such enhancements for vineyard properties. All actions to enhance ES have the potential to be used as marketing material to promote the grower's wine to consumers because of their sustainable or 'green' basis.

7.2.1 Biodiversity Conservation

Conserving biodiversity is generally critical for maintaining genetic diversity, which as an ES, has the potential to provide valuable resources such as medicines, products for materials in science, genes for resistance to crop pests and plant pathogens and

Fig. 7.3 A typical Australasian vineyard in the winter. Note bare earth and shortly mown rye grass floor

economically important ornamental species (Costanza et al. 1997). Enhancing biodiversity in an agricultural system is known to promote greater sustainability because of an enhanced stability of the system (Norberg et al. 2001). Greater biodiversity of such systems is thought to enhance the capacity to absorb or recover from disturbances (Fischer et al. 2006) and leads to a reduced reliance on external inputs to maintain production (Milestad and Darnhofer 2003). Biodiversity conservation also has huge potential to provide opportunities for recreation and tourism (New Zealand Biodiversity 2011). When one considers that vineyards largely occupy Mediterranean biomes and that these are biodiversity "hotspots", the preservation or restoration of biodiversity within them is particularly important (MEA 2005). Compounding this issue for Australasian vineyards is that they predominantly occur within lowland habitats which have been extensively modified resulting in many of the natural ecosystems being considered highly threatened (Norton and Miller 2000; Walker et al. 2005; MFE 2007; de Lange et al. 2010). Conserving (or restoring) biodiversity within vineyards therefore offers to create more resilient vineyard systems which generate cultural ES, as well as preserve genetic resources for current or future use.

7.2.2 *Nutrient Management and Soil Erosion Control*

Soil health is an important consideration of vineyard management as it directly contributes to vine growth and grape quality (Jackson and Schuster 2002; Reeve et al. 2005)

Table 7.1 Potential actions to enhance ecosystem services within vineyards and associated studies

Ecosystem service to enhance (consumed)	Action to enhance ecosystem service within vineyard	Ecosystem services enhanced (produced)	Studies exploring this action in vineyards
Biodiversity conservation	Restore or protect biodiversity within the vineyard property e.g. native plantings and other habitat	All cultural services and genetic resources	1, 2, 3
Biological control	Establish non-crop plants to provide beneficial arthropods (natural enemies) with shelter, nectar, alternative food and prey	Wine (quantity and quality)	1, 3, 4, 5, 6
Nutrient management	Establish cover crop(s) to moderate soil nutrient levels available to grape vines	Wine (quantity and quality)	2, 3, 5, 6
Erosion control	Reduce soil exposure to wind and rain erosion with cover crops	Aesthetics and wine	1, 3
Greenhouse gas sequestration	Establish vegetation in non-productive areas of the vineyard property	Climate regulation	1, 3
Winery wastewater filtration	Utilise land or natural wetland technologies to filter winery wastewater, e.g., establish native wetlands next to winery	Waste treatment	7, 8
Weed suppression	Establish mulch or groundcover beneath vines which suppress/ out compete weeds but which do not adversely effect vine performance	Aesthetics and wine	1, 3, 5

[1]Tompkins (2008)
[2]Whitelaw-Weckert et al. (2007)
[3]Sandhu and Nidumolu (2009)
[4]Jacometti et al. (2008)
[5]Danne et al. (2010)
[6]Nicholls et al. (2008)
[7]Vymazal et al. (2006)
[8]Banrock Station (2011)

and is seen by many to underpin the sustainability of agroecosystems in general (Altieri 1999; van Bruggen and Semenov 1999). The supporting services of the soil may be enhanced through continuous cover crop(s) which are known to increase the soil organic matter and lead to improved soil structure, water infiltration, water holding capacity, nutrient storage capacity and microbial density (Gulick et al. 1994; Bugg and van Horn 1997). These attributes may improve wine quality and quantity (Jackson and Schuster 2002; Tesic et al. 2007) while reducing the need for external fertilizer inputs.

Fig. 7.4 Buckwheat sown between vine rows to enhance ES of biological control of grape pests

7.2.3 Biological Control of Pests

Simplified agroecosystems such as conventional vineyards (Nicholls et al. 2008) have been shown to host few natural enemies of pests and this is largely attributed to pesticide applications and a lack of non-crop vegetation (Altieri 1999; Gurr et al. 2004; Tscharntke et al. 2007). Many scientists have argued that conventional pesticides could be reduced by conserving natural enemies of crop pests by providing them with non-crop habitat (Sotherton 1984; Fry 1995; Altieri 1999; Thies and Tscharntke 1999; Thomas and Marshall 1999; Landis et al. 2000; Ponti et al. 2005; Pywell et al. 2005), inferring that natural enemies are able to increase in number within these unsprayed refuges, exploit resources and then move out into the crop to provide pest control services (Jonsson et al. 2009). Indeed, Australasian research within vineyards looking at ecologically-based pest control has found that if growers provide natural enemies of grape pests with non-crop refuges (often including flowering plants), increased levels of pest control can occur (Berndt et al. 2006; Jacometti et al. 2008; Danne et al. 2010). The enhancement of biological control within vineyards offers wine growers an opportunity to reduce operational costs by reducing pesticide use. Jacometti et al. (2008) found that sowing the floral resource buckwheat one in every 10 vineyard rows (Fig. 7.4) resulted in a reduction in leafroller numbers, ultimately saving the Marlborough winegrower $250/ha per year that would otherwise have been spent on controlling this pest. Such actions may however become less about cost and more about market access as consumers tolerance towards pesticide use declines. Stuart Smith, the chairman of New Zealand Winegrowers, stated in 2007 'The day is coming and I believe it is coming soon, that New Zealand will have to produce residue-free wine' while acknowledging the need for growers to find alternatives to the conventional use of pesticides. Enhancing the ES of biological control offers such an alternative.

7.2.4 Winery Wastewater Filtration

Vineyard operations produce a considerable amount of wastewater that requires treatment (often by law) prior to its discharge (EPA 2004; Cowey 2010). In Australasia, most treated winery wastewater is destined for land-based disposal upon vine rows, grazing pastures or woodlots (Kumar et al. 2008). Conventional wastewater treatments vary from direct discharge into septic tanks to capital intensive systems such as aerobic digesters and aeration ponds (Hamoudi-Viaud et al. 2004). These require substantial freshwater use, pretreatment with aqueous ammonia to adjust wastewater pH and careful monitoring of inorganic salts, organic compounds, yeast and bacteria (Szymanski et al. 2007; Kumar et al. 2008). Overall, they have high costs and energy usage. An alternative that has been put forward as a more sustainable treatment system is that of constructed wetlands which remove wastewater pollutants by natural self-purification processes. The high level of biodiversity present in constructed wetlands means more degradation mechanisms are at play compared to conventional treatment plants which only utilise a few families of specialised bacteria (Vymazal et al. 2006). Low construction and maintenance costs, minimal energy requirements, an ability to tolerate flow fluctuations, minimal sludge production, synergies with biodiversity conservation and high aesthetic appearance (Chague-Goff and Rosen 2001) arguably make wetlands a viable wastewater treatment option for small and medium scale winery operations seeking greater production sustainability (Grismer et al. 2003).

7.2.5 Weed Suppression

Weeds pose a significant risk to New Zealand's primary agricultural industries especially in high-value horticultural systems such as wine grapes (Sanguankeo et al. 2009) where weeds can compete with vines for nutrients and moisture (Ingles et al. 1998; Sullivan 2003; Sanguankeo et al. 2009) or increase the risk of frost damage (Evans 1999) threatening yields. Weed management techniques in New Zealand viticulture employ a variety of cultural, chemical and mechanical methods including cultivation, mulches, and pre- and post-emergence herbicides (Pool et al. 1990).

For conventional viticulture however, the use of glyphosate-based herbicide is by far the most commonly used technique (Tesic et al. 2007) and this is likely due to the high cost-effectiveness and broad-spectrum nature of these herbicides (Dastgheib and Frampton 2000; Manktelow et al. 2004).

In New Zealand an upward trend in herbicide sales alongside agricultural production has been attributed to the rapid growth of vineyard areas around the country where approximately 48 tonnes of herbicide active ingredients are applied each year (Manktelow et al. 2004). Sustainable production programmes such as kiwi fruit 'Kiwi Green' and wine grape 'Sustainable Winegrowing New Zealand' have greatly improved growers' understanding of efficient herbicide use (Manktelow et al. 2004). Concerns over chemical use such as toxicity issues (Powles et al. 1998), spray drift risk (Holland et al. 1995), negative impacts upon soil fauna (Cross et al. 1993; Whitelaw-Weckert et al. 2007), reduced plant disease resistance (Johal and Huber 2009) and herbicide resistance in target weeds (Tesic et al. 2007) are causing Australasian wine growers to seek alternative weed management practices (Whitelaw-Weckert et al. 2007; Tompkins 2010).

Cover crops can provide an alternative control method through their suppression of weed species. Suppression occurs by the cover crop outcompeting the weed for space, light or nutrients (Porter 1998) or in some cases allelopathic cover crops will restrict weed growth by chemical root exudates (Grundy et al. 1999; Delabays and Mermillod 2000). Adoption of alternative weed management, such as cover crops, would provide growers with an ecosystem service which would reduce herbicide usage and address other concerns for human and environmental health.

7.2.6 Greenhouse Gas Sequestration

Widespread concern about the effects of human induced climate change has brought about international efforts to reduce greenhouse gas (GHG) emissions (IPCC 2001). New Zealand's commitment to this goal is evident from its signing of the United Nations Framework Convention for Climate Change and its ratification of the Kyoto protocol (MFE 1997; NZCCP 2001). Half of New Zealand's emissions are from agriculture, predominantly of methane and nitrous oxide with the remainder mostly of carbon dioxide (Wicock et al. 2008). To mitigate emissions of these GHG, energy efficiencies, the development of renewable energy technologies and conservation initiatives are being undertaken (MFE 1998). The use of afforestation to offset greenhouse gas emissions under article 3.3 of the Kyoto protocol (UNFCCC 1998) acknowledges the service plants provide in the sequestration of GHG. New Zealand plans to meet approximately half of its emissions targets in this way (NZCCP 2001). Plants have the ability to sequester GHG through their photosynthetic activity and may also enhance the sequestration of GHG through soil biogeochemical processes. This may occur through plant mediation of soil characteristics, including alterations of soil fauna (Schlesinger 1997). Consequently, New Zealand, alongside other Australasian countries, sees re-forestation as a valuable carbon sink

(Stephens et al. 2005). Establishing woody non-crop vegetation within vineyard properties could therefore enhance the ES of carbon sequestration and thereby provide the service of climate regulation.

Two examples of projects which have helped Australasian wine growers undertake such actions as those described in Table 7.1 are the Greening Waipara project in New Zealand and the Vineyard Ecosystem Management Project in Australia.

7.2.7 The Greening Waipara Project

Initiated in 2005, the Greening Waipara project (http://bioprotection.org.nz/greening-waipara) seeks to motivate wine growers of the Waipara valley to re-establish native New Zealand plants within their properties. The incentive put before growers was that native plant establishment would deliver valuable ES. Collaboration between the Waipara valley Winegrowers Inc., local council and community, Lincoln University and Landcare Research meant that by 2010 over 60 properties of the Waipara valley had become involved. The establishment of native New Zealand plants within vineyard properties was proposed as a form of ecological engineering (Gurr et al. 2004) to enhance ES (Costanza et al. 1997; Daily 1997) which would have tangible values for growers and improve the sustainability of the area's wine production. Potential ES provided by native plants included erosion management, filtration of winery effluent, enhanced biological control of pests, biodiversity conservation, marketing and ecotourism (Figs. 7.5 and 7.6). Field trials determined which plant species may be recommended to wine growers for the provision of specific ES. For example the native plant species *Leptinella dioica* Hooker (Asteraceae) and *Acaena inermis* Hooker (Rosaceae) (Fig. 7.7) were found to be capable of suppressing weeds beneath grapevines, providing vineyard managers with an alternative form of weed control, reducing their use of conventional mechanical and chemical methods. Furthermore, Waipara winegrowers that responded to a survey overwhelmingly agreed that the practices to enhance ES within their vineyards provided them with point of difference in marketing opportunities and generated greater regional brand recognition (Tompkins 2010).

7.2.8 Vineyard Ecosystem Management Project (CSIRO)

Similar to the Greening Waipara project, this Australian government initiative run by CSIRO also investigates the provision of ES by native plants in and around vineyards and aims to quantify the economic value of these ES and other potential benefits. The project is teaching growers to use an 'Ecosystem Based Business Risks Analysis Tool' model to identify their impacts upon ecosystems. This model aims to develop practices to mitigate or reduce these impacts and consequently enhance the sustainability of their wine production (Sandhu et al. 2008, 2009). For example

Fig. 7.5 A planting of native New Zealand vegetation next to a winery, providing ES including Eco-tourism opportunities

Henschke, one of the longest established wine producers in the Barossa region and a keen advocate of this project, has identified several ecosystem based 'risks' where wine making operations impact on ecosystems. Impacts identified by Henschke include greenhouse gas emissions, chemical pollution (pesticides and herbicides) and soil erosion. The owners have since implemented tree planting, integrated pest management techniques and cover crops to avoid or reduce these impacts respectively. In doing so Henschke have improved ES which support their business and have consequently benefitted from enhanced ES (Sandhu and Nidumolu 2009).

7.3 Evaluating the Potential Economic Gain of Enhancing ES Within Vineyards

For agriculture in general many argue that ES need to be further incorporated into the economy to ensure their conservation within these systems (Tilman et al. 2002; Gutman 2007; Kroeger and Casey 2007; Sandhu et al. 2008). For vineyards, as for other crops, this would first require an ability to measure ES so that a market evaluation is possible (Dale and Polasky 2007; Swinton et al. 2006). If this was achieved,

Fig. 7.6 A planting of native New Zealand vegetation has created a 'biodiversity trail' near to a Canterbury winery and its cellar door. This action by the grower has enhanced ES of biodiversity conservation and in doing so has provided a tourism opportunity

wine growers could receive financial incentives to preserve the ES that their vineyard properties generated. The reduction of operational vineyard costs through the enhancement of ES provides economic incentives to growers to undertake some of those activities listed in Table 7.1. One example mentioned earlier is presented by Jacometti et al. (2008) who found savings in pest management costs could be achieved by enhancing the service of biological control. The study gave growers a clear protocol, also termed a Service Providing Unit (SPU) (Luck et al. 2003), for enhancing an ES, instructing growers to sow buckwheat in every tenth inter-row at 45 kg/ha. More such SPUs for other ES within vineyards would no doubt increase grower action to enhance them. However, the specific financial benefits of actions to enhance ES within vineyards are poorly known, requiring further research (Tompkins 2010).

Fig. 7.7 Native New Zealand groundcover established beneath grapevines to enhance multiple ES delivered by the vineyard

Where ES generated within vineyard properties are public goods, it is likely that government-generated incentives, rather than market-based payments, will be necessary as growers will have little financial interest in maintaining such services (Dale and Polasky 2007; Kroeger and Casey 2007). For example, growers that restore native plant biodiversity on their properties may find market-based incentives such as marketing opportunities that exist for such actions. However, the restored biodiversity may provide public goods (i.e., ES) which lack any direct financial incentive for the grower. Conservation of species, cultural value or aesthetics would be examples of public goods. It may be that the marketing opportunities fail to provide adequate incentive to restore biodiversity. However, if the accumulated value of all ES is compensated for, including the public ES, growers may be satisfactorily motivated (and appropriately compensated) to restore biodiversity within their vineyards. Compensation for ES which are public goods would probably entail government generated incentives such as subsidies or tax reductions (Kroeger and Casey 2007) and could be delivered via agri-environment schemes similar to those of the United States,

United Kingdom and Europe, although these have achieved mixed results (Kleijn and Sutherland 2003; Kleijn et al. 2006). In Australia market-based approaches to enhance ES (beyond provisioning services) within agricultural land was implemented through the National Market-Based Instruments Pilot Program (DEWHA 2009; Windle and Rolfe 2008; Yang et al. 2010). This program, along with other independent environmental initiatives, has promoted cost effective ways of encouraging growers to address ES degradation providing support for restoration of biodiversity (Stoneham et al. 2003; ten Kate et al. 2004; Carroll et al. 2007), wetland construction (Robertson 2004), water purification (Bjornlund 2003), and many other ES (Tallis et al. 2009).

7.4 Conclusion

At present ES are of interest to agriculture because their enhancement may improve the productivity and sustainability of agroecosystems (Tilman et al. 2002). This applies to many global crops including wine grapes. ES which may be enhanced within vineyard properties include soil formation and structure, nutrient management, biodiversity, biological pest control, winery wastewater filtration, weed suppression and greenhouse gas sequestration. Through their enhancement, it has been argued that other ES which are of direct benefit to growers and also society at large can also be enhanced. Not only can the quality and quantity of the wine be improved but also additional ES such as aesthetic benefits, waste treatment, climate regulation, genetic resources and opportunities for recreation, education and tourism can be generated.

Government or grower initiated projects may achieve such enhancements. However, further work needs to investigate the economic gain of enhancing ES within vineyards which may lead to a more widespread adoption of such practice by growers. While growers may be motivated by personal financial gain (or savings) via ES enhancements, a shortfall in expenditure to undertake ES enhancement may occur which would require government or industry compensation schemes. This could be justified to the public when the ES being enhanced have value to the public.

References

Altieri MA (1999) The ecological role of biodiversity in agroecosystems. Agric Ecosyst Environ 74:19–31

Australia21 (2011) National ecosystem services strategy. http://www.australia21.org.au/pdf/final.doc

Banrock Station (2011) http://www.banrockstation.com

Berndt LA, Wratten SD, Scarratt SL (2006) The influence of floral resource subsidies on parasitism rates of leafrollers (Lepidoptera: Tortricidae) in New Zealand vineyards. Biol Control 37:50–55

Bisson LF, Waterhouse AL, Ebeler SE, Walker ME, Lapsley JT (2002) The present and future of the international wine industry. Nature 414:696–699

Bjornlund H (2003) Farmer participation in markets for temporary and permanent water in south-eastern Australia. Agric Water Manag 63:57–76

Boller E (1992) The role of integrated pest management in integrated production of viticulture in Europe. In: Proceedings, Brighton crop protection conference, pests and diseases, British Crop Protection Council, Brighton, Surrey, UK, 23–26 Nov, pp 499–506

Bugg RL, van Horn M (1997) Ecological soil management and soil fauna: best practices in California vineyards. In: Hamilton R, Tassie L, Hayes P (eds) Proceedings, Australian society viticulture and oenology, viticulture seminar: viticultural best practice, Mildura, Australia, 1 Aug, pp 23–34

Campbell HC (1999) Green protectionism and organic food exporting from New Zealand: crisis experiments in breakdown of fordist trade and agricultural policies. Rural Sociol 64:302–319

Carroll N, Fox J, Bayon R (2007) Conservation and biodiversity banking: a guide to setting up and running biodiversity credit trading systems. Earthscan, Oxford

Chague-Goff C, Rosen MR (2001) Using sediment chemistry to determine the impact of treated wastewater discharge on a natural wetland in New Zealand. Environ Geol 40:1411–1423

Charles H, Godfray J, Beddington JR, Crute IR, Haddad L, Lawrence D et al (2010) Food security: the challenge of feeding 9 billion people. Science 327:812–818

Costanza R, d'Arge R, de Groot R, Farber S, Grasso M, Hannon B et al (1997) The value of the world's ecosystem services and natural capital. Nature 387:253–260

Cowey G (2010) When it all goes wrong: the treatment and disposal of juice, wine and lees waste material. Aust N Z Grapegrow Winemak 554:51–54

Cross JV, Marks MJ, Greenfield A (1993) Weed control in integrated fruit production. Acta Hortic (ISHS) 347:169–177

Daily GC (1997) Nature's services: societal dependence on natural ecosystems. Island Press, Washington, DC

Dale VH, Polasky S (2007) Measures of the effects of agricultural practices on ecosystem services. Ecol Econ 64:286–296

Danne A, Thomson LJ, Sharley D, Penfold C, Hoffmann AA (2010) Effects of native grass cover crops on beneficial and pest invertebrates in Australian vineyards. Environ Entomol 39:970–978

Dastgheib F, Frampton C (2000) Weed management practices in apple orchards and vineyards in the South Island of New Zealand. N Z J Crop Hortic Sci 28:53–58

de Lange P, Heenan P, Norton D, Rolfe J, Sawyer J (2010) Threatened plants of New Zealand. Canterbury University Press, Christchurch

Delabays N, Mermillod G (2000) Searching for allelopathic species to be used as cover crop or ground cover. In: Proceedings, 13th international scientific conference, IFOAM & FIBL, Basel, Switzerland

DEWHA (Department of the Environment, Water, Heritage and the Arts) (2009) Ecosystem services: key concepts and applications, Occasional paper No 1. Department of the Environment, Water, Heritage and the Arts, Canberra

EPA (Environment Protection Authority) (2004) EPA guidelines for wineries and distilleries. Environment Protection Authority, South Australia

Evans RG (1999) Frost protection in orchards and vineyards. http://www.sidney.ars.usda.gov/personnel/pdfs/Frost%20Protection%20in%20Orchards%20and%20Vineyards.pdf

FAO (2007) The state of food and agriculture. FAO, Rome

Fiedler AK, Landis DA, Wratten SD (2008) Maximising ecosystem services from conservation biological control: the role of habitat management. Biol Control 45:254–271

Fischer J, Lindenmayer D, Manning A (2006) Biodiversity, ecosystem function and resilience: ten guiding principles for commodity production landscapes. Front Ecol Environ 4:80–86

Forbes SL, Cohen DA, Cullen R, Wratten SD, Fountain J (2009) Consumer attitudes regarding environmentally sustainable wine: an exploratory study of the New Zealand marketplace. J Clean Prod 17:1195–1199

Fry GLA (1995) Landscape ecological principles and sustainable agriculture. In: McKinlay RG, Atkinson D (eds) Proceedings, integrated crop protection: towards sustainability? British Crop Protection Council, Edinburgh, Surrey, UK, 11–14 Sept, pp 247–254

Grismer ME, Carr MA, Shephard HL (2003) Evaluation of constructed wetland treatment performance for winery wastewater. Water Environ Res 75:412–421

Grove Mill (2011) http://www.grovemill.co.nz

Grundy A, Bond W, Burston S, Jackson L (1999) Weed suppression by crops. In: Marshall G (ed) Proceedings, the 1999 Brighton conference – Weeds, British Crop Protection Council, Brighton, Surrey, UK, 15–18 Nov, pp 957–962

Gulick SH, Grimes DW, Munk DS, Goldhamer DA (1994) Cover-cropped-enhanced water infiltration of slowly permeable fine sandy loam. Soil Sci Soc Am J 58:1539–1546

Gurr GM, Wratten SD, Altieri MA (2004) Ecological functional structural plant models for pest management: advances in habitat manipulation for arthropods. CSIRO Publishing, Melbourne

Gutman P (2007) Ecosystem services: foundations for a new rural–urban compact. Ecol Econ 62:383–387

GWRDC (Grape and Wine Research and Development Corporation) (2007) Five year R and D plan 2007–2012. http://www.gwrdc.org.au

Hall MC, Mitchell R (2008) Wine marketing: a practical guide. Butterworth Heinemann, Oxford

Hamoudi-Viaud MN, Berthoumieux F, DescotesA (2004) Traditional winery waste water techniques. http://www.infowine.com/default.asp?scheda=1149

Holland PH, May B, Maber J (1995) Spray drift from orchards. http://www.hortnet.co.nz/publications/science/sprydrft.htm

Hubbert MK (1981) The world's evolving energy system. Am J Phys 49:1007–1023

Ingles C, Bugg R, Glenn T, McGourty G, Christensen P (1998) Cover cropping in vineyards; a grower's handbook, vol 3338. University of California/Division of Agriculture and Natural Resources, Oakland

IPCC (Intergovernmental Panel on Climate Change) (2001) Third assessment report, climate change. Cambridge University Press, Cambridge

Jackson D, Schuster D (2002) Canterbury grapes and wine 1840–2002. Shoal Bay Press, Christchurch

Jacometti M, Scarratt S, Wratten SD (2008) Buckwheat means no sprays are needed. New Zealand Grower (April): 76–78

Johal G, Huber D (2009) Glyphosate effects on diseases of plants. Eur J Agron 31:144–152

Jonsson M, Wratten SD, Robinson KA, Sam SA (2009) The impact of floral resources and omnivory on a four trophic level food web. Bull Entomol Res 99:275–285

Kassam A, Friedrich T, Shaxson F, Pretty J (2009) The spread of Conservation Agriculture: justification, sustainability and uptake. Int J Agric Sustain 7:292

Kleijn D, Sutherland WJ (2003) How effective are European agri-environment schemes in conserving and promoting biodiversity? J Appl Ecol 40:947–969

Kleijn D, Baquero RA, Clough Y, Diaz M, De Esteban J, Fernandez F et al (2006) Mixed biodiversity benefits of agri-environment schemes in five European countries. Ecol Lett 9:243–254

Kremen C (2005) Managing ecosystem services: what do we need to know about their ecology? Ecol Lett 8:469–479

Kroeger T, Casey F (2007) An assessment of market-based approaches to providing ecosystem services on agricultural lands. Ecol Econ 64:321–332

Kumar A, Correll R, Kookana R (2008) Winery wastewater workshops translate research into practice. Aust N Z Wine Ind J 23:60–63

Landis DA, Wratten SD, Gurr GM (2000) Habitat management to conserve natural enemies of arthropod pests in agriculture. Annu Rev Entomol 45:175–201

Luck GW, Daily GC, Ehrlich PR (2003) Population diversity and ecosystem services. Trends Ecol Evol 18:331–336

Manktelow D, Stevens P, Walker J, Gurnsey S, Park N, Zabkiewicz J et al (2004) Trends in pesticide use in New Zealand. Ministry for the Environment, Wellington

MEA (Millennium Ecosystem Assessment) (2005) Ecosystems and human well-being: biodiversity synthesis. World Resources Institute, Washington, DC
MFE (Ministry for the Environment) (1997) The New Zealand response. New Zealand's second national communication under the convention for climate change. Ministry for the Environment, Wellington
MFE (Ministry for the Environment) (1998) Climate change, more than just carbon dioxide: significance, sources and solutions for non-CO_2 greenhouse gases in New Zealand. Ministry for the Environment, Wellington
MFE (Ministry for the Environment) (2007) Land cover database: Canterbury region. http://www.maf.govt.nz/statistics/primaryindustries/regions/tables/landcover/lc-cant.htm
Milestad R, Darnhofer I (2003) Building farm resilience: the prospects and challenges of organic farming. J Sustain Agric 22:81–97
New Zealand Biodiversity (2011) Why we value biodiversity. http://www.biodiversity.govt.nz/picture/biodiversity/why.html
Nicholls CI, Parrella M, Altieri MA (2001) The effects of a vegetational corridor on the abundance and dispersal of insect biodiversity within a northern California organic vineyard. Landsc Ecol 16:133–146
Nicholls CI, Altieri MA, Ponti L (2008) Enhancing plant diversity for improved insect pest management in northern California organic vineyards. Acta Hortic 785:263–278
Norberg J, Swaney DP, Dushoff J, Lin J, Casagrandi R, Levin SA (2001) Phenotypic diversity and ecosystem functioning in changing environments: a theoretical framework. Proc Natl Acad Sci 98:11376–11381
Norton BDA, Miller CJ (2000) Some issues and options for the conservation of native biodiversity in rural New Zealand. Ecol Manag Restor 1:26–34
NZCCP (New Zealand Climate Change Policy) (2001) Climate change: the Government's preferred policy package. Department of Prime Minister and Cabinet, Wellington
PCE (The Parliamentary Commissioner for the Environment) (2004) Growing for good: intensive farming, sustainability and New Zealand's environment. Parliamentary Commissioner for the Environment, Wellington
Pinstrup-Andersen P, Pandya-Lorch R (1998) Food security and sustainable use of natural resources: a 2020 Vision. Ecol Econ 26:1–10
Ponti L, Ricci C, Veronesi F, Torricelli R (2005) Natural hedges as an element of functional biodiversity in agroecosystems: the case of a central Italy vineyard. Bull Insectol 58:19–23
Pool RM, Dunst RM, Lako AN (1990) Comparison of sod, mulch, cultivation, and herbicide floor management practices for grape production in nonirrigated vineyards. J Am Soc Hortic Sci 115:872–877
Porter R (1998) Weed suppression using cover crops. Aust Grapegro Winemak 414:30–33
Powles SB, Lorraine-Colwill DF, Dellow JJ, Preston C (1998) Evolved resistance to glyphosate in rigid ryegrass (*Lolium rigidum*) in Australia. Weed Sci 46:604–607
Pretty J, Hine R (2001) Reducing food poverty with sustainable agriculture: a summary of new evidence. http://www2.essex.ac.uk/ces/ResearchProgrammes/CESOccasionalPapers/SAFErep SUBHEADS.htm
Pywell RF, James KL, Herbert I, Meek WR, Carvell C, Bell D, Sparks TH (2005) Determinants of overwintering habitat quality for beetles and spiders on arable farmland. Biol Conserv 123:79–90
Reeve JR, Carpenter-Boggs L, Reganold JP (2005) Soil and winegrape quality in biodynamically and organically managed vineyards. Am J Enol Vitic 56:367–376
Reid WV, Mooney HA, Cropper A, Capistrano D, Carpenter SR, Watson KR et al (2005) Ecosystems and human well-being: synthesis. Island Press, Washington, DC
Robertson MM (2004) The neoliberalization of ecosystem services: wetland mitigation banking and problems in environmental governance. Geoforum 35:361–373
Robertson GP, Swinton SM (2005) Reconciling agricultural productivity and environmental integrity: a grand challenge for agriculture. Front Ecol Environ 3:38–46

Sachs JD, Baillie JEM, Sutherland WJ, Armsworth PR, Ash N, Beddington J et al (2009) Biodiversity conservation and the millennium development goals. Science 325:1502
Sandhu HS, Nidumolu U (2009) Identifying ecosystem based risks and opportunities: case of Henschke wines. CSIRO Publishing, Melbourne
Sandhu HS, Wratten SD, Cullen R (2008) The future of farming: the value of ecosystem services in conventional and organic arable land. An exploratory approach. Ecol Econ 64:835–848
Sandhu HS, Nidumolu U, Sandhu SK (2009) Modelling risks and opportunities arising from ecosystem change in food and fibre value chains. In: Proceedings of the 10th international Ecology Congress, Brisbane, Australia, 16–21 Aug
Sanguankeo P, Leon R, Malone J (2009) Impact of weed management practices on grapevine growth and yield components. Weed Sci 57:103–107
Schlesinger WH (1997) Biogeochemistry, an analysis of global change. Academic, San Diego
Sotherton NW (1984) The distribution and abundance of predatory arthropods overwintering on farmland. Ann Appl Biol 105:423–429
Stephens JMC, Molan PC, Clarkson BD (2005) A review of *Leptospermum scoparium* (Myrtaceae) in New Zealand. N Z J Bot 43:431–449
Stoneham G, Chaudhri V, Ha A, Strappazzon L (2003) Auctions for conservation contracts: an empirical examination of Victoria's Bush Tender trial. Aust J Agric Resour Econ 47:477–500
Sullivan P (2003) Overview of cover crops and green manures. http://www.attra.ncat.org
Swinton SM, Lupi GP, Robertson GP, Landis DA (2006) Ecosystem services from agriculture: looking beyond the usual suspects. Am J Agric Econ 88:1160–1166
SWNZ (Sustainable Winegrowers New Zealand) (2010) Sustainable winegrowers New Zealand: introduction. http://www.nzwine.com/swnz
Szymanski N, Lott S, Gulifer L, Binns P, Glastonbury R (2007) Re-engineering of waste management systems at De Bortoli Winery, Bilbul, NSW. Aust N Z Grapegrow Winemak 523:65–69
Tallis H, Goldman R, Uhl M, Brosi B (2009) Integrating conservation and development in the field: implementing ecosystem service projects. Front Ecol Environ 7:12–20
ten Kate K, Bishop J, Bayon R (2004) Biodiversity offsets: views, experience and the business case. IUCN/Insight Investment, Gland/Cambridge/London
Tesic D, Keller M, Hutton R (2007) Influence of vineyard floor management practices on grapevine vegetative growth, yield, and fruit composition. Am J Enol Vitic 58:1–11
Thies C, Tscharntke T (1999) Landscape structure and biological control in agro ecosystems. Science 285:893–895
Thomas CFG, Marshall EJP (1999) Arthropod abundance and diversity in differently vegetated margins of arable fields. Agric Ecosyst Environ 72:131–144
Tilman D, Cassman KG, Matson PA, Naylor R, Polasky S (2002) Agricultural sustainability and intensive production practices. Nature 418:671–677
Tompkins JM (2008) Endemic New Zealand plants for pest management in vineyards. In: Mason PG, Gillespie DR, Vincent C (eds) 3rd International Symposium on Biological Control of Arthropods. USDA Forest Service, Christchurch, New Zealand, pp 234–245
Tompkins JM (2010) Ecosystem services provided by native New Zealand plants in vineyards. PhD dissertation, Lincoln University, New Zealand. http://www.researchassrchive.lincoln.ac.nz
Tscharntke A, Bommarco R, Clough Y, Crist T, Kleijn D, Rand T et al (2007) Conservation biological control and enemy diversity on a landscape scale. Biol Control 43:294–309
UNFCCC (United Nations Framework Convention on Climate Change) (1998) The Kyoto protocol to the convention on climate change. UN Climate Change Secretariat, Bonn
van Bruggen AHC, Semenov AM (1999) A new approach to the search for indicators of root disease suppression. Aust Plant Pathol 28:4–10
Vymazal J, Greenway M, Tonderski K, Brix H, Mander U (2006) Constructed wetlands for wastewater treatment. In: Verhoeven JTA, Beltman B, Bobbink R, Whigham DF (eds) Wetlands natural resource management, vol 190. Springer, Berlin, pp 69–96
Walker S, Price R, Rutledge D (2005) New Zealand's remaining indigenous cover: recent changes and biodiversity protection needs. Department of Conservation, Wellington

Whitelaw-Weckert MA, Rahman L, Hutton RJ, Coombes N (2007) Permanent swards increase soil microbial counts in two Australian vineyards. Appl Soil Ecol 36:224–232
Wicock R, Elliott S, Hudson N, Parkyn S, Quinn J (2008) Climate change mitigation for agriculture: water quality benefits and costs. Water Sci Technol 58:2093–2099
Windle J, Rolfe J (2008) Exploring the efficiencies of using competitive tenders over fixed price grants to protect biodiversity in Australian rangelands. Land Use Policy 25:388–398
Yang W, Bryan B, MacDonald D, Ward J, Wells G, Crossman N, Connor J (2010) A conservation industry for sustaining natural capital and ecosystem services in agricultural landscapes. Ecol Econ 69:680–689

Chapter 8
Habitat Diversity at the Field and Landscape Level: Conservation Biological Control Research in California Viticulture

Albie Miles, Houston Wilson, Miguel Altieri, and Clara Nicholls

8.1 Introduction: The Need for Ecologically Based Viticulture in California

The intensification of viticulture in California has led to the creation of grape monocultures characterized by an absence of non-crop plant diversity in and around vineyards. The continued expansion of vineyards into California native plant communities has also led to an aggregate reduction of non-crop habitats at the landscape scale (Heaton and Merenlender 2000). Such increased concentration of plant host resources and the reduction of non-crop habitats supporting natural enemies have been shown to increase pest densities, with associated crop losses and reduced overall crop productivity (Root 1973; Russell 1989; Corbett and Rosenheim 1996a; Altieri and Nicholls 2004). To manage recurring pest problems, California grape growers rely principally on the use of synthetic pesticides, including organophosphate and carbamate insecticides, known to pose a range of environmental quality and human health risks (Bentley 2009; CDPR 2009; UC IPM 2010b; Eskenazi et al. 2010).

With increasing concern over the environmental impacts of viticulture, rising production costs, and increased regulation of pesticides, the demand for research driven by ecologically-based pest management (EBPM) strategies has steadily grown (Broome and Warner 2008; Meadows 2008; Ross and Golino 2008; Brodt et al. 2009). In addition to the use of insecticides accepted under the United States Department of Agriculture, National Organic Program, California grape growers have sought to use EBPM strategies, including on-farm diversification to promote

A. Miles (✉) • H. Wilson • M. Altieri • C. Nicholls
Department of Environmental Science, Policy and Management, University of California, 130 Mulford Hall #3114, Berkeley, CA 94720-3114, USA
e-mail: albiemiles@berkeley.edu; houston@berkeley.edu; agroeco3@berkeley.edu; nicholls@berkeley.edu

N.J. Bostanian et al. (eds.), *Arthropod Management in Vineyards: Pests, Approaches, and Future Directions*, DOI 10.1007/978-94-007-4032-7_8,

biological control (Altieri et al. 2005; Ross and Golino 2008). Despite growing interest and adoption, few field and farm-scale EBPM strategies in use today have been scientifically evaluated for their ability to consistently regulate pest populations below economic thresholds. With the exception of the general principles (Altieri et al. 2011), California grape growers lack specific guidelines on how to successfully diversify their vineyards or conserve non-crop habitats in the surrounding landscapes to ensure biological control of important arthropod pests.

8.2 Key Hypotheses Informing Research in Vineyard Diversification in California: Natural Enemies and Resource Concentration

Two main hypotheses have been used for evaluating the effect of on-farm vineyard diversification strategies on biological control in California: (1) the Natural Enemies Hypothesis (NEH) (Andow 1991a), and (2) the Resource Concentration Hypothesis (RCH) (Root 1973). The NEH predicts a positive correlation between plant species richness, natural enemy abundance and the regulation of herbivore pests through increased predation and parasitism. The RCH predicts that herbivore pests are more likely to find and remain on agricultural host plants grown in pure stands (monocultures) than in more biologically diversified (polycultures) cropping systems. The RCH predicts that most specialized herbivore species are likely to attain the highest relative densities in monocultures when compared to diversified farming systems (Root 1973). In more complex agroecosystems, the dilution of plant host resources, interspecific competition, and more favorable environmental conditions for natural enemies are understood to be complementary factors that serve to regulate herbivore pest densities (Russell 1989; Altieri and Nicholls 2003; Andow 1991b; Costello and Daane 2003).

8.3 Vineyard Diversification Studies in California: Field-Level Research

Multiple on-farm diversification studies have measured the impact of overwintering and summer cover crops on biological control of *Erythroneura* leafhoppers (Table 8.1).

In a 2-year study, Flaherty (1969) measured the impact of the weedy Johnson grass (*Sorghum halepense* (L.) Persoon) on population densities of the Willamette mite (*Eotetranychus willamettei* Ewing) in a Tulare County ‘Thompson Seedless’ vineyard. Researchers measured population densities of predators and pests in both weed infested and grass-free vines, concluding that the Johnson grass supported populations of alternate prey (the twospotted spider mite) which moved between the weedy vegetation and the vine canopy. Provided with an alternate food source, the

Table 8.1 Summary of research on effects of vineyard diversification on biological control in California vineyards: 1998–2005

Description of study	Reference	Main effect	Economically important level of biocontrol?	Reported mechanism influencing biocontrol
Effect of weedy vegetation (Johnson grass) on population densities of Willamette mites (*Eotetranychus willamettei*)	Flaherty (1969)	98% reduction in Willamette mites	Yes	Natural enemies: weedy grasses support populations of alternate prey used by predatory mites (*Metaseiulus occidentalis*). Predatory mites were maintained at high density and well dispersed in system thus able to respond rapidly to increased abundance of pest mites, controlling pest at lower densities
Effect of cover crops (vetch and oats) on population densities of *Erythroneura variabilis*	Roltsch et al. (1998)	40–50% reduction in 2nd and 3rd generation leafhopper density (year 1), 2nd generation only (year 2)	No	Natural enemies: spider densities negatively correlated with leafhopper densities in presence of cover crop
Effect of cover cropping (vetch and barley) on *Erythroneura* leafhoppers	Costello and Daane (1998)	15–20% reduction in nymph leafhopper density	No	Resource concentration: reduced vine vigor influencing plant host quality. No significant difference in natural enemy abundance
Effect of cover cropping (vetch and barley) on *Erythroneura* leafhoppers.	Costello and Daane (2003)	15–20% reduction in nymph leafhopper density	No	Resource concentration and natural enemies: reduced vine vigor influencing plant host quality. Predation by spiders suggested as contributing factor to enhanced biocontrol
Effect of cover cropping (vetches and oats) on *Erythroneura* (variegated) leafhoppers	Hanna et al. (2003)	35% higher density of 1st generation of leafhopper nymphs in exclusion sub-plots	No	Natural enemies: 35% higher density of 1st generation leafhopper nymphs were found in exclusion sub-plots, suggesting role of spiders in regulating leafhoppers. Higher density of spiders in cover crop plots, but no significant effect on biocontrol. Suggested as resulting from intraguild predation and low seasonal density of leafhoppers
Effect of floral resource provisioning (*Helianthus annuus* and *Fagopyrum esculentum*) on *Erythroneura* leafhoppers and thrips	Nicholls et al. (2000) and Altieri et al. (2005)	15% reduction in nymph leafhopper densities (for periods). 32% reduction in thrips	No (leafhoppers) Yes (thrips)	Natural enemies: spiders correlated with low leafhopper densities, and *Orius* spp. correlated with reduced thrips densities in treatment plots

predatory mites (*Metaseiulus occidentalis* (Nesbitt)) were maintained at higher densities and were better dispersed throughout the vineyard area influenced by the weedy vegetation when compared to plots with no Johnson grass. Predatory mites were thus able to respond more rapidly to an increased abundance of the Willamette mite pest and control them at lower densities resulting in significantly lower Willamette mite densities in the diversified (with Johnson grass) versus the simple (no Johnson grass) plots. To further substantiate that enhanced predation by predatory mites was the cause of lower pest mite densities, researchers evaluated the impact of insecticide applications (thus reduced predatory mite populations) on population densities of Willamette mites. Plots with and without Johnson grass that were treated with insecticide showed both lower densities of predatory mites and consistent and significantly higher populations densities of Willamette mite, indicating ecological release of herbivore mites from predation (Flaherty 1969).

Roltsch et al. (1998) conducted several experiments to determine the effect of resident weedy vegetation and cover crops on spider densities and biological control of the variegated leafhopper (*Erythroneura variabilis* Beamer) in a San Joaquin Valley 'Thompson Seedless' vineyard. Building upon prior studies suggesting that vineyard spiders could be influenced by the ground cover habitats, researchers sought to evaluate the impact of the planted cover crops, common vetch (*Vicia sativa* L.), purple vetch (*Vicia benghalensis* L.) and oat (*Avena sativa* L.) on leafhopper densities and spider abundance and diversity. Agelenid (*Holonena nedra* Chamberlin & Ivie), and theridiid (*Theridion* spp.) spiders were found to be more abundant in the vine canopy in ground cover plots. A corresponding inverse relationship was found for leafhopper densities, with the highest densities found in control plots (no cover). Further corroboration of spiders playing a key role in regulating leafhopper densities was found with a strong positive correlation between high late-season leafhopper densities and low spider abundance in insecticide (dimethoate) treated vineyards, indicating an ecological release of leafhoppers from predation by spiders (Roltsch et al. 1998). No further mechanistic studies were conducted to empirically validate enhanced predation by spiders in the presence of ground covers.

A field diversification study examining the impact of overwintering cover crop mixtures and resident weedy vegetation on the variegated leafhopper (*E. variabilis)* was reported by Hanna et al. (2003). Prior research had established that spiders are the most abundant generalist natural enemy in vineyards and other agroecosystems. Furthermore, they are the only natural enemy, other than *Anagrus* spp. (Mymaridae), present in sufficient densities to regulate *Erythroneura* leafhoppers (Costello and Daane 1999, 2003). Researchers thus set out to evaluate the impact of cover cropping on spider and leafhopper abundance using a fall planted mixture of purple vetch (*V. benghalensis*), common vetch (*V. sativa*) and 'Cayuse' oat (*A. sativa*). Using the cover crop mixture and bare-ground as main plots and vine exclusion as sub-plots (to restrict spiders), researchers evaluated the relative impact of each treatment on spider and *E. variabilis* densities. Parasitism rates by *Anagrus* spp. were found to be similar in all plots throughout the year and other generalist natural enemies were found to be rare. Spider exclusion resulted in an average 35% increase in the density of first generation *E. variabilis* nymphs only. Yet, despite a 1.6-fold increase in spider densities on vines with cover crops (no exclusion), the cover crop did not significantly affect the density of *E. variabilis* on grape vines. Researchers suggest that this was due to insufficient spider enhancement from the cover crop and low overall leafhopper abundance during the study period. Interestingly, the cover crop mix had no significant impact on vine vigor/nutrient status, in contrast to the findings of Costello and Daane (1999, 2003). While this study provided support for the hypothesis that in-field diversification can enhance spider abundance, it does not always lead to lower pest densities, perhaps because of the complexity and variability of trophic interactions (e.g. inter and intraguild predation) in agroecosystems (Hanna et al. 2003).

Nicholls et al. (2000) conducted a 2-year comparative study of the effect of floral resource provisioning on biological control in an organic wine grape vineyard in Hopland, California. Comparing two 1-ha vineyard blocks (with and without flowers), researchers measured the impact of the summer cover crops, annual sunflower (*Helianthus annuus* L.), and annual buckwheat (*Fagopyrum esculentum* Moench), on population densities of western grape leafhopper (*Erythroneura elegantula* Osborn), western flower thrips (*Frankliniella occidentalis* (Pergande)), and key natural enemies (parasitoids and generalist predators). Researchers reported an estimated 15% lower density from mid to late season (July–August) leafhopper nymphs in cover cropped vineyards when compared to monocultures and a significantly lower density of thrips (32%) for both years of the study. The study also found a greater abundance and richness of generalist natural enemies (*Orius* spp., Coccinellid beetles, and thomisid spiders) in the treated vs. control plot. Although researchers found a higher density of *Anagrus* spp.[1] wasps in the control plots, no significant difference in rates of parasitism were found between treatment and control plots. Lower density of leafhopper nymphs in the treatment plot (with cover crops) was attributed to impacts of generalist predators, namely spiders and *Orius* spp. anthocorids. Lower density of thrips in treatment plots was attributed to the impact of the generalist *Orius* spp. predators. The researchers also studied the impact of mid-season mowing of the flowering cover crops on pest and beneficial insects, reporting a significant but temporary increase in density (18%) of both generalist predators and *Anagrus* parasitoids, and a subsequent lower (27%) leafhopper nymph density in the vine canopy after mowing (Nicholls et al. 2000; Altieri et al. 2005).

Daane and Costello (1998) assessed the influence of purple vetch (*V. benghalensis*) and barley (*Hordeum vulgare* L.) cover crops and resident weedy vegetation on vine vigor, natural enemy and leafhopper abundance in four San Joaquin Valley vineyards. They found that season-long cover cropping reduced late season leafhopper

[1]Early research referred to all species of *Anagrus* wasps, a key egg parasitoid of *Erythroneura* leafhoppers, found in vineyard as '*Anagrus epos* Girault.' Recent taxonomic revisions of *Anagrus epos* by Triapitsyn (1998) have revealed a complex of species, including the two most common grape leafhopper parasitoids in California: *A. erythroneurae* and *A. daanei*. As such, *Anagrus* spp. will hereafter be referred to as simply '*Anagrus*.'

nymph densities by 15–20%. Though a treatment effect was clearly determined, the level of leafhopper reduction was not considered economically important and the mechanisms leading to pest reduction were not clearly established. No significant differences in the density of leafhopper predators or *Anagrus* spp. parasitoids were found on vines in cover cropped versus control plots. Additionally, no consistent differences in parasitism rates by *Anagrus* spp. wasps were observed between treatments and control plots, leading researchers to conclude that natural enemy fitness, behavior and density were not significantly enhanced by cover cropping and therefore did not play an important role in regulating leafhopper densities. Assessments of the impact of cover cropping on vine vigor (indicated by petiole nitrogen and vine shoot biomass) however, showed significantly lower vigor and the lowest late season leafhopper density on vines with season long cover crops and resident weedy grasses. Additionally, researchers found the lowest total number of leafhopper eggs on grape vines in cover cropped plots (Daane and Costello 1998). In a follow-up study, Costello and Daane (2003) re-evaluated the influence of the same cover crops (purple vetch and barley) on leafhopper abundance to determine how their presence had reduced leafhopper density, and to isolate the relative influence of cover crops on the nutrient status of vines (i.e. plant host quality) from the impact of cover crops on natural enemy fitness on biological control. Three treatments were established and compared in the 2-year study: ground cover (vetch and barley), no-cover (tilled control) and ground cover with exclusion (i.e. with barriers limiting arthropods and spiders moving into the vine canopy). They showed mid- and late season leafhopper densities were significantly reduced in plots with the ground cover compared with the no-cover. Neither leafhopper egg parasitism by *Anagrus* spp. nor spider density (on vines or ground) could explain differences in leafhopper density. Vine vigor, however, was determined to be significantly lower in cover crop than in the no-cover plots, and late season leafhopper density was highest in ground cover/exclusion plots. Grapevine vigor had the strongest correlation with leafhopper density, with low vigor resulting from the apparent competition between the cover crops, resident weedy vegetation and grapevines and not from the impact of natural enemies. Higher late season leafhopper density in the cover/exclusion plots was, however, attributed to the reduced predation by spiders. The study suggests that cover crops may have a significant impact on soil quality and vine growth, complementing any function they serve in enhancing the natural enemies of vineyard pests (Costello and Daane 2003; Daane et al. 2005).

8.4 Landscape Ecology and Conservation Biological Control in California Vineyards

The intensification of production has not only produced simplified individual cropping systems (i.e., monocultures), but in addition the regional adoption of such practices has led to the aggregate simplification of entire agricultural landscapes (Tscharntke et al. 2005). The process of agroecosystem simplification is particularly acute in wine grape regions as the geographic branding of wine (e.g. premiums paid for wine produced in Napa County) further encourages regional land use conversion from natural habitat to high-value wine grape production. This loss of both agro-biodiversity and natural habitats that surround agroecosystems can lead to the loss of multiple ecosystem services, including biological control (Kremen et al. 2002; Altieri and Nicholls 2004).

The term landscape 'heterogeneity' (alternately landscape 'complexity' or 'diversity') has been used in the ecology and conservation literature to describe the area, arrangement and/or composition of natural habitats surrounding agroecosystems (Bianchi et al. 2006). Studies of landscape effects on ecosystem services typically quantify ecological features within a 1–3 km radius around a crop field, although some studies have measured landscape features at scales ranging from as little as 0.4 km to at most 25 km (Thies and Tscharntke 1999; Östman et al. 2001; Steffan-Dewenter et al. 2002). Landscapes are generally quantified in terms of the relative proportion of various habitat types within a given area (e.g., 32% oak woodland within a 1.5 km radius of a crop field), although some studies simply utilize categorical terms to describe a landscape (e.g., 'complex' and 'simple' landscapes) (Thies and Tscharntke 1999).

While researchers previously hypothesized that landscape heterogeneity could have a significant impact on biological control (van Emden 1965), it is only more recently that they have begun to address this relationship empirically. Bianchi et al. (2006) conducted a review of the ecological literature measuring the influence of landscape heterogeneity on arthropod populations and biological control in agriculture. Their analysis showed that in 74% of the cases studied, increased natural enemy diversity and abundance were correlated positively with increased landscape heterogeneity. However, in only 45% of the studies reviewed, increased landscape heterogeneity correlated positively with decreased pest densities, reduced crop damage or increased yield. While landscape heterogeneity has been shown to have a significant and positive influence on natural enemy diversity and abundance at the field level, meta-analyses conducted to date have shown that landscape heterogeneity does not consistently result in enhanced biological control (Bianchi et al. 2006; Chaplin-Kramer et al. 2011). The relationship between landscape heterogeneity and enhanced pest regulation in agriculture is therefore considered to be specific to the cropping system and life-history characteristics of key pests and their natural enemies (With et al. 2002; Hunter 2002; Tscharntke et al. 2007). A more detailed understanding of how specific biophysical features of landscapes influence arthropod populations will be essential for the development of cost-effective habitat enhancement strategies aimed at improving biological control and other ecosystem services to agriculture.

8.4.1 Research on Overwintering Habitat for **Anagrus** *spp.*

Several studies have evaluated the contribution of natural enemy refuges to pest regulation in California grape systems. A majority of the existing work has focused on the effect of *Anagrus* overwintering habitat and whether its proximity to vineyards influences biological control of *Erythroneura* leafhoppers. This is because this parasitoid must locate alternate leafhopper host eggs to complete winter diapause. Although *Anagrus* can complete multiple generations by parasitizing *Erythroneura* eggs during the spring and summer, these pest leafhoppers overwinter as adults while *Anagrus* overwinters as larvae (UC IPM 2010a). Overwintering habitat that supports alternate leafhopper host(s) may be limited (due to plant community composition) or lie at a great distance from vineyards. Low quality or distant overwintering habitat for *Anagrus* may lead to delayed spring colonization of vineyards, allowing early grape leafhopper populations to develop unchecked. This can result in leafhopper damage to young grape shoots and/or large populations of adult leafhoppers at the end of the growing season, which can interfere with harvest activities (UC IPM 2010a).

Researchers have attempted to address this management problem by investigating how habitat patches that serve as natural enemy refuge can contribute to early-season control of grape leafhoppers. Studies primarily evaluate the use of blackberry and prune refuges (*Rubus* spp. and *Prunus* spp., respectively) around California vineyards. Although some of this section draws from the broader North American literature, many of the known alternate host plants for overwintering *Anagrus* can be found in California. An overview of known alternate host plants (and associated leafhoppers) for overwintering *Anagrus* wasps is included in Table 8.2.

8.4.1.1 Studies of Wild Blackberry Refuges

A 1966 study of blackberry refuges revealed a gradient of parasitoid activity that declined with increasing distances from the refuges. Leafhopper egg parasitism was observed up to 6.4 km away from the blackberry stands. Beyond this distance egg parasitism rates declined substantially. Researchers concluded that the observed trend was likely due to *Anagrus* dispersing outward from the blackberry refuge. The study did not include any direct measurements of dispersal (e.g. mark/recapture) or quantitative assessments of *Anagrus* densities (Doutt et al. 1966).

In a related survey of *Anagrus* dispersal, Doutt and Nakata (1973) monitored vineyards for parasitoid activity at increasing distances from a large riparian area. It was assumed that the riparian habitat harbored a high density of wild blackberry, although no formal information on plant species composition was reported for the riparian area. Sampling vineyards at increasing distances from the riparian habitat (up to 32 km), researchers observed leafhopper egg parasitism 3–4 weeks earlier in vineyards located at closer proximity (<8 km) to the riparian forest. This finding again led researchers to conclude that *Rubus* spp. were harboring overwintering populations of *Anagrus* wasps and that these parasitoids were dispersing into nearby vineyards earlier in the spring. In addition, researchers observed earlier leafhopper egg parasitism in vineyards located downwind from the riparian ecosystem when compared with vineyards upwind at similar or closer distances. This finding led to the suggestion that dominant wind direction also plays an important role in *Anagrus* dispersal (Doutt and Nakata 1973).

Table 8.2 Summary of plant and leafhopper host associations for *Anagrus* spp.

Plant species	Common name	*Anagrus* species	Host species	Region	Reference
Aceraceae					
Acer glabrum	Douglas maple	*A. atomus*	?	BC	1
Acer saccharum	Sugar maple	*A. daanei*	?	NY	2, 3
		Anagrus spp.	?	NY	3
Betulaceae					
Alnus sp.	Alder	*A. erythroneurae*	?	BC	1
Betula occidentalis	Water birch	*A. atomus*	?	BC	1
		A. avalae	?	BC	1
		A. erythroneurae	?	BC	1
Betula pendula	European white birch	*A. atomus*	?	BC	1
Ostrya virginiana	Hophornbeam	*A. atomus*	?	NY	3
Cornaceae					
Cornus racemosa	Gray dogwood	*A. yawi*	?	NY	3
Cornus stolonifera	Red osier dogwood	*A. daanei* (?)	?	BC	1
		A. erythroneurae	?	BC	1
Fabaceae					
Robinia pseudo-acacia	Black locust	*A. daanei*	?	NY	3
		A. epos	?	NY	3
		A. nigriventris	?	NY	3
		Anagrus spp.	?	NY	3
Fagaceae					
Quercus rubra	Northern red oak	*Anagrus* spp.	?	NY	3
Juglandaceae					
Juglans nigra	Black walnut	*Anagrus* spp.	?	NY	3
Lamiaceae					
Lavendula angustifolia	Lavender	*A. atomus*	?	BC	1

(continued)

Table 8.2 (continued)

Plant species	Common name	*Anagrus* species	Host species	Region	Reference
Mentha spp.	Garden mint	*A. atomus*	?	BC	1
		A. erythroneurae	?	BC	1
	Mint	*A. atomus*	?	BC	1
		A. erythroneurae	?	BC	1
Nepeta cataria	Catnip	*A. atomus*	?	BC	1
		A. erythroneurae	?	BC	1
Nepeta x *mussinii*	Persian catnip	*A. atomus*	?	BC	1
		A. erythroneurae	?	BC	1
Salvia officinalis	Garden sage	*A. atomus*	?	BC	1
		A. erythroneurae	?	BC	1
Oleaceae					
Fraxinus americana	White ash	*Anagrus* spp.	?	NY	3
Rosaceae					
Crateagus sp.	Hawthorn	*Anagrus* spp.	?	NY	3
Fragaria x *ananassa*	Strawberry	*A. atomus* or *A. erythroneura*	?	BC	1
Malus domestica	Apple	*A. atomus*	?	BC	1
		A. avalae	?	BC	1
		A. erythroneurae	?	BC	1
Malus pumila	Apple	*Anagrus* spp.	?	NY	3
Malus spp.	Apple	*A. atomus*	*Typhlocyba pomaria*	BC, ON	2
			Empoasca maligna	MI	2
			?	BC	2
		A. avalae	*T. pomaria*	ON	2
		A. daanei	*T. pomaria*	MI	2
			?	MI	2

		A. erythroneurae	*T. pomaria*	CA, WA	2
		Anagrus sp.	*T. pomaria*	MI	2
		A. tretiakovae	?	MI	2
Prunus avium	Sweet cherry	*A. erythroneurae*	?	BC	1
Prunus domestica	Cultivated plum	*A. erythroneurae*	?	BC	1
Prunus dulcis	Almond	*A. atomus*	?	CA	2
		A. daanei	?	CA	2
		A. erythroneurae	?	CA	2
Prunus persica	Peach	*A. atomus*	*Edwardsiana prunicola*	CA	2
			Typhlocyba quercus (?)	CA	2
			?	CA	2
		A. avalae	*E. prunicola*	CA	2
			?	CA	2
		A. erythroneurae	*E. prunicola*	CA	2
			?	CA	2
		A. tretiakovae	*Erythroneura plena*	MD	2
Prunus serotina	Black cherry	*Anagrus* spp.	?	NY	3
Prunus virginiana	Choke cherry	*A. atomus*	?	BC	1
		A. avalae	?	BC, OR, WA	1, 4
		A. daanei	?	BC	1
		A. tretiakovae	?	OR, WA	4
Purshia tridentata	Antelope bitterbrush	*A. atomus*	?	OR, WA	4
Rosa eglanteria	Sweetbrier	*A. atomus*	?	OR, WA	4
		A. erythroneurae	?	OR, WA	4
Rosa multiflora	Multiflora rose	*A. atomus*	?	NY	3
		A. daanei	?	NY	3
Rosa rugosa	Rugose Rose	*A. atomus*	?	OR, WA	4
		A. erythroneurae	?	OR, WA	4
		A. tretiakovae	?	OR, WA	4

(continued)

Table 8.2 (continued)

Plant species	Common name	*Anagrus* species	Host species	Region	Reference
Rosa spp.	Rose	*A. atomus*	*Edwardsiana rosae*	CA, NY,	2
			?	BC, NY, OR, WA	1, 4
		A. avalae	*E. rosae*	BC, OR(?)	2
			?	BC, OR, WA	1, 4
		A. daanei	*E. rosae*	CA, NY	2
		A. erythroneurae	?	BC, OR, WA	1, 4
		A. nr. sp. *daanei*	?	OR, WA	4
		A. tretiakovae	?	OR, WA	4
Rosa woodsii	Wood's rose	*A. atomus*	?	OR, WA	4
		A. erythroneurae	?	OR, WA	4
Rubus armeniacus (*R. discolor, R. procerus*)	Himalayan blackberry	*A. atomus*	?	OR, WA	4
		A. avalae	?	OR, WA	4
		A. erythroneurae	?	OR, WA	4
		A. nigriventris	?	OR, WA	4
		A. nr. sp. *columbi*	?	OR, WA	4
		A. tretiakovae	?	OR, WA	4
Rubus laciniatus	Evergreen blackberry	*A. atomus*	?	OR, WA	4
Rubus spp.	Blackberry	*A. atomus*	*Dikrella* spp.	CA	2
			?	BC, CA	1, 2
		A. daanei	*Dikrella* spp. (?)	CA	2
			?	CA	2
		A. erythroneurae	*D. californica* (?)	CA	2
			D. cruentata (?)	CA	2
			Dikrella spp.	CA	2
			?	BC, CA	1, 2
		A. nigriventris	?	CA	2
Rubus spp.	Tayberry	*A. atomus*	?	BC	1
		A. erythroneurae	?	BC	1

Rutaceae					
Zanthoxylum americanum	Common prickly ash	*A. daanei*	?	NY	3
		Anagrus spp.	?	NY	3
Salicaceae					
Salix babylonica	Weeping willow	*A.* nr. sp. *avalae*	?	OR, WA	4
Salix nigra	Willow	*A. erythroneurae*	?	NY	3
Salix spp.	Willow	*A. atomus*	?	OR, WA	4
		A. erythroneurae	?	OR, WA	4
		A. nr. sp. *nigriventris*	?	BC, OR, WA	1, 4
		Anagrus sp.	?	BC	1
Ulmaceae					
Ulmus pumila	Siberian elm	*A. atomus*	?	BC	1
Vitaceae					
Parthenocissus quinquefolia	Virginia creeper	*A. daanei*	?	BC	1
Vitis cv. Castel	Grape	*A. daanei*	*Erythroneura* spp. (?)	NY	3
		Anagrus spp.	*Erythroneura* spp. (?)	NY	3
		A. tretiakovae	*Erythroneura* spp. (?)	NY	3
Vitis cv. GR-7	Grape	*A. daanei*	*Erythroneura* spp. (?)	NY	3
		Anagrus spp.	*Erythroneura* spp. (?)	NY	3
		A. tretiakovae	*Erythroneura* spp. (?)	NY	3
Vitis cv. Seyval blanc	Grape	*A. daanei*	*Erythroneura* spp. (?)	NY	3
Vitis labrusca Bailey cv. Concord	Grape	*A. daanei*	*Erythroneura* spp. (?)	NY	3
		A. nigriventris	*Erythroneura* spp. (?)	NY	3
		A. tretiakovae	*Erythroneura* spp. (?)	NY	3
Vitis labrusca Bailey cv. Delaware	Grape	*A. tretiakovae*	*Erythroneura* spp. (?)	NY	3

(continued)

Table 8.2 (continued)

Plant species	Common name	*Anagrus* species	Host species	Region	Reference
Vitis labrusca Bailey cv. Niagara	Grape	*A. daanei*	*Erythroneura* spp. (?)	NY	3
		A. erythroneurae	*Erythroneura* spp. (?)	NY	3
		A. tretiakovae	*Erythroneura* spp. (?)	NY	3
Vitis riparia	Grape	*A. tretiakovae*	*Erythroneura* spp. (?)	NY	3
Vitis spp.	Grape	*A. daanei*	*Erythroneura bistrata*	NY	2
			E. comes	NY	2
			E. elegantula	CA	2
			Erythroneura spp.	CA, NY	2
			E. variabilis	CA	2
			E. ziczac	BC, NY	2
			?	BC, CA, NY	2
		A. epos	*Dikrella*. sp.	NM	2
			E. comes	NY	2
			Erythroneura spp.	BCA, SON	2
			E. variabilis	SON	2
			E. vulnerata	CO	2
			?	NM, SON	2
		A. erythroneurae	*D. cockerellii*	NM	2
			Dikrella sp.	NM	2
			E. bistrata	NY	2
			E. comes	NY	2
			E. elegantula	CA	2
			Erythroneura spp.	BCA, CA	2
			E. variabilis	BCA, CA	2
			?	BCA, CA, NY, SON	2
		A. flaveolus	?	SON	2
		A. nigriventris	*E. comes*	NY	2

		A. tretiakovae	*D. cockerellii* (?)	COA	2
			E. bistrata	NY	2
			E. comes	DE, NY	2
	Grape	*A. tretiakovae*	*Erythroneura* spp.	AZ, COA, NY	2
			E. variabilis	AZ, SON	2
			E. ziczac	NY	2
			?	AZ, BCA, MI, NM, NY, SON	2
		A. yawi	*E. comes*	NY	2
Vitis vinifera	Grape	*A. daanei*	?	BC	1
Vitis vinifera L. cv. Chardonnay	Grape	*A. daanei*	*Erythroneura* spp. (?)	NY	3
		A. epos	*Erythroneura* spp. (?)	NY	3
		A. erythroneurae	*Erythroneura* spp. (?)	NY	3
		A. tretiakovae	*Erythroneura* spp. (?)	NY	3

State and province abbreviations used: Canada: *BC* British Columbia, *ON* Ontario; United States: *AZ* Arizona, *CA* California, *CO* Colorado, *DE* Delaware, *MD* Maryland, *MI* Michigan, *NM* New Mexico, *NY* New York, *OR* Oregon, *WA* Washington; Mexico: *BCA* Baja California, *COA* Coahuila, *SON* Sonora
[1]Lowery et al. (2007)
[2]Triapitsyn (1998)
[3]Williams and Martinson (2000)
[4]Wright and James (2007)
(?) Unverified species or region of collection

Although none of the studies above measured whether early season parasitism significantly influenced pest densities, the findings led to the development of recommendations that growers establish blackberry refuges around their vineyards to promote early season biological control of leafhoppers. More than a decade after the recommendations were made, further scientific evaluation of the plantings showed that the blackberry refuges did not consistently enhance biological control (Flaherty et al. 1985). Researchers posited that the on-farm blackberry refuges were unsuccessful because many were planted outside of their native riparian habitats, and that reduced canopy cover and lower soil moisture levels reduced the quality of the refuges which contributed to lower populations of both blackberry leafhopper (*Dikrella cruentata* (Gillette)) and *Anagrus*. Flaherty et al. (1985) attempted to substantiate this hypothesis by providing shade structures to *Rubus* plantings. Findings suggest that while the shade treatment did enhance *D. cruentata* populations on the blackberry, *Anagrus* densities were not significantly increased.

Due to its inability to consistently control leafhopper populations, California growers largely abandoned the planting of blackberry around Central Valley vineyards by the late 1980s. Additionally, identification of *Rubus* spp. as a systemic host of Pierce's Disease (*Xylella fastidiosa* Wells et al.) led to its removal from many riparian habitats of the Northern and Central Coast grape growing regions (Purcell and Saunders 1999; Baumgartner and Warren 2005).

8.4.1.2 Experiments Involving Prune Refuges

Counter to previous findings, Kido et al. (1983) reported high early season leafhopper parasitism in vineyards adjacent to prune orchards and revealed an additional alternate host for overwintering *Anagrus*, the prune leafhopper (*Edwardsiana prunicola* (Edwards)), which was reproducing in French prune (*Prunus* spp.) orchards neighboring vineyards. Following this discovery, Kido et al. (1984) conducted a non-replicated 2-year study quantifying population densities of *E. prunicola* and *Anagrus* in two vineyards adjacent to prune orchards. Only one vineyard-orchard pair was studied each year. Based on observations of leafhopper egg parasitism 3–4 weeks earlier in vineyards adjacent (<30 m) to the prune orchards, researchers concluded that *Anagrus* populations remained active in the prune trees throughout the growing season and dispersed into the nearby vineyards to parasitize grape leafhoppers eggs in the spring. Kido et al. (1984) concluded that French prunes could be used like *Rubus* spp. to enhance overwintering habitat for *Anagrus* wasps and thereby increase biological control of leafhoppers.

Building upon the above assessments, Wilson et al. (1989) monitored *Anagrus* activity in two vineyards, one adjacent and the other located at some distance away from a prune orchard (exact distance not reported). The study showed that the prune orchard harbored high densities of *Anagrus* wasps, and that leafhopper egg parasitism occurred approximately 3–4 weeks earlier in the nearby vineyard. Like others, the study concluded that *Anagrus* could successfully overwinter in French prune refuges and potentially contribute to early-season control of grape leafhoppers.

The effect of wind speed was also measured on *Anagrus* colonization. Prune trees subjected to low velocity winds were found to have a higher abundance of *Anagrus*. Based on these and the findings of Doutt and Nakata (1973), researchers advised growers to plant French prune trees upwind from their vineyards to augment populations of *Anagrus* and enhance biological control of leafhoppers.

Prior to 1990, all research conducted on the *Anagrus*-leafhopper system had been based on non-replicated comparisons and indirect assessments of *Anagrus* dispersal from overwintering refuges. While early season leafhopper egg parasitism was reported to be enhanced with proximity to *Rubus* spp. and *Prunus* spp. refuges, no assessment of vineyards pest densities were conducted.

The first direct assessment of *Anagrus* movement was carried out by Corbett and Rosenheim (1996a) using rubidium (Rb) to mark prune refuges adjacent to two vineyard sites over a 2-year period. Early season *Anagrus* populations were monitored at increasing distances away from the refuges to quantify the proportion of the *Anagrus* population found in the vineyard that originated in the Rb-marked prune trees. Consistent with the previous prune refuge studies, a higher density of *Anagrus* was found in vine rows directly adjacent (10–20 m) and downwind from prune trees. Only a small percentage of these parasitoids, however, were positively marked with Rb. Given the conflicting evidence, the authors concluded that although prune trees did directly contribute to vineyard *Anagrus* populations, the presence of regional riparian habitats appeared to be a greater source of *Anagrus*. The apparent 'prune tree effect' was partially the result of the prune stands acting as windbreaks for aerially dispersing *Anagrus* assumed to be coming from nearby riparian habitats (Corbett and Rosenheim 1996a).

Murphy et al. (1996, 1998a, b) completed a more comprehensive evaluation of the effect of prune refuges on biological control. In these studies, 18–24 pairs of vineyard blocks with and without nearby prune orchards were evaluated over 2 years. Researchers tested the hypotheses suggested in previous studies by evaluating the influence of prune trees on *Anagrus* abundance in vineyards, measuring leafhopper parasitism rates, and quantifying the abundance of leafhopper nymphs at increasing distances away from the prune refuge plantings. These studies again showed that *Anagrus* was more abundant approximately 3–4 weeks earlier in vineyards adjacent to prune orchards (Murphy et al. 1996). A similar effect was seen in parasitism rates, where leafhopper egg parasitism was significantly higher and occurred approximately 3–4 weeks earlier in vineyards adjacent to the prune orchards (Murphy et al. 1998a). Although significant differences in *Anagrus* density and parasitism rates between treatment and control blocks diminished later in the growing season (second and third leafhopper generations), early season effects of prune refuges could potentially influence late-season leafhopper populations. The final component of this study showed, however, that leafhopper nymph densities were not found to be significantly different between treatment and control sites (Murphy et al. 1998b). These results raised additional questions about the source habitats for *Anagrus* and highlighted the need to carry out more thorough evaluations of alternate overwintering habitat, and conduct further mark-recapture studies to better understand *Anagrus* dispersal across the landscape.

Corbett and Rosenheim (1996b) conducted another mark/recapture study of *Anagrus*, this time using fluorescent dust to mark wasps emerging from vineyard grape leaves. While not a study of *Anagrus* dispersal from any type of refuge *per se*, this study did provide new information on *Anagrus* movement within a vineyard. The study revealed that *Anagrus* appeared to disperse up to 24.5 m/day and contrary to all prior evidence, it had a significant tendency to disperse upwind. However, these novel findings are not definitive. In their discussion, Corbett and Rosenheim (1996b) suggested that, because their data on dominant wind speed and direction was from a nearby weather station, it may not have been representative of wind characteristics within the study vineyard itself. Like the prune refuges, the vineyard canopy structure may have altered wind speed and direction within the vineyard and this might have subsequently influenced *Anagrus* dispersal.

Prune orchards can still be found near some commercial vineyards in California. While these orchards could potentially provide a patchwork of overwintering habitat for *Anagrus* wasps, their area relative to the vineyards is small and their contribution to biological control is likely negligible. Researchers have suggested that small refuges (prune, blackberry or otherwise) may not be viable over the long term, as their entire population of alternate overwintering hosts risk being eliminated by overwhelming populations of *Anagrus* produced in large vineyards during the summer (Mills and Daane 2005).

8.4.1.3 North American Research on Alternate Overwintering Habitat for *Anagrus* spp.

Studies evaluating the impact of habitat patches on biological control of leafhoppers examined only two plant genera, *Rubus* spp. and *Prunus* spp., both in the Rosaceae. This limited range of known overwintering host plants for *Anagrus* has encouraged researchers to seek out new alternate host plants that could be utilized for habitat enhancement in proximity to vineyards. This work is especially important in light of recent taxonomic revisions to the *Anagrus* complex, which revealed that not all *Anagrus* species overwintering near vineyards are necessarily the same as those found parasitizing *Erythroneura* leafhoppers in vineyards during the growing season (Triapitsyn 1998).

Overwintering habitat assessments have been conducted in various viticulture regions in North America, revealing a range of new plant and host associations for *Anagrus* (Table 8.2). While *Anagrus* is consistently encountered on plants in the Rosaceae, this parasitoid also appears to be associated with plants in many other families. At present, *Anagrus* appears to reproduce exclusively on eggs from species in the Ciccadellidae. A summary of known plant and leafhopper host associations for *Anagrus* spp. is presented in Table 8.2.

8.4.2 Measuring the Effect of Plant Corridors, Flower Islands, and Native Vegetation

Nicholls et al. (2001) evaluated the influence of non-crop habitat on biological control in a northern California organic vineyard. They focused on two separate

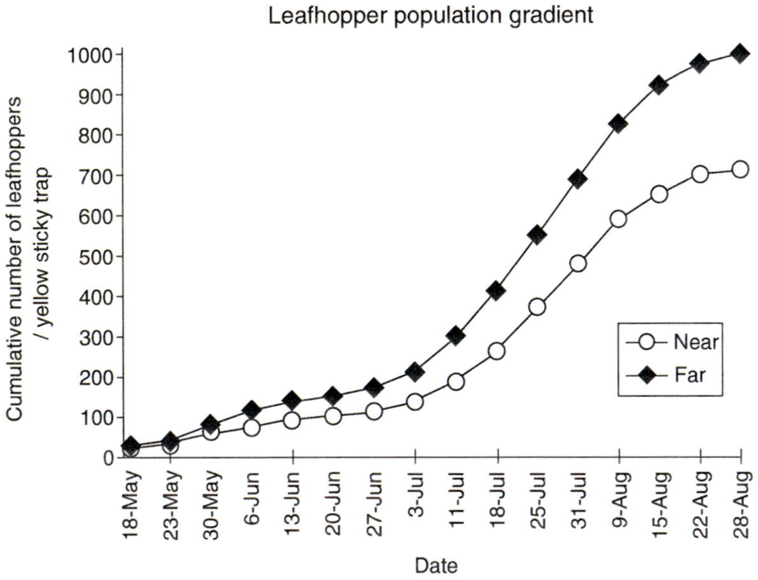

Fig. 8.1 Season patterns of adult leafhoppers in a vineyard 'Near' and 'Far' from a corridor of flowering vegetation (Hopland, California 1996) (From Altieri et al. 2005)

non-crop habitats: (1) a vegetational corridor bisecting the vineyard, and (2) a riparian forest abutting the vineyard. The corridor consisted of 65 different species of flowering plants. No description of plant species composition was provided for the riparian forest. Natural enemy and pest populations were monitored at increasing distances away from the corridor and the riparian forest. The study found that the abundance of generalist predators decreased at increasing distances away from both the forest and corridor, while *Anagrus* densities increased towards the center of the vineyard plots. Leafhopper egg parasitism rates did not exhibit any significant spatial trends relative to the two non-crop habitats, although parasitism rates were generally higher towards the center of the vineyard blocks. Thrips, *F. occidentalis*, and grape leafhopper adult densities both increased at greater distances away from the two non-crop habitats (Fig. 8.1). *Anagrus* dispersal was evaluated through an indirect assessment of movement, and no clear information was provided about dominant wind direction relative to non-crop habitats. In that study, Nicholls et al. (2001) concluded that the distribution of *Anagrus* was likely following that of the leafhoppers resulting from a density-dependent relationship between the parasitoid and host rather than any influence from non-crop habitat.

To further understand the spatial patterns of biological control in vineyards, Altieri et al. (2005) evaluated population densities of pest and beneficial insects at increasing distances away from a 0.25 ha on-farm 'flower island' in a Northern California vineyard. The island was composed of 33 species of flowering shrubs

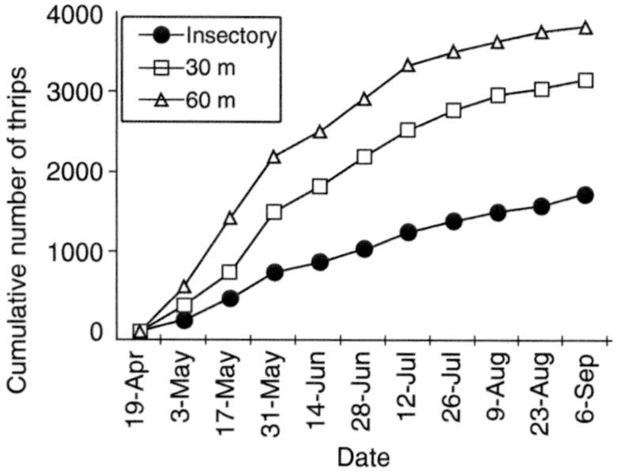

Fig. 8.2 Cumulative number of thrips per yellow sticky trap in 2004 at Benziger vineyard (Glen Ellen, California)

and herbs predominantly from the Asteraceae, Agavaceae, and Lamiaceae. The assemblage of flowering plants was selected to provide floral resources from April to late September. Natural enemy populations (*Orius* spp., Coccinellidae, Syrphidae, *Anagrus* wasps), leafhopper egg parasitism, and thrips density were recorded at increasing distances away from the island (10, 30 and 60 m) over a single growing season. Results showed that natural enemy densities, and leafhopper egg parasitism rates both decreased and abundance of thrips increased at greater distances away from the island (Fig. 8.2). The researchers thus suggested that the flower island may have served as a source of pollen, nectar or alternate prey for natural enemies which led to the observed changes in leafhopper egg parasitism and thrips densities.

While much conservation biological control research in California viticulture has focused on the *Anagrus*-leafhopper system, other research has been conducted to evaluate the relationship between natural habitats and vineyard spider populations. Spiders are known to be the most abundant generalist predator in vineyards and natural habitats could be contributing to these vineyard populations (Costello and Daane 1995, 1999; Roltsch et al. 1998). Hogg and Daane (2010, 2011) evaluated how oak woodland-chaparral, and riparian habitats contributed to vineyard spider populations. Spiders were sampled throughout the growing season in natural habitats and at multiple distances into vineyards. They reported that spider dispersal into vineyards appeared to occur later in the growing season (July and August) and that spider species diversity and abundance significantly differed between natural and vineyard habitats. The observed differences in species composition became more pronounced with increasing distance away from the natural habitats (up to 250 m), and vineyards were found to be dominated by just a few spider species. Researchers suggested that natural habitats serve as an important source of vineyard spider populations. While no assessment of pest densities was conducted in these two studies, the researchers noted that the observed changes in vineyard spider species composition relative to distance away from adjacent source habitats likely has implications for biological control.

8.4.3 Landscape Restoration to Enhance Ecosystem Services to California Vineyards

Habitat restoration in California agriculture is characterized by the establishment of mixed-use hedgerows intended to promote biological control, pollination and other ecosystem services. Hedgerows typically consist of combinations of annuals, herbaceous, and woody perennial shrubs and trees. These plantings have been found to attract populations of important natural enemies of vineyard pests, including *Orius* spp., *Geocoris* spp., Coccinellidae, Chrysopidae, Nabidae, Syrphidae and various spiders (Dufour and Appropriate Technology Transfer for Rural Areas Organization 2000; Robins et al. 2001; Earnshaw 2004). Despite the limited data on the aggregation of natural enemies, the impact on biological control of vineyard pests remains largely unexplored. Given the lack of scientific data on the impact of such plantings, growers and government programs supporting such efforts may not be fully realizing the outcomes they are intended to achieve.

Additionally, the inconsistent findings of many of the previous vineyard diversification studies described above may be in part due to a failure of researchers to adequately account for the influence of the surrounding landscape on biological control. Identifying the key qualities and quantities of non-crop habitats that support natural enemies will be an essential step in developing scientifically based landscape restoration programs that effectively enhance biological control and other ecosystem services to vineyards. Despite the many important contributions of ecologists and biological control specialists to date, many research gaps remain. Filling these gaps will be essential in providing the empirical evidence needed to define the specific types of habitat enhancement that leads to cost-effective regulation of important vineyard pests.

8.5 Current Diversification Research at UC Berkeley: Field-Scale Analysis

Building upon the prior field and landscape-scale studies in conservation biological control in vineyards discussed above, researchers at the university of California, Berkeley, have recently initiated the first comprehensive, multi-scalar study of the impact of floral resource provisioning (FRP) and landscape complexity in Napa, Sonoma, San Joaquin and Fresno County wine grape systems.

The floral resource provisioning theory predicts that the addition of flowering plants to simplified agroecosystems improves biological control by providing insect parasitoids or predators with key food sources (e.g., nectar, pollen) that would otherwise limit fitness of natural enemies (Barbosa 1998; Landis et al. 2000; Altieri and Nicholls 2004; Heimpel and Jervis 2005; Lee and Heimpel 2008). The floral resource provisioning systems attract the interests of researchers and growers because of its theoretical appeal and success in some cropping systems

(Tonhasca and Byrne 1994; Gurr and Wratten 2000; Letourneau et al. 2010). The floral resource provisioning schemes also attract some skepticism in the scientific community as the outcomes of hundreds of on-farm diversification studies have been mixed (Andow 1991a; Lavandero et al. 2005; Wäckers et al. 2005; Straub et al. 2008). However, in a recent meta-analysis, Letourneau et al. (2010) showed that on-farm diversification strategies consistently supported a greater abundance, and diversity of natural enemies and increased pest control. Further, when FRP strategies do appear successful, the ecological processes underlying enhanced pest regulation often remains unsubstantiated or not fully understood (Gurr et al. 2000, 2004; Landis et al. 2000; Nicholls et al. 2000). Finally, the relationship between FRP and pest densities in vineyards and other cropping systems may also be explained by multiple alternative hypotheses (Wratten et al. 1998; Corbett 1998; Costello and Daane 2003; Gurr et al. 2004; Heimpel and Jervis 2005; Bianchi et al. 2006). The current scientific consensus is that FRP can enhance biological control, but its success is both context and system specific (Altieri and Nicholls 2004; Tscharntke et al. 2007). Moreover, while FRP programs have the potential to decrease reliance on pesticides, the uncertainty of the effectiveness of this and other diversification schemes restricts large-scale implementation.

Prior studies in field-scale diversification in California vineyards were limited by a number of key factors. First, in Costello and Daane (1998, 2003), non-flowering cover crops (i.e., barley) were used, and less consideration was given to seasonal availability of floral resources, flower morphology and accessibility, and/or the quality of floral resources needed to enhance the fitness of natural enemies (Wäckers 2004; Begum et al. 2006; Vattala et al. 2006). The findings of Nicholls et al. (2000) were limited to a comparative analysis of two large vineyard blocks without full substantiation of the cause of enhanced biological control. Additionally, in all the prior on-farm diversification research in California, the landscape context (i.e., the area and diversity of non-crop habitat) was not fully taken into account (Tscharntke et al. 2005, 2007; Chaplin-Kramer et al. 2011).

To address some of the limitations and build upon prior studies in vineyard diversification, the current UC Berkeley conservation biological control project will assess the impact of floral resource provisioning and landscape complexity in several key grape producing regions. At the field level, the study will measure the impact of five flowering ground covers (annual buckwheat (*F. esculentum*), lacy phacelia (*Phacelia tanacetifolia* Bentham), sweet alyssum (*Lobularia maritima* (L.) Desvaux), bishop's weed (*Ammi majus* L.), and wild carrot (*Daucus carota* L.)) on biological control of *Erythroneura* leafhoppers (*E. elegantula* and *E. variabilis*) and vine mealybug (*Planococcus ficus* (Signoret)) by the parasitoid wasps *Anagrus* spp. and *Anagyrus pseudococci* (Girault) (Hymenoptera: Encyrtidae) in California vineyards. The research includes eight split-block trials on commercial vineyards in Napa and Sonoma County and two fully replicated research designs, one located in Lodi and the other at the UC Kearney Agriculture Center in Fresno County. The research will test multiple hypotheses (i.e., natural enemies and resource concentration) of biological control in vineyards to advance scientific knowledge of cost-effective and ecologically-based pest management. The study will quantify the impacts of FRP on population densities of pest and beneficial insects and analyze the biological mechanisms (e.g., longevity, fecundity, parasitism rates) theorized to be enhanced through FRP. Comparative cost-benefit analyses (FRP vs. conventional practices) will evaluate the cost-effectiveness of the tested strategies. In addition, the study will measure natural enemy movement from flowering cover crops to the vine canopy and substantiate nectar feeding through laboratory studies and anthrone testing. In a separate replicated and complementary study, researchers are testing the effect of methyl salicylate lures (a beneficial insect attractant) on natural enemies, pest densities and biological control (James 2003, 2006; Cook et al. 2006; James and Price 2004). Data from laboratory studies indicate that FRP has a significant positive impact on the longevity of *A. pseudococci* females (A. Miles et al., unpubl. data).

8.6 Current Diversification Research at UC Berkeley: Landscape Analysis

The landscape component of the UC Berkeley conservation biological control project will evaluate the influence of landscape heterogeneity on the effectiveness of a field-scale FRP treatment to enhance biological control of grape leafhopper (*E. elegantula*) in northern California wine grape vineyards. A field FRP treatment plot will be compared to a control plot in 20 separate vineyards situated along a gradient of landscape heterogeneity. The FRP treatment will consist of three annual flowering plant species: lacy phacelia (*P. tanacetifolia*), bishop's weed (*Ammi majus* L.) and wild carrot (*D. carota*). This combination of species was selected to provide floral bloom throughout the entire growing season. These species are also drought tolerant, require no additional irrigation and can readily be integrated with standard vineyard management practices in northern California. Populations of the leafhopper pest and its key natural enemies will be monitored along with parasitism rates, crop damage and yield. Vine vigor will also be assessed in order to evaluate the influence of plant nutrient status on pest densities. Additionally, an assessment of *Anagrus* dispersal from natural habitats into adjacent vineyards will be conducted. Finally, *Anagrus* overwintering habitat will be assessed. Plant species commonly found in northern California vineyard landscapes will be sampled and evaluated for overwintering parasitoids. Plant material found to support significant *Anagrus* populations will be further evaluated to determine the associated insects that serve as alternate-hosts for the parasitoid. In combination, these studies are intended to generally evaluate how vineyard landscape composition influences the ability of field-scale FRP to enhance biological control of key wine grape pests. The goal of this research is to determine thresholds of landscape heterogeneity within which the use of field-scale FRP is most cost-effective for enhancing biological control.

8.7 Conclusion: Field and Landscape Level Diversification for Conservation Biological Control

Results of the California studies reviewed above show a pattern consistent with the larger national and international conservation biological control literature: treatment effects from diversification strategies are discernable, yet cost-effective biological control is not consistently achieved (Andow 1991a) (English-Loeb et al. 2003; Begum et al. 2006; Berndt et al. 2006; Straub et al. 2008). Nevertheless, meta analyses showed that diversification had a moderate effect on the abundance of plant herbivores (Tonhasca and Byrne 1994; Letourneau et al. 2010). Other meta-analyses of landscape factors have also shown that while natural enemy abundance, richness, predation and parasitism rates do increase significantly with landscape heterogeneity, pest densities are not found to be consistently lower (Bianchi et al. 2006; Chaplin-Kramer et al. 2011). Despite the large body of existing research, significant gaps remain in the conservation biological control literature. Findings from the research proposals outlined herein will help provide the necessary information for advancing the science of conservation biological control and developing more cost-effective ecologically-based pest management strategies for California vineyards.

8.8 Proposals and Considerations for Future Research: Conservation Biological Control in California Vineyards

The following are guidelines and specific proposals for research that would serve to advance the science and practice of conservation biological control in California viticulture. Proposals include both natural and social science studies.

As the effect of field-scale habitat enhancement strategies can be influenced by features in the surrounding landscape, future research must consider the influence of non-crop habitats that lie beyond the individual field or vineyard boundary. Broad correlative studies of landscape heterogeneity, natural enemy and pest density must be conducted along with detailed evaluations of the ecological processes theorized to influence biological control. To provide reliable data for use in developing effective pest management strategies, studies must be conducted for a minimum of 2 years and include full replication at the field and landscape scale. Field-scale evaluations of diversified cropping systems should assess both natural enemies and pest densities along with empirical tests of parasitism and predation (Bianchi et al. 2006). Measures should be taken to determine the impact of treatments on herbivore population densities, crop yield and quality. Studies measuring the impact of intercropping must account for the influence of non-crop vegetation on plant nutrient status along with impacts on the fitness of natural enemies (Daane and Costello 1998; Altieri and Nicholls 2003). Multi-trophic interactions must also be considered as increased diversity, and abundance of natural enemies in complex agricultural habitats can lead to intraguild predation and subsequent release of pests from biological control (Finke and Denno 2004; Straub et al. 2008).

Studies involving habitat manipulation should evaluate both the natural enemies and resource concentration hypotheses. Invertebrate response to landscape heterogeneity should be evaluated in a way that can address both of these hypotheses. At a minimum, this would require separately examining insect response to the relative area, diversity and connectivity of both natural habitat and agricultural land at the landscape scale. The high probability of idiosyncratic and species specific response to the landscape will require that observed trends be evaluated relative to a number of alternate measures of landscape heterogeneity, including perimeter to area ratio, mean patch size, and distance away from natural habitats (for details see Concepción et al. 2008). As non-crop habitats cannot be assumed to benefit only predators and parasitoids, studies should simultaneously measure the impact of non-crop vegetation on the fitness of insect pests (van Emden 1965; Baggen et al. 1999; Roschewitz et al. 2005).

As habitat diversity will influence insect movement at both the field and landscape scale, researchers are encouraged to consider the movement and distribution of arthropods in relation to the elements of heterogeneity under study (Corbett 1998; Dover and Settele 2009). The results of Corbett and Rosenheim (1996a) demonstrate the importance of empirical assessments of parasitoid dispersal from non-crop habitats. Quantifying insect movement between in-field habitat and crop and from non-crop habitats into cropping systems will be critical to developing a more nuanced understanding of the impact of heterogeneity at multiple spatial scales. Recent advances with relatively inexpensive marking systems (Hagler and Jones 2010) will help make this a reality.

Controlled field and laboratory trials are essential for determining the physiological influence of non-crop vegetation on key pests and natural enemies (Wäckers et al. 2005). Quantifying the influence of multiple species of flowering plants on parasitoid longevity, fecundity, parasitism rates and sex ratios of key biological control agents can help form the empirical basis for understanding enhanced biological control in field trials. To further substantiate nectar feeding, researchers should consider anthrone or HPLC testing to determine changes in parasitoid gut-sugar levels in the presence of flowers (Steppuhn and Wäckers 2004; Heimpel and Jervis 2005). Ideally, such work would be conducted under conditions most resembling the vineyard environment (Lee and Heimpel 2008).

It is important for applied research in conservation biological control to include on-farm and participatory trials in commercial vineyard settings. Such dialog with growers encourages the development of practices suitable for large-scale implementation and facilitates a social learning process between researchers, and growers that may improve the relevancy of research, and advance grower adoption of successful ecologically-based pest management practices (Röling and Wagemakers 2000; Warner 2007a, b). Cost-benefit analysis, including data on impact to other ecosystem services (e.g., soil quality, etc.) will provide a more holistic basis for grower decision making regarding the true costs and benefits of vineyard diversification (Fiedler et al. 2008; Gurr et al. 2003; Jackson et al. 2007).

Habitat enhancement tactics may also be successfully combined with the many new chemical ecology approaches (e.g., pheromones) to further enhance biological control (Daane et al. 2008). 'Attract and reward' strategies, for example, combine the use of herbivore-induced plant volatile (HIPV) compounds with in-field FRP and has shown much promise in enhancing the effectiveness of diversification schemes (James 2006; Khan et al. 2008). One such HIPV compound, methyl salicylate (MeSA) has been shown to increase abundance of some natural enemies in grape vineyards as well as in other cropping systems (James and Price 2004; James 2006; Lee 2010).

Future research must also include relevant economic and social assessments which may assist in developing ecologically-based pest management practices suitable for commercial adoption and provide a sound basis for the formulation of public policy (Cullen et al. 2008). To date, little work has been done to evaluate the impacts of public policy on vineyard habitat management or the ability of public institutions to adequately respond to grower research needs and coordination of agricultural restoration efforts at the regional scale. Finally, it will be critical to gather information on consumer perceptions of product quality and value associated with agricultural goods produced using ecologically-based farming practices (Forbes et al. 2009; Zucca et al. 2009; Howard and Allen 2010; Delmas and Grant 2010).

References

Altieri MA, Nicholls CI (2003) Soil fertility management and insect pests: harmonizing soil and plant health in agroecosystems. Soil Tillage Res 72:203–211

Altieri MA, Nicholls CI (2004) Biodiversity and pest management in agroecosystems. Food Products Press, Binghamton

Altieri MA, Ponti L, Nicholls CI (2005) Manipulating vineyard biodiversity for improved insect pest management: case studies from northern California. Int J Biodivers Sci Ecosyst Serv Manag 1:191–203

Altieri MA, Nicholls CI, Wilson H, Miles A (2011) Habitat management in vineyards: a growers manual for enhancing natural enemies of pests. Laboratory of Agroecology/University of California, Berkeley

Andow DA (1991a) Vegetational diversity and arthropod population response. Annu Rev Entomol 36:561–586

Andow D (1991b) Yield loss to arthropods in vegetationally diverse agroecosystems. Environ Entomol 20:1228–1235

Baggen L, Gurr G, Meats A (1999) Flowers in tritrophic systems: mechanisms allowing selective exploitation by insect natural enemies for conservation biological control. Entomol Exp Appl 91:155–161

Barbosa P (1998) Conservation biological control. Academic, San Diego

Baumgartner K, Warren JG (2005) Persistence of *Xylella fastidiosa* in riparian hosts near northern California vineyards. Plant Dis 89:1097–1102

Begum M, Gurr GM, Wratten SD, Hedberg PR, Nicol HI (2006) Using selective food plants to maximize biological control of vineyard pests. J Appl Ecol 43:547–554

Bentley WJ (2009) The integrated control concept and its relevance to current integrated pest management in California fresh market grapes. Pest Manag Sci 65:1298–1304

Berndt LA, Wratten SD, Scarratt SL (2006) The influence of floral resource subsidies on parasitism rates of leafrollers (Lepidoptera: Tortricidae) in New Zealand vineyards. Biol Control 37:50–55

Bianchi F, Booij C, Tscharntke T (2006) Sustainable pest regulation in agricultural landscapes: a review on landscape composition, biodiversity and natural pest control. Proc R Soc B 273:1715

Brodt S, Klonsky K, Thrupp A (2009) Market potential for organic crops in California: almonds, hay, and winegrapes. University of California/Agriculture and Natural Resources, Oakland

Broome JC, Warner KD (2008) Agro-environmental partnerships facilitate sustainable wine-grape production and assessment. Calif Agric 62:133–141

CDPR (California Department of Pesticide Regulation) (2009) Department of pesticide regulation 2007 Annual pesticide use report indexed by commodity (Napa County). California Department of Pesticide Regulation. http://www.cdpr.ca.gov/docs/pur/purmain.htm

Chaplin-Kramer R, O'Rourke ME, Blitzer EJ, Kremen C (2011) A meta-analysis of crop pest and natural enemy response to landscape complexity. Ecol Lett 14:922–932

Concepción ED, Diaz M, Baquero RA (2008) Effects of landscape complexity on the ecological effectiveness of agri-environment schemes. Landsc Ecol 23:135–148

Cook SM, Khan ZR, Pickett JA (2006) The use of push-pull strategies in integrated pest management. Annu Rev Entomol 52:375–400

Corbett A (1998) The importance of movement in the response of natural enemies to habitat manipulation. In: Pickett CH, Bugg RL (eds) Enhancing biological control: habitat management to promote natural enemies of agricultural pests. University of California Press, Berkeley, pp 25–48

Corbett A, Rosenheim JA (1996a) Impact of a natural enemy overwintering refuge and its interaction with the surrounding landscape. Ecol Entomol 21:155–164

Corbett A, Rosenheim JA (1996b) Quantifying movement of a minute parasitoid, *Anagrus epos* (Hymenoptera: Mymaridae), using fluorescent dust marking and recapture. Biol Control 6:35–44

Costello MJ, Daane KM (1995) Spider (Araneae) species composition and seasonal abundance in San Joaquin Valley grape vineyards. Environ Entomol 24:823–831

Costello MJ, Daane KM (1998) Influence of ground cover on spider populations in a table grape vineyard. Ecol Entomol 23:33–40

Costello MJ, Daane KM (1999) Abundance of spiders and insect predators on grapes in central California. J Arachnol 27:531–538

Costello MJ, Daane KM (2003) Spider and leafhopper (*Erythroneura* spp.) response to vineyard ground cover. Environ Entomol 32:1085–1098

Cullen R, Warner KD, Jonsson M, Wratten SD (2008) Economics and adoption of conservation biological control. Biol Control 45:272–280

Daane KM, Costello MJ (1998) Can cover crops reduce leafhopper abundance in vineyards? Calif Agric 52:27–33

Daane KM, Smith RJ, Klonsky KM, Bentley WJ (2005) Research article. Organic vineyard management in California. In: IPM in organic systems, XXII International Congress of Entomology, Brisbane, Australia, 16 Aug 2004, pp 37N–55N

Daane KM, Cooper ML, Triapitsyn SV, Walton VM, Yokota GY, Haviland DR et al (2008) Vineyard managers and researchers seek sustainable solutions for mealybugs, a changing pest complex. Calif Agric 62:167–176

Delmas MA, Grant LE (2010) Eco-labeling strategies and price-premium: the wine industry puzzle. Bus Soc 20(10):1–39

Doutt RL, Nakata J (1973) The Rubus leafhopper and its egg parasitoid: an endemic biotic system useful in grape-pest management. Environ Entomol 2:381–386

Doutt R, Nakata J, Skinner F (1966) Dispersal of grape leafhopper parasites from a blackberry refuge. Calif Agric 20:14–15

Dover J, Settele J (2009) The influences of landscape structure on butterfly distribution and movement: a review. J Insect Conserv 13:3–27

Dufour R, Appropriate Technology Transfer for Rural Areas (Organization) (2000) Farmscaping to enhance biological control. ATTRA. https://attra.ncat.org/attra-pub/summaries/summary.php?pub=145

Earnshaw S (2004) Hedgerows for California agriculture: a resource guide. Community Alliance with Family Farmers, Davis

English-Loeb G, Rhainds M, Martinson T, Ugine T (2003) Influence of flowering cover crops on *Anagrus* parasitoids (Hymenoptera: Mymaridae) and *Erythroneura* leafhoppers (Homoptera: Cicadellidae) in New York vineyards. Agric For Entomol 5:173–181

Eskenazi B, Huen K, Marks A, Harley KG, Bradman A, Barr DB, Holland N (2010) PON1 and Neurodevelopment in children from the CHAMACOS study exposed to organophosphate pesticides in utero. Environ Health Perspect 118:1775

Fiedler AK, Landis DA, Wratten SD (2008) Maximizing ecosystem services from conservation biological control: the role of habitat management. Biol Control 45:254–271

Finke DL, Denno RF (2004) Predator diversity dampens trophic cascades. Nature 429:407–410

Flaherty DL (1969) Ecosystem trophic complexity and densities of the Willamette mite, *Eotetranychus willamettei* Ewing (Acarina: Tetranychidae). Ecology 50:911–916

Flaherty D, Wilson L, Stern V, Kido H (1985) Biological control in San Joaquin valley vineyards. In: Hoy MA, Herzog DC (eds) Biological control in agricultural IPM systems. Academic, New York, pp 501–520

Forbes SL, Cohen DA, Cullen R, Wratten SD, Fountain J (2009) Consumer attitudes regarding environmentally sustainable wine: an exploratory study of the New Zealand marketplace. J Clean Prod 17:1195–1199

Gurr G, Wratten S (2000) Biological control: measures of success. Kluwer Academic Publishers, Dordrecht

Gurr G, Wratten S, Barbosa P (2000) Success in conservation biological control of arthropods. In: Gurr G, Wratten S (eds) Biological control: measures of success. Kluwer Academic Publishers, Dordrecht, pp 105–132

Gurr GM, Wratten SD, Luna JM (2003) Multi-function agricultural biodiversity: pest management and other benefits. Basic Appl Ecol 4:107–116

Gurr GM, Wratten SD, Altieri MA (2004) Ecological engineering for pest management. CSIRO Publishing, Collingwood

Hagler JR, Jones VP (2010) A protein-based approach to mark arthropods for mark-capture type research. Entomol Exp Appl 135:177–192

Hanna R, Zalom FG, Roltsch WJ (2003) Relative impact of spider predation and cover crop on population dynamics of *Erythroneura variabilis* in a raisin grape vineyard. Entomol Exp Appl 107:177–191

Heaton E, Merenlender A (2000) Modeling vineyard expansion, potential habitat fragmentation. Calif Agric 54:12–19

Heimpel GE, Jervis MA (2005) Does floral nectar improve biological control by parasitoids? In: Wäckers FL, Rijn PCJ, Bruin J (eds) Plant-provided food for carnivorous insects: a protective mutualism and its applications. Cambridge University Press, Cambridge, pp 267–304

Hogg BN, Daane KM (2010) The role of dispersal from natural habitat in determining spider abundance and diversity in California vineyards. Agric Ecosyst Environ 135:260–267

Hogg BN, Daane KM (2011) Ecosystem services in the face of invasion: the persistence of native and nonnative spiders in an agricultural landscape. Ecol Appl 21:565–576

Howard PH, Allen P (2010) Beyond organic and fair trade? An analysis of ecolabel preferences in the United States. Rural Sociol 75:244–269

Hunter MD (2002) Landscape structure, habitat fragmentation, and the ecology of insects. Agric For Entomol 4:159–166

Jackson LE, Pascual U, Hodgkin T (2007) Utilizing and conserving agrobiodiversity in agricultural landscapes. Agric Ecosyst Environ 121:196–210

James DG (2003) Synthetic herbivore-induced plant volatiles as field attractants for beneficial insects. Environ Entomol 32:977–982

James DG (2006) Methyl salicylate is a field attractant for the goldeneyed lacewing, *Chrysopa oculata*. Biocontrol Sci Technol 16:107–110

James DG, Price TS (2004) Field-testing of methyl salicylate for recruitment and retention of beneficial insects in grapes and hops. J Chem Ecol 30:1613–1628

Khan ZR, James DG, Midega CAO, Pickett JA (2008) Chemical ecology and conservation biological control. Biol Control 45:210–224

Kido H, Flaherty D, Bosch D, Valero K (1983) Biological control of grape leafhopper. Calif Agric 37:4–6

Kido H, Flaherty D, Bosch D, Valero K (1984) French prune trees as overwintering sites for the grape leafhopper egg parasite. Am J Enol Vitic 35:156

Kremen C, Williams NM, Thorp RW (2002) Crop pollination from native bees at risk from agricultural intensification. Proc Natl Acad Sci USA 99:16812–16816

Landis DA, Wratten SD, Gurr GM (2000) Habitat management to conserve natural enemies of arthropod pests in agriculture. Annu Rev Entomol 45:175–201

Lavandero B, Wratten S, Shishehbor P, Worner S (2005) Enhancing the effectiveness of the parasitoid *Diadegma semiclausum* (Helen): movement after use of nectar in the field. Biol Control 34:152–158

Lee JC (2010) Effect of methyl salicylate-based lures on beneficial and pest arthropods in strawberry. Environ Entomol 39:653–660

Lee JC, Heimpel GE (2008) Floral resources impact longevity and oviposition rate of a parasitoid in the field. J Anim Ecol 77:565–572

Letourneau DK, Armbrecht I, Salguero Rivera B, Montoya Lerma J, Jiménez Carmona E et al (2010) Does plant diversity benefit agroecosystems? A synthetic review. Ecol Appl 21:9–21

Lowery DT, Triapitsyn SV, Judd GJR (2007) Leafhopper host plant associations for *Anagrus* parasitoids (Hymenoptera: Mymaridae) in the Okanagan Valley, British Columbia. J Entomol Soc Br Columbia 104:9–15

Meadows R (2008) Research news: research fuels sustainable viticulture revolution. Calif Agric 62:127–131

Mills N, Daane K (2005) Biological and cultural controls: non-pesticide alternatives can suppress crop pests. Calif Agric 59:23–28

Murphy BC, Rosenheim JA, Granett J (1996) Habitat diversification for improving biological control: abundance of *Anagrus epos* (Hymenoptera: Mymaridae) in grape vineyards. Environ Entomol 25:495–504

Murphy BC, Rosenheim JA, Dowell RV, Granett J (1998a) Habitat diversification tactic for improving biological control: parasitism of the western grape leafhopper. Entomol Exp Appl 87:225–235

Murphy B, Rosenheim J, Granett J, Pickett C, Dowell R (1998b) Measuring the impact of a natural enemy refuge: the prune tree/vineyard example. In: Pickett CH, Bugg RL (eds) Enhancing biological control: habitat management to promote natural enemies of agricultural pests. University of California Press, Berkeley, pp 297–309

Nicholls CI, Parrella MP, Altieri MA (2000) Reducing the abundance of leafhoppers and thrips in a northern California organic vineyard through maintenance of full season floral diversity with summer cover crops. Agric For Entomol 2:107–113

Nicholls CI, Parrella M, Altieri MA (2001) The effects of a vegetational corridor on the abundance and dispersal of insect biodiversity within a northern California organic vineyard. Landsc Ecol 16:133–146

Östman Ö, Ekbom B, Bengtsson J (2001) Landscape heterogeneity and farming practice influence biological control. Basic Appl Ecol 2:365–371

Purcell A, Saunders S (1999) Fate of Pierce's disease strains of *Xylella fastidiosa* in common riparian plants in California. Plant Dis 83:825–830

Robins P, Holmes RB, Laddish K (2001) Bring farm edges back to life! Yolo County Resource Conservation District, Woodland

Röling NG, Wagemakers M (2000) Facilitating sustainable agriculture: participatory learning and adaptive management in times of environmental uncertainty. Cambridge University Press, Cambridge

Roltsch W, Hanna R, Zalom F, Shorey H, Mayse M (1998) Spiders and vineyard habitat relationships in central California. In: Pickett CH, Bugg RL (eds) Enhancing biological control: habitat management to promote natural enemies of agricultural pests. University of California Press, Berkeley, pp 311–318

Root RB (1973) Organization of a plant-arthropod association in simple and diverse habitats: the fauna of collards (*Brassica Oleracea*). Ecol Monogr 43:95–124

Roschewitz I, Hucker M, Tscharntke T, Thies C (2005) The influence of landscape context and farming practices on parasitism of cereal aphids. Agric Ecosyst Environ 108:218–227

Ross K, Golino D (2008) Wine grapes go green: the sustainable viticulture story. Calif Agric 62:125–126

Russell EP (1989) Enemies hypothesis: a review of the effect of vegetational diversity on predatory insects and parasitoids. Environ Entomol 18:590–599

Steffan-Dewenter I, Münzenberg U, Bürger C, Thies C, Tscharntke T (2002) Scale-dependent effects of landscape context on three pollinator guilds. Ecology 83:1421–1432

Steppuhn A, Wäckers F (2004) HPLC sugar analysis reveals the nutritional state and the feeding history of parasitoids. Funct Ecol 18:812–819

Straub CS, Finke DL, Snyder WE (2008) Are the conservation of natural enemy biodiversity and biological control compatible goals? Biol Control 45:225–237

Thies C, Tscharntke T (1999) Landscape structure and biological control in agroecosystems. Science 285:893

Tonhasca A, Byrne DN (1994) The effects of crop diversification on herbivorous insects: a meta-analysis approach. Ecol Entomol 19:239–244

Triapitsyn SV (1998) *Anagrus* (Hymenoptera: Mymaridae) egg parasitoids of *Erythroneura* spp. and other leafhoppers (Homoptera: Cicadellidae) in North American vineyards and orchards: a taxonomic review. Trans Am Entomol Soc 124:77–112

Tscharntke T, Klein AM, Kruess A, Steffan-Dewenter I, Thies C (2005) Landscape perspectives on agricultural intensification and biodiversity – ecosystem service management. Ecol Lett 8:857–874

Tscharntke T, Bommarco R, Clough Y, Crist TO, Kleijn D, Rand TA et al (2007) Conservation biological control and enemy diversity on a landscape scale. Biol Control 43:294–309

UC IPM (2010a) How to manage pests: grape. http://www.ipm.ucdavis.edu/PMG/selectnewpest.grapes.html

UC IPM (2010b) Relative toxicities of insecticides and miticides used in grapes to natural enemies and honey bees. http://www.ipm.ucdavis.edu/PMG/r302900111.html#REFERENCE

Van Emden H (1965) The role of uncultivated land in the biology of crop pests and beneficial insects. Sci Hortic 17:121–136

Vattala HD, Wratten SD, Phillips CB, Wäckers FL (2006) The influence of flower morphology and nectar quality on the longevity of a parasitoid biological control agent. Biol Control 39:179–185

Wäckers FL (2004) Assessing the suitability of flowering herbs as parasitoid food sources: flower attractiveness and nectar accessibility. Biol Control 29:307–314

Wäckers FL, Rijn PCJ, Bruin J (2005) Plant-provided food for carnivorous insects: a protective mutualism and its applications. Cambridge University Press, Cambridge

Warner K (2007a) Agroecology in action: extending alternative agriculture through social networks. The MIT Press, Cambridge

Warner KD (2007b) The quality of sustainability: agroecological partnerships and the geographic branding of California wine grapes. J Rural Stud 23:142–155

Williams L III, Martinson TE (2000) Colonization of New York vineyards by *Anagrus* spp. (Hymenoptera: Mymaridae): overwintering biology, within-vineyard distribution of wasps, and parasitism of grape leafhopper, *Erythroneura* spp. (Homoptera: Cicadellidae), eggs. Biol Control 18:136–146

Wilson LT, Pickett CH, Flaherty D, Bates T (1989) French prune trees: refuge for grape leafhopper parasite. Calif Agric 43:7–8

With KA, Pavuk DM, Worchuck JL, Oates RK, Fisher JL (2002) Threshold effects of landscape structure on biological control in agroecosystems. Ecol Appl 12:52–65

Wratten S, Van Emden H, Thomas M (1998) Within-field and border refugia for the enhancement of natural enemies. In: Pickett CH, Bugg RL (eds) Enhancing biological control: habitat management to promote natural enemies of agricultural pests. University of California Press, Berkeley, pp 375–403

Wright LC, James DG (2007) *Anagrus* spp. (Hymenoptera: Mymaridae) reared from plants collected during winter in south central Washington and north central Oregon. J Entomol Soc Br Columbia 104:17–24

Zucca G, Smith DE, Mitry DJ (2009) Sustainable viticulture and winery practices in California: What is it, and do customers care? Int J Wine Res 2:189–194

Chapter 9
Management of Phytophagous Mites in European Vineyards

Carlo Duso, Alberto Pozzebon, Serge Kreiter, Marie-Stéphane Tixier, and Marco Candolfi

9.1 Introduction: Problems with Mites in Vineyards

The first infestations of spider mites in European vineyards were detected in the second half of the nineteenth century after the invasion of powdery mildew, downy mildew and phylloxera from North America. Serious problems associated with *Tetranychus urticae* Koch were detected in Italy and Austria and local outbreaks of *Panonychus ulmi* (Koch) were recorded. At that time the eriophyid *Colomerus vitis* (Pagenstecher) was known but not considered important, probably because sulfur was largely used to control grape diseases. Two additional species were described at the beginning of the twentieth century: the spider mite *Eotetranychus carpini* (Oudemans) and the eriophyid *Calepitrimerus vitis* (Nalepa). At that time, problems with mites injurious to grapes were negligible but the situation suddenly changed after World War II. Most of these problems were immediately associated with the extensive use of chlorinated hydrocarbon insecticides and later, by ethylene-bis-dithiocarbamate fungicides (EBDC). Three spider mite species were involved in the outbreaks: *E. carpini*, *P. ulmi* and *T. urticae* (Rambier 1958; Rota 1962; Zangheri and Masutti 1962). Most researchers of the time thought that mite outbreaks were due to the detrimental effects of pesticides on natural enemies of spider mites.

C. Duso (✉) • A. Pozzebon
Department of Agronomy, Food, Natural resouces, Animals and Environment, University of Padova, viale dell'Università, 16, 35020 Legnaro (PD), Italy
e-mail: carlo.duso@unipd.it

S. Kreiter • M.-S. Tixier
Montpellier SupAgro, UMR CBGP, Campus International de Baillarguet, CS 30016-34988 Montferrier sur Lez cedex, France

M. Candolfi
Innovative Environmental Services (IES) Ltd., Benkenstrasse 260, CH-4108 Witterswil, Switzerland

N.J. Bostanian et al. (eds.), *Arthropod Management in Vineyards: Pests, Approaches, and Future Directions*, DOI 10.1007/978-94-007-4032-7_9,

However, there was little evidence to support this fact. Meanwhile, according to Chaboussou (1965) infestations were caused by changes in plant physiology induced by organic pesticides, favorable to mite demographic parameters ('trophobiosis'). Chaboussou (1965) and others also tried to classify the most common pesticides according to their effect on spider mites. However, contradictory results were reported and no scientific proof could be obtained, and trophobiosis remained a matter of great debate among entomologists.

Studies carried out by Ivancich Gambaro (1973) supported the hypothesis that mite outbreaks were caused by the disruption of the balance between predatory and phytophagous mites exerted by organic pesticides. Infestations by *E. carpini* were prevalent in vineyards treated with some EBDC fungicides. She compared copper fungicides and EBDC fungicides and obtained a huge increase of predatory mite densities (i.e. *Kampimodromus aberrans* (Oudemans)) in the plots treated with the copper fungicides with a subsequent decline of *E. carpini* densities. Unfortunately, these findings were not appreciated immediately by the scientific community. Ad hoc experiments provided additional evidence for the role of fungicides in mite outbreaks (Girolami 1981). These and other experiments emphasized the need to elucidate the side effects of the most common pesticides on beneficials as a first step in the implementation of IPM strategies. Apart from EBDC fungicides, a number of insecticides (organophosphates, carbamates and pyrethroids) have also been associated with mite outbreaks because of their adverse effects on predatory mites (Girolami 1981). Resistance of predatory mites to organophosphates (OPs) and carbamates was reported by Duso et al. (1992) and Vidal and Kreiter (1995) and to pyrethroids by Bonafos et al. (2007). Moreover, these insecticides were repellent to spider mites, favored their dispersal and affected their demographic parameters.

Mechanisms involved in mite outbreaks in the nineteenth century remain unclear, since organic pesticides were developed later and only copper and sulfur fungicides were used. Why had mite problems declined in the subsequent decades? Copper fungicides have negligible effect on mites, whereas sulfur has acaricidal properties. It is likely that resistance to sulfur was first developed in spider mites, favoring their outbreaks, and later in predatory mites, re-establishing the balance between phytophagous and predatory mites. Finally, the appearance of organic pesticides led to tremendous changes in mite communities (Girolami 1981).

In the late 1980s and early 1990s, the reduced use of broad-spectrum pesticides, the adoption of action threshold (AT) levels for spider mites, and especially the increased knowledge of the potential of macropredators and predatory mites, allowed a reduction of spider mite outbreaks. Currently, the damage caused by *P. ulmi* and *T. urticae* in grapes seems to be low, whereas damage caused by *E. carpini* is increasing in some countries (e.g., Italy, Spain, Greece). Factors promoting this trend are not clear, and problems with *Cal. vitis* are more frequent in young than in old vineyards. It has been suggested that strains resistant to sulfur and other fungicides with acaricidal properties have been selected in nurseries, but no clear evidence for this phenomenon has been presented.

9.2 Biology, Ecology and Economic Importance of Spider Mites

9.2.1 *Panonychus ulmi*

The European red mite, *P. ulmi* is considered the most important spider mite in European vineyards with *T. urticae* and *E. carpini*. It is widespread in temperate regions and less damaging in hot regions. It overwinters as eggs laid on 1- or 2-year old branches (usually at the insertion of shoots) from late summer onwards. Eggs hatch from April to May and the first generation develops at sprouting. The first generation can be completed in about 20 days. The number of generations per year is from four to six in central Europe to six to seven in southern Europe (Girolami 1981). Demographic parameters of *P. ulmi* have been estimated using apple leaves as a plant substrate. At 15 and 21°C, the intrinsic rates of population increase (r_m) are respectively 0.056 and 0.134 and the reproductive rate (R_0) ranges from 10.1 to 17.4 (Herbert 1981a). Mean generation time ranges from 18 to 41 days depending on temperature (Ramsdell and Jubb 1979; Herbert 1981a).

The spatio-temporal distribution of *P. ulmi* has been thoroughly studied. The head of the trunk (Double Guyot system) is the preferred oviposition site, nevertheless a significant percentage of winter eggs are also laid in the basal part of branches (Baillod et al. 1989b; Candolfi et al. 1992a). At the beginning of the growing season, the spider mites prefer basal leaves and later, leaves located at the middle of shoots. Spider mite densities are higher on main leaves than on lateral leaves (Candolfi et al. 1992a). *Panonychus ulmi* seasonal dynamics seem to follow a general pattern despite differences between regions. After winter egg hatch, the spider mite population remains at low levels until June and begins to increase in early summer, peaking in mid- or late summer (Wermelinger et al. 1992; Duso and Pasqualetto 1993). Its development is positively affected by high temperature (Ramsdell and Jubb 1979; Wermelinger et al. 1992), leaf nitrogen and iron content, and carbohydrate and amino acid levels (Schreiner 1984). Rainfall, high relative humidity (RH) (Wermelinger et al. 1992), as well as high K^+ leaf phenolic content (Schreiner 1984) reduce the rate of population increases. Infestation by *P. ulmi* can vary significantly among grapevine cultivars, and leaf hairiness promotes spider mite population increases (Schreiner 1984; Rilling 1989).

Panonychus ulmi feeds on the spongy mesophyll and palisade cells causing leaf discoloration, and at high densities causes a decrease in CO_2 exchange rates (Rilling and Düring 1990). However, net photosynthesis of field grown grapevines was not affected at infestation levels of up to 3,500 mite-days per leaf (peak of about 60 mites per leaf) (Candolfi et al. 1993b). Kast (1989) did not observe a negative effect of *P. ulmi* on berry yield on cv Müller except for a slight decrease on must soluble solids at peak mite densities at about 56 mites per leaf. In Switzerland, yield, berry quality, and plant growth of three cultivars (Müller x Thurgau, Gewürztraminer, and Pinot Noir) were not affected by spider mite infestation levels of up to 36,000 mite-days per leaf (Candolfi et al. 1993a). Candolfi et al. (1993a, b) stated that *P. ulmi* is less harmful than *T. urticae* to grapes. It should be stressed that most of these experiments were conducted on potted vines and thus the response of plants to mites in field conditions may be different. Moreover, the effect of an infestation should be evaluated even in the subsequent growing season. Girolami (1981) proposed an AT of 10–20 motile forms per leaf (depending on phenological stage) based on a series of field trials.

Baillod et al. (1989b) suggested sampling 5–10 segments (between the fifth and the eighth internodes) of 1-year old branches during the winter to forecast the risk of *P. ulmi* infestations. Baillod et al. (1979) proposed a sequential sampling plan to monitor *P. ulmi* abundance during the growing season by using a presence-absence statistic. However, at spider mite densities approaching the AT, approximately 100% of the leaves were infested, and thus this method was unable to differentiate between plots that required treatment from those that did not. Girolami (1981) showed that the dispersion of *P. ulmi* could be described by the negative binomial distribution and proposed a sampling strategy for spider mite based on the cumulative number of spider mites on leaves.

9.2.2 *Tetranychus urticae*

The twospotted spider mite, *T. urticae*, is a polyphagous species that colonizes grapes as well as several weeds occurring in vineyards (Schruft et al. 1979; Arias and Nieto 1981; Boller et al. 1985). Demographic parameters have been calculated on apple leaf substrates by Herbert (1981b) at 15°C, 18°C, and 21°C. At these temperatures, the intrinsic rates of increase (r_m) are respectively 0.069, 0.156, and 0.372; the net reproduction rates (R_0) 20.8, 38.4, and 58.1; and the mean generation times 44.0, 23.4, and 10.9 days. The life table of *T. urticae* was constructed by rearing mites on grape leaves at 21–28°C (Wang and DaHan 2006). The intrinsic rate of increase, net reproduction rate, mean generation time were 0.207, 39.59, 17.80 days, respectively. These data would suggest a lower performance of *T. urticae* on grape than on apple leaves. The potential of this mite to increase its population level increased with high temperatures (at 33–38°C the egg to egg period is about 5.5 days) associated with low RH rates (Van de Vrie et al. 1972).

Females are usually the overwintering stage. However, in dry, warm climatic conditions (e.g. southwestern Spain), juveniles can also be found overwintering on weeds at protected sites (Arias and Nieto 1980). In spring, females can resume their activity before bud swell, and mites have to colonize weeds where the first generation appears (Arias and Nieto 1981). In central Europe (e.g. Switzerland) females come out of diapause in March and April and most of them move to weeds (Baillod et al. 1989a). In Spain mites disperse from weeds to grapevine leaves in spring (Arias and Nieto 1991). In Switzerland and France, dispersal from weeds to grapevine occurs in early summer (Baillod et al. 1989a; Kreiter et al. 1991). In Spain, *T. urticae* can develop up to 15 generations per year (Arias and Nieto 1991). In late summer, the spider mites enter into diapause, depending on temperature and photoperiod, and move to overwintering sites.

In southwestern Spain, most of overwintering females can be found on vines (Arias and Nieto 1980), whereas in Germany most of them occur on fallen leaves and weeds (Schruft et al. 1979). The migration from weeds to grapevines depends on natural weed senescence, weed distribution and diversity. The use of some herbicides favors the migration of spider mites to vines (Boller et al. 1985; Kreiter et al. 1991).

The spatio-temporal distribution of *T. urticae* depends on the time at which mites colonize vines. In Spain, *T. urticae* infests the first leaves of the shoot in spring and then moves to the upper part of the shoot. After berry set, the spider mites are located on the top main leaves of the primary shoots and on the first lateral leaves (Arias and Nieto 1978). In Switzerland, where *T. urticae* migrates to the vine in June, most of the spider mites stay on the lower six to seven main leaves from the shoot base (Baillod et al. 1989a).

Tetranychus urticae feeds predominantly on the spongy mesophyll and palisade cells on the leaf undersurface, causing a loss of chlorophyll, leaf discoloration, and leaf drop (Candolfi 1991). On defoliated vines, *T. urticae* can feed on any green tissue including shoots and berries (Arias and Nieto 1981). In semi-dry climatic conditions, serious damage can be observed from July onwards, and the level of grapevine defoliation is related to early intense feeding (Arias and Nieto 1978, 1981). Feeding intensity of *T. urticae* dramatically increases from 10 to 35°C (Candolfi et al. 1991) which explains why spider mite population cause more damage in hot climate areas. Additional factors affecting the degree of damage are the time and duration of infestation, cultivar, and soil moisture. Some grape cultivars can compensate for feeding damage by increasing the lateral leaf area (Candolfi 1991). Under water stress, a compensatory growth of the lateral shoots does not occur. Increasing mite densities result in significant reduction of the net photosynthesis, transpiration, as well as stomatal and mesophyll conductance (Candolfi et al. 1992b). At 6,000 mite-days per leaf, net CO_2 assimilation of cv Pinot Noir leaves was reduced by 21.3–52.2%, depending on leaf position and phenological stage. The effect of *T. urticae* on gas exchange was more pronounced during bloom. Growth of cv Pinot Noir grapevines was only slightly affected by *T. urticae* feeding (Candolfi 1991). Total plant dry weight was reduced by 12.6% when 7,000 mite-days per leaf had been accumulated during the growing season. These cumulative densities caused adverse effects to 1–2-year old wood and roots but not to yield and berry quality (Candolfi 1991). The effects of *T. urticae* on grape physiology were not affected by cultivar.

In Spain, Arias and Nieto (1983) reported an average reduction in sugar content of 0.05° Brix for each 10% defoliation. Yield was reduced by 0.3 kg/vine during an experimental year. Arias and Nieto (1981) also showed that sugar content of the juice declined by about 0.05° Brix for each week of defoliation. On the other hand, studies conducted in the Czech Republic showed that a density of 16.8 *T. urticae* per leaf in August caused a reduction in yield (0.2 kg/vine on cv Sauvignon) but not in sugar content of the must (Hluchy and Pospisil 1992). The variability found in these studies is reflected by the different ATs suggested for *T. urticae* on grapevines, varying from 20% to 40% of infested leaves (Baillod et al. 1993) to 2,000–3,000 mite-days per leaf under non-stress conditions (Candolfi 1991). Arias and Nieto (1983) proposed an economic injury level based on the knowledge of the relationship between damage symptoms and defoliation, the effects of spider mites on yield and quality, the cost of protection, and the efficacy of control.

Two sampling strategies for *T. urticae* have been proposed to forecast the risk of infestation. Baillod et al. (1989a) suggested using the percentage of infested leaves. However, at spider mite densities higher than 12 motile forms per leaf, 90–100% of the leaves become infested, making this method inaccurate for predicting high spider mite densities. It may be more appropriate to monitor the percentage of leaves with feeding symptoms (Arias and Nieto 1978). The percentage of infested vines can be used as an alternative sampling method because *T. urticae* colonize vines after spending some time on weeds that may change their spatial distribution in the field from year to year (Arias and Nieto 1980).

9.2.3 *Eotetranychus carpini*

The yellow spider mite, *E. carpini*, is an important mite pest of southern Europe including southern Switzerland (Mathys and Tencalla 1960; Zangheri and Masutti 1962; Villaronga et al. 1991). The life cycle of *E. carpini* is completed on grapevine, where females overwinter under bark crevices. In April, females move to the new vegetation. Populations persist on basal leaves in spring and spread along the shoots after bloom. The mite can complete from four to six (Switzerland) to seven to eight generations (Italy and France) (Rambier 1958; Mathys and Tencalla 1960; Zangheri and Masutti 1962). In late summer, females migrate from the leaves to overwintering sites.

Laboratory studies showed that the lower thermal threshold for development is about 7 ± 1°C and the optimal temperature for development and reproduction is 26 ± 1°C (Bonato et al. 1990). Life history and demographic parameters were studied by Castagnoli et al. (1989) at 25 ± 1°C. Developmental times (egg to adult) require 12.03 days. Mean longevity is about 34.72 days, and total fecundity is 53.11 eggs. Bonato et al. (1990) conducted similar studies at 15°C, 19.8°C, 22.7°C, 26°C and 30.3°C, and 60% RH. Developmental times decreased from 28.4 to 9.7 days as the temperature increased from 15 to 30.3°C. Total fecundity ranged from 28.6 (15°C) to 46.4 eggs (26°C), but declined to 29.3 at 30°C. The intrinsic rate of natural increase went from 0.058 at 15°C to 0.153 at 26°C but diminished to 0.130 at 30.3°C. Relative humidity also affected demographic parameters: a RH of 30% was optimal for the species (O. Bonato, unpubl. data).

Little is known about the spatial distribution of overwintering females. In spring *E. carpini* colonizes leaf undersurfaces, in particular leaf portions located along the main veins. Therefore, most eggs are laid at the conjunction of the midrib and veins or along the veins. During the growing season, the mites are more concentrated on leaves located in midshoots (Baillod et al. 1979).

In southern France, populations peak generally between mid-July and the beginning of August, with a second peak in September (Laurent and Agulhon 1987).

This pattern can be observed in northern Italy where spring infestations can be also very serious. Moreover, a summer decline of mite populations has been recorded in vineyards when *E. carpini* coexists with *P. ulmi* (Duso and Pasqualetto 1993). Population growth is promoted by moderate temperature, low relative humidity, and pubescent leaf undersurfaces (Duso and Vettorazzo 1999).

Eotetranychus carpini feeds on spongy mesophyll and palisade cells. Yellow (on some cultivars reddish) spots appear on leaves, mainly along the main veins. As these symptoms spread to the entire surface, leaves progressively dry and abscise. This symptom is easily observed in mid- and late summer and involves first basal leaves, and then leaves located in the middle part of the shoots. The influence of *E. carpini* on vine physiology, plant growth, yield, and berry quality has not been studied. Nervertheless, empirical ATs of 60–70% infested leaves in spring, and 30–45% infested leaves in summer has been proposed by Baillod et al. (1979). Girolami (1981) proposed an AT of 5–10 motile forms per leaf.

Sequential sampling plans based on the % of leaves infested by one or more mites were proposed by Baillod et al. (1979). A method taking into account the % of leaves showing spider mite feeding injury was proposed by Laurent and Agulhon (1987).

9.2.4 *Tetranychus mcdanieli*

The McDaniel mite, *Tetranychus mcdanieli* McGregor, is a serious pest of deciduous fruit trees and grapes in North America (Jeppson et al. 1975). In Europe, it was first detected in 1981 in the Champagne region in France, where it seemed to be localized (Rambier 1982). Females overwinter under the bark, and they feed in spring first on vine buds and then on leaves. In North America the development from egg to adult requires about 8 days and seven to nine generations can be completed within a year (Jeppson et al. 1975). Using bean leaves as substrate, Tanigoshi et al. (1975) reported the net reproductive rate and intrinsic rate of natural increase as 75.1 and 0.201 at 25°C. In France, depending on climatic conditions, populations peak in July and August. High temperature, low humidity, and cultivar attributes are key factors promoting the growth of mite populations (Rambier 1982).

Mites spin profuse webbing, and at high infestation levels, leaves can mat together. The first adverse effect of feeding is chlorophyll loss, followed by leaf discoloration and leaf drop. Leaf discoloration and leaf drop are more pronounced in hot and dry seasons. The impact of this spider mite on grapevine physiology and yield parameters is unknown. In France, ATs proposed for *T. urticae* are used to keep Mcdaniel mite populations at non-damaging levels.

9.2.5 *Tetranychus turkestani*

The strawberry spider mite, *Tetranychus turkestani* Ugarov & Nikolski, is a serious pest of annual crops, as well as of deciduous fruits in several regions (Jeppson et al. 1975).

In Western Europe, it has been recorded in France, Spain, and Portugal. Overwintering females can be detected in litter under loose bark. In spring, mites feed on weeds before infesting grapevines. In vineyards, population densities reach moderate levels in July. Mites feed mainly on the leaf undersurfaces, inducing typical damage. Webbing produced by large populations cause leaves and stems to mat together. The influence of *T. turkestani* on grapevine physiology and productivity has not been investigated. Action thresholds reported for *T. urticae* are used in France.

9.3 Biology, Behavior and Economic Importance of Eriophyoid Mites

9.3.1 Calepitrimerus vitis

The grape rust mite, *Cal. vitis*, is a serious pest in Europe and elsewhere (Duso and de Lillo 1996; Bernard et al. 2005). The biology of *Cal. vitis* has been intensively studied in Portugal, Italy and Spain (Carmona 1973; Liguori 1988; Perez-Moreno and Moraza 1998). Females (so-called deutogynes) overwinter under the bud scales or under the bark at the insertion between 1- and 2-year old branches. In spring they lay eggs at the base of the shoots, and the progeny remains on this substrate. Protogynes appear in May, and later mites migrate along shoots to colonize leaf undersurfaces. At the end of June, mites can be found in the new buds at the leaf axils. Population densities increase in July to peak in midsummer. From August onwards, deutogynes appear again and their proportion over protogynes increase progressively. This phase depends on environmental conditions and infestation levels. Studies carried out in Italy confirmed trends in seasonal abundance reported above.

This species develops from two to four generations per year in central Europe, but its potential increases in southern Europe and in Moldavia (Duso and de Lillo 1996). Perez-Moreno and Moraza (1998) emphasized the significance of dry and hot summers on mite abundance.

Calepitrimerus vitis feeding may cause death of the growth point of buds, stunted shoot growth, shortened shoot internodes, development of lateral shoots, leaf deformation, reduced cluster size, and flower drop (Carmona 1973, 1978). In Swiss vineyards, large infestations occurring in summer caused the bronzing of leaves and prevented berries to ripen. The berries appeared brown and cracked (Baillod and Guignard 1986). Since the 1980s severe infestations have been reported from Italy, France, Switzerland, and Germany, especially in young vineyards where sprouting had been dramatically affected. In older vineyards, the incidence of damage has been much milder (Duso and de Lillo 1996).

Erroneous diagnoses of grape rust mite infestations have also been made, e.g., the so-called ‘Restricted Spring Growth’ (RSG) in Australia. Bernard et al. (2005) showed that leaf and shoot distortions, as well as retarded shoot growth in early spring were caused by large populations of overwintered *Cal. vitis* females. They also showed that shoot length reduction depended on the cultivar. Another syndrome, was the so-called Short Shoot Syndrome (SSS). It was associated with high grape rust mite population densities in the US (Walton et al. 2007). The symptoms were malformed leaves, short and angled shoots in spring, and scar tissues with bronzed leaves in summer. However, the real cause was bunch necrosis in early season.

The increase in summer temperatures in recent years seems to favor *Cal. vitis* outbreaks in central Europe. Under these conditions, the pest can develop more generations and find more opportunities to overwinter. Pesticide use is likely a major factor promoting *Cal. vitis* outbreaks. Eriophyoid mites can show resistance to fungicides with acaricidal properties and possibly to acaricides. However, this phenomenon has not been investigated in depth (Van Leeuwen et al. 2010a). Furthermore, a number of fungicides and insecticides used in viticulture can be detrimental to predators that prey on *Cal. vitis*. In this context, the use of sulfur (especially of dust applications) needs to be evaluated (Bernard et al. 2005; Walton et al. 2007).

Wind and human activities are the most important factors in mite dispersal (Duffner et al. 2001). A high number of Eriophyidae (32.1% *Cal. vitis*) were trapped in a wind chamber during experiments carried out in Germany. A significant number of *Cal. vitis* were detected on the clothes and hands of people working in vineyards and nurseries. Cultural practices can also assume a great significance in mite dispersal.

In a vineyard comprising several cultivars, Castagnoli et al. (1997) reported *Cal. vitis* to be the dominant phytophagous mite species on cv Canaiolo. This cultivar is characterized by high pubescence on the underside of leaves. Predatory mites, mainly *Typhlodromus exhilaratus* Ragusa, were also more abundant on this cultivar. Duso and Vettorazzo (1999) reported that the predatory mites *K. aberrans* and *Typhlodromus pyri* Scheuten were more abundant on grape cultivars with hairy leaves. Consequently, eriophyid mites decreased to lower densities on cultivars with hairy leaves. The susceptibility of a cultivar to *Cal. vitis* may be related to temperature requirements. Cultivars requiring relatively high temperatures to grow in spring (e.g. cv Cabernet Sauvignon) are frequently regarded more susceptible (Bernard et al. 2005).

The effect of *Cal. vitis* on grapevine physiology and performance has not been studied in depth. According to Carmona (1978) a density of 20–25 mites per bud during winter result in severe symptoms on shoots the following spring. Therefore, she considered a density of 20 overwintering females per bud as the AT in Portugal. In Switzerland, the presence of overwintering females is considered sufficient to apply acaricides (Baillod and Guignard 1986). In the Czech Republic an AT of 280 *Cal. vitis* per leaf had been proposed for the summer (Hluchy and Pospisil 1992). In Australia, high mite densities in spring did not affect significantly vine fruitfulness and yield parameters at berry set (Bernard et al. 2005). These highly varied results reported in experiments aimed at assessing the effect of infestation on yield parameters stress the lack of experimental data on the impact of *Cal. vitis* on vine physiology.

Forecasting the risk associated to *Cal. vitis* is fundamental, especially in newly planted vineyards. Baillod and Guignard (1986) proposed to sample buds located at the base (first to second) and in the middle (fourth to eighth) of 1-year old branches, as they are the most infested. Perez-Moreno and Moraza (1998) reported that the central buds were the most infested in Portugal. Washing techniques to estimate more accurately *Cal. vitis* populations in summer and winter have been described by Perez-Moreno and Moraza (1998).

9.3.2 Colomerus vitis

The grape erineum mite, *Col. vitis*, occurs in the most important viticultural areas of the world. Three strains of this species, the erineum, bud, and leaf curl strains, have been distinguished (e.g. Smith and Stafford 1948). The erineum strain is very common in Europe while records on the presence of the other strains need to be confirmed (Duso and de Lillo 1996). Carew et al. (2004) used molecular markers to gain further insight into the identity of the erineum and the bud strains in Australia. Patterns of genetic variation observed using PCR-RFLP of ITS 1 revealed the existence of two distinct species. Microsatellite markers showed an extensive genetic differentiation between the two populations (species) even at micro-geographical scales.

The biology and ecology of the erineum strain has been investigated in Switzerland (Mathez 1965), as well as in California (Smith and Stafford 1948). Females overwinter under the external bud scales, less frequently under the bark crevices of branches. Their activity starts with bud swell, when females induce the first felty patches ('erinea') on leaves, where reproduction takes place and the progeny find optimal food. In early spring the infestations are concentrated on basal leaves. Later, mite populations migrate along the shoots reaching the apical leaves, where erinea are induced in large numbers. Lateral shoots are easily infested. In late summer, eriophyoids start to migrate to overwintering sites but a part of the population can persist on recently developed leaves. Little is known about the biological and demographic parameters of *Col. vitis*. According to Mathez (1965), the first generation is completed in about 25 days and up to seven generations per year can be completed.

The biology of the bud strain has been studied in California (Kido and Stafford 1955; Smith and Schuster 1963), while there is little information for Europe. The life cycle is completed within the buds. Females overwinter inside the buds, and at bud swelling, they feed and reproduce on the primordia. Following shoot growth, mites move under the buds at the leaf axils or crawl from bud to bud. The mites penetrate into the new buds and feed on the primordia. Primordial clusters are infested from July onwards. A generation requires about 20 days to be completed. In California, oviposition starts in early spring, peaks in July and declines in late summer.

In South Africa, mite reproduction reaches high levels a few weeks after bud burst (Dennill 1986). The number of generations fluctuates among different countries.

The leaf curl strain has been reported occasionally in Europe. Its life cycle is thought to be similar to that of the erineum strain (Duso and de Lillo 1996).

Fig. 9.1 Injury on leaf caused by *Colomerus vitis* (Photo by Carlo Duso)

The erineum mite produces blisters on upper surface of the leaf, which correspond to patches of trichomes on the under surface of the leaf (Fig. 9.1). The blister color can vary from light green to yellow or dark red. Young leaves are more susceptible to damage than older leaves. Blisters can be also induced on inflorescences. The distribution of erinea along the shoots of vines not pruned during the growing season is concentrated on two areas: the first three to four leaves in spring and from the 9th and 10th leaves onwards in summer (Mathez 1965). On vines pruned after blossom, lateral shoots are frequently infested probably because of their susceptibility. The erineum strain is not considered economically important. In the 1960s, infestations increased in Switzerland, because of changes in pruning techniques, that allowed a large development of lateral shoots (Mathez 1965). A detailed study on the impact of the erineum strain on yield and berry quality was carried out in Switzerland (Linder et al. 2009). When an infestation was at maximal abundance, 3% of leaves had more than 60% of the leaf area damaged. However, no significant effects on transpiration rates were measured. The chlorophyll index was not affected, while photosynthesis and stomatal conductance rates slightly decreased on heavily infested leaves. Plant damage was not correlated with overwintering mite populations. The authors concluded that chemical control was not needed against the erineum mite, confirming empirical observations carried out in several countries.

Feeding of buds by mites causes cell hypertrophy (polyps) and the development of scar tissues. Mite infestations cause alteration of shoots and inflorescences, death of terminal buds, development of lateral shoots, death of overwintering buds, delayed berry maturation, and reduction of sugar content (Smith and Stafford 1948;

Kido and Stafford 1955; Smith and Schuster 1963). More recently, Bernard et al. (2005) observed 100–500 mites per bud on dead apical meristems of primary and secondary buds. The infestation of the first 10 buds were more pronounced and these were suggested as sampling units (Kido and Stafford 1955).

A yield reduction of about 50% had been associated with high infestations in California (Kido and Stafford 1955). However, the systematics in a number of studies and the related yield losses in California were controversial (Barnes 1958, 1992; Smith and Schuster 1963). In Europe infestations are occasional.

The curl mites feed on the leaf veins causing alteration in leaf growth (Smith and Stafford 1948). The economic importance of this strain is considered to be negligible, probably because mites are vulnerable to predators and pesticides (Schwartz 1986).

9.4 Another Mite Injurious to Grapes

Citrus flat mite, *Brevipalpus lewisi* (McGregor) (Tenuipalpidae) feeds on more than 30 host plants, including the grapevine (Jeppson et al. 1975). Problems with *B. lewisi* have been recorded in Spanish vineyards where four to five overlapping generations can be completed (Arias and Nieto 1985; Rodríguez et al. 1987). At 22°C and 40% RH, a generation is completed in 47.7 days and fecundity is about 14.2 eggs per female (Buchanan et al. 1980). Females overwinter on branches, under the bark and on shed leaves. They leave their overwintering sites after bud burst and lay eggs on the basal internodes of the developing shoots where the first generation takes place. Later, the mites disperse from the shoots to stems, inflorescences and leaves. In summer most of the population is located on leaves. Population density peaks in September (Arias and Nieto 1985; Rodríguez et al. 1987). Mite feeding causes weakening of the shoots and sometimes their death (Raikov and Nachev 1965). Injury to inflorescences results in abnormal berry development and berry shatter (Arias et al. 1986). Raikov and Nachev (1965) have reported yield losses amounting to 30%. Arias et al. (1986) have reported a correlation between the number of internodes showing feeding symptoms on shoots and a decrease in cluster weight of a table grape cultivar. A better correlation was determined between feeding symptoms on the inflorescences and final cluster weight. No economic injury levels have yet been established for *B. lewisi*. If the internodes in the cluster region and the inflorescences show feeding symptoms at blossom, then chemical control is recommended.

9.5 Biological Control

Spider mites and rust mites are classic examples of pesticide-induced pests. Their occurrence is negligible in minimally disturbed vineyards and markedly reduced in organic vineyards. This is because of the presence of several macropredators and

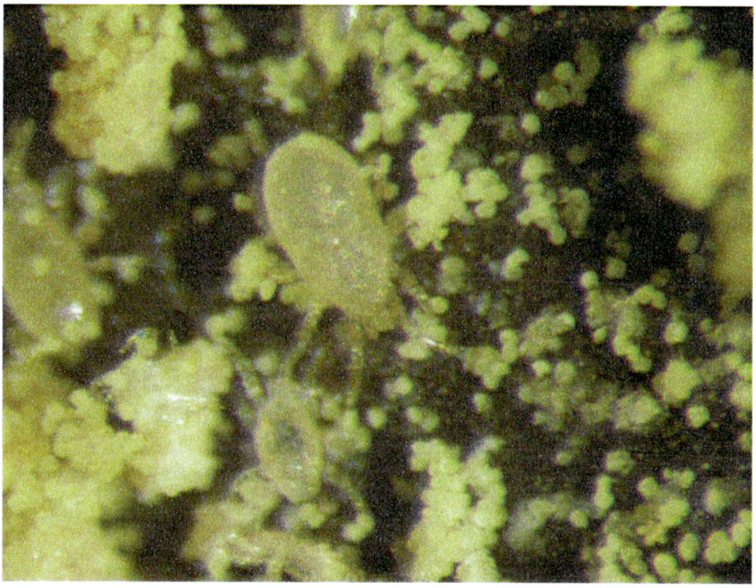

Fig. 9.2 The predatory mite *Kampimodromus aberrans* (Photo by Stefano Vettore)

predatory mites, especially those belonging to the family Phytoseiidae. The most important macropredators in European vineyards belong to the Thysanoptera (Aelothripidae), Heteroptera (Anthocoridae and Miridae), Coleoptera (Coccinellidae and Staphilinidae), and Neuroptera (Chrysopidae). Macropredators have relatively long developmental times, high reproductive potential and voracity, their impact is often significant at high mite infestation levels, but they do not persist when prey are scarce (Girolami 1981; Duso and Pasqualetto 1993). Predatory mites can build stable populations in vineyards representing the most important component of biocontrol resources (Ivancich Gambaro 1973). The Stigmaeidae have a potential in controlling eriophyids (Duso and de Lillo 1996) but their response to spider mite build up is slower when compared to that exhibited by phytoseiids. Therefore, the latter have attracted most interest of researchers.

Surveys have revealed a great diversity of phytoseiids in European vineyards. As an example, more than 20 phytoseiid species have been recorded in Italian and French vineyards. Among them, *T. pyri* dominates in central Europe, *K. aberrans* (Fig. 9.2) in southern Europe where *Amblyseius andersoni* (Chant), *T. exhilaratus*, and *Phytoseius finitimus* Ribaga are also present (Castagnoli 1989; Villaronga et al. 1991; Papaioannou-Souliotis et al. 1999; Tixier et al. 2000b; Kreiter et al. 2000; Ragusa di Chiara and Tsolakis 2001). It should be stressed that these species are generalist predators, type III, after McMurtry and Croft (1997). They can persist when prey densities decline by surviving on alternative foods. Some of them show a narrow association with the host plant, a capacity to regulate their densities and to compete with other predators (McMurtry and Croft 1997). Knowledge of their feeding habits and relationships with plants is fundamental for conservation biological control tactics.

The above mentioned phytoseiid species develop and reproduce on spider mites (e.g. *P. ulmi* and *E. carpini*) and show comparable demographic parameters when fed with eriophyoids (e.g. *Col. vitis*) and pollen (e.g. Ragusa 1979; Castagnoli and Liguori 1986; Castagnoli et al. 1989; Duso and Camporese 1991). *Amblyseius andersoni* exhibits shorter developmental times and higher oviposition rates than *T. pyri* and *K. aberrans* (Duso and Camporese 1991; Schausberger 1992). However, these two species respond better than *A. andersoni* to spider mite population increases (Duso 1989). In this respect, Ivancich Gambaro (1973), Duso and Vettorazzo (1999) have reported *T. pyri* and *K. aberrans* to be effective in controlling *E. carpini*. Some ecological factors negatively affect *A. andersoni* and *T. pyri* in vineyards. *Amblyseius andersoni* populations decline when RH levels and prey availability decrease. In contrast, *T. pyri* is affected by high temperatures and interspecific competition. The ability of *K. aberrans* to out-compete *T. pyri* and *A. andersoni*, has been repeatedly observed in northeastern Italy and it is likely due to its capacity to survive in conditions of prey scarcity, as well as its tolerance to high temperatures and low RH rates (Girolami et al. 1992; Duso and Pasqualetto 1993; Duso and Vettorazzo 1999). *Typhlodromus pyri* and *K. aberrans* are favored on grape cultivars with hairy leaf undersurfaces. In contrast, *A. andersoni* shows an opposite trend (Duso 1992; Kreiter et al. 2002; Loughner et al. 2008). Leaf morphology strongly affects colonization patterns when these species co-occur and can mediate interspecific competition (Duso and Vettorazzo 1999).

Typhlodromus pyri, *K. aberrans* and *A. andersoni* have been proven to be effective in preventing rust mite infestations (Duso and de Lillo 1996; Perez-Moreno and Moraza 1998; Duso and Vettorazzo 1999).

Typhlodromus exhilaratus exhibited a higher intrinsic rate of increase on *E. carpini* and pollen than on *P. ulmi* (Castagnoli and Liguori 1986; Castagnoli et al. 1989). The adaptation of *T. exhilaratus* to low RH is a fundamental requirement to colonize vineyards of southern Europe (Liguori and Guidi 1995). This species also proved to have potential to manage *Cal. vitis* (Liguori 1988).

The economic importance of *P. finitimus* is controversial. *Phytoseius finitimus* may have some potential to manage *P. ulmi* (Duso and Vettorazzo 1999) but its ability has not been compared with that of other phytoseiids. Its populations reach high densities on grape cultivars with hairy leaf undersurfaces, where it competes successfully with other predatory mites (Duso and Vettorazzo 1999).

Generalist phytoseiids occurring in vineyards can consume tenuipalpids, eriophyids, tydeids, winterschmidtiids, and young stages of thrips or coccids as alternative prey. The presence of these preys can enhance phytoseiid performance, and the role of eriophyids in this context is crucial (Liguori 1988; Engel and Ohnesorge 1994a, b). However, it should be stressed that some alternative prey (e.g., *Cal. vitis*) is difficult to manage. Tydeids are very common in vineyards and their role as alternative prey for phytoseiids found in Europe has been suggested. Unfortunately tydeids are preyed upon, notably by *Paraseiulus talbii* (Athias-Henriot), which is not a biocontrol agent of phytophagous mites in vineyards (Camporese and Duso 1995). Generalist phytoseiids can also consume immature or adult phytoseiid stages (McMurtry and Croft 1997).

Pollen is a fundamental food source for generalist predatory mites. Grape leaves are excellent pollen traps, and their analysis shows definite trends in pollen fluctuations. In northern Italy, pollen densities are relatively high at sprouting, after bloom and in late summer (Duso et al. 1997). When pollen is abundant on grape leaves, the population size of *T. pyri*, *K. aberrans* and *A. andersoni* increases (Engel and Ohnesorge 1994b; Duso et al. 1997, 2004b).

Plant pathogenic fungi can constitute additional food resources for generalist phytoseiids. Grape downy mildew (GDM) (*Plasmopara viticola* (Berkeley and Curtis ex. de Bary) Berlese and De Toni), and grape powdery mildew (GPM) (*Uncinula necator* (Schweinitz) Burrill), are the most important worldwide grape pathogens. The spread of GDM foliar symptoms can promote population increases of *A. andersoni* and *T. pyri* (Duso et al. 2003) as these species can develop and reproduce on GDM in the laboratory (Pozzebon and Duso 2008). Demographic parameters of *A. andersoni* on GDM reached higher values when compared to those related to *T. pyri* and thus GDM can mediate interspecific competition between the two species. GDM spread has positive effects on tydeid populations and consequently to their predator *P. talbii* (Duso et al. 2005). The implications of GDM spread on biological control are unclear. In contrast with GDM, GPM is a supplementary food for *A. andersoni* and *T. pyri* (Pozzebon et al. 2009). Interactions between powdery mildew and phytoseiids have been poorly documented, despite the economic importance of this pathogen in the Mediterranean region.

The management of non-prey foods for generalist phytoseiids is fundamental for conservation biocontrol tactics. Hedgerows can provide pollen (and phytoseiids) for contiguous vineyards. In an experimental farm comprising a hedgerow and a contiguous vineyard, elders produced large amounts of pollen allowing for population increases of the predatory mite *Euseius finlandicus* (Oudemans). However, the importance of this strategy for grapes is unclear (Duso et al. 2004b).

Pollen produced by species belonging to the Poaceae family is a major component of wind-borne pollen in vineyards of various regions, and Poaceae are widely used as cover crops. Experiments conducted in northern Italy showed that a reduction in grass mowing increased pollen densities on grape canopy and consequently phytoseiid densities (Girolami et al. 2000).

The management of plant pathogenic fungi is more risky and controversial. GDM infections benefit *A. andersoni* over *T. pyri*, but *T. pyri* is more effective than *A. andersoni* at managing tetranychids on grapes. On the other hand, *A. andersoni* is unique in colonizing new vineyards potentially exposed to rust mite infestations.

9.6 Predatory Mite Augmentation: Should We Prefer to Release Resistant Strains?

Phytoseiids may require a long time to colonize vineyards. Predatory mites occurring on natural vegetation surrounding vineyards can find problems to settle in vineyards if certain pesticides are used (Tixier et al. 1998, 2000b). In large-scale field experiments conducted in the 1980s, phytoseiids (i.e., *K. aberrans*, *T. pyri*, and

A. andersoni) were released in Italian, French, and Swiss vineyards with varied results. In some situations, phytoseiids settled successfully, but were eradicated by OPs and EBDC pesticides. In the 1990s, the grape protection scenario was dramatically affected in some European countries by the spread of grapevine yellows (Flavescence dorée) and a consequent increase of insecticide used to control the leafhopper vector *Scaphoideus titanus* Ball. These changes in pesticide use suggested releasing OP resistant phytoseiid strains, mainly of *T. pyri* and *K. aberrans*. The release of a strain of *K. aberrans*, collected in a vineyard in northern Italy treated with EBDC fungicides and organophosphates, gave excellent results in terms of spider mite control and predatory mite persistence.

Implications of genetic polymorphism on the success of phytoseiid releases have been poorly researched. The above mentioned strain of *K. aberrans* has been the subject of multiple releases in Italy and is associated with a restricted polymorphism. More studies are required on this topic for a better long term biological control program.

9.7 Natural Vegetation and Phytoseiid Mite Management in Neighboring Vineyards

Plant diversity in uncultivated areas surrounding crops is assumed to increase natural enemy density and diversity. These uncultivated areas are supposed to provide alternative, stable and durable food and habitat resources for natural enemies. The role of uncultivated areas on Phytoseiidae communities has been investigated in Europe from the 1980s (e.g. Boller et al. 1988). Since the late 1980s, densities of phytoseiid mites have increased in vineyards with integrated pest management programs (e.g., Girolami et al. 1992; Kreiter et al. 2000). In such vineyards, only pesticides known to be innocuous to natural enemies are used in order to avoid phytoseiid extinction and promote their abundance. The increase in phytoseiid mite densities observed in such vineyards was so high and rapid, that it could not only be explained by a development of the phytoseiid mites naturally present in the vineyards, and it was assumed that phytoseiid mites immigrated from the surroundings and 'colonized' the vine plots. Nevertheless, many questions were raised. Where did these phytoseiid mites come from? Does the uncultivated vegetation surrounding vineyards shelter phytoseiid mites? Which species? Do these phytoseiid mites disperse in the vineyards? At what density? How do they perform such dispersal? Do all the species found in uncultivated areas disperse? To what distance is dispersal possible? What happens once the mites arrive in the crops? Do they settle well? Only a few studies have dealt with the process of colonization of phytoseiid mites. The most important and complete studies were conducted in France, and focused on the two prevailing species *K. aberrans* in southern France, and *T. pyri* in the rest of France.

Phytoseiid mites have been reported from uncultivated areas surrounding crops, especially vineyards (e.g., Boller et al. 1988; Tixier et al. 1998; Ragusa di Chiara and Tsolakis 2001; Duso et al. 2004a; Barbar et al. 2005). However, in a crop protection framework, the objective would be to promote the phytoseiid species commonly encountered in vine fields and known for their ability to efficiently control mite pests. Depending on each geographical and climatic situation, crop management conditions and plant diversity, the phytoseiid mite density and diversity could be different. An overview of studies dealing with phytoseiid mite abundance, in vegetation surrounding crops, emphasizes two key factors that explain phytoseiid mite abundance and diversity.

The first factor is plant composition in the neighboring areas. Although Phytoseiidae are predators, their abundance is affected by plant leaf characteristics (Walter 1996; Kreiter et al. 2002). High trichome and domatia densities are correlated with high predator abundances (Karban et al. 1995; Walter 1996). This applies especially for *K. aberrans* (Kreiter et al. 2002). Indeed, if some phytoseiids prefer pubescent leaves, others (e.g., *Euseius* spp.) prefer smooth plant substrates (Duso et al. 2004a, b). Furthermore, on the same leaf or the same plant, different leaf structures could be present, providing different favorable habitats for different phytoseiid species, thus enhancing phytoseiid diversity. The presence of pollen, liquids (nectars) on plants could affect phytoseiid mite, abundance, and diversity. Plants with pubescent leaves retain more pollen and consequently harbor higher phytoseiid mite populations (Kreiter et al. 2002). Because of these narrow relationships between phytoseiid development and their plant niche, their diversity and densities would be directly correlated to floristic diversity and to the abundance of suitable plants for their development.

The second factor is that the main species found on natural vegetation are often prevalent in the neighboring vine crops (Tixier et al. 1998, 2006; Barbar et al. 2005). One of the hypotheses proposed to explain such observations is that the agricultural practices applied on vine crops could affect and 'select' the neighboring fauna of phytoseiids. However, there is no study focusing on the factors that could affect such a biodiversity homogenization in agroecosystems.

Phytoseiid mites can disperse by aerial and ambulatory means. Dispersal is known to be linked to declining habitat conditions, i.e. overcrowding, poor quality and quantity of food, intraspecific and interspecific competition, and plant senescence. In some experiments, phytoseiid mites were caught both in ambulatory and aerial traps within vineyards (Tixier et al. 1998, 2000a, 2002a). However, aerial dispersal contributed to a greater extent to mite movement into crops (Tixier et al. 1998, 2000a). In these experiments, the species mainly found in uncultivated vegetation (i.e. *K. aberrans*) was also the prevalent species dispersing in the vine crops. Males, females and immatures were found in the traps, and the main vector of this seemingly random dispersal was found to be the wind. Dry north and northwest winds, ranging from 14 to 31 km/h, were found to be favorable for dispersal of phytoseiid mites. Lastly, the phytoseiid mites were mainly caught in areas close to the greatest neighboring reservoir (higher phytoseiid density). Traps located over 90 m from this reservoir captured a few phytoseiid mites (Tixier et al. 1998, 2000a). Thus, an environment rich in phytoseiid mites, especially *K. aberrans,* would constitute a reservoir source from which random aerial dispersal supported by wind would occur. Furthermore, it seems that the number of mites trapped represents only a very small portion of the population present in the source area, confirming the low dispersal ability of *K. aberrans*. Generalizations about phytoseiid dispersal are difficult to make. Phytoseiid mites would disperse from reservoir source areas in high density if the surrounding vegetation would provide them with food and shelter. The proximity of reservoir sources to the surrounding vegetation seems to be a key factor to ensure high dispersal rates. The mechanisms of dispersal (airbone and ambulatory) differ from one species to another. In southern France, the arrival and settlement of *K. aberrans* and exchanges between uncultivated and cultivated vineyards were investigated by molecular typing. The study showed genetic similarities for females originating from (1) *Quercus pubescens* Willdenow and *Celtis australis* L. in adjacent woody areas, and (2) various parts of the experimental plot. However, differentiation in genetic patterns of predatory mites from these two groups was observed both in spring and summer (Tixier et al. 2002a, b). Thus, despite the great number of migrants that arrived in the experimental plots and originated from woody areas, it seemed that the gene flow between these two agroecosystem compartments was low (Tixier et al. 1998, 2000b). This molecular typing helps to explain the results obtained from earlier artificial releases of *K. aberrans* that showed that released mites rarely settle well. Several hypotheses have been proposed to explain such results: (i) low dispersal ability (distance or frequency) of *K. aberrans* (Tixier et al. 1998), (ii) highly aggregated distribution of this species, and (iii) mortality after arrival (Tixier et al. 2000b). Pesticide applications, for instance, can affect the settlement of phytoseiid mites arriving from uncultivated surrounding areas. Management of pesticides within the plots, based on knowledge on their side effects (Sentenac et al. 2002) is thus one of the key factors to enhance colonization success. However, it has been shown that populations present in the neighboring vegetation, even if more susceptible than the ones present in vineyards, present also a high level of resistance to some insecticides and fungicides (Tixier and Kreiter 2003; Barbar et al. 2008). Thus, other factors would affect phytoseiid mite settlement in the vine plots. Could reproduction incompatibilities between migrants and specimens already present in the plot explain the low gene flow observed? Do host plant shifts during dispersal of phytoseiids affect settlement in the vineyards? Finally, what about the survival of mites during dispersal? Answers to these questions are still needed to enhance the management of phytoseiids naturally occurring in vineyard surroundings. For this, sampling schemes along with molecular typing have to be developed. Currently, molecular typing is difficult because of the absence of adequate molecular markers for phytoseiids especially microsatellites.

Vineyard surroundings are not the unique source of floristic and faunistic diversity in agroecosystems. Phytoseiid mites can also be present on co-planted trees or in the inter-row vegetation (ground cover vegetation). Thus, studies are needed to determine the impact of the presence of such floristic biodiversity inside vine plots on phytoseiid mite density, diversity and the success of biological control programs (Barbar et al. 2006; Liguori et al. 2011).

Lastly, all these studies require knowledge of phytoseiid mite taxonomy to ensure a correct identification of the species present in order to assess their real control efficiency. Morphologically related species could be present in the same environment requiring an in-depth taxonomic expertise (Tixier et al. 2006).

9.8 Chemical Control

Since biological control is a successful strategy to manage mites injurious to grapes, chemical control would be necessary only when predatory mites have been eradicated and their re-colonization is difficult. Chemical control may be required even in young vineyards to control *Cal. vitis* populations. The following technical aspects should be considered when selecting and applying pesticides for the control of phytophagous mites: (1) action thresholds based or economic injury levels (EILs), (2) pesticide spectrum (ovicide, larvicide, nymphicide and/or adulticide), (3) modes of action taking into account resistance management strategies, and (4) side effects on beneficial arthropods.

A number of acaricides (e.g. dicofol, propargite, clofentezine, bromopropylate, hexythiazox, abamectin) and insecticides with acaricidal activity (e.g. endosulfan and flufenoxuron) have been extensively used in the past. More recently these compounds have been replaced by mitochondrial electron transport inhibitors (METIs) (e.g. pyridaben and fenazaquin) and mite growth inhibitors (e.g. spirodiclofen, spiromesifen, acequinocyl, bifenazate and etoxazole). Additional information on acaricides can be found in Dekeyser (2005), Van Leeuwen et al. (2010b), and Bostanian et al. (Chap. 4). Updated information on the registration status of a specific pesticide in Europe may be obtained at the EU Pesticides Database (2009).

Acaricide application schedule can be followed according to grapevine phenological stages. The use of acaricides to control overwintering forms was frequent in the past but no longer in recent protocols. From bud swell to bud burst or later, *Cal. vitis* can be reduced by using mineral oils or sulfur. From bud burst to two- or three-unfolded leaves, acaricides may be applied to control eggs or juveniles of *P. ulmi* (mite growth regulators should be preferred). When inflorescences are clearly visible different developmental stages of mites occur and compounds active against motile stages and mite growth inhibitors can be applied. Similar strategies are followed in early or midsummer.

The development of resistant strains is an important issue for chemical control. This is of particular importance for spider mites (EU Pesticide Database 2009). Spider mites rapidly develop resistant strains and the principal genetic and ecological factors involved are: (1) arrhenotokous reproduction, (2) high reproductive rate, (3) inbreeding, (4) a very short life cycle, and (5) lack of dispersal of the phytophagous mites from treated areas and a low level of immigration from untreated areas (Cranham and Helle 1985; Croft and Van de Baan 1988). In Europe, resistance in *P. ulmi* and *T. urticae* has been reported in fruit orchards (e.g. Nauen et al. 2001) and other crops (Van Leeuwen et al. 2010b). To our knowledge, only one scientific report on resistance of spider mites occurring in European vineyards has been published. Rambier (1964) reported that a *P. ulmi* strain from vineyards located in Bas-Languedoc (France) was resistant to demeton-methyl. With the exception of this report, the hypothesis of the occurrence of resistant strains of spider mites has been drawn from field observations, but no characterization of resistance has been carried out.

9.9 Side Effects of Pesticides on Mite Communities

In the last three decades increasing interest has been channeled to the development of IPM tactics in vineyards and the study of side-effects of pesticides on mite communities.

In 1974, the International Organization of Biological Control (IOBC) created the working group 'Pesticides and Beneficial Arthropods', which focused its activities on the evaluation of the effects of pesticides on beneficial arthropods. Since 1996, data on the effects of plant protection products to non-target arthropods are included in mandatory procedures for registration of every active ingredient, and authorization of use in the European Union (Directive 91/414/EEC). This directive requires that the effects of plant protection products have to be tested on *T. pyri,* a predatory mite of major importance in viticulture. So far, many studies have been published on pesticide effects to non-target arthropods. Information on specific pesticides used in European viticulture can be obtained in various publications (e.g., Hassan et al. 1994; Kreiter et al. 1997; Sterk et al. 1999; Sentenac et al. 2002), and several more recent references in Bostanian et al. (Chap. 4).

The compatibility of pesticides with the conservation of beneficial mites is a key aspect in IPM, so that only harmless pesticides should be used in the program. If no harmless products are available, products that are slightly or moderately harmful to beneficials can be used with some restrictions (fewer applications per season). Harmful products should never be considered in IPM.

Historically, chlorinated hydrocarbons, carbamates, and pyrethroids proved to be detrimental to predatory mites at different levels. Pyrethrins are also highly toxic to predatory mites, but because of their low persistence, re-colonization of treated plants often takes place. IGRs, neonicotinoids and molt accelerating compounds are characterized by low to moderate toxicity on predatory mites (Kreiter et al. 1997; Sentenac et al. 2002; Tosi et al. 2006). Among fungicides, several studies highlighted the detrimental effects of EBDCs and dinocap (Angeli and Ioriatti 1994; Kreiter et al. 1998; Pozzebon et al. 2002). Wettable sulfur is generally less toxic than dust sulfur (Papaioannou-Souliotis et al. 1998). The more recently developed fungicides are generally characterized by good selectivity (Sentenac et al. 2002; Nicotina et al. 2004; Bostanian et al. 2009).

Field observations suggested the occurrence of resistant strains among predatory mites, and laboratory studies have definitely shown resistance levels (e.g. Posenato 1994; Auger et al. 2004, 2005; Bonafos et al. 2007).

Leaf morphology can reduce negative effects of non-selective pesticides. *Typhlodromus pyri* appeared to tolerate repeated applications of mancozeb in varieties characterized by heavy pubescent leaf undersurfaces (Pozzebon et al. 2002). Grape downy and powdery mildews provide alternative food for generalist predatory mites. Fungicides can exert a direct impact on phytoseiids due to their toxicity, but they can also have an indirect effect reducing food availability for predatory mites. For example, the presence of foliar symptoms of downy mildew is associated with faster colonization by predators of grapevines treated with non-selective insecticides than with those not showing foliar symptoms (Pozzebon and Duso 2010).

References

Angeli G, Ioriatti C (1994) Susceptibility of two strains of *Amblyseius andersoni* Chant (Acari: Phytoseiidae) to dithiocarbamate fungicides. Exp Appl Acarol 16:669–679

Arias GA, Nieto JC (1978) Observaciones sobre la biología de la "araña amarilla" (*Tetranychus urticae* Koch) en las viñas de "Tierra de Barros" (Badajoz) durante 1976 y 1977. Ministerio de agricultura, Dirección general de la producción agraria. Serv Def Plagas Insp Fitopatol 31/78:1–45

Arias GA, Nieto JC (1980) Observaciones sobre la biología de la "araña amarilla (*Tetranychus urticae* Koch) en los viñedos de "Tierra de Barros" (Badajoz) durante 1978 y 1979. Ministerio de Agricultura, Dirección general de la producción agraria. Serv Def Plagas Insp Fitopatol 4/80:1–39

Arias GA, Nieto JC (1981) Observaciones sobre la biología de la "araña amarilla" (*Tetranychus urticae* Koch) y correlación entre sintomas y pérdidas en una viña de "Tierra de Barros" (Badajoz) durante 1980. Serv Def Plagas Insp Fitopatol 9/81:1–41

Arias GA, Nieto JC (1983) Estimación de las pérdidas producidas por la "araña amarilla común" (*Tetranychus urticae* Koch) en "Tierra de Barros" (Badajoz) y propuesta de un umbral de tolerancia económica. Bol Serv Def Plagas Insp Fitopatol 9:227–252

Arias GA, Nieto JC (1985) El "ácaro de la roña" (*Brevipalpus lewisi* McGregor), nuevo parásito de la vid en España: invernación, colonización de las cepas y prospección en la comarca de Guareña (Badajoz). Bol Serv Def Plagas Insp Fitopatol 11:193–203

Arias GA, Nieto JC (1991) La "araña amarilla común" (*Tetranychus urticae* Koch) en "Tierra de Barros" – I: Bioecología. *Vitivinicultura* II:38–41

Arias GA, Nieto JC, Merino RS (1986) Estimación de las pérdidas que produce el "ácaro de la roña" (*Brevipalpus lewisi* McGregor). In: Proceedings of the VIII Jornadas de Viticultura y Enología de "Tierra de Barros", Almendralejo, Spain, p 7, 5–9 May 1986

Auger P, Bonafos R, Kreiter S (2004) Mancozeb resistance patterns among *Kampimodromus aberrans* and *Typhlodromus pyri* (Acari: Phytoseiidae) strains from French vineyards. Can Entomol 136:663–673

Auger P, Bonafos R, Kreiter S, Delorme R (2005) A genetic analysis of mancozeb resistance in *Typhlodromus pyri* (Acari: Phytoseiidae). Exp Appl Acarol 37:83–91

Baillod M, Guignard E (1986) Nouveaux dégâts de l'acariose bronzée et du court-noué parasitaire dus à *Calepitrimerus vitis* (Nalepa) (Acari, Eriophyidae) en 1984 et 1985. Rev Suisse Vitic Arboric Hortic 18:285–288

Baillod M, Bassino JP, Piganeau P (1979) L'estimation du risque provoqué par l'acarien rouge (*Panonychus ulmi* Koch) et l'acarien des charmilles (*Eotetranychus carpini* Oudemans) en viticulture. Rev Suisse Vitic Arboric Hortic 11:123–130

Baillod M, Antonin P, Mittaz C (1989a) Migrations, estimation des populations et nuisibilité de l'acarien jaune commun *Tetranychus urticae* Koch dans la viticulture valaisanne. Rev Suisse Vitic Arboric Hortic 21:179–183

Baillod M, Guignard E, Mischler M, Vautier P (1989b) Contrôle des oeufs de l'acarien rouge (*Panonychus ulmi* Koch) sur le bois de taille en viticulture et prévision du risque. Rev Suisse Vitic Arboric Hortic 21:7–14

Baillod M, Charmillot PJ, Jermini M, Meylan A, Vallotton R, Antonin P et al (1993) Protection intégrée et stratégies de lutte contre les ravageurs de la vigne. Rev Suisse Vitic Arboric Hortic 25:23–29

Barbar Z, Tixier M-S, Kreiter S, Cheval B (2005) Diversity of Phytoseiid mites in uncultivated areas adjacent to vineyards: a case study in south of France. Acarologia 45:145–154

Barbar Z, Tixier M-S, Cheval B, Kreiter S (2006) Effects of agroforestry on phytoseiid mite communities (Acari: Phytoseiidae) in vineyards in the south of France. Exp Appl Acarol 40:175–188

Barbar Z, Tixier M-S, Kreiter S (2008) Assessment of pesticide susceptibility in *Typhlodromus exhilaratus* and *Typhlodromus phialatus* strains in a vineyard in the south of France. Exp Appl Acarol 42:95–105

Barnes MM (1958) Relationships among pruning time response, symptoms attributed to grape bud mite, and temporary early season boron deficiency in grapes. Hilgardia 28:193–224

Barnes MM (1992) Grape erineum mite. In: Flaherty DL, Christensen LP, Lanini WT, Marois JJ, Philips PA, Wilson LT (eds) Grape pest management, vol Publication no. 3343. University of California, Division of Agriculture and National Resources, Oakland, pp 262–264

Bernard MB, Horne PA, Hoffmann AA (2005) Eriophyoid mite damage in *Vitis vinifera* (grapevine) in Australia: *Calepitrimerus vitis* and *Colomerus vitis* (Acari: Eriophyidae) as the common cause of the widespread 'Restricted Spring Growth' syndrome. Exp Appl Acarol 35:83–109

Boller EF, Janser E, Zahner S, Potter C (1985) Kann Herbizideinsatz im Weinbau Spinnmilbenprobleme verursachen? Schweiz Z Obst Weinbau 121:527–531

Boller EF, Remund U, Candolfi PM (1988) Hedges as a potential sources of *Typhlodromus pyri*, the most important predatory mite in vineyards of northern Switzerland. Entomophaga 33:249–255

Bonafos R, Serrano E, Auger P, Kreiter S (2007) Resistance to deltamethrin, lambda-cyhalothrin and chlorpyriphos-ethyl in some populations of *Typhlodromus pyri* Scheuten and *Amblyseius andersoni* (Chant) (Acari: Phytoseiidae) from vineyards in the south west of France. Crop Prot 26:169–172

Bonato O, Cotton D, Kreiter S, Gutierrez J (1990) Influence of temperature on the life-history parameters of the yellow grapevine mite *Eotetranychus carpini* (Oudemans) (Acari: Tetranychidae). Int J Acarol 16:241–245

Bostanian NJ, Thistlewood HMA, Hardman JM, Racette G (2009) Toxicity of six novel fungicides and sulphur to *Galendromus occidentalis* (Acari: Phytoseiidae). Exp Appl Acarol 47:63–69

Buchanan GA, Bensgston M, Exley EM (1980) Population growth of *Brevipalpus lewisi* McGregor (Acarina: Tenuipalpidae) on grapevines. Aust J Agric Res 31:957–965

Camporese P, Duso C (1995) Life history and life table parameters of the predatory mite *Typhlodromus talbii*. Entomol Exp Appl 77:149–157

Candolfi MP (1991) Einfluss von *Tetranychus urticae* Koch und *Panonychus ulmi* Koch (Acari) auf Gaswechsel. PhD dissertation, Wachstum, Ertrag und Traubenqualität der Weinrebe

Candolfi MP, Keller M, Boller EF (1991) Mite-load function improves precision of feeding damage estimation in *Tetranychus urticae*. Entomol Exp Appl 58:289–293

Candolfi MP, Boller EF, Wermelinger B (1992a) Spatio-temporal distribution of *Panonychus ulmi* Koch (Acari, Tetranychidae) on Guyot-trained grapevines. J Appl Entomol 114:244–250

Candolfi MP, Boller EF, Wermelinger B (1992b) Influence of the twospotted spider mite, *Tetranychus urticae*, on the gas exchange of Pinot noir grapevine leaves. Vitis 31:205–212

Candolfi MP, Wermelinger B, Boller EF (1993a) Influence of the red mite (*Panonychus ulmi* KOCH) on yield, fruit quality and plant vigour of three *Vitis vinifera* varieties. Vitic Enol Sci 48:161–164

Candolfi MP, Wermelinger B, Boller EF (1993b) Photosynthesis and transpiration of "Riesling x Sylvaner" grapevine leaves as affected by the European red mite (*Panonychus ulmi* Koch) (Acari: Tetranychidae) feeding. J Appl Entomol 115:233–239

Carew ME, Goodisman MAD, Hoffmann AA (2004) Species status and population genetic structure of grapevine eriophyoid mites. Entomol Exp Appl 111:87–96

Carmona MM (1973) Notes on the bionomics of *Calepitrimerus vitis* (Nal.) (Acarina: Eriophyidae). In: Proceedings of the 3rd International Congress of Acarology, Prague, Czechoslovakia, pp 197–199, 31 Aug–6 Sept 1971

Carmona MM (1978) *Calepitrimerus vitis* (Nalepa), responsàvel pela «Acariose da videira». 1- Notas sobre a morfologia, biologia e sintomatologia. Agron Lusit 39:29–56

Castagnoli M (1989) Recent advances in knowledge of the mite fauna in the biocenoses of grapevine in Italy. In: Cavalloro R (ed) Influence of environmental factors on the control of grape pest, diseases and weeds. A. A. Balkema, Rotterdam, pp 169–180

Castagnoli M, Liguori M (1986) Tempi di sviluppo e ovideposizione di *Typhlodromus exhilaratus* Ragusa (Acarina: Phytoseiidae) allevati con vari tipi di cibo. Redia 69:361–368

Castagnoli M, Amato F, Monagheddu M (1989) Osservazioni biologiche a parametri demografici di *Eotetranychus carpini* (Oud.) (Acarina: Tetranychidae) e del suo predatore *Typhlodromus exhilaratus* Ragusa (Acarina: Phytoseiidae) in condizioni di laboratorio. Redia 72:545–557

Castagnoli M, Liguori M, Nannelli R (1997) Le popolazioni degli Acari nei vigneti inerbiti del Chianti: confronto tra cultivar. Redia 80:15–31

Chaboussou F (1965) La multiplication par voie trophique des tétranyques à la suite des traitements pesticides. Relations avec les phénomènes de résistance acquise. Boll Zool Agrar Bachic 7:144–184

Cranham JE, Helle W (1985) Pesticide resistance in Tetranychidae. In: Helle W, Sabelis MW (eds) Spider mites: their biology, natural enemies and control, vol 1B. Elsevier, Amsterdam, pp 405–421

Croft BA, Van de Baan HE (1988) Ecological and genetic factors influencing evolution of pesticide resistance in Tetranychid and Phytoseiid mites. Exp Appl Acarol 4:277–300

Dekeyser M (2005) Acaricide mode of action. Pest Manag Sci 61:103–110

Dennill GB (1986) An ecological basis for timing control measures against the grape vine bud mite *Eriophyes vitis* Pgst. Crop Prot 5:12–14

Duffner K, Schruft G, Guggenheim R (2001) Passive dispersal of the grape rust mite *Calepitrimerus vitis* Nalepa 1905: (Acari, Eriophyoidea) in vineyards. J Pest Sci 74:1–6

Duso C (1989) Role of *Amblyseius aberrans* (Oud.), *Typhlodromus pyri* Scheuten and *Amblyseius andersoni* (Chant) (Acari, Phytoseiidae) in vineyards. I. The effects of single or mixed phytoseiid population releases on spider mite densities (Acari, Tetranychidae). J Appl Entomol 107:474–492

Duso C (1992) Role of *Amblyseius aberrans* (Oud.), *Typhlodromus pyri* Scheuten and *Amblyseius andersoni* (Chant) (Acari, Phytoseiidae) in vineyards. III: influence of variety characteristics on the success of *A. aberrans* and *T. pyri* releases. J Appl Entomol 114:455–462

Duso C, Camporese P (1991) Developmental times and oviposition rates of predatory mites *Typhlodromus pyri* and *Amblyseius andersoni* (Acari: Phytoseiidae) reared on different foods. Exp Appl Acarol 13:117–128

Duso C, de Lillo E (1996) 3.2.5. Grape. In: Lindquist EE, Sabelis MW, Bruin J (eds) Eriophyoid mites: their biology, natural enemies and control, vol 6, World crop pests. Elsevier Science, Amsterdam, pp 571–582

Duso C, Pasqualetto C (1993) Factors affecting the potential of phytoseiid mites (Acari: Phytoseiidae) as biocontrol agents in north Italian vineyards. Exp Appl Acarol 17:241–258

Duso C, Vettorazzo E (1999) Mite population dynamics on different grape varieties with or without phytoseiids released (Acari: Phytoseiidae). Exp Appl Acarol 23:741–763

Duso C, Camporese P, van der Geest LPS (1992) Toxicity of a number of pesticides to strains of *Typhlodromus pyri* and *Amblyseius andersoni* (Acari: Phytoseiidae). Entomophaga 37:363–372

Duso C, Malagnini V, Paganelli A (1997) Indagini preliminari sui rapporti tra polline e *Kampimodromus aberrans* (Oudemans) (Acari: Phytoseiidae). Allionia 35:229–239

Duso C, Pozzebon A, Capuzzo C, Bisol PM, Otto S (2003) Grape downy mildew spread and mite seasonal abundance in vineyards: evidence for the predatory mites *Amblyseius andersoni* and *Typhlodromus pyri*. Biol Control 27:229–241

Duso C, Fontana P, Malagnini V (2004a) Diversity and abundance of phytoseiid mites (Acari: Phytoseiidae) in vineyards and in the surrounding vegetation in northeastern Italy. Acarologia 64:31–47

Duso C, Malagnini V, Paganelli A, Aldegheri L, Bottini M (2004b) Pollen availability and phytoseiid abundance (Acari: Phytoseiidae) on natural and secondary hedgerows. BioControl 49:397–415

Duso C, Pozzebon A, Capuzzo C, Malagnini V, Otto S, Borgo M (2005) Grape downy mildew spread and mite seasonal abundance in vineyards: effects on *Tydeus caudatus* and its predators. Biol Control 32:143–154

Engel R, Ohnesorge B (1994a) Die Rolle von Ersatznahrung und Mikroklima im System *Typhlodromus pyri* Scheuten (Acari, Phytoseidae) – *Panonychus ulmi* Koch (Acari, Tetranychidae) auf Weinreben I. Untersuchungen im Labor. J Appl Entomol 118:129–150

Engel R, Ohnesorge B (1994b) Die Rolle von Ersatznahrung und Mikroklima im System *Typhlodromus pyri* Scheuten (Acari, Phytoseidae) – *Panonychus ulmi* Koch (Acari, Tetranychidae) auf Weinreben I. Freilandversuche. J Appl Entomol 118:224–238

EU Pesticides Database (2009) Active substances. http://www.ec.europa.eu/sanco_pesticides/public/index.cfm

Girolami V (1981) Danni, soglie di intervento, controllo degli acari della vite. In: Proceedings of the III incontro su La difesa integrata della vite, Latina, Italy, pp 111–143, 3–4 Dec 1981

Girolami V, Picotti P, Coiutti C (1992) Ruolo determinante del fitoseide *Amblyseius aberrans* (Oud.) nel controllo degli acari fitofagi. L'Informatore Agrario 68:65–69

Girolami V, Borrella E, Di Bernardo A, Malagnini V (2000) Influenza positiva sui Fitoseidi della fioritura del cotico erboso. L'Informatore Agrario 51:71–73

Hassan SA, Bigler F, Bogenschütz H, Boller E, Brun J, Calis JNM et al (1994) Results of the sixth joint pesticide testing programme of the IOBC/WPRS-Working Group "Pesticides and Beneficial Organisms". Entomophaga 39:107–119

Herbert HJ (1981a) Biology, life tables, and intrinsic rate of increase of the European red mite, *Panonychus ulmi* (Acarina: Tetranychidae). Can Entomol 113:65–71

Herbert HJ (1981b) Biology, life tables and innate capacity for increase of the twospotted spider mite, *Tetranychus urticae* (Acarina: Tetranychidae). Can Entomol 113:371–378

Hluchy M, Pospisil Z (1992) Damage and economic injury levels of eriophyid and tetranychid mites on grapes in Czechoslovakia. Exp Appl Acarol 14:95–106

Ivancich Gambaro P (1973) Il ruolo del *Typhlodromus aberrans* Oudemans (Acarina Phytoseiidae) nel controllo biologico degli Acari fitofagi del Veronese. Boll Zool Agrar Bachic 11:151–165

Jeppson L, Keifer H, Baker E (1975) Mites injurious to economic plants. University of California Press, Berkeley/Los Angeles

Karban R, English-Loeb G, Walker MA, Thaler J (1995) Abundance of phytoseiid mites on *Vitis* sp.: effects of leaf hairs, domatia, prey abundance and plant phylogeny. Exp Appl Acarol 19:189–197

Kast WK (1989) Untersuchungen zur Befall-Verlust-Relation und Bekämpfungsschwelle bei der Obstbaumspinnmilbe (*Panonychus ulmi* Koch) an Reben. Dtsch Weinbau Jahr 40:199–209

Kido H, Stafford EM (1955) The biology of the grape bud mite *Eriophyes vitis* (Pgst). Hilgardia 24:119–141

Kreiter S, Brian F, Magnien C, Sentenac G, Valentin G (1991) Spider mites and chemical control of weeds: interactions. In: Dusbabek F, Bukva V (eds) Modern acarology, vol 2. Academia, Prague and SPB Academic Publishing bv, The Hague, pp 725–736

Kreiter S, Sentenac G, Weber M, Rinville C, Barthes D, Auger P (1997) Effets non intentionnels de quelques produits phytopharmaceutiques sur *Typhlodromus pyri, Kampimodromus aberrans* et *Phytoseius plumifer*. Phytoma Def Cult 493:51–58

Kreiter S, Sentenac G, Barthes D, Auger P (1998) Toxicity of four fungicides to the predaceous mite *Typhlodromus pyri* (Acari: Phytoseiidae). J Econ Entomol 91:802–811

Kreiter S, Tixier MS, Auger P, Muckensturm N, Sentenac G, Doublet B, Weber M (2000) Phytoseiid mites of vineyards in France (Acari: Phytoseiidae). Acarologia 61:77–96

Kreiter S, Tixier MS, Croft BA, Auger P, Barret D (2002) Plants and leaf characteristics influencing the predaceous mite *Kampimodromus aberrans* (Acari: Phytoseiidae) in habitats surrounding vineyards. Environ Entomol 31:648–660

Laurent JC, Agulhon R (1987) Les tétranyques de la vigne. Evolution des populations estivales et conséquences de leurs attaques sur la qualité de la récolte dans le vignoble méditerranéen. Conférence Internationale sur les Ravageurs en Agriculture, Paris, An. ANPP 6: 229–234, 1–3 décembre 1987

Liguori M (1988) Effetto di trattamenti antiparassitari diversi sulle popolazioni del fitoseide predatore *Typhlodromus exhilaratus* Ragusa e su quelle degli acari fitofagi in un vigneto del senese. Redia 71:455–462

Liguori M, Guidi S (1995) Influence of different constant humidities and temperatures on eggs and larvae of a strain of *Typhlodromus exhilaratus* Ragusa (Acari Phytoseiidae). Redia 78:321–329

Liguori M, Tixier MS, Hernandes AF, Douin M, Kreiter S (2011) Agroforestry management and phytoseiid communities in vineyards of the south of France. Exp Appl Acarol 54:1–15

Linder C, Jermini M, Zufferey V (2009) Nuisibilité de l'érinose sur le cépage Muscat. Rev Suisse Vitic Arboric Hortic 41:177–181

Loughner R, Goldman K, Loeb G, Nyrop J (2008) Influence of leaf trichomes on predatory mite (*Typhlodromus pyri*) abundance in grape varieties. Exp Appl Acarol 45:111–122

Mathez F (1965) Contribution à l'étude morphologique et biologique d'*Eriophyes vitis* Pgst., agent de l'Erinose de la vigne. Bull Soc Entomol Suisse 37:233–283

Mathys G, Tencalla Y (1960) L'acarien des charmilles (*Eotetranychus carpini* Oudemans) dans le vignoble tessinois. Rev Rom Agric Vitic Arboric 16:29–31

McMurtry JA, Croft BA (1997) Life-styles of phytoseiid mites and their roles in biological control. Annu Rev Entomol 42:291–321

Nauen R, Stumpf N, Elbert A, Zebitz CPW, Kraus W (2001) Acaricide toxicity and resistance in larvae of different strains of *Tetranychus urticae* and *Panonychus ulmi* (Acari: Tetranychidae). Pest Manag Sci 57:253–261

Nicotina M, Capone GC, Prisco A (2004) Selectivity of some fungicides used against powdery mildew in a vineyard in south Italy for phytoseiid mites (Parasitiformes, Phytoseiidae). Phytophaga 14:557–561

Papaioannou-Souliotis P, Markoyiannaki Printziou D, Tsagkarakou A, Rumbos I, Adamopoulos I (1998) Effects of different fungicides and insecticides on populations of *Phytoseius finitimus* (Ribaga) in vineyard in four regions of Greece. Redia 81:17–35

Papaioannou-Souliotis P, Markoyaiannaki-Printziou D, Rumbos I, Adamopoulos I (1999) Phytoseiid mites associated with vine in various provinces of Greece: a contribution to faunistics and biogeography, with reference to eco-ethological aspects of *Phytoseius finitimus* (Ribaga) (Acari: Phytoseiidae). Acarologia 40:113–125

Perez-Moreno I, Moraza MML (1998) Population dynamics and hibernation shelters of *Calepitrimerus vitis* in the vineyards of Rioja, Spain, with a description of a new eriophyid extraction technique (Acari: Eriophyidae). Exp Appl Acarol 22:215–226

Posenato G (1994) Popolazioni di *Amblyseius aberrans* (Oud.) resistenti ad esteri fosforici e ditiocarbammati. L'Informatore Agrario 50:41–43

Pozzebon A, Duso C (2008) Grape downy mildew *Plasmopara viticola*, an alternative food for generalist predatory mites occurring in vineyards. Biol Control 45:441–449

Pozzebon A, Duso C (2010) Pesticide side-effects on predatory mites: the role of trophic interactions. In: Sabelis MW, Bruin J (eds) Trends in acarology. Springer, Dordrecht, pp 465–469

Pozzebon A, Duso C, Pavanetto E (2002) Side effects of some fungicides on phytoseiid mites (Acari, Phytoseiidae) in north Italian vineyards. Anz Schaedlkd J Pest Sci 75:132–136

Pozzebon A, Loeb G, Duso C (2009) Grape powdery mildew as a food source for generalist predatory mites occurring in vineyards: effects on life-history traits. Ann Appl Biol 155:81–89

Ragusa S (1979) Laboratory studies on the food habits of the predacious mite *Typhlodromus exhilaratus*. In: Rodriguez JG (ed) Recent advances in acarology, vol I. Academic, London, pp 485–490

Ragusa di Chiara S, Tsolakis H (2001) Phytoseiid faunas of natural and agricultural ecosystems in Sicily. In: Halliday RB, Walter DE, Proctor HC, Norton RA, Colloff MJ (eds) Proceedings of the 10th International Congress of Acarology, CSIRO Pub, Collingwood, Australia, pp 522–529

Raikov E, Nachev P (1965) Winter control of the vine Phytoptipalpid. Rastitelna Zashchita 13:6–8

Rambier A (1958) Les tétranyques nuisibles à la vigne en France continentale. Rev Zool Agric 57:1–20

Rambier A (1964) Essais acaricides réalisés en Bas-Languedoc dans un foyer de *Panonychus ulmi* Koch (Tetranychidae) résistant au déméton. C R Acad Agric Fr 50:267–278

Rambier A (1982) Un acarien, sur vigne en Champagne, nouveau en France: *Tetranychus mcdanieli* McGregor 1931, du groupe *pacificus*. Prog Agric Vitic 99:261–266

Ramsdell DR, Jubb GL (1979) Seasonal life history of European red mite on concord grapes in Erie country, Pennsylvania. Environ Entomol 8:1018–1020

Rilling G (1989) Differential response of grapevine cultivars to European red mite (*Panonychus ulmi* Koch) – elaboration of a screening method. Vitis 28:97–110

Rilling G, Düring H (1990) Structure and function of grapevine leaves (*Vitis vinifera* L.) as affected by the European red mite (*Panonychus ulmi*). Vitis 29:27–42

Rodríguez JA, Arias A, Santiago R, Nieto JC (1987) Observaciones sobre la biología de *Brevipalpus lewisi* McGregor en viñedos de la Comarca de Guareña (Badajoz), 1984–1986. Bol San Veg Plagas 13:249–259

Rota P (1962) Osservazioni sugli Acari Tetranichidi dannosi alle piante coltivate ed ornamentali in Italia. Boll Zool Agrar Bachic 4:31–136

Schausberger P (1992) Vergleichende untersuchungen über den einfluß unterschiedlicher nahrung auf die präimaginalentwicklung und die reproduktion von *Amblyseius aberrans* Oud. und *Amblyseius finlandicus* Oud. (Acarina, Phytoseiidae). J Appl Entomol 113:476–486

Schreiner U (1984) Untersuchungen zur Anfälligkeit verschiedener Rebsorten, Pfropfkombinationen und Unterlagen gegenüber Tetranychus urticae und Panonychus ulmi. PhD dissertation, Universität Kaiserslautern, Germany

Schruft G, Mittenmüller K, Stärk OJ (1979) Die Ueberwinterung der Gemeine Spinnmilbe *Tetranychus urticae* Koch an der Rebe und ihr Auftreten im Frühjahr. Wein-Wiss 34:55–60

Schwartz A (1986) Leaf curl mite in vineyards. Viticulture and Oenology F, pp 28

Sentenac G, Bonafos R, Ruelle B, Coulon T, Escaffre P, Auger P, Kreiter S (2002) Effets non intentionnels de certains produits phytopharmaceutiques sur *Typhlodromus pyri*, *Kampimodromus aberrans* et *Phytoseius plumifer*. Phytoma 555:50–55

Smith LM, Schuster RO (1963) The nature and extent of *Eriophyes vitis* injury to *Vitis vinifera* L. Acarologia 5:530–539

Smith LM, Stafford EM (1948) The bud mite and the erineum mites of grapes. Hilgardia 18:317–334

Sterk G, Hassan SA, Baillod M, Bakker F, Bigler F, Blumel S et al (1999) Results of the seventh joint pesticide testing programme carried out by the IOBC/WPRS-Working Group 'Pesticides and Beneficial Organisms'. BioControl 44:99–117

Tanigoshi LK, Hoyt SC, Browne RW, Hogan JA (1975) Influence of temperature on population increase of *Tetranychus mcdanieli*. Ann Entomol Soc Am 68:972–986

Tixier MS, Kreiter S (2003) The dispersal of *Kampimodromus aberrans* between uncultivated areas and grape fields: does pesticide applications affect the settlement of migrants? Insect Sci Appl 23:21–29

Tixier MS, Kreiter S, Auger P, Weber M (1998) Colonization of Languedoc vineyards by phytoseiid mites (Acari: Phytoseiidae): influence of wind and crop environment. Exp Appl Acarol 22:523–542

Tixier MS, Kreiter S, Croft BA, Auger P (2000a) Colonization of vineyards by phytoseiid mites: their dispersal patterns in plot and their fate. Exp Appl Acarol 24:191–211

Tixier MS, Kreiter S, Auger P, Sentenac G, Salva G, Weber M (2000b) Phytoseiid mite species located in uncultivated areas surrounding vineyards in three French regions. Acarologia 41:127–140

Tixier MS, Kreiter S, Auger P (2002a) How can molecular data contribute to the analysis of the colonization of vineyards by *Kampimodromus aberrans*. In: Bernini F (ed) Acarid phylogeny and evolution; adaptation in mites and ticks. Kluwer Academic Publisher, Dordrecht, pp 331–340

Tixier MS, Kreiter S, Croft BA, Auger P (2002b) Colonization of vineyards by *Kampimodromus aberrans* (Oudemans) (Acari: Phytoseiidae): dispersal from surrounding plants as indicated by random amplified polymorphism DNA typing. Agric For Entomol 4:255–264

Tixier MS, Kreiter S, Cheval B, Guichou S, Auger P, Bonafos R (2006) Immigration of phytoseiid mites from surrounding uncultivated areas into a newly planted vineyard. Exp Appl Acarol 39:227–242

Tosi L, Farinazzo E, Posenato G, Girolami V (2006) Effetti collaterali di insetticidi su *Kampimodromus aberrans*. L'Informatore Agrario 62:54–56

Van de Vrie M, McMurtry JA, Huffaker CB (1972) Ecology of Tetranychid mites and their natural enemies: a review. III. Biology, ecology and pest status and host-plant relations of tetranychids. Hilgardia 41:343–432

Van Leeuwen T, Witters J, Nauen R, Duso C, Tirry L (2010a) The control of eriophyoid mites: state of the art and future challenges. Exp Appl Acarol 51:205–224

Van Leeuwen T, Vontas J, Tsagkarakou A, Dermauw W, Tirry L (2010b) Acaricide resistance mechanisms in the two-spotted spider mite *Tetranychus urticae* and other important Acari: a review. Insect Biochem Mol Biol 40:563–572

Vidal C, Kreiter S (1995) Resistance to a range of insecticides in the predaceous mite *Typhlodromus pyri* (Acari: Phytoseiidae): inheritance and physiological mechanisms. J Econ Entomol 88:1097–1105

Villaronga P, Marques J, Casanovas S, Ferragut F (1991) Les acariens phytophages et prédateurs dans les vignobles de l'Alt-Emporda (Girona, Espagne). Prog Agric Vitic 108:519–523

Walter DE (1996) Living on leaves: mites, tomenta and leaf domatia. Annu Rev Entomol 41:101–114

Walton VM, Dreves AJ, Gent DH, James DG, Martin RR, Chambers U, Skinkis PA (2007) Relationship between rust mites *Calepitrimerus vitis* (Nalepa), bud mites *Colomerus vitis* (Pagenstecher) (Acari: Eriophyidae) and short shoot syndrome in Oregon vineyards. Int J Acarol 33:307–318

Wang WJ, DaHan H (2006) Life table of *Tetranychus urticae* feeding on grape leaf in the laboratory. Chin Bull Entomol 43:851–853

Wermelinger B, Candolfi MP, Baumgärtner J (1992) A model of the European red mite (Acari: Tetranychidae) population dynamics and its linkage to grapevine growth and development. J Appl Entomol 114:155–166

Zangheri S, Masutti L (1962) Osservazioni e considerazioni sul problema degli acari della vite nelle Venezie. Riv Vitic Enol Conegliano 15:75–89

Chapter 10
A Holistic Approach to Future Management of Grapevine Phylloxera

Kevin S. Powell

10.1 Introduction: Taxonomy and Distribution

Grapevine phylloxera, *Daktulosphaira vitifoliae* (Fitch), belongs to the Phylloxeridae family. Phylloxerids are a group of gall-forming sap-sucking insects including the minor pests pecan phylloxera (*Phylloxera devastatrix* Pergande) and pear phylloxera (*Aphanostigma piri* (Cholodovski)), which live on deciduous trees and perennial fruit crops (Powell 2008) and are related to the superfamily Aphidioidea. Grapevine phylloxera, the main economically important phylloxerid, is monophagous to *Vitis* spp. (Vitaceae) and is widely recognized as the most significant insect pest of commercial European grapevines, *Vitis vinifera* L. Grapevine phylloxera was first described in 1855 on native *Vitis* spp. (Granett et al. 2001a) but its devastating effect on *V. vinifera* was not recognized until the accidental introduction of this pest to Europe in the early 1860s. After the widely reported economic impact on the European wine industry (Ordish 1972) and removal of over two million ha of grapevines (Jackson 2008), it spread to South Africa, the Middle East, Asia, and Australasia later in the nineteenth century (Boehm 1996; Campbell 2004).

Phylloxera is widely distributed in most grape-growing regions of the world (EPPO 1990). However, in some countries, including China (Du et al. 2011), Australia (Powell 2008), Russia (Frolov and David'yan 2009), and Armenia (ARD 2007), its distribution appears restricted only to some grape-growing regions. More recent detections include (1) the Yarra Valley, Australia in 2006, (2) the Ararat Valley, Armenia in 2009 (V. Keushguerian, pers. comm.), and (3) the Hunan, Shaanxi and Liaoning provinces of China in 2006–2007 (Du et al. 2011).

K.S. Powell (✉)
Department of Primary Industries, Biosciences Research Division, Rutherglen Centre, RMB 1145, Rutherglen, VIC 3685, Australia
e-mail: kevin.powell@dpi.vic.gov.au

N.J. Bostanian et al. (eds.), *Arthropod Management in Vineyards: Pests, Approaches, and Future Directions*, DOI 10.1007/978-94-007-4032-7_10,

10.2 Life Cycle

The grapevine phylloxera life cycle has recently been reviewed (Forneck and Huber 2009). Its life cycle varies depending on the genetic characteristics of both the insect and its *Vitis* host parentage, with some phylloxera genetic strains being root feeders (*radicicolae)*, some being leaf feeders (*gallicoae*), and some feeding on both parts of the grapevine. The 'classical' description of the life cycle is regarded as cyclic parthenogenesis (alternating between asexual and sexual life phases on the same host) and holocyclic (sexual) reproduction (Granett et al. 2001a). However, in some grape-growing environments, for example California, North America; northeast and central Victoria, Australia; and parts of Europe, only anholocyclic (asexual) reproduction has been reported (Corrie et al. 2003; Forneck and Huber 2009).

As the anholocyclic form causes significant economic damage, only its characteristics will be described in the rest of this chapter. Grapevine phylloxera is an abundant egg layer (Fig. 10.1a), producing up to several hundred eggs per adult female but the fecundity of the female adult will depend on genetic strain, rootstock food source, and environmental conditions. Eggs hatch within 10 days into first instar nymphs (Fig. 10.1b), commonly referred to as 'crawlers' due to their dispersive

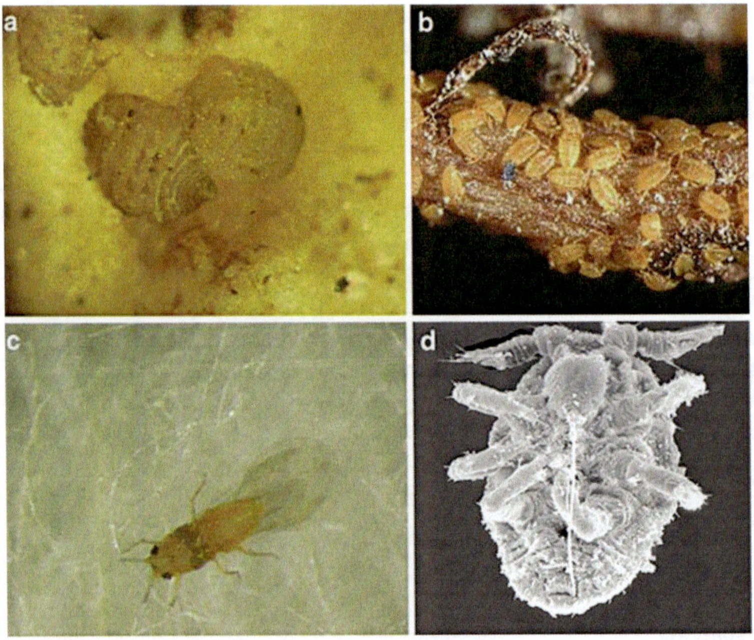

Fig. 10.1 (**a**) Eggs being oviposited by adult apterous *Daktulosphaira vitifoliae*. (**b**) First instar *Daktulosphaira vitifoliae* nymphs or 'crawlers' on lignified root of *Vitis vinifera*. (**c**) Winged (alate) adult *D. vitifoliae*. (**d**) Scanning electron micrograph of first instar *D. vitifoliae* showing stylet

Fig. 10.2 Three types of grapevine phylloxera induced galls: (**a**) leaf galls on 'resistant' grapevine rootstock foliage, (**b**) *yellow hook*-shaped nodosities produced on both non-lignified roots of *Vitis vinifera* and *Vitis* sp., (**c**) swellings or tuberosities produced on *V. vinifera*

characteristics, which under suitable conditions establish primary feeding sites. Once established on a feeding site, either leaf or root, crawlers settle and develop a further three more 'intermediate' nymphal stages until they develop into either alate (winged, Fig. 10.1c) or apterous (wingless) female asexual adults. All nymphs and apterous adult stages feed using a stylet (needle-like mouthpart, Fig. 10.1d) on individual parenchymal (non-vascular) cell contents (Forneck et al. 2002) of the host plant. Depending on their primary feeding site, they cause either leaf galls (Fig. 10.2a) or root galls (Fig. 10.2b, c). Leaf galls are more commonly found on American *Vitis* spp. and some hybrids with this parentage, and are relatively rare on *V. vinifera* (Remund and Buller 1994). The propensity to form leaf galls is most probably linked to phylloxera genetics, environmental conditions and host plant physiology. Although relatively uncommon in Hungary, leaf galls have been increasingly observed on *V. vinifera* (Molnár et al. 2009). In Australia fully developed leaf galls have never been reported on *V. vinifera*, although partially-formed leaf galls have been observed in northeast Victoria in a vineyard containing multiple phylloxera genetic strains (K. Powell, unpubl. data). Although visually intriguing, and a potential risk of transfer of phylloxera via leaf or canopy material, leaf-galling is more commonly observed on rootstock suckers derived from American *Vitis* spp. or hybrids, or nursery vines. It can be managed through standard desuckering techniques (Powell 2001a).

Root-galling phylloxera genetic strains are much more of an economic threat and quarantine concern than leaf-galling strains. On American *Vitis* rootstocks under field conditions, *radicicolae* appear to be limited to non-lignified roots and are rarely economically important, unless the rootstocks have partial *V. vinifera* parentage. In contrast, on ungrafted *V. vinifera*, root feeding by phylloxera is generally observed on both lignified and non-lignified roots and it can, depending on phylloxera genotype and environment, cause significant economic damage to the root system. This root damage results directly in yield decline and eventual vine death by disrupting nutrient and water uptake by the vine root system, and it has indirect negative effects through entry of fungal pathogens causing root necrosis (Omer et al. 1995; Granett et al. 1998; Omer and Granett 2000; Edwards et al. 2006).

10.3 Genetic Diversity

Phylloxera biotypes have been reported in many grape-growing countries (Stevenson 1970a; De Klerk 1979a; King and Rilling 1985; Song and Granett 1990; Corrie et al. 1998) and resistance-breaking biotypes have affected rootstock hybrids with partial *V. vinifera* parentage (Granett et al. 1985), but until relatively recently the genetic variability within phylloxera populations has been poorly characterised. In Australia for example, using RAPD DNA typing, only three phylloxera biotypes were initially reported (Corrie et al. 1997, 1998). However, the identification and utilization of mitochondrial microsatellite markers combined with more extensive ground surveys (Corrie et al. 2002, 2003) have led to the characterization of at least 83 phylloxera genotypes in Australia alone (Umina et al. 2007). Similarly, in Europe over 80 distinct genotypes originating from leaf-galling populations have been identified using microsatellite markers (Vorwerk and Forneck 2006). The existence of 'superclones' (Korosi et al. 2007; Herbert et al. 2010) has highlighted the importance of distinguishing among phylloxera clones while screening for both phylloxera resistance and susceptibility. In addition, molecular typing of newly detected phylloxera incursions can be useful in a traceback situation to determine links to the original source of an infestation.

10.4 Seasonal Abundance and Population Dynamics

The seasonal activity and relative abundance of grapevine phylloxera on ungrafted *V. vinifera* has been studied under commercial field conditions in several grape-growing countries including North America (Omer et al. 1997), Europe (Porten and

Fig. 10.3 Methods for assessing grapevine phylloxera population abundance and dispersal using (**a**) root examination, (**b**) emergence traps, (**c**) adhesive aerial sticky traps, or (**d**) adhesive trunk traps

Huber 2003), the Middle East (Al-Antary et al. 2008), Australia (Powell et al. 2003; Herbert et al. 2006), and New Zealand (King and Buchanan 1986). The population dynamics and dispersive potential of grapevine phylloxera are dependent on which genetic strain of phylloxera is considered. This will influence the relative risks of transfer and quarantine breakdown and thereby determine the timing of any rootstock replanting strategy.

Several monitoring methods have been developed to assess phylloxera population dynamics, below- and above-ground, during the grapevine growing season (Powell et al. 2000). Below-ground monitoring of phylloxera is generally conducted by examination of a section of the grapevine root, either *in situ* or *ex situ*, using a hand lens (Fig. 10.3a). This system clearly allows examination of all life stages of live phylloxera, but the technique itself is labour intensive and unsuitable for regular monitoring due to potential damage to the host plant root system by excessive digging and/or root excision.

Alternative methods of monitoring include pitfall, sticky (both trunk and aerial), suction, and emergence traps (Fig. 10.3b–d) which, although only capturing dispersive life stages of the insect, allow regular, relatively simple, and cost-effective monitoring during the growing season. These above-ground monitoring methods have been used extensively to quantify relative risks of transfer of different phylloxera genetic strains, both in grafted and ungrafted vineyards (Powell et al. 2003; Herbert et al. 2006; Trethowan and Powell 2007). Ultimately, in the future, improved monitoring systems will be developed, which may then allow the calculation of economic thresholds to determine strategic use of control measures.

In all grape-growing regions where *radicicolae* phylloxera have been monitored (De Klerk 1972; King and Buchanan 1986; Omer et al. 1997; Powell et al. 2000; Porten and Huber 2003; Al-Antary et al. 2008), using either traps or root examination, seasonal activity follows a similar pattern. Insects overwinter on the root system as crawlers and when soil temperatures increase in the spring, crawlers develop through four nymphal stages to adulthood. The adults reproduce in the spring and reach peak abundance in the summer months, with the populations then declining in autumn. Phylloxera has a short life cycle with a high reproductive capacity and a single asexual adult can, under ideal conditions, produce up to several hundred viable eggs. The fecundity, relative abundance and generational period of phylloxera populations can be influenced by phylloxera genetics and can vary depending on environmental factors, particularly precipitation, air and soil temperature, as well as soil textural and chemical characteristics. For example, of the 83 different phylloxera genetic strains characterized in Australia so far, 49 feed exclusively on the grapevine root system and consequently a proportion of these populations is present below-ground all year (Umina et al. 2007). Therefore, planting material and soil from infested vineyards represent a considerable quarantine risk. In addition, even though they do not feed on the plant canopy, their dispersive stages, the crawlers and alate adults, can be detected above-ground in spring and summer on the grapevine trunk, soil surface, foliage and grape bunches (Powell et al. 2000). This highlights additional quarantine risks during this busy period of the viticulture season.

In Australia, leaf-galling genetic strains of phylloxera, which predominate above-ground, have to date only been detected on rootstock suckers or rootstock nursery plantings and not on *V. vinifera* leaves. In contrast, in areas of Europe such as Hungary (Molnár et al. 2009), leaf-galling has been observed on *V. vinifera* and is therefore considered more of a quarantine risk.

10.5 Feeding Physiology and Anatomy

Radicicolae phylloxera feed on susceptible *V. vinifera* roots and cause the initiation of nodosities. There are yellow fleshy hook-like structures on non-lignified immature roots (Fig. 10.2b). Nodosities can also be seen on 'resistant' rootstocks. In contrast, tuberosities (Fig. 10.2c) can develop on lignified roots of *V. vinifera* only. The level of damage to the grapevine root system is closely related to the phylloxera virulence level and host plant physiological status, and the resultant damage generally only becomes clear after several seasons post-infestation in the form of canopy decline (Fig. 10.4).

At the cellular level, the stylet of phylloxera enters the outer region of the cortex and feeding is from single parenchymal cells, rather than on vascular tissue (Kellow et al. 2004). No clear evidence of phloem feeding has been observed using electrophysiological monitoring techniques (Kingston et al. 2007a). The mechanism of

Fig. 10.4 Visual symptoms of canopy decline and premature yellowing of phylloxera-infested ungrafted *Vitis vinifera*

susceptibility and resistance of grapevine roots is a relatively unexplored research area for phylloxera, but it is feasible that primary and secondary metabolites may be involved and metabolomic or genomic approaches may assist in identifying whether any induced defense responses occur below-ground (van Dam 2009).

The internal morphology of the phylloxera digestive system has recently been characterized as a compartmentalized midgut with the posterior region having a storage role prior to digestion activities in the anterior region (Kingston 2007). This may aid the insect in producing enough nutritional reserves to survive for up to 7 days away from its host plant and act as an energy source for egg production (Kingston et al. 2007b). This ability to survive for around a week in the absence of a food source exacerbates the risk of quarantine breakdown.

10.6 Environmental Conditions and Climate Change

Several biotic and abiotic factors influence phylloxera establishment, dispersal and population abundance, as well as the level of economic damage inflicted on infested grapevines. Yet, no studies have been conducted to date or models developed to determine the effect of predicted climate change on grapevine phylloxera distribution. However, elevated temperature, reduced water availability, extreme weather events, and elevated CO_2 are likely to have a major impact on phylloxera distribution and its establishment directly or indirectly as they will influence the host plant 'health' and nutritional status. Indirectly, changes in root biomass, root phenology, and nutritional quality may also influence the development of phylloxera, as has been suggested for root-feeding aphids (Salt et al. 1996).

10.6.1 Temperature

One factor which is rarely considered but has a major impact on grapevine phylloxera establishment and the subsequent level of associated grapevine damage, is the combined effect of soil environment and climate. Soil and air temperature directly affect nymphal development, with the optimum range being 21–28°C (Granett and Timper 1987; Fisher and Albrecht 2003). Root gall formation, as induced by phylloxera feeding, commences at 18°C (Turley et al. 1996). Degree-days can be used as an indicator for optimizing the efficacy of phylloxera monitoring and detection systems (Herbert et al. 2006). However, life tables of different genetic strains of phylloxera have not yet been developed and this is an important research gap which will need to be addressed. Temperature range (particularly minimum and maximum) and extreme temperature events (e.g. frosts or extreme heat) are likely to directly influence phylloxera and associated vine damage by affecting nymphal development, fecundity, feeding site establishment, feeding behavior, population dynamics, and available ecogeographical range. In a recent Australian study, two phylloxera genetic strains differed in their response to temperature and humidity (Korosi et al. 2009). Temperatures exceeding 40°C combined with 30% relative humidity cause complete mortality of phylloxera dispersive stages within 2 h. This would suggest that in viticultural regions where the number of extreme heat days is predicted to increase under future climate change scenarios, the likelihood of phylloxera establishment and/or survival would be reduced. The lower thermal limits for phylloxera are poorly defined but cooler temperatures are unlikely to have a major impact, as the phylloxera may respond to frost events by moving down the soil profile.

10.6.2 Water

Water availability, relative humidity and soil moisture can also directly influence insect pest and host plant interactions by affecting the environment both above and below-ground. Reduced soil moisture and increased soil temperature combined have the potential to adversely affect grapevine phylloxera. However, because phylloxera has the ability to move down to cooler and more humid soil strata, it may avoid these changes. Increased soil moisture from using mulches increases the risk of phylloxera transfer above-ground (Powell et al. 2007).

10.6.3 Carbon Dioxide

Carbon dioxide is the main source of carbon that plants use to produce energy during photosynthesis. Elevated levels of CO_2 are expected under future climate change scenarios. A variable range of responses to elevated CO_2, including reduced or increased fecundity, change in population density, and faster rates of development have been observed in a broad range of aphid species (Holopainen 2002). Elevated CO_2 levels can modify the C:N ratios in the plant, which can then impact secondary metabolites and concentrations of carbon-based and nitrogen-based compounds (Bezemer and Jones 1998) such as carbohydrates (e.g., sucrose) and amino acids which are both essential dietary components for phylloxera (Kingston et al. 2007b). Elevated CO_2 can increase cell production and cell expansion in plant roots which will have lower total nitrogen, higher ratios of carbon to nitrogen, and increased carbohydrate levels (Lawler et al. 1997). This could potentially affect grapevine phylloxera establishment and development since it responds to slight modifications in diet (Kingston 2007).

Feeding behavior could also be modified as sap-sucking insects can detect changes in the plant sap C:N ratio and may use this to assess the suitability of plants as a food source (Holopainen 2002). *Radicicolae* phylloxera may be more adaptable to elevated CO_2 because concentrations of this gas in the soil are generally higher than in the atmosphere (Staley and Johnson 2008). *Gallicoae* phylloxera, having a predominantly above-ground habitat and are likely to be more sensitive to changes in atmospheric CO_2.

10.6.4 Soil Chemistry and Texture

Soil chemistry and structure have long been regarded as key drivers of phylloxera establishment, rate of spread, and risk of dispersal. Anecdotal evidence often suggests that phylloxera prefer clay soils and consequently sandy soils are not widely regarded as at risk for phylloxera establishment. The theory is that cracked clay soils facilitate easier underground population migration, whereas sandy-loam soils do not allow this easy passage for the crawlers to disperse. Ermolaev (1990) suggests that the amount of silicon in sandy soil could also reduce phylloxera abundance. In studies conducted in South Africa, De Klerk (1972) showed that there was a higher risk of phylloxera in soils with less than 65% sand content. In Canada, phylloxera infestations were less common on loam and sandy-loam soils than clay soils (Stevenson 1964). Nevertheless, in all the vineyards with predominantly sandy soils, phylloxera was still detected, albeit in lower abundance. However, in both studies there were survey limitations: (1) they were primarily conducted on grafted vines which naturally support lower levels of phylloxera on roots than ungrafted vines, (2) the examination was conducted to a soil depth of only 30 cm for all soil types, and (3) sample number was limited. These limitations could lead to the misleading notion that vineyards on sandy soils are not at risk from phylloxera infestation. Observations in ungrafted *V. vinifera* vineyards in Australia have shown that distribution of phylloxera and rate of spread do differ spatially and temporally (Bruce et al. 2009). This may be caused by differences in soil chemical characteristics (Powell et al. 2003).

There is some debate over the influence of biophysical and biochemical soil characteristics on phylloxera development and dispersal. Although textural properties such as relative sand and clay content are cited as key determinants of the level of phylloxera abundance (Nougaret and Lapham 1928), a 4-year field study in Austrian vineyards indicated that a combination of pH, organic carbon, and soil texture all influence phylloxera populations (Reisenzein et al. 2007). Clay and inorganic content were positively correlated with phylloxera abundance. In contrast, soil characteristics had minimal impact on phylloxera establishment in the Pacific Northwest region of North America (Chitkowski and Fisher 2005). Under Australian soil conditions, some chemical factors have been highlighted which may influence phylloxera abundance and dispersal. In this respect, Powell et al. (2003) reported that toxic levels of aluminum exchange capacity are related to higher phylloxera abundance. Aluminum is known to stress plant root systems inhibiting root growth and affecting root chemistry (Delhaize and Ryan 1995; Dipierro et al. 2005). The rate of spread and establishment of phylloxera has also been linked to higher levels of electrical conductivity in the soil (Bruce et al. 2009, 2011) and this has highlighted the potential for use of remote soil sensing as a tool in targeted phylloxera detection.

Worldwide, current rootstock recommendations rarely consider the impact of soil type when screening rootstocks for phylloxera resistance, but field studies have shown that the population abundance of some phylloxera genetic strains on rootstocks may be affected by environmental conditions (Trethowan and Powell 2007).

10.7 Fungal Interactions

When feeding, phylloxera create stylet insertion points where pathogenic soil microbes may enter the root system to cause additional root damage through necrosis. Several ubiquitous fungal pathogens including *Fusarium* spp., *Cylindrocarpon destructans* (Zins), *Pythium ultimum* Trow and *Phaeoacremonium* spp. have been implicated as secondary pathogens (Granett et al. 1998; Omer and Granett 2000; Edwards et al. 2006). However, the extent of this secondary damage, and the future potential to control such damage using fungicides, will be dependent on other edaphic factors that include the diversity and inoculum levels of beneficial soil microbes, available soil moisture, and relative feeding damage caused by different phylloxera genetic strains.

10.8 Management Options

Although the use of resistant rootstocks is widely advocated as the main form of phylloxera management, several alternative or supplementary long- and short-term options exist. Selection of control measures will be determined largely by whether the grower is dealing with pre- or post-incursion scenarios and the relative extent of economic damage. Pre-incursion management will include the use of surveillance and detection tools and imposition of protective quarantine protocols. Post-incursion options will place greater emphasis on the use of short term and long term management options such as the use of resistant rootstocks and other control options.

10.8.1 Detection and Surveillance

The first stage in any phylloxera management program is to detect an incursion. This may seem obvious but it is relatively difficult in the case of phylloxera because of its small size, genetic variability, and unpredictable below-ground spatial distribution. Therefore, currently and in the future, a range of detection options need to be considered and developed further.

In its most dispersive stage, the crawler is relatively small (around 0.3 mm in length, Kingston 2007) and hence not easily visible to the naked eye (Fig. 10.5) without examining the grapevine root system with a hand lens. The genetic diversity of phylloxera also creates challenges with detection sensitivity as those genetic strains which are less virulent may only be present in low numbers (Herbert et al. 2003)

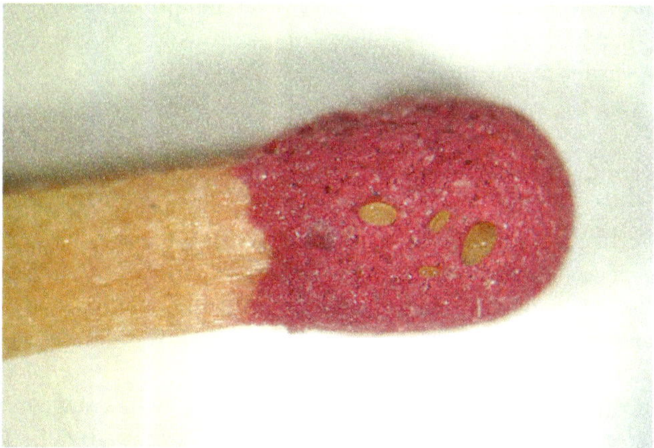

Fig. 10.5 *Daktulosphaira vitifoliae* life stages (egg, crawler, intermediate and apterous adult-shown here) are relatively difficult to detect with the naked eye due to their small size and are shown here on a match head

even on susceptible *V. vinifera*. Generally, the first indication that phylloxera may be present in a vineyard is seen when grapevines show stress symptoms in the foliage or canopy (Fig. 10.4). This occurs several seasons after the initial infestation. On ungrafted *V. vinifera* these above-ground symptoms become visible as premature senescence in autumn, stunting of lateral shoot growth, reduced grape yields, and reduced overall vigor.

The predominant conventional method used worldwide to detect phylloxera is to conduct a ground survey whereby a systematic visual inspection of the grapevine root system is carried out with the aid of a shovel and a hand lens (Fig. 10.3a). This is regarded as a 'primary' detection method as it entails looking for the insect and associated root damage (galling) directly. In grape-growing regions of countries where both phylloxera and the use of phylloxera resistant rootstocks are not yet widespread, such as Australia, new and more sensitive approaches to its detection are being developed which include primary non-destructive techniques such as detecting the DNA of the insect in the soil (Herbert 2005) or using trapping techniques (Powell et al. 2007). Secondary techniques may also be used, such as examining stress symptoms in the grapevine itself (Blanchfield et al. 2006) through chemical or spectral 'fingerprinting' techniques.

10.8.1.1 Primary Detection: Molecular Fingerprinting and Trapping

One of the most recent innovations in phylloxera detection has been the development of a phylloxera-specific DNA probe. This has been validated under both laboratory (Herbert 2005) and field conditions (Herbert et al. 2008a; Bruce et al. 2011). Under field conditions, the approach appears significantly more sensitive compared with conventional ground surveying and on par with emergence trapping in ungrafted *V. vinifera* vineyards. The technique does, however, require further validation under a range of soil conditions at different times of the growing season and in vineyards in which low virulence genetic strains predominate.

10.8.1.2 Secondary Detection: Photosynthetic Response to Phylloxera

When phylloxera attacks the roots of susceptible vines it reduces the uptake of water and nutrients. The imposed water or nutrient stress partially explains the reduced growth of vines following infestation, and provides an opportunity to examine the grapevine canopy for above-ground symptoms of phylloxera presence even when the insect is actually below-ground.

Leaf pigment composition is sensitive to plant stress, with a range of abiotic and biotic factors responsible for either loss or reduction of photosynthetic pigments (e.g., chlorophylls) or the production of photoprotective pigments (such as zeaxanthin and ß-carotene). Pigment composition therefore correlates with plant stress indicators such as declines in photosynthetic efficiency as measured by chlorophyll fluorescence (Gamon et al. 1997; Lovelock and Robinson 2002). A novel approach

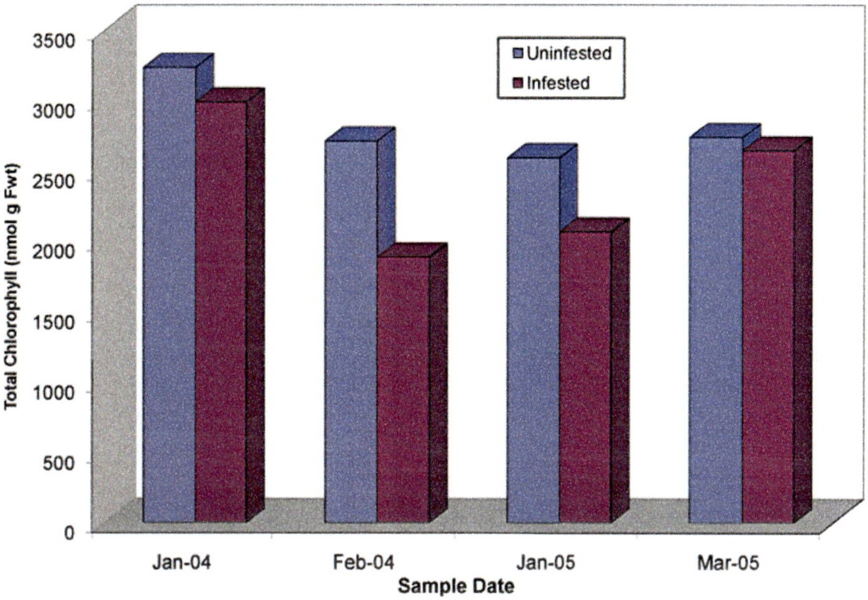

Fig. 10.6 Total chlorophyll content of ungrafted *Vitis vinifera* cv. Pinot Noir leaves from phylloxera infested and phylloxera free grapevines

to phylloxera detection is to examine plant pigments to identify the specific physiological changes associated with phylloxera infestation (Fig. 10.6, Blanchfield et al. 2006). Using HPLC analysis, plant photosynthetic pigment changes can be matched with changes in stress parameters and ultimately with spectral changes in the leaves (Renzullo et al. 2006a). If changes extend to the spectral level, there is a real possibility that spectral-specific airborne sensors could identify early infestation (Renzullo et al. 2006b). However, chlorophyll fluorescence and pigment changes would allow further verification of infection, and might be needed if spectral changes are unable to resolve infestation from the impacts of abiotic stressors.

10.8.1.3 Secondary Detection: Host Plant Chemical Response to Phylloxera

The search also continues for the identification of a chemical compound or group of compounds which may be specifically upregulated in the leaves of phylloxera infested grapevines when phylloxera is feeding on the root system. Should such a unique 'chemical fingerprint' be determined, this would dramatically improve our ability to detect relatively low levels of phylloxera when visual symptoms are absent or difficult to distinguish. Two techniques have been explored to identify such chemical fingerprint(s). They are nuclear magnetic resonance (NMR) and mass spectroscopy (MS). The nuclear magnetic resonance spectroscopy is particularly suitable for this type of investigation for the following reasons: (1) all compounds containing hydrogen atoms (i.e. essentially all organic compounds) produce a signal in the host plant, ensuring that all components of any leaf extracts examined will be detected, (2) the intensity of the signals produced is directly proportional to the number of molecules of the compound present in the mixture, allowing quantification components to be made, and (3) in favourable cases, the information available in the form of a chemical shift (frequency), coupling and integration can allow the structure (or at least some structural information) of the components to be deduced, even in complex mixtures of compounds.

Two recent metabolomic studies have used NMR and MS techniques (Tucker et al. 2007; Rochfort et al. 2009) to compare grapevine leaves from phylloxera infested and uninfested grapevines under both field and controlled conditions. These studies collectively highlighted potential flavonol, terpenoid, and fatty acid compounds for further investigation. Several flavonol compounds including isorhamnetin glycoside, rutin, kaempherol glycoside and quercetin glycoside appear to be markedly upregulated in the leaves when phylloxera is present on the root system of *V. vinifera* (Rochfort et al. 2009). In addition, α-linalool and the linolenic:linoleic acid ratio were also important indicators of infestation (Tucker et al. 2007).

Without further validation to determine if these biochemical responses are actually phylloxera-specific, and not just general disease indicators, it remains to be seen whether any of these specific compounds will offer a new approach to phylloxera detection in the future. Furthermore, the rapid advancement of metabolite analytic platforms and methods may eventually lead to the identification of other novel chemicals which may indeed be unique to grapevine phylloxera infestations (Benheim et al. 2011).

To date, chemical markers have been examined only in leaf material. However, should a chemical marker of infestation be found in grapes or grape juice and a satisfactory method of its analysis developed, juice could be collected for testing when it is produced during the crushing process of winemaking. This would simplify the sampling procedure and avoid the quarantine issues involved with transferring grapevine leaves to a quarantine laboratory. One potential drawback is that any chemical markers would likely be present in very low concentrations because the juice from infested grapevines would be diluted by juice from healthy grapevines and so would require a highly sensitive analytical technique such as liquid chromatography-mass spectrometry (LCMS) for the trace detection of the compounds originally identified by the NMR procedure.

Ultimately, the detection of chemical markers will need to be validated under a range of stress conditions to ensure that marker compounds detected are truly phylloxera infestation-specific and not induced by other diseases or stresses.

10.8.1.4 Secondary Detection: Ground, Airborne or Satellite Remote Sensing for Phylloxera

Remote sensing technologies have been used widely in the viticulture industry worldwide to map grapevine variability in canopy development and maturation at

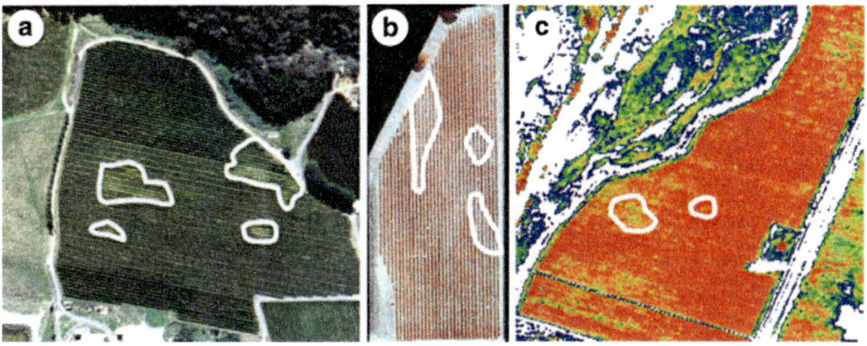

Fig. 10.7 Remotely-sensed imagery from three vineyards in northeastern Victoria, Australia (**a**) a true-colour composite from a Compact Airborne Imaging Spectrometer (CASI), (**b**) false-colour composite of a Digital Multi-Spectral (DMS) Imager image, and (**c**) a Normalised Difference Vegetation Index image (NDVI) derived from the imagery of a multispectral QuickBird satellite sensor. All images show regions of low vigour (i.e. weak spots, identified by *white polygons*) along the rows of grapevines. Weak spots, however, may be a result of phylloxera infestation just as easily as it may be dehydration or nutrient deficiency

the vineyard block level (Proffitt et al. 2006). The imagery obtained has been used to support selective harvesting, the quantification of missing grapevines, and broad regional assessment of grape-growing areas. The advantage of this technology is that it can rapidly survey a much wider area of vineyard than is feasible with on-ground methods.

Studies have previously demonstrated that some airborne and satellite remote-sensing techniques, such as infrared photography, multispectral and hyperspectral imaging spectroscopy, have identified 'weak spots' in vine canopies that are a result of phylloxera infestation (Bell 1995; Frazier et al. 2004; Renzullo et al. 2006a). However, spectral discrimination (Fig. 10.7a–c) between phylloxera infestation and other stresses is still far from an exact science. Validation, particularly at the canopy scale, is required to determine whether phylloxera-specific spectral characteristics can be distinguished from other stress signals. These attributes should be detectable for a range of phylloxera genetic strains, in a range of soil conditions, and for a range of grapevine cultivars and rootstock types. Integration of other field collected data such as soil conductivity spatial maps (using EM38 sensors) may supplement remote canopy imaging systems and allow a targeted approach to phylloxera surveillance (Bruce et al. 2009).

10.8.1.5 Integrated Targeted Detection

Electromagnetic induction sensors measure bulk electrical conductivity which is determined by relative proportions of clay, salinity, and moisture in the soil (Proffitt et al. 2006). The EM38 sensor collects data from 0 to 1.5 m, reflecting the typical depth of grapevine roots and the known depths at which grapevine phylloxera is found on the root system (De Klerk 1972; Buchanan 1990). In the vertical dipole position, 65% of its response comes from the 25 to 150 cm soil depth. The data are collected in two passes. Typically in the vineyard environment, EM38 data is collected by mounting the equipment on a rubber sled and pulling via an all-terrain vehicle through the centre of the grapevine midrow. A differential global positioning system is used in conjunction with the EM38 to accurately map a vineyard area.

EM38 soil survey techniques measure and subsequently map underground soil conductivity in the vicinity of grapevine root zone. Soil conductivity is also related to soil texture and chemistry. Mapping of infested sites and linking spectral data with phylloxera dispersal information from infested sites should improve current detection methods, and may allow the use of soil maps and canopy imagery to predict the rate of underground population migration in phylloxera-infested vineyards (Fig. 10.8, Bruce et al. 2009). This approach has the potential to be utilized in detection of other soil-borne pests of grapevines.

The data obtained from such an integrated detection approach would reduce reliance on (1) historical quarantine status, (2) perceived soil textural risk based on anecdotal evidence, and (3) relative proximity to existing phylloxera infested quarantine zones (PIZs) (DPI 2010).

10.8.2 Eradication

Attempts to eradicate grapevine phylloxera after initial detection have been conducted at both a vineyard and regional scale but they have been unsuccessful to date. One of the few partial 'successes' of this approach was the detection of phylloxera, for the first time, in the Geelong region of Victoria, Australia, in 1877. Within 2 months of its detection in this region, a Vine Disease Eradication Bill was introduced and within 8 years of the initial discovery all grapevines in the region were apparently destroyed as part of an eradication program. However, within 3 years phylloxera was discovered in other parts of the state (Buchanan 1990).

These attempts to eliminate phylloxera from Australia ultimately failed but led to the introduction of quarantine measures in 1890 along with recommendations on the use of phylloxera-resistant rootstocks. In 2010, despite the relative lack of resistant rootstock use in Australia, representing only around 5% of the grape-growing regions, phylloxera is still not widespread and is known to occur only in certain regions of Victoria and New South Wales (DPI 2010). However, phylloxera-specific quarantine protocols (NVHSC 2009) are enforced in Australia to restrict the spread of phylloxera from known PIZs to areas considered free from phylloxera (Table 10.1).

In contrast to Australia, most other major grape-producing countries manage phylloxera by area-wide use of resistant rootstock. One of the common misconceptions in the viticulture industry is that, if phylloxera is detected in an ungrafted *V. vinifera* vineyard, removal of the infested grapevines and replacement with 'resistant' rootstocks will eradicate the insect from the vineyard. By planting rootstocks which have no *V. vinifera* parentage in their lineage, economic damage can certainly

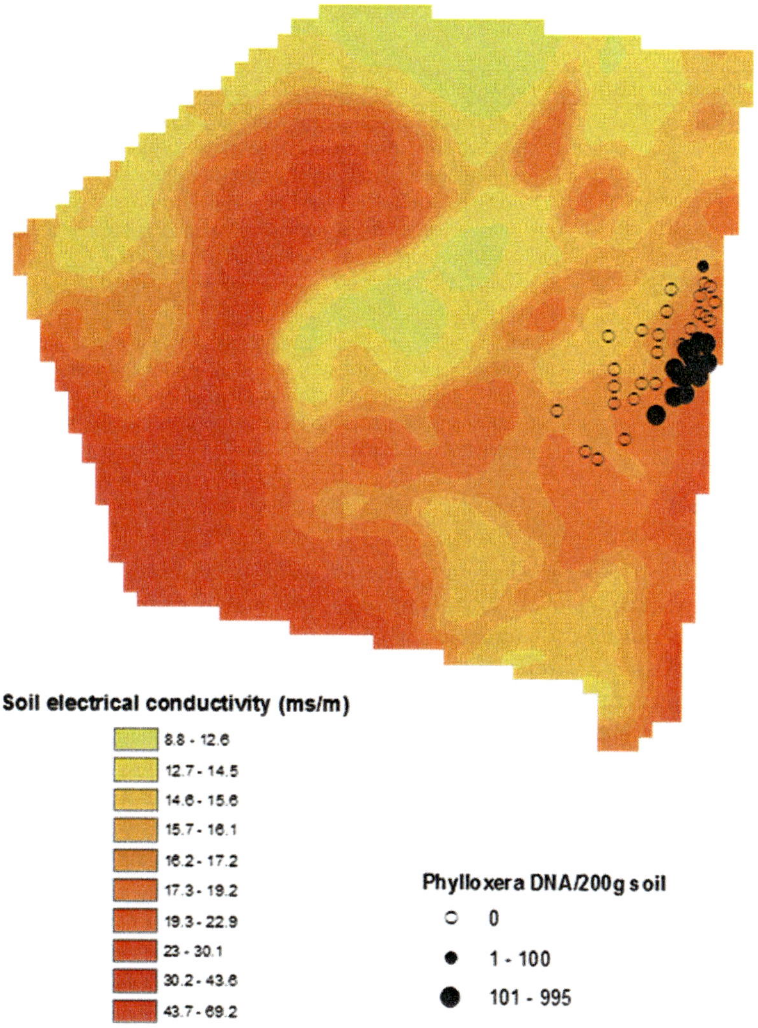

Fig. 10.8 EM38 imagery of phylloxera infested vineyard and soil DNA probe collection points as part of a targeted integrated phylloxera detection approach based on soil electrical conductivity values

be avoided, but the rootstocks will still support or tolerate low levels of some phylloxera strains.

10.8.3 Quarantine

Grapevine phylloxera is not widespread in some countries, such as China and Australia, because its natural dispersal is limited to 100 m per year (King and

Table 10.1 Definitions of grapevine phylloxera quarantine zones in Australia (modified after NVHSC 2008)

Quarantine zone status	Description
Phylloxera infested zone (PIZ)	A PIZ contains vineyards known to be infested with phylloxera or to have been infested with phylloxera. Its boundary must be a minimum of 5 km from any infested vineyard The boundary of the PIZ may be defined by local government boundaries or other landmarks, provided they are no closer than 5 km from the edges of any infested vineyards
Phylloxera risk zone (PRZ)	The boundaries of a PRZ are determined by default. They include all areas not defined as a PIZ or PEZ
Phylloxera exclusion zone (PEZ)	A PEZ is an area that has been established by historical information and/or a survey program as not being infested by phylloxera, and is governed by appropriate regulations to control the movement of grapevine material, specified grape products and vineyard equipment into the area

Buchanan 1986). It spends a large proportion of its life cycle underground and some genotypes produce relatively low populations. It is mainly distributed by human-assisted vectors such as viticulture machinery, infested clothing, footwear, and grape products (Deretic et al. 2003).

Phylloxera have two active dispersive life stages, the alates (winged adults) and the first instar nymphs (crawlers), both of which can move above and below-ground in both ungrafted and grafted vineyards. Movement of both life stages from the root system is seasonally dependent, with migration above-ground occurring in the spring and summer months (Powell et al. 2003). Alates are considered a relatively low quarantine risk because the sexual component of the life cycle is short (Corrie et al. 2002; Vorwerk and Forneck 2006). Crawlers can be extremely abundant, however, and represent a relatively high risk of transfer (Herbert et al. 2006) because they can disperse onto the grapevine foliage and grape bunches (Powell et al. 2000). This represents a risk of phylloxera transfer on grapes, in grape bins, on grape harvesters, and on the clothing of vineyard personnel at harvest time. In Australia quarantine protocols are in place to minimize the risk of phylloxera transfer by such routes (Table 10.2). Post-harvest grape material such as grapes or unfermented pomace (a mixture of grape seeds, skins and stalks) and other winery waste products have also been shown to carry viable phylloxera crawlers (Deretic et al. 2003). Post-harvest fermentation (Deretic et al. 2003), sulfur dioxide fumigation (Buchanan 1990), and composting (Bishop et al. 2002) are all used as disinfestation protocols for phylloxera. Concentrated sodium hypochlorite can also be an effective disinfestation treatment for footwear (Dunstone et al. 2003). Cold water alone is ineffective as crawlers and eggs can survive for several days submerged in water (Korosi et al. 2009).

Like most insects, phylloxera has a relatively limited temperature range and this factor has been exploited through the development of a dry heat disinfestation method for viticultural machinery (Korosi et al. 2012). Because phylloxera has been

Table 10.2 Summary of some of the potential vectors by which grapevine phylloxera could be transferred from a potentially infested vineyard property, and some of the recommended methods for quarantine and disinfestation used in Australia

Risk vector	Quarantine and/or disinfestation protocol	Reference
Whole fresh winegrapes	Annual ground surveys of the source vineyard, to determine phylloxera infestation status	NVHSC (2009)
	Secure packaging and cleaning of grape transport bin	
	Inspection of grape bins and transport vehicles	
	Wash down facilities for grape bins and transport	
Grapevine cuttings and rootlings	Hot water treatment at 50 ± 1°C for 30 min or 54 ± 1°C for 5 min	NVHSC (2009)
Diagnostic samples[a]	One of the following:	NVHSC (2009)
	Freezing to −18°C for 24 h	
	Freezing and transport at −196°C	
	Oven drying at 45°C for 2 h minimum	
	Hot water treatment at 50 ± 1°C for 30 min, or 54 ± 1°C for 5 min	
	Fixative, e.g. 70% ethanol	
Must or juice	Minimum of 72 h fermentation	NVHSC (2009)
	Filtered to achieve maximum particle size of 50 µm	
Marc (pomace)	72 h fermentation composting as per Australian Standard AS 4454	NVHSC (2009) Bishop et al. (2002)
Table grapes	Sulphur pads with a minimum of 970 g/kg sodium metabisulphite or fumigation with methyl bromide	Buchanan (1990) NVHSC (2009)
Vineyard equipment	Grape bins: immerse in water at 70°C for a minimum 2 min	NVHSC (2009) Korosi et al. (2009)
	Harvesters: dry heat treatment for 75 min at 45°C or 2 h at 40°C	
Footwear	Scrub footwear in 2% sodium hypochlorite (a.i.) for 30 min	Dunstone et al. (2003)
Clothing	Disposable overalls	NVHSC (2009)
Vineyard vehicles	Avoid entry to grapevine areas where possible	NVHSC (2009)
	Plan the route to travel on sealed roads	

[a]For example petioles, soil, insects, etc.

detected on grapevine harvesters (King and Buchanan 1986) and is known to be present in the grapevine canopy during harvest time (Powell et al. 2000), dry heat sheds are used in Australia to disinfest machinery between vineyards, and sometimes even between infested blocks within a vineyard to avoid transferring the insect around the property. Hot water treatment is also an effective recommended disinfestation treatment for vine cuttings and grape bins (Table 10.2) (NVHSC 2009).

The inherent ability of phylloxera to disperse naturally, without human assistance, is limited. Several vectors of phylloxera transfer have been identified (Table 10.2), and in some instances, quantified (King and Buchanan 1986; Deretic et al. 2003). This has led to the development of a range of quarantine measures in some countries (EPPO 1990) to restrict phylloxera spread. One of the most comprehensive sets of protocols is the Australian National Phylloxera Management Protocol (NVHSC 2009),

Table 10.3 Common rootstock hybrid combinations tested for grapevine phylloxera resistance and some commercially available rootstocks (modified after Whiting 2003)

Parentage	Rootstock varieties
V. berlandieri x *V. riparia*	5BB Kober
	5A Teleki
	5C Teleki
	420A Millardet
	S04
V. riparia x *V. rupestris*	Schwarzmann
	101-14
	3309 Courdec
V. cinerea x *V. riparia*	Börner
V. berlandieri x *V. rupestris*	1103 Paulsen
	99 Richter
	110 Richter
	140 Ruggeri
V. rupestris x *V. candicans*	Ramsey

which not only highlights the diverse range of transfer risks, but also highlights a range of disinfestation options to prevent or reduce the risk of transfer from infested to uninfested vineyards or regions.

10.8.4 Resistant Rootstocks

The native range of grapevine phylloxera is restricted to eastern North America in regions where *Vitis riparia* Michaux is distributed. Grapevine phylloxera was inadvertently transferred to Europe and other grape-growing regions on wild American *Vitis* spp. which were introduced as breeding material to develop hybrids to control the fungal disease powdery mildew (Gale 2003). The natural tolerance of American *Vitis* spp. has been exploited for over 150 years in breeding programs where *V. vinifera*, which possesses characteristics important for fruit quality, is grafted on to the root system of the phylloxera-tolerant hybrids.

A range of rootstock hybrid combinations are commercially available (Table 10.3) and new hybrids are under development (Korosi et al. 2011) along with different approaches around the world to screen for phylloxera resistance. Unfortunately, there is no consistent approach to screening for phylloxera resistance. Available methods include laboratory (both excised roots and tissue culture) (Forneck et al. 2001; Kellow et al. 2002; Makee et al. 2004), glasshouse, and field screening. While the genotype of the resistant hybrid is usually well characterized, the genetic characteristics of phylloxera, with few notable exceptions, are rarely considered in any screening program.

A three-phase screening system (Korosi et al. 2007) has been developed, but it is not internationally adopted, whereby single clonal lineages of phylloxera are screened to determine resistance levels to both 'susceptible' *V. vinifera* and grafted

American *Vitis* hybrids. This triphasic approach is used rather than a single system to account for variability in the host plant response to phylloxera infestation in the three phase screening systems used. The three screening systems are: (1) excised root bioassays (Granett et al. 1983), (2) potted grapevine (Korosi et al. 2007), and (3) monitoring field grown grapevines (Trethowan and Powell 2007). Granett et al. (2001b) have shown that in excised root bioassays population levels appear higher than on field grown vines, indicating that if only excised root bioassays were used for resistance screening an overestimate of tolerance could be inadvertently assumed. Glasshouse trials allow a whole plant response to be assessed under controlled conditions and also allow for testing under different soil conditions. Both excised root laboratory and whole plant glasshouse trials allow a relatively rapid assessment of rootstock tolerance to different phylloxera clonal lines (Figs. 10.9 and 10.10a, b). Field trials take longer to assess but do allow a more realistic evaluation of grapevine response to phylloxera infestation under commercial vineyard conditions (Fig. 10.11).

An electrophysiological technique called the electrical penetration graph (EPG), originally developed for aphids, has been adapted to study the feeding behavior of root-feeding grapevine phylloxera on susceptible and resistant grapevines (Kingston et al. 2007a). With further refinement, this approach may dramatically improve the rate of future rootstock screening against specific phylloxera clonal lineages.

10.8.5 Genetic Modification of Host Plant Resistance

There has been very limited focus on the potential to develop genetically modified grapevines with phylloxera resistance. This is because of the natural resistance in American *Vitis* spp. has been harnessed very successfully through conventional plant breeding techniques. However, because phylloxera is essentially monophagous, the potential to identify antimetabolites, which may be used in genetically modified grapevines, is intriguing. For example, secondary metabolites such as cyanogenic glucosides, when expressed in *V. vinifera* roots, affect phylloxera fecundity under laboratory conditions (Franks et al. 2006).

An artificial diet rearing system has been developed for both *radicicolae* (Kingston et al. 2007a) and *gallicoae* phylloxera (Forneck and Wöhrle 2003) which with further modification, could be used to screen potential antimetabolites. These could include plant-derived lectins which have been shown to be effective against a broad range of hemipteran pests (Hilder et al. 1995; Powell 2001b; Trębicki et al. 2009).

10.8.6 Cultural Management

Opportunities for controlling phylloxera through soil management such as tillage or through soil amendments such as organic or physical mulches have received

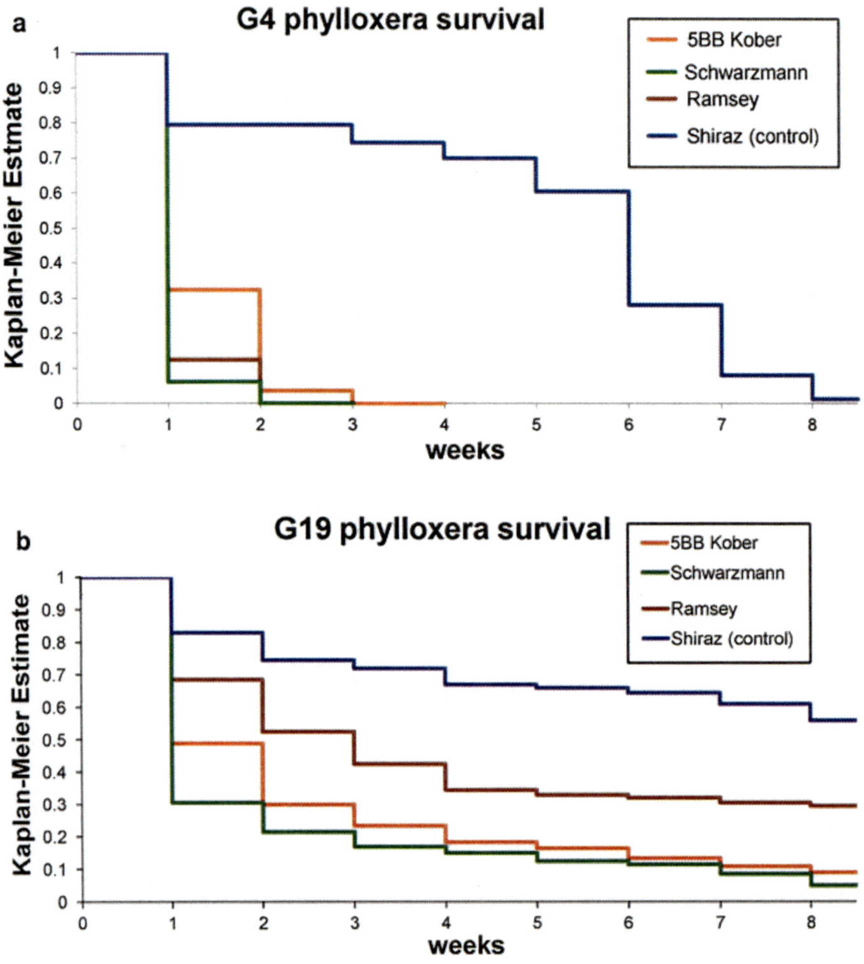

Fig. 10.9 Survival of G4 and G19 clonal lineages of *Daktulosphaira vitifoliae* on excised roots of four rootstock types

surprisingly limited research attention. Lotter et al. (1997, 1999) conducted soil surveys in organically managed vineyards as well as conventionally managed phylloxera-infested vineyards, and reported a lower incidence of phylloxera-associated damage in the organically managed vineyards.

Organic mulch applications have the potential to change the physical and textural properties of the soil environment, making it either more or less conducive to phylloxera survival on the root system or they may directly affect the insects' mobility through the soil. However, the formulation of the mulch application is likely to be a key determinant of the efficacy of this type of management. In a 3-year study

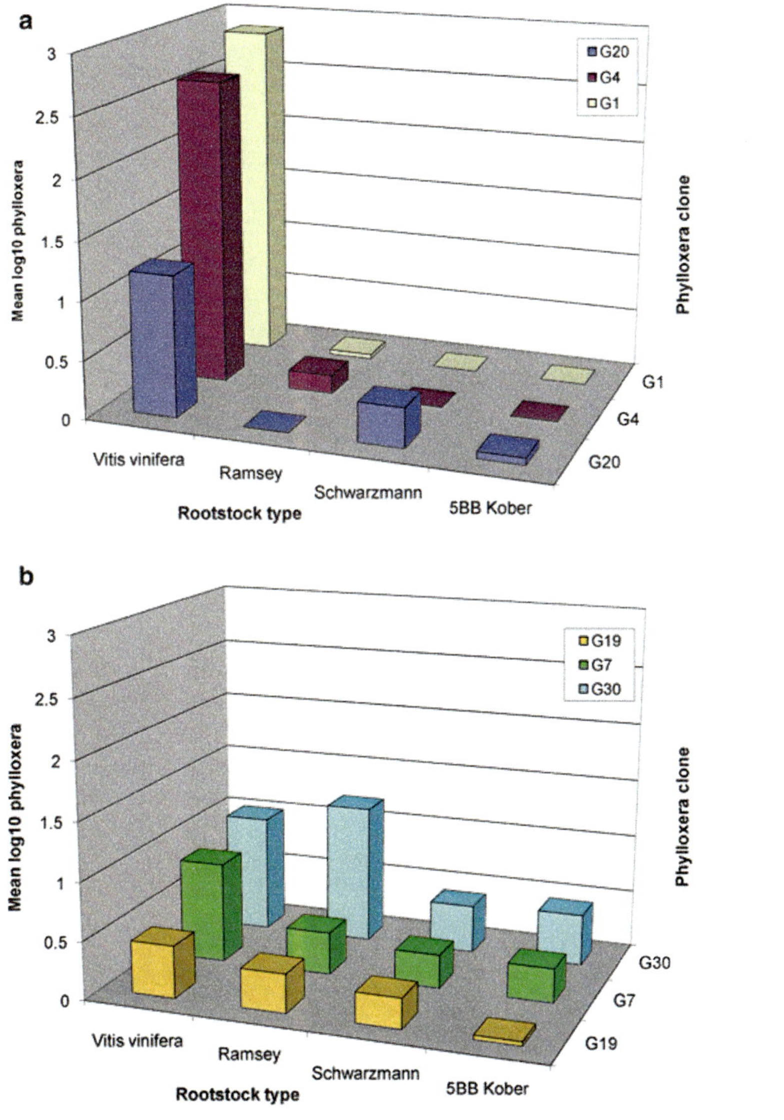

Fig. 10.10 Glasshouse bioassay data from six phylloxera clonal lineages (**a**) G1, G4 and G20, and (**b**) G7, G19 and G30 screened against four rootstock types: ungrafted *Vitis vinifera,* Ramsey, Schwarzmann and 5BB Kober

conducted in the cool climate grape-growing region of the King Valley, northeast Victoria, Australia, composted green waste mulch applications increased the abundance and dispersal of phylloxera above ground, thereby increasing the quarantine risk (Powell et al. 2006). In contrast, in a similar study conducted in central Victoria, Australia, some grape-pomace mulch formulations actually reduced

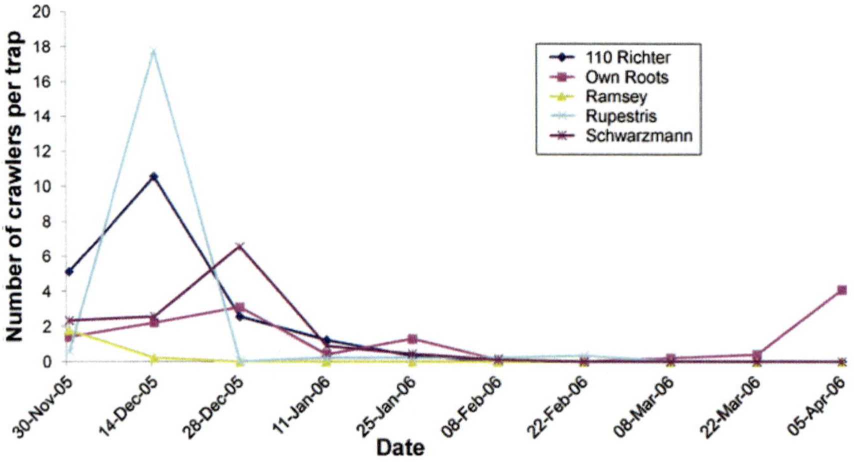

Fig. 10.11 Seasonal above-ground dispersal of grapevine phylloxera, in a vineyard with mixed phylloxera clonal lineages, from five rootstock types in a commercial vineyard in northeast Victoria, Australia, as assessed using emergence traps

phylloxera population levels under field conditions (Powell et al. 2007). In Germany, reduced phylloxera abundance was observed when spruce sawdust was applied as mulch, but whether this was due to changed physical or changed microbial properties of the soil is unclear (Huber et al. 2003). The use of water or nitrogen management of grapevines has also been described as a short term option for phylloxera management (Vega 1956; Kopf 2000), but several factors would preclude widespread use of this technique, including relative soil permeability (i.e., leaching) and availability of irrigation water.

10.8.7 Biological Control

Options for the use of biological agents, either natural or introduced, for control of grapevine phylloxera are surprisingly poorly explored despite the potential opportunities that might exist for either a classical or inundative biological control approach. The use of natural enemies to control phylloxera has rarely been discussed and this may in part be due to the somewhat complex life cycle of this pest that includes both root-feeding and foliar feeding habitats (depending on genetic strain and host plant). Some natural predators of phylloxera have been reported (Rondani cited in Dalmasso 1956; Gorkavenko 1976; Jubb and Masteller 1977; Wheeler and Henry 1978; Wheeler and Jubb 1979), but warrant further investigation of their potential efficacy.

Biocontrol agents for potential phylloxera control include the nematode *Heterorhabditis bacteriophora* Poinar (English-Loeb et al. 1999), the fungi

Beauveria bassiana (Balsamo) Vuillemin (Granett et al. 2001a), *Metarhizium anisopliae* (Metshnikoff) Sorokin (Kirchmair et al. 2004; Huber and Kirchmair 2007), and *Cephalosporium* spp. (Vega 1956) and the predatory mite *Tyroglyphus phylloxerae* Riley (van Driesche and Bellows 1996).

10.8.8 Chemical Control

The search for a 'silver bullet' to control grapevine phylloxera by chemical means has received sporadic attention over the last 150 years. However, a broad range of chemicals, some of which have detrimental phytotoxic and environmental impacts, have been tested against phylloxera. Insecticide groups including organochlorines (Williams 1979), organophosphates (Manojlovic 1989), pyrethroids (Johnson et al. 2008), carbamates (Rammer 1980; Loubser et al. 1992) and neonicotinoids (Al-Antary et al. 2008; Herbert et al. 2008b) have been tested (Table 10.4) against both leaf-galling and root-galling forms with limited success to date on a commercial scale.

Each of these insecticide groups has a different mechanism of action that requires different timing, application rate and application method. Furthermore, their efficacy can be influenced by soil, climate and host plant conditions and the ability of phylloxera to develop overlapping generations. Although some of these chemicals may indeed reduce above-ground population levels of phylloxera in the short term, and hence minimize the risk of phylloxera transfer on viticulture machinery, the control of below-ground root-galling grapevine phylloxera is inherently more difficult. One difficulty is the vertical distribution of the insect on the grapevine root system. On ungrafted *V. vinifera*, phylloxera has been found on grapevine roots to a depth of up to 1.2 m into the soil profile (De Klerk 1972).

Because phylloxera is predominantly a root feeder, a downwardly mobile systemic action is important for targeted delivery of the insecticide. The neonicotinoids are downwardly mobile systemics and both imidacloprid and thiamethoxam have been shown to be effective *in vitro* and *in planta* under controlled environment studies (Herbert et al. 2008b) and field conditions (Al-Antary et al. 2008). The secondary impacts of potential chemical control agents are also important to consider. For example, imidacloprid has an extremely long half-life in the soil, up to 365 days, and is toxic to common predators in vineyards such as lacewings, predatory bugs (Bernard et al. 2007) and predatory mites (Bostanian et al. 2010).

To overcome some of the challenges of reaching roots with insecticides, a relatively novel insecticide, spirotetramat, has recently been developed, which specifically targets hemipteran pests (Nauen et al. 2008). It has a two-way systemic movement profile enabling movement between the leaves and the roots that will provide the potential to target all genetic strains of phylloxera. Recent trials indicate it has some effect on grapevine phylloxera (van Steenwyk et al. 2009; Sleezer et al. 2011).

Table 10.4 Selected chemical insecticides tested against gallicolae and *radicicolae Daktulosphaira vitifoliae*

Active ingredient	Phylloxera type	Country	Reference
Aldicarb	*Radicicolae*	Australia	Buchanan and Godden (1989)
	Gallicoae	South Africa	Loubser et al. (1992)
Carbofuran	*Radicicolae*	USA	Rammer (1980)
		USA	Granett et al. (1986)
		Australia	Buchanan and Godden (1989)
		New Zealand	King et al. (1983)
Endosulfan	*Gallicoae*	Canada	Stevenson (1970b)
Fenamiphos	*Radicicolae*	Australia	Buchanan and Godden (1989)
		South Africa	De Klerk (1979b)
Hexachlorobutadiene		South Africa	De Klerk (1979b)
		Russia	Litvinov (1982)
		Russia	Gorenshtein (1983)
Hexachlorocyclohexane	*Gallicoae*	Bulgaria	Kostadinov (1995)
		Yugoslavia	Manojlovic (1989)
Imidacloprid	*Radicicolae*	Australia	Herbert et al. (2008b)
		Jordan	Nazer et al. (2006)
		Jordan	Al-Antary et al. (2008)
	Gallicoae	Brazil	Botton et al. (2004)
Oxamyl	*Radicicolae*	Australia	Buchanan & Godden (1989)
		New Zealand	King et al. (1983)
		Jordan	Nazer et al. (2006)
Sodium tetrathiocarbonate	*Radicicolae*	USA	Weber et al. (1996)
Spirotetramat	*Gallicoae*	USA	van Steenwyk et al. (2009)
		USA	Johnson et al. (2008)
Thiamethoxam	*Radicicolae*	Australia	Herbert et al. (2008b)
		Jordan	Nazer et al. (2006)
		Jordan	Al-Antary et al. (2008)
	Gallicoae	Brazil	Botton et al. (2004)

10.9 Future Phylloxera Management

The future for management of grapevine phylloxera will be determined by advances in research, changes in the world geographic distribution of commercial grapevines due to climate change, and the continued academic interest in this insect. In the last decade there has been a resurgence in phylloxera research activity, particularly in areas of detection and genetic characterization.

The use of resistant rootstocks plays a predominant role in the management of phylloxera and it will be significantly enhanced by advances in phylloxera and grapevine genetics. Recently, the genome for *V. vinifera* cv. Pinot Noir was mapped (Jaillon et al. 2007) and mapping of the grapevine phylloxera genome has recently been proposed (Delmotte et al. 2011). If the genome of phylloxera is mapped, this will provide major advances in our knowledge of the genetic basis of the interaction between phylloxera and its host, and will enhance our current limited knowledge of leaf and root gall formation. As *Vitis* spp. is the only host of phylloxera, the potential exists for breeding immune rootstocks through genetic manipulation. However, if genetic modification is not widely accepted by the grape and wine consumer, the focus may change to the use of alternative forms of phylloxera management, such as the use of classic biocontrol which has received minimal attention to date.

Advances in detection techniques for phylloxera are likely to focus on the integration of secondary detection methods (such as the use of remote sensing methods for area-wide surveillance of either soil or vegetation), which could be enhanced by the use of satellite imagery with targeted primary detection methods (either molecular or chemical fingerprinting). As detection methods become more sensitive, it is also likely that the distribution maps of phylloxera will be modified. This could affect the use of quarantine protocols both nationally and internationally. A change in distribution of regions suitable for growing grapes (Jones 2007; Webb et al. 2010) is likely to directly affect phylloxera establishment and distribution. The potential impacts of climate change on phylloxera has received very little attention but further research in this area could lead to the development of predictive modelling tools to determine relative risks of phylloxera establishment and development. Fundamental knowledge of the effect of environmental factors on different phylloxera strains could also provide improvements in both quarantine protocols and rootstock recommendations.

References

Al-Antary TM, Nazer IK, Qudeimat EA (2008) Population trends of grapevine phylloxera, *Daktulosphaira (Viteus) vitifoliae* Fitch. (Homoptera: Phylloxeridae) and effect of two insecticides on its different stages in Jordan. Jordan J Agric Sci 4:343–349

ARD (Agriculture and Rural Development) (2007) Armenia managing food safety and agricultural health: an action plan. Agriculture and Rural Development Department and Europe and Central Asia Region/The World Bank/Zangak-97 Publishing House, Washington/Yerevan

Bell C (1995) GRAPES project prepares California vineyard managers for insect infestation. Geogr Info Syst 5:44–47

Benheim D, Rochfort S, Ezernieks V, Robertson E, Potter ID, Korosi GA, Powell KS (2011) Early detection of grape phylloxera (*Daktulosphaira vitifoliae* Fitch) infestation through identification of chemical biomarkers. Acta Hortic 904:17–24

Bernard M, Horne PA, Papacek D, Jacometti MA, Wratten SJ, Evans K et al (2007) Guidelines for environmentally sustainable wine grape production in Australia: IPM adoption self-assessment guide for growers. Aust N Z Grapegrow Winemak 518:26–36

Bezemer TM, Jones TH (1998) Plant-insect herbivore interactions in elevated atmospheric CO_2: quantitative analyses and guild effects. Oikos 82:212–222

Bishop A, Powell KS, Gibson T, Barchia IM, Wong PTW (2002) Mortality of grape phylloxera in composting organics. Aust J Grape Wine Res 8:48–55

Blanchfield AL, Robinson SA, Renzullo LJ, Powell KS (2006) Phylloxera-infested grapevines have reduced chlorophyll and increased photo-protective pigment content – can leaf pigment composition aid pest detection? Funct Plant Biol 33:507–514

Boehm W (1996) The phylloxera fight: protecting south Australia from the phylloxera threat. The Phylloxera and Grape Industry Board of south Australia, Adelaide

Bostanian NJ, Hardman JM, Thistlewood HMA, Racette G (2010) Effects of six selected orchard insecticides on *Neoseiulus fallacis* (Acari: Phytoseiidae) in the laboratory. Pest Manag Sci 66:1263–1267

Botton M, Ringenberg R, Zanardi OZ (2004) Controle químico da forma galícola da filoxera *Daktulosphaira vitifoliae* (Fitch, 1856) (Hemiptera: Phylloxeridae) na cultura da videira. Cien Rural 34:1327–1331

Bruce RJ, Lamb DW, Mackie AM, Korosi GA, Powell KS (2009) Using objective biophysical measurements as the basis of targeted surveillance for detection of grapevine phylloxera *Daktulosphaira vitifoliae* Fitch: preliminary findings. Acta Hortic 816:71–80

Bruce RJ, Lamb DW, Hoffmann AA, Runting J, Powell KS (2011) Towards improved early detection of grapevine phylloxera (*Daktulosphaira vitifoliae* Fitch) using a risk-based assessment. Acta Hortic 904:123–131

Buchanan GA (1990) The distribution, biology and control of grape phylloxera, *Daktulosphaira vitifoliae* (Fitch), in Victoria. Ph.D. dissertation, La Trobe University, Melbourne, Australia

Buchanan GA, Godden GD (1989) Insecticide treatments for control of grape phylloxera (*Daktulosphaira vitifolii*) infesting grapevines in Victoria, Australia. Aust J Exp Agric 29:267–271

Campbell C (2004) Phylloxera: how wine was saved for the world. Harper Collins, London

Chitkowski RL, Fisher JR (2005) Effect of soil type on the establishment of grape phylloxera colonies in the Pacific Northwest. Am J Enol Vitic 56:207–211

Corrie AM, Buchanan G, van Heeswijck R (1997) DNA typing of populations of phylloxera (*Daktulosphaira vitifoliae* (Fitch)) from Australian vineyards. Aust J Grape Wine Res 3:50–56

Corrie AM, Kellow A, Buchanan G, van Heeswijck R (1998) Phylloxera biotypes in Australia. Aust Grapegrow Winemak 417:26–32

Corrie AM, Crozier RH, van Heeswijck R, Hoffmann AA (2002) Clonal reproduction and population genetic structure of grape phylloxera, *Daktulosphaira vitifoliae,* in Australia. Heredity 88:203–211

Corrie AM, van Heeswijck R, Hoffmann AA (2003) Evidence for host-associated clones of grape phylloxera *Daktulosphaira vitifoliae* (Hemiptera: Phylloxeridae) in Australia. Bull Entomol Res 93:193–201

Dalmasso G (1956) Lutte contre le phylloxéra (à l'exclusion de l'emploi des porte-greffes résistants). Rapport général pour l'Europe. Bull de l'Off Int du Vin 29:5–30

De Klerk CA (1972) Occurrence and distribution of the vine phylloxera, *Phylloxera vitifolia* (Fitch), in the Olifants River Irrigation Area, Northwest Cape Province. Phytophylactica 4:25–26

De Klerk CA (1979a) An investigation of two morphometric methods to test for the possible occurrence of morphologically different races of *Daktulosphaira vitifoliae* (Fitch) in South Africa. Phytophylactica 11:51–52

De Klerk CA (1979b) Chemical control of the vine phylloxera with hexaclorobutadiene. Phytophylactica 11:83–85

Delhaize E, Ryan PR (1995) Aluminum toxicity and tolerance in plants. Plant Physiol 107:315–321

Delmotte F, Forneck A, Powell KS, Rispe C, Tagu D (2011) Proposal to sequence the genome of the grape phylloxera (*Daktulosphaira vitifoliae* Fitch). http://www.aphidbase.com/var/aphidbase/storage/htmlarea/2965/file/White-Paper-Phylloxera_25may2011-1.pdf

Deretic J, Powell KS, Hetherington SL (2003) Assessing the risk of phylloxera transfer during post-harvest handling of wine grapes. Acta Hortic 617:61–66

Dipierro N, Mondelli D, Paciolla C, Brunetti G, Dipierro S (2005) Changes in the ascorbate system in the response of pumpkin (*Cucurbita pepo* L.) roots to aluminium stress. J Plant Physiol 162:529–538

DPI (Department of Primary Industries) (2010) Phylloxera management zones in Victoria. http://www.dpi.vic.gov.au/DPI/nrenfa.nsf/fid/34C3DAAA05EC6348CA2577520020304F

Du Y-P, Wang Z-S, Zhai H (2011) Grape root cell features related to phylloxera resistance and changes of anatomy and endogenous hormones during nodosity and tuberosity formation. Aust J Grape Wine Res 17:291–297

Dunstone RJ, Corrie AM, Powell KS (2003) Effect of sodium hypochlorite on first instar phylloxera (*Daktulosphaira vitifoliae* Fitch) mortality. Aust J Grape Wine Res 9:107–109

Edwards J, Powell KS, Granett JA (2006) Tritrophic interactions between grapevines, phylloxera and pathogenic fungi – establishing the root cause of grapevine decline. Aust N Z Grapegrow Winemak 513:33–37

English-Loeb G, Villani M, Martinson T, Forsline A, Consolie N (1999) Use of entomophagic nematodes for control of grape phylloxera (Homoptera: Phylloxeridae): a laboratory evaluation. Biol Control 28:890–894

EPPO (European Plant Protection Organisation) (1990) Data sheets on quarantine pests. *Viteus vitifoliae*. EPPO quarantine pest. Prepared by CABI and EPPO for the EU under Contract 90/399003

Ermolaev AA (1990) Resistance of grape phylloxera on sandy soils. Agrokhimiya 2:141–142

Fisher JR, Albrecht MA (2003) Constant temperature life tables of populations of grape phylloxera from Washington and Oregon. Acta Hortic 617:43–48

Forneck A, Wöhrle A (2003) A synthetic diet for phylloxera (*Daktulosphaira vitifoliae* Fitch). Acta Hortic 617:129–134

Forneck A, Huber L (2009) (A)sexual reproduction – a review of life cycles of grape phylloxera, *Daktulosphaira vitifoliae*. Entomol Exp Appl 131:1–10

Forneck A, Walker MA, Blaich R (2001) An *in vitro* assessment of phylloxera (*Daktulosphaira vitifoliae* Fitch) (Hom., Phylloxeridae) life cycle. J Appl Entomol 125:443–447

Forneck A, Kleinmann S, Blaich R, Anvari SF (2002) Histochemistry and anatomy of phylloxera (*Daktulosphaira vitifoliae*) nodosities on young roots of grapevine (*Vitis* spp). Vitis 41:93–97

Franks TK, Powell KS, Choimes S, Marsh E, Iocco P, Sinclair BJ et al (2006) Consequences of transferring three sorghum genes for secondary metabolite (cyanogenic glucoside) biosynthesis to grapevine hairy roots. Transgenic Res 15:181–195

Frazier P, Whiting J, Powell KS, Lamb D (2004) Characterising the development of grape phylloxera infestation with multi-temporal near-infrared aerial photography. Aust N Z Grapegrow Winemak 485a:133–142

Frolov AN, David'yan GE (2009) Pests *Viteus vitifolii* fitch – grape Phylloxera. Interactive agricultural ecological atlas of Russia and neighboring countries. Economic plants and their diseases, pests and weeds. http://www.agroatlas.ru/en/content/pests/Viteus_vitifolii

Gale G (2003) Saving the vine from Phylloxera: a never-ending battle. In: Sandler M, Pinder R (eds) Wine: a scientific exploration. Taylor & Francis, London, pp 70–91

Gamon JA, Serrano L, Surfus JS (1997) The photochemical reflectance index: an optical indicator of photosynthetic radiation use efficiency across species, functional types, and nutrient levels. Oecologia 112:492–501

Gorenshtein RS (1983) Modelling the behaviour in soil of hexachlorobutadiene, used for phylloxera control in vineyards. Khim Sel'sk Khoz 4:55–57

Gorkavenko EB (1976) Entomophages of grape phylloxera and their role in reducing of the pest population in the southern areas of Ukraine. Tr Vses Nauchno-issled Inst Zashch Rast 46:88–97

Granett J, Timper P (1987) Demography of grape phylloxera *Daktulosphaira vitifoliae* (Homoptera: Phylloxeridae), at different temperatures. J Econ Entomol 80:327–329

Granett J, Bisabri-Ershadi B, Carey J (1983) Life-tables of phylloxera on resistant and susceptible grape rootstocks. Entomol Exp Appl 34:13–19

Granett J, Timper P, Lider LA (1985) Grape phylloxera (*Daktulosphaira vitifoliae*) (Homoptera: Phylloxeridae) biotypes in California. J Econ Entomol 78:1463–1467

Granett J, Timper P, White J (1986) Grape phylloxera, *Daktulosphaira vitifoliae* (Homoptera: Phylloxeridae), susceptibility to carbofuran: stage and clonal variability. J Econ Entomol 79:1096–1099

Granett J, Omer AD, Pessereau P, Walker MA (1998) Fungal infections of grapevine roots phylloxera-infested vineyards. Vitis 37:39–42

Granett J, Walker MA, Kocsis L, Omer AD (2001a) Biology and management of grape phylloxera. Annu Rev Entomol 46:387–412

Granett J, Omer AD, Walker MA (2001b) Seasonal capacity of attached and detached vineyard roots to support grape phylloxera (Homoptera: Phylloxeridae). J Econ Entomol 94:138–144

Herbert KS (2005) The early detection and alternative management of phylloxera in ungrafted vineyards. Ph.D. dissertation, La Trobe University, Melbourne, Australia

Herbert KS, Powell KS, Hoffmann AA, Parsons Y, Ophel-Keller K, van Heeswijck R (2003) Early detection of phylloxera – present and future directions. Aust N Z Grapegrow Winemak 473a:93–96

Herbert KS, Hoffmann AA, Powell KS (2006) Changes in grape phylloxera abundance in ungrafted vineyards. J Econ Entomol 99:1774–1783

Herbert KS, Powell KS, McKay A, Hartley D, Herdina H, Ophel-Keller K (2008a) Developing and testing a diagnostic probe for grape phylloxera applicable to soil samples. J Econ Entomol 101:1934–1943

Herbert KS, Hoffmann AA, Powell KS (2008b) Assaying the potential benefits of thiamethoxam and imidacloprid for phylloxera suppression and improvements to grapevine vigour. Crop Prot 27:1229–1236

Herbert KS, Umina PA, Mitrovski PJ, Powell KS, Viduka K, Hoffmann AA (2010) Clone lineages of grape phylloxera differ in their performance on *Vitis vinifera*. Bull Entomol Res 19:1–8

Hilder VA, Powell KS, Gatehouse AMR, Gatehouse JA, Gatehouse LN, Shi Y et al (1995) Expression of snowdrop lectin in transgenic tobacco plants results in added protection against aphids. Transgenic Res 4:18–25

Holopainen JK (2002) Aphid response to elevated ozone and CO_2. Entomol Exp Appl 104: 137–142

Huber L, Kirchmair M (2007) Evaluation of efficacy of entomopathogenic fungi against small-scale grape-damaging insects in soil – experiences with grape phylloxera. Acta Hortic 633:167–171

Huber L, Eisenbeis G, Porten M, Ruhl EH (2003) The influence of organically managed vineyard soils on the phylloxera populations and the vigour of grapevines. Acta Hortic 617:55–59

Jackson RS (2008) Wine science: principles and applications, 3rd edn. Academic, San Diego

Jaillon O, Aury JM, Noel B, Policriti A, Clepet C, Casagrande A et al (2007) The grapevine genome sequence suggests ancestral hexaploidization in major angiosperm phyla. Nature 49:463–467

Johnson DTB, Lewis B, Sleezer S (2008) Chemical evaluation and timing of applications against foliar form of grape phylloxera, 2006. Arthropod Manag Test 33:C11

Jones GV (2007) Climate change: observations, projections, and general implications for viticulture and wine production. In: Essick E, Griffin P, Keefer B, Miller S, Storchmann K (eds) Economics department working paper no. 7. Whitman College, Department of Geography, southern Oregon University, Ashland, OR, pp 1–14

Jubb GL, Masteller EC (1977) Survey of arthropods in grape vineyards of Erie County, Pennsylvania: Neuroptera. Environ Entomol 6:419–428

Kellow AV, McDonald G, Corrie AM, van Heeswijck R (2002) *In vitro* assessment of grapevine resistance to two populations of phylloxera from Australian vineyards. Aust J Grape Wine Res 8:109–116

Kellow AV, Sedgley M, van Heeswijck R (2004) Interaction between *Vitis vinifera* and grape phylloxera: changes in root tissue during nodosity formation. Ann Bot 93:581–590

King PD, Rilling G (1985) Variations in the galling reaction of grapevines: evidence of phylloxera biotypes and clonal reaction to phylloxera. Vitis 24:32–42

King PD, Buchanan GA (1986) The dispersal of phylloxera crawlers and spread of phylloxera infestations in New Zealand and Australian vineyards. Am J Enol Vitic 37:26–33

King PD, Meekings JS, Smith SM, Lauren SM (1983) Insecticidal control of phylloxera on grapes. In: Hartley MJ (ed) Proceedings, 36th New Zealand weed and pest control conference. New Zealand Weed and Pest Control Society, Palmerston North, New Zealand, pp 140–144

Kingston KB (2007) Digestive and feeding physiology of grape phylloxera (*Daktulosphaira vitifoliae* Fitch). Ph.D. dissertation, Australian National University, Canberra, Australia

Kingston KB, Powell KS, Cooper PD (2007a) Characterising the root-feeding habits of grape phylloxera using electrical penetration graph. Acta Hortic 733:159–166

Kingston KB, Powell KS, Cooper PD (2007b) Grape phylloxera: new investigations into the biology of an old grapevine pest. Aust N Z Grapegrow Winemak 521a:12–17

Kirchmair M, Huber L, Rianer J, Strasser H (2004) *Metarhizium anisopliae*, a potential biological control agent against grape phylloxera. BioControl 49:295–303

Kopf A (2000) Untersuchungen zur abundanz der reblaus (*Dactylosphaera vitifolii* Shimer) und zur nodositätenbildung in abhängigkeit von umweltfaktoren. Ph.D. dissertation, University of Hohenheim, Germany

Korosi GA, Trethowan CJ, Powell KS (2007) Screening for rootstock resistance to grapevine phylloxera genotypes from Australian vineyards under controlled conditions. Acta Hortic 733:159–166

Korosi GA, Trethowan CJ, Powell KS (2009) Reducing the risk of phylloxera transfer on viticultural waste and machinery. Acta Hortic 816:53–61

Korosi GA, Powell KS, Clingeleffer PR, Smith B, Walker RR, Wood J (2011) New hybrid rootstock resistance screening for phylloxera under laboratory conditions. Acta Hortic 904:53–58

Korosi GA, Mee P, Powell KS (2012) Influence of temperature and humidity on mortality of grapevine phylloxera *Daktulosphaira vitifoliae* clonal lineages: a scientific validation of a disinfestation procedure for viticultural machinery. Aust J Grape Wine Res 18:43–47

Kostadinov A (1995) The control of grapevine leaf form phylloxera. Seskostopanska Nauka i Proizvodstvo. Agric Sci Prod 33:25–26

Lawler IR, Foley WJ, Woodrow IE, Cork SJ (1997) The effects of elevated CO_2 atmospheres on the nutritional quality of *Eucalyptus* foliage and its interaction with soil nutrient and light availability. Oecologia 109:59–68

Litvinov PI (1982) Use of hexachlorobutadiene for phylloxera control in vineyards. Khim Sel'sk Khoz 1:29–31

Loubser JT, van Aarde IMF, Hoppner GFJ (1992) Assessing the control potential of aldicarb against grapevine phylloxera. S Afr J Enol Vitic 13:84–86

Lotter DW, Granett J, Rizzo D (1997) Soil ecology of grape Phylloxera and the potential for biological control: differences in root damage caused by grape phylloxera in organic vs. conventionally managed northern California vineyards. Organic farming research foundation project report, Santa Cruz, CA

Lotter DW, Granett J, Omer AD (1999) Differences in grape phylloxera-related grapevine root damage in organically and conventionally managed vineyards in California. Hortic Sci 34:1108–1111

Lovelock CE, Robinson SA (2002) Surface reflectance properties of Antarctic moss and their relationship to plant species, pigment composition and photosynthetic function. Plant Cell Environ 25:1239–1250

Makee H, Charbaji T, Ayyoubi Z, Idris I (2004) Evaluating resistance of some rootstocks to grape phylloxera with *in vitro* and excised root testing systems. In Vitro Cell Dev Biol Plant 40:225–229

Manojlovic B (1989) Possibility of chemical control of gall midges *Daktulosphaira vitifoliae* Fitch (Homoptera: Phylloxeridae) on American grapevine. Zast Bilja 40:73–87

Molnár JG, Németh CS, Májer J, Jahnke GG (2009) Assessment of phylloxera leaf galling incidence on European grapevines in Badacsony Hungary. Acta Hortic 816:97–104

Nauen R, Reckmann U, Thomzik J, Thielert W (2008) Biological profile of spirotetramat (Movento®) – a new two-way systemic (ambimobile) insecticide against sucking pest species. Bayer Crop Sci J 61:245–278

Nazer IK, Al-Antary TM, Abu Jbara R (2006) Chemical control of grape phylloxera *Daktulosphaira (Viteus) vitifoliae* Fitch. (Homoptera: Phylloxeridae) sing three chemical soil treatments. Jordan J Agric Sci 2:338–347

NVHSC (National Vine Health Steering Committee) (2008) National phylloxera management protocol: definitions of phylloxera management zones. http://www.gwrdc.com.au/webdata/resources/files/DefinitionsOfZonesForProtocol.pdf

NVHSC (National Vine Health Steering Committee) (2009) National phylloxera management protocol. http://www.gwrdc.com.au/nvhscphylloxera.htm
Nougaret RL, Lapham MH (1928) A study of phylloxera infestation in California as related to types of soils. U S Dep Agric Tech Bull 20:1–39
Omer AD, Granett J (2000) Relationship between grape phylloxera and fungal infections in grapevine roots. Z Pflanzenkrankh Pflanzenschutz 107:285–294
Omer AD, Granett J, De Benedictus JA, Walker MA (1995) Effect of fungal root infections on the vigour of grapevines infested by root-feeding grape phylloxera in California vineyards. Vitis 34:165–170
Omer AD, Granett J, Downie DA, Walker MA (1997) Population dynamics of grape phylloxera in California vineyards. Vitis 36:199–205
Ordish G (1972) The great wine blight. Sidgewick & Jackson, London
Porten M, Huber RL (2003) An assessment method for the quantification of *Daktulosphaira vitifoliae* (Fitch) (Hem., Phylloxeridae) populations in the field. J Appl Entomol 127:157–162
Powell KS (2001a) Taking aim at phylloxera. VHS video, Department of Primary Industries, Rutherglen, Victoria, Australia
Powell KS (2001b) Antimetabolic effects of plant lectins towards nymphal stages of the planthoppers *Tarophagous proserpina* and *Nilaparvata lugens*. Entomol Exp Appl 99:71–77
Powell KS (2008) Grape phylloxera: an overview. In: Johnson SN, Murray PJ (eds) Root feeders: an ecosystem perspective. CAB International, Wallingford, pp 96–114
Powell KS, Brown D, Dunstone R, Hetherington S, Corrie AM (2000) Population dynamics of phylloxera in Australian vineyards and implications for management. In: Powell KS, Whiting J (eds) Proceedings, international symposium on grapevine phylloxera management, Melbourne, Department of Natural Resources & Environment, Victoria, Australia, pp 7–20
Powell KS, Slattery WF, Deretic J, Herbert K, Hetherington S (2003) Influence of soil type and climate on the population dynamics of grapevine phylloxera in Australia. Acta Hortic 617:33–41
Powell KS, Burns A, Norng S, Granett J, McGourty G (2006) Influence of composted green waste on the population dynamics and dispersal of grapevine phylloxera *Daktulosphaira vitifoliae*. Agric Ecosyst Environ 119:33–38
Powell KS, Trethowan CJ, Blanchfield AL, Norng S (2007) Composted winery waste and its influence on grape phylloxera in ungrafted vineyards. Acta Hortic 733:143–150
Proffitt A, Bramley R, Lamb D, Winter E (2006) Precision viticulture – a new era in vineyard management and wine production. Winetitles Pty. Ltd., Ashford
Rammer I (1980) Fields studies with carbofuran for the control of the root form of the grape phylloxera. J Econ Entomol 73:327–331
Reisenzein H, Baumgarten A, Pfeffer M, Aust G (2007) The influence of soil properties on the development of grape phylloxera populations in Austrian viticulture. Acta Hortic 733:3–23
Remund U, Buller E (1994) Die Reblaus – wieder aktuell? Schweiz Z Obst Weinbau 130:242–244
Renzullo LJ, Blanchfield AL, Powell KS (2006a) A method of wavelength selection and spectral discrimination of hyperspectral reflectance spectrometry. IEEE Trans Geosci Remote Sens 44:1986–1994
Renzullo LJ, Blanchfield AL, Guillermin R, Powell KS, Held AA (2006b) Comparison of prospect and HPLC estimates of leaf chlorophyll contents in a grapevine stress study. Int J Remote Sens 27:817–823
Rochfort S, Ezernieks V, Trenerry C, Jones R, Imsic M, Panozzo J, Powell KS (2009) MS Metabolomics – from biomarker discovery to rapid class targeted analysis. In: Proceedings ANZSMS 22-22nd biennial mass spectrometry conference in Australia and New Zealand, Sydney, Australia, 27–30 Jan 2009, p 42
Salt DT, Fenwick P, Whittaker JB (1996) Interspecific herbivore interactions in a high CO_2 environment: root and shoot aphids feeding on *Cardamine*. Oikos 77:326–330
Sleezer S, Johnson DT, Lewis B, Goggin F, Rothrock C, Savin M (2011) Foliar grape phylloxera, *Daktulosphaira vitifoliae* (Fitch) seasonal biology, predictive model, and management in the Ozarks region of the United States. Acta Hortic 904:151–156

Song G-C, Granett J (1990) Grape phylloxera (Homoptera: Phylloxeridae) biotypes in France. J Econ Entomol 83:489–493

Staley JT, Johnson SN (2008) Climate change impacts on root feeders. In: Johnson SN, Murray PJ (eds) Root feeders: an ecosystem perspective. CAB International, Wallingford, pp 192–214

Stevenson AB (1964) Seasonal history of root-infesting *Phylloxera vitifoliae* (Fitch) (Homoptera: Phylloxeridae) in Ontario. Can Entomol 96:79–987

Stevenson AB (1970a) Strains of the grape phylloxera with different effect on the foliage of certain grape cultivars. J Econ Entomol 63:135–138

Stevenson AB (1970b) Endosulfan and other insecticides for control of the leaf form of the grape phylloxera in Ontario. J Econ Entomol 63:125–128

Trębicki P, Harding RM, Powell KS (2009) Anti-metabolic effects of *Galanthus nivalis* agglutinin and wheat germ agglutinin on nymphal stages of the common brown leafhopper using a novel artificial diet system. Entomol Exp Appl 131:99–105

Trethowan CJ, Powell KS (2007) Rootstock-phylloxera interactions under field conditions. Acta Hortic 733:115–122

Tucker DJ, Lamb DL, Powell KS, Blanchfield AL, Brereton IM (2007) Detection of phylloxera infestation in grapevines by NMR methods. Acta Hortic 733:173–181

Turley M, Granett J, Omer AD, De Benedictus JA (1996) Grape phylloxera (Homoptera: Phylloxeridae) temperature threshold for establishment of feeding sites and degree-day calculations. Environ Entomol 25:842–847

Umina PA, Corrie AM, Herbert KS, White VL, Powell KS, Hoffmann AA (2007) The use of DNA markers for pest management: clonal lineages and population biology of grape phylloxera. Acta Hortic 733:183–195

van Dam NM (2009) Below ground herbivory and plant defenses. Annu Rev Ecol Evol Syst 40:373–391

van Driesche RG, Bellows TS (1996) Biological control. Chapman and Hall, New York

van Steenwyk RA, Varela LG, Ehlhardt M (2009) Insecticide evaluations for grape phylloxera with foliar applications of Movento. In: Abstracts 83rd orchard pest and disease management conference, Pullman, Portland, Oregon, Washington State University, WA, 14–16 Jan 2009, p 24

Vega J (1956) Lutte contre le phylloxera (à l'exclusion de l'emploi des porte-greffes résistants). Rapport général pour l'Amérique latine. Bull de l'Off Int du Vin 29:31–42

Vorwerk S, Forneck A (2006) Reproductive mode of grape phylloxera (*Daktulosphaira vitifoliae*, Homoptera: Phylloxeridae) in Europe: molecular evidence for predominantly asexual populations and a lack of gene flow between them. Genome 49:678–687

Webb L, Dunn GM, Barlow EWR (2010) Winegrapes. In: Stokes C, Howden M (eds) Adapting agriculture to climate change. CSIRO, Canberra, pp 101–118

Weber E, De Benedictis J, Smith RJ, Granett J (1996) Enzone does little to improve health of phylloxera-infested vineyards. Calif Agric 50:19–23

Wheeler AG, Henry TJ (1978) *Ceratocapsus modestus* (Hemiptera: Miridae), a predator of grape phylloxera: seasonal history and description of fifth instar. Melsheimer Entomol Ser 25:6–10

Wheeler AG, Jubb GL (1979) *Scymnus cervicalis* Mulsant, a predator of grape phylloxera with notes on *S. brullei* Mulsant as a predator of woolly aphids on elm (Coleoptera: Coccinellidae). Coleopt Bull 33:199–204

Whiting J (2003) Selection of grapevine rootstocks and clones for greater Victoria. Department of Primary Industries, Melbourne

Williams R (1979) Foliar and subsurface insecticidal applications to control aerial form of the grape phylloxera. J Econ Entomol 79:407–409

Chapter 11
Leafhoppers and Planthoppers: Their Bionomics, Pathogen Transmission and Management in Vineyards

Chrystel Olivier, Charles Vincent, Julien Saguez, Brian Galka, Phyllis G. Weintraub, and Michael Maixner

11.1 Introduction

Auchenorrhyncha is the hemipteran suborder that includes cicadas, leafhoppers, froghoppers or spittlebugs, planthoppers and treehoppers. Leafhoppers (Cicadellidae) are cosmopolitan and one of the largest insect families with approximately 22,000 described species (Forero 2008). Planthoppers (infra-order Fulgoromorpha) are mainly tropical with approximately 20 described families (Urban and Cryan 2007). Leafhoppers and planthoppers have piercing-sucking mouthparts that cause direct damage to plants by feeding in mesophyll cells or on xylem and/or phloem sap, and indirect damage by transmitting pathogens. As pathogens are not easily managed in plants, the most

C. Olivier (✉) • B. Galka
Saskatoon Research Centre, Agriculture and Agri-Food Canada, 107 Science Place, Saskatoon, SK, Canada S7N 0X2
e-mail: chrystel.olivier@agr.gc.ca; brian.galka@agr.gc.ca

C. Vincent
Horticultural Research and Development Centre, Agriculture and Agri-Food Canada, 430 Gouin Blvd., Saint-Jean-sur-Richelieu, QC, Canada J3B 3E6
e-mail: charles.vincent@agr.gc.ca

J. Saguez
Co-Lab R&D, Division de Ag-Cord. Inc., 655 Delorme, Granby, QC, Canada J2J 2H4
e-mail: saguezj@yahoo.com

P.G. Weintraub
Agricultural Research Organization, Gilat Research Center, D.N. Negev 85280, Israel
e-mail: phyllisw@volcani.agri.gov.il

M. Maixner
Institute for Plant Protection in Fruit Crops and Viticulture, Julius Küln-Institut (JKI), Federal Research Centre for Cultivated Plants, Brüningstr. 84, 54470 Bernkastel-Kues, Germany
e-mail: michael.maixner@jki.bund.de

N.J. Bostanian et al. (eds.), *Arthropod Management in Vineyards: Pests, Approaches, and Future Directions*, DOI 10.1007/978-94-007-4032-7_11,

common control methods rely on the use of insecticides to manage insect populations. This chapter will provide an overview of the life cycles, feeding behavior and vector abilities of leafhoppers and planthoppers causing damage in vineyards. Present and future management methods will be presented in a viticultural context.

11.2 Bionomics

11.2.1 Feeding Strategies

Grapevines are grown worldwide in several environmental conditions and are attacked by a large number of pests including leafhoppers and planthoppers. Most of the Auchenorrhyncha pests feed from phloem tissues. The leafhopper subfamily Cicadellinae (sharpshooters) feed on xylem tissues, and most species of the sub-family Typhlocibynae feed on mesophyll cells and secondarily in the vascular bundle.

Leafhoppers and planthoppers have piercing-sucking mouthparts to feed on plant tissues and extract sap fluids. The paired mouthparts form needle-like stylets with two canals, one to ingest food and one to release saliva. There are two types of saliva, a gel-like form that hardens to produce the salivary sheath and an enzymatic watery form that serves multiple purposes depending on the species, such as dissolving cell walls or cell contents (Backus et al. 2005b). All leafhoppers choose a feeding substrate and pierce the cell layers. The stylets penetrate intracellularly into the tissues by extending and retracting. Salivary sheath feeders secrete sheath saliva around the stylets to form a covering that physically protects the stylets and absorbs defensive compounds secreted by the plant. Fluids are then pumped into the insect directly from the plant or after release of watery saliva (Backus et al. 2005a, b). Cell rupture feeders do not produce a salivary sheath, but instead use one of three feeding tactics (Backus et al. 2005b). Mesophyll, phloem or xylem cells are ruptured and drained either slowly or quickly with copious watery saliva and very little secretion of sheath saliva.

11.2.2 Life Cycles

Leafhopper eggs are laid singly or a few at a time under the plant epidermis, usually in stems or leaves or under the bark. Some planthoppers deposit their eggs in the soil close to their natural herbaceous host plants (Sforza et al. 1999). Eggs may remain in dormancy during winter or hatch after 1–4 weeks of embryonic development. Leafhoppers are paurometabolous; nymphs undergo five instars before molting into the sexually mature adults. Adult males and females seek each other out for mating through specialized vibrational courtship calls (Hunt and Morton 2001; Mazzoni et al. 2009).

Leafhopper longevity and abundance depend on geographic area, climate, plant species and tissues fed upon (Nielson 1968; McClure 1980; Olsen et al. 1998).

Leafhoppers can overwinter in any life stage. Sharpshooters and most species of grape leafhoppers feed on many types of green vegetation during winter. They enter vineyards soon after bud break to feed and reproduce (Daane and Costello 2000; Hopkins and Purcell 2002). The cicadellids *Scaphoideus titanus* Ball and *Oncopsis alni* (Schrank) overwinter as eggs under the bark of their host plants (Vidano 1964; Claridge and Howse 1968), whereas the cixiid *Hyalesthes obsoletus* Signoret hibernates as nymphs on plant roots (Sforza et al. 1999). Typhlocybinae are usually multivoltine, the number of generations average 2 to 4. However, in hot areas such as Arizona or the San Joaquin Valley (Daane and Costello 2000), there can be 5–7 generations. In comparison, *S. titanus* is univoltine (Vidano 1964).

Leafhopper dispersion occurs over short or long distances (Carlson et al. 1992) and usually depends on wind (Larsen and Whalon 1988), food sources (Zhou et al. 2003), breeding sites (McClure 1980) or canopy density (Lessio and Alma 2004). Only a few species of leafhoppers are known to migrate over long distances. Among them, in North America, populations of *Macrosteles quadrilineatus* Forbes and *Empoasca fabae* (Harris) migrate into the Great Lakes region and southern Canada from the Gulf Coast regions of North America (Taylor and Reling 1986; Valk and Stevenson 1994). Most grape Typhlocybinae cover short distances, migrating from their overwintering sites to the vineyards (Daane and Costello 2000; Böll and Herrmann 2004), with first populations increasing at the edges of the vineyards for *Erythroneura ziczac* Walsh (Zimmerman et al. 1996). In France, studies showed that populations of *Empoasca vitis* (Göthe) were re-dispersing after immigration in vineyards in search of oviposition sites (Decante and Van Helden 2008). In Europe, *H. obsoletus* feeds on bindweed (*Convolvulus arvensis* L.) and stinging nettles (*Urtica dioica* L.) and its dispersion is dependent on the distribution of its food source in the vineyards and the surrounding areas (Bressan et al. 2007). Plant host and canopy densities can also affect leafhopper dispersal. The movement of the monophagous *S. titanus* in vineyards is greatly reduced in high density vineyards (Lessio and Alma 2004).

11.2.3 Economically Significant Pest Species

Bentley et al. (2007) listed 17 species of leafhoppers considered to be grape pests worldwide because of their feeding habits. Within the group of phloem-sap feeders, two leafhopper species are recognized to be vectors of phytoplasma diseases in grape, *S. titanus*, the vector of Flavescence dorée (FD) and *O. alni*, the vector of Palatinate grapevine yellows (Maixner et al. 2000). The planthopper *H. obsoletus* is recognized to be the vector of Bois noir (BN) (Maixner 1994). *Scaphoideus titanus* and *H. obsoletus* are considered to be of high economic importance (Boudon-Padieu and Maixner 2007; Laimer et al. 2009). Among the xylem-sap feeders, numerous species of sharpshooters are known to be vectors of *Xylella fastidiosa* Wells et al., the bacterial causal agent of Pierce's disease. The blue-green sharpshooter, *Graphocephala atropunctata* (Signoret), the red-headed sharpshooter *Carneocephala fulgida* Nottingham, the green sharpshooter *Draeculacephala minerva* Ball and the glassy-winged sharpshooter *Homalodisca vitripennis* (formally = *coagulata*) (Germar), are among those most frequently mentioned as being a vector of Pierce's disease in grapevine (Myers et al. 2007; Hopkins and Purcell 2002). The Typhlocybinae cause direct damage to the grapes by emptying mesophyll cell content or by cell rupture feeding. This subfamily comprises many serious grape pests such as *Erythroneura comes* (Say), *Erythroneura vitis* (Harris), *Erythroneura variabilis* Beamer, *E. ziczac*, *Erythroneura vulnerata* Fitch, and *Erythroneura elegantula* Osborn as well as *Em. fabae* in North America (Zimmerman et al. 1996; Daane and Costello 2000), *Arboridia adanae* (Dlabola) in Eastern Mediterranean regions (Yigit and Erckle 1992), *Jacobiasca lybica* (Bergevin & Zanon), *Em. vitis*, and *Zygina rhamni* Ferrari in Europe (Martinson et al. 1997; Mazzoni et al. 2008; Lenz et al. 2009).

The number of recognized leafhopper pests is expected to increase due to constant environmental changes and the development of new techniques to sample, detect and identify pathogens and their vectors (Boudon-Padieu and Maixner 2007; Foissac and Wilson 2010). For instance, many leafhoppers considered to be potential pathogen vectors are awaiting investigations to confirm their vector status (Orenstein et al. 2003; Beanland et al. 2006).

11.3 Damage to Grapevines

11.3.1 Direct Plant Damage

When continuous and heavy feeding occurs by mesophyll feeders, the damaged leaves have reduced photosynthetic capacity and eventually fall off, causing a reduction in vine vigor and berry sugar content (Olsen et al. 1998). Some phloem feeding Typhlocybinae, namely of the tribe Empoascini, seriously damage plants and cause symptoms of rolling and discoloration of leaf margins, commonly referred to as 'hopperburn'. Hopperburn symptoms are caused by the plant responses to the wounds created by the stylets and exacerbated by the leafhopper saliva (reviewed by Sōgawa 1982; Backus et al. 2005b).

The effects of leafhopper infestation vary depending on geographical locations leafhopper densities, vine cultivars, cultural practices, and the location of damaged leaves (Martinson et al. 1994, 1997; Daane and Costello 2000; Daane et al. 2005). In North America, vineyards located in irrigated warm areas (e.g., California) can tolerate a loss of 20% of foliar area due to damage by *Erythroneura* spp. without affecting yield or maturity (Flaherty et al. 1992). However, vineyards in cool climates have a much lower tolerance for leafhopper injury (Martinson et al. 1994). Sugar content of berries can decrease to 8–9% (normal = 20–25%), but crop losses can go up to 100%. Strong negative effects on bud fruitfulness, with reduced grape production, have also been observed. These effects were carried over for several years after an infestation, despite low subsequent leafhopper infestations (Martinson et al. 1997).

In Europe, economic thresholds for *Empoasca* spp. are considered lower in the southern regions than in the North. Nevertheless, *J. lybica* may cause severe damage and even complete yield losses in vineyards of the Mediterranean regions (Alma 2002).

Plant damage to grapes is also caused by excretion of honeydew by leafhoppers. Honeydew wastes act as excellent substrates for the development of sooty black mold (Weiss 2006) on the berries and leaves, decreasing the quality of the grapes and reducing the photosynthetic capacities (Daane and Costello 1998). *Metcalfa pruinosa* (Say), a planthopper introduced to Europe from North America, is able to cause severe damage to grapes by honeydew and wax excretion (Della Giustina and Navarro 1993).

11.3.2 Indirect Plant Damage

Leafhoppers and planthoppers are known to transmit two important groups of phytopathogenic agents in vineyards: phloem-limited phytoplasmas, the causal agents of plant yellows diseases and xylem-limited bacteria such as *X. fastidiosa*, the Pierce's disease agent. Both groups of phytopathogens are acquired when leafhoppers feed on infected plants. However, phytoplasmas are transmitted in a circulative, propagative mode, whereas *X. fastidiosa* is transmitted in a non-circulative yet propagative mode (Purcell and Finlay 1979). There are many biological, physiological and physical barriers to pathogen acquisition, circulation and multiplication (Wayadande and Fletcher 1995; Wayadande et al. 1997; Bressan et al. 2006). Moreover, pathogen vector relationships tend to be rather specific. Phytoplasmas and *X. fastidiosa* have a dual host transmission cycle, requiring both insect and plant (Chatterjee et al. 2008; Hogenhout et al. 2008).

11.3.3 Grapevine Yellows

Phytoplasmas are non-culturable, wall-less, prokaryotes that cause several hundred diseases to various plants (Weintraub and Jones 2010). They live and reproduce in the phloem of their host plants and are transmitted by phloem-sap feeding leafhoppers, planthoppers and psyllids in a complex circulative, propagative transmission mode (Weintraub and Beanland 2006). Briefly, after ingestion, phytoplasmas cross the digestive epithelium, where they may multiply. Thereafter, they move into the hemolymph and eventually reach the salivary glands, where they multiply intensively. Vectors become infectious after a latent period varying from about 2 weeks to several months. The minimum time required for acquisition and inoculation is influenced by the particular plant species and the feeding habits of the vectors, and the latency period depends on the specific phytoplasmas-insect interactions (Weintraub and Beanland 2006; Hogenhout et al. 2008, 2009).

Different grapevine yellows diseases cause almost identical symptoms in grapevines. Typical yellows disease symptoms include leaf chlorosis and downwards rolling, flower abortion or berry withering, dieback of shoot tips, uneven or total lack of lignification of canes and reduced vitality (Caudwell and Martelli 1992), as well as lower wine quality (Matus et al. 2008). Physiological symptoms include collapse of tissues (Meignoz et al. 1992), callose deposition in phloem (Hren et al. 2009), and impaired photosynthesis (Bertamini et al. 2002). Plant physiology, metabolism and gene expression are deregulated in infected plants (Albertazzi et al. 2009; Hren et al. 2009). Grapevine cultivars, rootstocks, plant age, and health status are determinant factors in symptom severity (Cousin and Boudon-Padieu 2002; Sharon et al. 2003; Gajardo et al. 2009). Remissions of infected grapevines have been observed but the mechanisms are not well understood (Musetti et al. 2007; Constable 2010).

11.3.3.1 Flavescence Dorée (FD)

Flavescence Dorée is the most important grapevine yellows disease. It is caused by phytoplasmas belonging to groups 16SrV-C and -D (Angelini et al. 2003). It has reached epidemic proportions in the affected areas of Europe (Laimer et al. 2009). The FD vector, *Scaphoideus titanus* (Schvester et al. 1961), is a monophagous leaf-hopper that feeds and reproduces only on *Vitis* spp. native to North America, *S. titanus* was introduced into Europe presumably in the beginning of the twentieth century and was reported for the first time in southwestern France in 1958. The European origin of FD and the presence of European plants and vectors that could act as FD reservoirs have been tentatively linked to its epidemiology on grapevine (Filippin et al. 2009). The shift from broad-spectrum insecticides to selective insecticides, such as insect growth regulators or *Bacillus thuringiensis* Berliner, is also suspected to have led to unexpected establishment of *S. titanus* and therefore have contributed to the spread of FD (Belli et al. 2000). FD has been registered as a quarantine disease since 1993 (EPPO A2 list N° 94; EU Annex designation II/A2) and mandatory control measures have been developed to manage FD propagation (Mannini 2007; Steffek et al. 2007) but the vectors and the pathogens are still spreading (Bertin et al. 2007). In contrast, although *S. titanus* is native to North America, FD has not been formally reported in the US and Canada (Maixner et al. 1993).

11.3.3.2 Bois Noir (BN)

Bois Noir is the second most important grapevine yellows disease. It is caused by phytoplasmas of the stolbur (16SrXII-A) group (Daire et al. 1997). BN is present all over Europe. Stolbur phytoplasmas cause problems not only in grapevines but also in various solanaceous crops and corn. They are endemic to Europe and widespread in different herbaceous plant species that belong to the natural vegetation. Different strains of the pathogens that are specific to plant hosts have been identified. They are mainly associated with field bindweed, hedge bindweed (*Calystegia sepium* (L.)), and stinging nettle (Langer and Maixner 2004; Bressan et al. 2007). These plants are natural hosts of *H. obsoletus*, which is the only confirmed vector of stolbur phytoplasmas to grapevines (Maixner 1994), and infections occur only through erratic feeding as grape is not a suitable host. Since not only the phytoplasma strains but also the vector populations are specifically associated with the different host plants, distinct plant host specific epidemiological cycles of stolbur phytoplasmas occur in the field (Johannesen et al. 2008). Due to the feeding preferences of *H. obsoletus*, grapevine is a dead end host for the pathogens (Cousin and Boudon-Padieu 2002) and is therefore insignificant for the spread of BN.

11.3.3.3 Other Grapevine Yellows Diseases

In Virginia and New York State, North American Grapevine Yellows (NAGY) have been described and associated with Aster Yellows (AY) (16SrI-A) and X-disease phytoplasmas (16SrIII-I), but the vectors are unknown (Wolf et al. 1994; Beanland et al. 2006). They cause severe damage on Chardonnay and Riesling cultivars (Pearson et al. 1985; Wolf 2000). In Canada, AY (16SrI-A & -B) was detected in grapevines for the first time in 2006 (Olivier et al. 2009) and potential vectors were identified. Symptoms of Australian Grapevine Yellows (AGY) are associated with phytoplasma strains 16SrXII-B (stolbur group), 16Sr-II (faba bean phyllody group) and 16SrI-D (AY group) detected in single or mixed infections (Constable 2010).

11.3.3.4 Pierce's Disease (PD)

Pierce's Disease is caused by *Xylella fastidiosa*, a γ-proteobacteria which infects over 100 plant species, including grapes and stone fruits (Hopkins and Purcell 2002; Myers et al. 2007). In infected grapevines, leaves become slightly yellow or red along the margins and eventually fall. Even though their petiole remains attached, fruit clusters shrivel, wood on new canes matures irregularly with the appearance of patches of green surrounded by bark, shoots are stunted and their growths are delayed (Davis et al. 1978). *Xylella fastidiosa* has been divided into several subspecies based on molecular analysis and grape can be infected by strains belonging to different subspecies (Galvez et al. 2010). The disease occurs in North, Central, and South America. Winter temperatures seem to limit the geographical range of PD (Feil and Purcell 2001), but global warming might favour the northern expansion of the disease that has been found in trees as far north as southern Ontario and Alberta in Canada (Goodwin and Zhang 1997). Climatic differences between regions can affect the timing and severity of symptoms (Galvez et al. 2010).

Xylella fastidiosa is transmitted by xylem-sap feeding Auchenorrhyncha, mainly spittle bugs and sharpshooters, in a non-circulative yet propagative foregut-borne transmission mode. Its biology and vectors are known (Redak et al. 2004; Chatterjee et al. 2008). *Xylella fastidiosa* is passively acquired during a short feeding period and there is no latent period. Mechanisms of acquisition-transmission are well described (Chatterjee et al. 2008; Backus et al. 2009).

11.4 Leafhopper Management

Knowledge of leafhopper biology (overwintering sites, life cycles, dispersal, plant host range, vector status, etc.) and disease epidemiology is a major step in developing sustainable management programs to control leafhoppers (Bentley et al. 2007; Weintraub and Wilson 2010). The development of reliable sampling and monitoring techniques and the establishment of economic thresholds are essential components of IPM programs (Daane and Costello 2000). Chemical control methods remain the most common technique used to manage leafhopper populations (Daane and Costello 2000; Weintraub and Wilson 2010). Nevertheless, combinations of biological, cultural and chemical controls are being developed, and promoted, and are being more widely used, due to the lower associated economic and ecological costs.

11.4.1 General Control Measures for Grape Leafhoppers

11.4.1.1 Insecticides

For growers, insecticide applications remain the key technology to manage leafhopper populations. However, their effectiveness is often variable, particularly when leafhoppers are highly mobile or have a wide host range. Application timing and insecticide choice are important for successful leafhopper population reduction. Care should be taken in evaluating the economic and ecological costs of the insecticide applications that can have undesirable effects in vineyards when they are carried out at inappropriate times. For example, emergence of adult parasitoids from leafhopper eggs is adversely affected by imidacloprid residues (Byrne and Toscano 2007). Furthermore, insecticide treatments may become very costly when they cause outbreaks of other arthropod pests that need to be controlled following chemical treatments against leafhoppers (Flaherty et al. 1992).

11.4.1.2 Cultural Practices

Ground cover, watering, and pruning are some cultural practices that can affect both leafhoppers and natural enemies. In some cases, it is recommended to maintain ground cover to favor beneficial insect populations (Daane and Costello 2000). In other cases, growers are encouraged to treat or remove all plants in or around vineyards which could constitute reservoirs for phytoplasmas or alternative host plants for leafhoppers (McClure 1980; Weintraub and Beanland 2006;

Maixner 2007). Lack of vegetation or wide grapevine inter-rows can prevent monophagous (e.g., *S. titanus*) leafhopper species movements and phytoplasma spreading (Lessio and Alma 2004).

Irrigation practices can influence plant canopy and physiology, and leafhopper densities. Water stress induces modifications of leaf structure and causes lower densities of leafhoppers (Trichilo et al. 1990), by affecting feeding behaviors and oviposition (Daane and Williams 2003), as well as increasing egg and nymph mortality (Costello and Daane 2003). Regulated deficit irrigation was suggested to be used as a cultural practice to manage leafhopper populations (Chaves et al. 2007).

11.4.1.3 Biological Control

Natural enemies of leafhoppers are generally present in vineyards. Their abundance and effectiveness can vary considerably depending on different parameters, such as cultural practices, vineyard location, climatic conditions, and soil characteristics. Parasitoids may parasitize leafhopper adults, nymphs, or eggs and reduce populations of primary vectors over a wide area (Moya-Raygoza et al. 2006). Although leafhopper management by parasitoids is less effective than by insecticides, natural enemy conservation can significantly affect leafhopper populations (Daane and Costello 2000). Parasitoid populations can be favored by the presence of host plants (Daane and Costello 2000).

The dryinid *Aphelopus albopictus* Ashmead attacks the western grape leafhopper, but controls only 10–40% of the leafhopper population. The use of *Anagrus* spp. egg parasitoids to control different leafhopper species is being explored (Krugner et al. 2009).

Predators including green lacewings, tiger flies, nabid bugs, and ladybird beetles can play a significant role in controlling leafhopper populations (Daane and Costello 2000). Entomopathogenic fungi have been identified in leafhopper populations, some of them inducing high mortality levels (Langer et al. 2005; Boucias et al. 2007). However, these approaches gave variable results under field conditions (Laimer et al. 2009).

11.4.2 Control of Phytoplasma and Xylella fastidiosa *Vectors*

Different grapevine yellows (GY) diseases display similar symptoms but differ considerably in epidemiology, which mainly depends on the biology of the respective vectoring species. The expansion of Pierce's disease is dependent on the expansion of the host vector range which is highly dependent on climate. Beside cultural practices to prevent infection and to promote recovery of vines, control measures against GY aim to reduce infection pressure through interruption of the specific transmission cycles. This can include the direct control of the particular vectors as well as indirect measures to deprive vectors of their food sources or to prevent them from feeding on grapevines.

11.4.2.1 Control of *Scaphoideus titanus*

In all countries where FD occurs, insecticide treatments against the vector are mandatory in vector-affected vine nurseries and in municipalities with infested vineyards, as well as in any area of the affected municipality growing plants belonging to the species *Vitis* genus (Klinger 2003). Mandatory control measures include: control of plant movements from Europe, eradication programs (insecticide spraying, vineyard uprooting if >20% of infected grapevines are infected), certification schemes for commercialization, and hot water treatments (Mannini 2007). Hot water treatment routinely used in certified nurseries also kills the eggs of *S. titanus*, as well as several other pathogens and pest eggs (Waite et al. 2001). The shift from broad-spectrum insecticides to selective insecticides such as insect growth regulators or *B. thuringiensis* var. *kurstaki* is also suspected to have led to unexpected establishment of *S. titanus* and therefore to have contributed to the spread of FD (Belli et al. 2000). *Scaphoideus titanus* populations are being managed by eliminating the eggs through burning of the pruned wood and by treating the vineyards before bud burst with oils, followed by two insecticide treatments 30 and 45 days after the first egg hatch (Caudwell and Martelli 1992; Rousseau 1997). The dates of the treatments are established on a regional basis. A potential third treatment might be applied depending on the vector abundance estimated by visual inspection after the second treatment. Trapping flying adults might help to reduce the number of sprays by confirming the near absence of adults of *S. titanus* following the first larvicidal treatment (Van Helden et al. 2007).

In organic vineyards, rotenone, pyrethrum, and azadirachtin control the FD vector (Steffek et al. 2007). Sulfur and paraffin oil applications after bud break might help reduce the population of *S. titanus* (Rousseau 1997). A component of current efforts to control the introduced FD vector *S. titanus* in France involves exploration for effective micro-Hymenoptera parasitoids in the Great Lakes region of North America where populations of the endemic *S. titanus* are low (Nusillard et al. 2003).

11.4.2.2 Control of *Hyalesthes obsoletus*

The spread of BN is more difficult to control and less effective than that of FD. This is due to the more complex epidemiology of BN, where alternative host plants of the pathogens and of *H. obsoletus* are involved.

H. obsoletus is able to inoculate grapevines successfully within a short time (Bressan et al. 2007) although only a very small proportion of the vector population is present on the grapevines. Therefore, the direct control of this vector by insecticide applications in vineyards is not effective. The natural host plants of *H. obsoletus* are widespread and determine the density and distribution of the planthoppers. The control of these plants in vineyards and their surrounding areas, either by herbicides or by non-chemical means is an effective measure to reduce the *H. obsoletus* population density (Maixner 2007). Weed control should cease from 3 weeks prior to adult emergence until the end of the flight activity to prevent an increased movement of infective vectors from their treated natural hosts to grapevines (Maixner 2006). A green cover should be established in vineyards and adjacent areas wherever possible. This not only suppresses the less competitive host plants but also reduces the attractiveness of the plots for the vectors which prefer open soil with sparse vegetation. Another technique to reduce the number of emerging adult vectors is plowing during winter which brings the hibernating nymphs up to the soil surface were they are killed by frost. Tamping the soil around vineyards decreases the survival of eggs and nymphs (Sforza and Boudon-Padieu 1998).

The role of soil inhabiting predators of nymphs of *H. obsoletus* is unknown. Parasitoids have only rarely been found (Sforza et al. 1999). *Hyalesthes obsoletus* was infected and killed by the entomophagous fungus *Metarhizium anisopliae* (Metschnikoff) Sorokin in the laboratory. However, field experiments did not effectively reduce the population density. In Israel, a push and pull approach to prevent *H. obsoletus* from entering vineyards has being evaluated (Zahavi et al. 2007) by placing plants of *Vitex agnus-castus* L. around vineyards, as this bush is a preferred host plant of *H. obsoletus.*

11.4.2.3 Control of *Xylella fastidiosa* Vectors

Several disease management strategies have been used with success in areas with heavy PD incidence. These include grapevine resistance, management of vector populations and changes in cultural practices.

Despite differences in PD susceptibility in a few cultivars such as 'Petit Sirah' or 'Chenin Blanc', the majority of the high quality cultivars of *Vitis vinifera* L. are susceptible to PD (Hopkins and Purcell 2002). Although the native *Vitis rotundifolia* Michaux shows resistance to PD, it is not favored by growers because the berries are of poor quality. Other native *Vitis* spp. have been used as sources of resistance in breeding programs (Walker et al. 2009). These native cultivars might be useful for gene introduction into high quality cultivars (Galvez et al. 2010).

The most efficient PD vector is considered to be *Homalodisca vitripennis* because of its mobility and wide host range. Managing the first generation of *H. vitripennis* with early insecticidal treatments in vineyards and its surrounding vegetation helps slow down the expansion of PD (Hopkins and Purcell 2002). Foliar and soil-applied insecticides have been tested (Bethke et al. 2001). Kaolin, a fine grained aluminosilicate mineral applied as a particle film, physically coats plants and protects the grapes from *H. vitripennis* (Germar) (Puterka et al. 2003), and consequently reduces grapevine infections (Tubajika et al. 2007).

11.4.2.4 Resistant Plants and Rootstocks

Grapevine cultivars directly influence leafhopper biodiversity in vineyards (Martinson and Dennehy 1995), and rootstocks seem to play a role in insect attraction (Sharon et al. 2003). Screening of rootstocks and resistant plants for both phytoplasmas associated with apple trees (Seemuller and Harries 2010) and *X. fastidiosa* (Krivanek and Walker 2005) has been on-going for years, although not particularly successful for grapevines (Laimer et al. 2009). Engineered grapevine cultivars can express proteins able to interfere with *X. fastidiosa* or phytoplasmas (Bruening et al. 2008; Laimer et al. 2009). Likewise, engineered rootstocks could target phytoplasmas in the roots during winter (Laimer et al. 2009).

11.4.2.5 Physical Control

Physical barriers are effective protection tactics because the vector is physically excluded from reaching the plant. However, this may not always be technically practical. Blua et al. (2005) erected a 5 m tall screen barrier to determine the movement of *H. vitripennis*. They found that the leafhopper was repelled by it and was deflected to surrounding vegetation. They concluded that a physical barrier is an effective management tool, especially in the case of high-value crops. In another study, insect exclusion screens which absorbed ultraviolet light were found to be more effective against phytoplasma-bearing leafhoppers in walk-in tunnels used to grow *Limonium* spp. flowers (Weintraub et al. 2008).

11.5 Future Management Methods

A number of different methods using symbiontic bacteria are currently being researched. Many arthropods carry a diverse assembly of symbiotic microorganisms that are maternally inherited and have major effects on their hosts (Duron et al. 2008). Bacteria can be genetically modified to prevent the transmission of pathogens. Symbionts can be genetically modified and successfully reintroduced into leafhoppers to compete with the pathogens (Bextine et al. 2005; Miller et al. 2006). Studies have also been conducted on symbionts that can interact with phytoplasmas in fat bodies and salivary glands of leafhoppers (Bigliardi et al. 2006; Marzorati et al. 2006). A very new and novel potential method of phytoplasma control involves the use of *Pseudomonas putida* (Trevisan) against the chrysanthemum yellows phytoplasma (Gamalero et al. 2010). When phytoplasma-infected plants were treated with these plant beneficial bacteria, the viability of the phytoplasma was reduced and in fully developed leaves the phytoplasma had a degenerate appearance.

11.6 Conclusion

Wine consumption is growing in popularity, especially in North America. Wine produced in California alone accounts for two-thirds of all of the wine sold in the US, and some 21 million people visited California vineyards and wineries in 2010

(Anonymous 2011). As agritourism is increasing in popularity, the potential for the movement of plants, pathogens and vectors by this massive population increases concomitantly. Furthermore, the increase in intensive agricultural vineyard practices could lead to the buildup of pathogen and vector populations which could be confounded by climate change (Boudon-Padieu and Maixner 2007), enabling increased winter survival (Anderson et al. 2004). Faced with these challenges, growers need an array of technologies to manage leafhoppers, planthoppers, and the pathogens they transmit. Sequencing and genomic/proteomic comparisons between pathogen strains will allow scientists to better understand the pathogen-host relationships, and may identify new management tactics. However, research must move forward in the realm of plant resistance, resistant rootstocks and agricultural techniques such as: appropriate cover crops and trap plants to enhance natural enemies and divert hopper pests from the crop. Other research avenues to be considered are pesticide timing and application technology, as well as mechanical/physical protection.

References

Albertazzi G, Milc J, Caffagni A, Francia E, Roncaglia E, Ferrari F, Tagliafico E, Stefani E, Pecchioni N (2009) Gene expression in grapevine cultivars in response to Bois Noir phytoplasma infection. Plant Sci 176:792–804

Alma A (2002) Auchenorrhyncha as pests on grapevine. Denisia 176:531–538

Anderson PK, Cunningham AA, Patel NG, Morales FJ, Epstein PR, Daszak P (2004) Emerging infectious diseases of plants: pathogen pollution, climate change and agrotechnology drivers. Trends Ecol Evol 19:535–544

Angelini E, Negrisolo E, Clair D, Borgo M, Boudon-Padieu E (2003) Phylogenetic relationships among Flavescence dorée strains and related phytoplasmas determined by heteroduplex mobility assay and sequence of ribosomal and nonribosomal DNA. Plant Pathol 52:663–672

Anonymous (2011) Agricultural Marketing Resource Center, Agritourism Profile. http://www.agmrc.org/commodities_products/agritourism/agritourism_profile.cfm

Backus EA, Habibi J, Yan F, Ellersieck M (2005a) Stylet penetration by adult *Homalodisca coagulata* on grape: electrical penetration graph waveform characterization, tissue correlation, and possible implications for transmission of *Xylella fastidiosa*. Ann Entomol Soc Am 98:787–813

Backus EA, Serrano MS, Ranger CM (2005b) Mechanisms of hopperburn: an overview of insect taxonomy, behavior, and physiology. Annu Rev Entomol 50:125–151

Backus EA, Holmes WJ, Schreiber F, Reardon BJ, Walker GP (2009) Sharpshooter X wave: correlation of an electrical penetration graph waveform with xylem penetration supports a hypothesized mechanism for *Xylella fastidiosa* inoculation. Ann Entomol Soc Am 102:847–867

Beanland L, Noble R, Wolf TK (2006) Spatial and temporal distribution of North American grapevine yellows disease and of potential vectors of the causal phytoplasmas in Virginia. Environ Entomol 35:332–344

Belli G, Bianco PA, Casati P, Scattini G (2000) Serious and widespread outbreaks of flavescence dorée in vines in Lombardy. L'Informatore Agrario 56:56–59

Bentley WJ, Varela L, Daane K (2007) Grapes, insects, ecology and control. In: Pimentel D (ed) Encyclopedia of pest management. Taylor and Francis, New York, NY, pp 1–8

Bertamini M, Nedunchezhian N, Tomasi F, Grando MS (2002) Phytoplasma [Stolbur-subgroup (Bois noir-BN)] infection inhibits photosynthetic pigments, ribulose-1, 5-bisphosphate carboxylase and photosynthetic activities in field Brown grapevine (*Vitis vinifera* L. cv. Chardonnay) leaves. Physiol Mol Plant Pathol 61:357–366

Bertin S, Giglielmino CR, Karam N, Gomulski LM, Malacrida AR, Gasperi G (2007) Diffusion of the Nearctic leafhopper *Scaphoideus titanus* Ball in Europe: a consequence of human trading activity. Genetica 131:275–285

Bethke JA, Blua MJ, Redak RA (2001) Effect of selected insecticides on *Homalodisca coagulata* (Homoptera: Cicadellidae) and transmission of oleander leaf scorch in a greenhouse study. J Econ Entomol 94:1031–1036

Bextine B, Lampe D, Lauzon C, Jackson B, Miller TA (2005) Establishment of a genetically marked insect-derived symbiont in multiple host plants. Curr Microbiol 50:1–7

Bigliardi E, Sacchi L, Genchi M, Alma A, Pajoro M, Daffonchio D, Marzorati M, Avanzati AM (2006) Ultrastructure of a novel *Cardinium* sp. symbiont in *Scaphoideus titanus* (Hemiptera: Cicadellidae). Tissue Cell 38:257–261

Blua MJ, Campbell K, Morgan DJW, Redak RA (2005) Impact of a screen barrier on dispersion behaviour of *Homalodisca coagulate* (Hemiptera: Cicadellidae). J Econ Entomol 98:1664–1668

Böll S, Herrmann JV (2004) A long-term study on the population dynamics of the grape leafhopper (*Empoasca vitis*) and antagonistic mymarid species. J Pest Sci 77:33–42

Boucias DG, Scharf DW, Breaux SE, Purcell DH, Mizell RE (2007) Studies on the fungi associated with the glassy-winged sharpshooter *Homalodisca coagulata* with emphasis on a new species *Hirsutella homalodiscae* nom. prov. BioControl 52:231–258

Boudon-Padieu E, Maixner M (2007) Potential effects of climate change on distribution and activity of insect vectors of grapevine pathogens. In: Proceedings of the international and multidisciplinary colloquium, global warming, which potential impact on the vineyards, Dijon, France, pp 1–8, 26–28 Mar 2007

Bressan A, Clair D, Sémétey O, Boudon-Padieu E (2006) Insect injection and artificial feeding bioassays to test the vector specificity of Flavescence dorée phytoplasma. Phytopathology 96:790–796

Bressan A, Turata R, Maixner M, Spiazzi S, Boudon-Padieu E, Girolami V (2007) Vector activity of *Hyalesthes obsoletus* living on nettles and transmitting a stolbur phytoplasma to grapevines: a case study. Ann Appl Biol 150:331–339

Bruening G, Feldstein PA, Civerolo EL (2008) Exploiting *Xylella fastidiosa* proteins for Pierce's disease control. In: Proceedings of the Pierce's disease research symposium, California Department of Food and Agriculture, Sacramento, CA, pp 142–148. http://ddr.nal.usda.gov/bitstream/10113/32825/1/IND44250853.pdf

Byrne FJ, Toscano NC (2007) Lethal toxicity of systemic residues of imidacloprid against *Homalodisca vitripennis* (Homoptera: Cicadellidae) eggs and its parasitoid *Gonatocerus ashmeadi* (Hymenoptera: Mymaridae). Biol Control 43:130–135

Carlson JD, Whalon ME, Landis DA, Gage SH (1992) Springtime weather patterns coincident with long-distance migration of potato leafhopper into Michigan. Agric For Meteorol 59:183–206

Caudwell A, Martelli GP (1992) Flavescence dorée. In: Martelli GP (ed) Transmissible diseases of grapevines. Handbook for detection and diagnosis. FAO, Rome, pp 97–102

Chatterjee S, Almeida RPP, Lindow S (2008) Living in two worlds: the plant and insect lifestyles of *Xylella fastidiosa*. Annu Rev Phytopathol 46:243–271

Chaves MM, Santos TP, Souza CR, Ortuño MF, Rodrigues ML et al (2007) Deficit irrigation in grapevine improves water-use efficiency while controlling vigour and production quality. Ann Appl Biol 150:237–252

Claridge MF, Howse PE (1968) Songs of some British *Oncopsis* species (Hemiptera: Cicadellidae). Proc R Entomol Soc Lond Ser A Gen Entomol 43:57–61

Constable FE (2010) Phytoplasma epidemiology: grapevines as a model. In: Weintraub PG, Jones P (eds) Phytoplasmas: genomes, plant hosts and vectors. CABI, Wallingford, pp 188–212

Costello MJ, Daane KM (2003) Spider and leafhopper (*Erythroneura* spp.) response to vineyard ground cover. Environ Entomol 32:1085–1098

Cousin M-T, Boudon-Padieu E (2002) Phytoplasma and phytoplasma diseases: vectors, control and research topics. Cahiers d'études et de recherches francophones 11:115–126

Daane KM, Costello MJ (1998) Can cover crops reduce leafhopper abundance in vineyard? Calif Agric 52:27–33

Daane KM, Costello MJ (2000) Variegated and Western grape leafhoppers. In: Christensen P (ed) Raisin production manual, vol 3393. University of California, ANR, Oakland, pp 173–181
Daane KM, Williams LE (2003) Manipulating vineyard irrigation amounts to reduce insect pest damage. Ecol Appl 13:1650–1666
Daane KM, Smith RJ, Klonsky KM, Bentley WJ (2005) Organic vineyard management in California. Organic-Researchcom 5:37–55
Daire X, Clair D, Reinert W, Boudon-Padieu E (1997) Detection and differentiation of grapevine yellows phytoplasmas belonging to the elm yellows group and the stolbur subgroup by PCR amplification of non-ribosomal DNA. Eur J Plant Pathol 103:507–514
Davis MJ, Purcell AH, Thompson SV (1978) Pierce's disease of grapevines: isolation of the causal bacterium. Science 199:75–77
Decante D, Van Helden M (2008) Spatial and temporal distribution of *Empoasca vitis* within a vineyard. Agric For Entomol 10:111–118
Della Giustina W, Navarro E (1993) *Metcalfa pruinosa*, un nouvel envahisseur? Phytoma 451:30–32
Duron O, Bouchon D, Boutin S, Bellamy L, Zhou L, Engelstädter J, Hurst GD (2008) The diversity of reproductive parasites among arthropods: Wolbachia do not walk alone. BMC Biol 6, Article no 27
Feil H, Purcell AH (2001) Temperature-dependent growth and survival of *Xylella fastidiosa* in vitro and in potted grapevines. Plant Dis 85:1230–1234
Filippin L, Jović J, Cvrković T, Forte V, Clair D et al (2009) Molecular characteristics of phytoplasma associated with Flavescence dorée in clematis and grapevine and preliminary results on the role of *Dictyophara europaea* as a vector. Plant Pathol 58:826–837
Flaherty D, Christensen L, Lanini WT, Marois J, Phillip P, Wilson L (1992) Major insect and mite pests. In: Flaherty DL, Christensen L, Lanini WT, Marois J, Phillip P, Wilson L (eds) Grape pest management, vol 3343. University of California, DANR, Oakland, pp 120–225
Foissac X, Wilson MR (2010) Current and possible future distribution of phytoplasma diseases and their vectors. In: Weintraub PG, Jones P (eds) Phytoplasmas: genomes, plant hosts and vectors. CABI, Wallingford, pp 294–324
Forero D (2008) The systematics of the Hemiptera. Rev Colomb Entomol 34:1–21
Gajardo A, Fiore N, Prodan S, Paltrinieri S, Botti S et al (2009) Phytoplasmas associated with grapevine yellows disease in Chile. Plant Dis 93:789–796
Galvez LC, Korus K, Fernandez J, Behn JL, Banjara N (2010) The threat of Pierce's disease to midwest wine and table grapes. http://www.apsnet.org/publications/apsnetfeatures/Pages/Pierces.aspx.
Gamalero E, D'Amelio R, Musso C, Cantamessa S, Pivato B et al (2010) Effects of *Pseudomonas putida* S1Pf1Rif against chrysanthemum yellows phytoplasma infection. Phytopathology 100:805–813
Goodwin PH, Zhang S (1997) Distribution of *Xylella fastidiosa* in Southern Ontario as determined by the polymerase chain reaction. Can J Plant Pathol 19:13–18
Hogenhout SA, Oshima K, Ammar E-D, Kakizawa S, Kingdom HN, Namba S (2008) Phytoplasmas: bacteria that manipulate plants and insects. Mol Plant Pathol 9:403–423
Hogenhout SA, Van der Hoorn RA, Terauchi R, Kamoun S (2009) Emerging concepts in effector biology of plant-associated organisms. Mol Plant Microbe Interact 22:115–122
Hopkins DL, Purcell AH (2002) *Xylella fastidiosa*: cause of Pierce's disease of grapevine and other emergent diseases. Plant Dis 86:1056–1066
Hren M, Nikolié P, Rotter A, Blejec A, Terrier N et al (2009) 'Bois noir' phytoplasma induces significant reprogramming of the leaf transcriptome in the field grown grapevine. BMC Genomics 10:460–477
Hunt RE, Morton TL (2001) Regulation of chorusing in the vibrational communication system of the leafhopper *Graminella nigrifrons*. Am Zool 41:1222–1228
Johannesen J, Lux B, Michel K, Seitz A, Maixner M (2008) Invasion biology and host specificity of the grapevine yellows disease vector *Hyalesthes obsoletus* in Europe. Entomol Exp Appl 126:217–227
Klinger T (2003) Arrêté du 9 juillet 2003 relatif à la lutte contre la flavescence dorée de la vigne et contre son agent vecteur. Journal Officiel de la République Française 167, 22 juillet 2003, p 12362, texte n°49

Krivanek AF, Walker MA (2005) *Vitis* resistance to Pierce's disease is characterized by differential *Xylella fastidiosa* populations in stems and leaves. Phytopathology 95:44–52

Krugner R, Johnson MW, Morgan DJW, Morse JG (2009) Production of *Anagrus epos* Girault (Hymenoptera: Mymaridae) on *Homalodisca vitripennis* (Germar) (Hemiptera: Cicadellidae) eggs. Biol Control 51:122–129

Laimer M, Lemaire O, Herrbach E, Golsdmith V, Minafra A et al (2009) Resistance to viruses, phytoplasmas and their vectors in the grapevine in Europe: a review. J Plant Pathol 91:7–23

Langer M, Maixner M (2004) Molecular characterisation of grapevine yellows associated phytoplasmas of the stolbur-group based on RFLP-analysis of non-ribosomal DNA. Vitis 43:191–199

Langer M, Maixner M, Kirchmair M, Huber L (2005) Efficacy of *Metarhizium anisopliae* against *Hyalesthes obsoletus* (Auchenorrhyncha: Cixiidae). Vitis 44:99–100

Larsen KJ, Whalon ME (1988) Dispersal of *Paraphlepsius irroratus* (Say) (Homoptera: Cicadellidae) in peach and cherry orchards. Environ Entomol 17:842–851

Lenz MS, Isaacs R, Flore JA, Howell GS (2009) Vegetative growth responses of Pinot gris (*Vitis vinifera* L.) grapevines to infestation by potato leafhoppers (*Empoasca fabae* Harris). Am J Enol Vitic 60:130–137

Lessio F, Alma A (2004) Dispersal patterns and chromatic response of *Scaphoideus titanus* Ball (Homoptera Cicadellidae), vector of the phytoplasma agent of grapevine flavescence dorée. Agric For Entomol 6:121–127

Maixner M (1994) Transmission of German grapevine yellows (Vergilbungskrankheit) by the planthopper *Hyalesthes obsoletus* (Auchenorrhyncha: Cixiidae). Vitis 33:103–114

Maixner M (2006) Grapevine yellows – current developments and unsolved questions. In: Proceedings of the 15th meeting of the international council for the study of virus and virus-like diseases of the grapevine, Stellenbosch, South Africa, pp 86–88, 3–7 Apr 2006

Maixner M (2007) Biology of *Hyalesthes obsoletus* and approaches to control this soil borne vector of Bois noir disease. IOBC/WPRS Bull 30:3–9

Maixner M, Pearson RC, Boudon-Padieu E, Caudwell A (1993) *Scaphoideus titanus*, a possible vector of grapevine yellows in New York. Plant Dis 77:408–413

Maixner M, Reinert W, Darimont H (2000) Transmission of grapevine yellows by *Oncopsis alni* (Schrank) (Auchenorrhyncha: Macropsinae). Vitis 39:83–84

Mannini F (2007) Hot water treatment and field coverage of mother plant vineyards to prevent propagation material from phytoplasma infections. Bull Insectol 60:311–312

Martinson TE, Dennehy TJ (1995) Varietal preference of *Erythroneura* leafhoppers (Homoptera: Cicadellidae) feeding on grapes in New York. Environ Entomol 24:550–558

Martinson TE, Dennehy TJ, Hoffman CJ (1994) Phenology, within-vineyard distribution, and seasonal movement of eastern grape leafhopper (Homoptera: Cicadellidae) in New York vineyards. Environ Entomol 23:236–243

Martinson TE, Dunst R, Lakso A, English-Loeb G (1997) Impact of feeding injury by Eastern grape leafhopper (Homoptera: Cicadellidae) on yield and juice quality of concord grapes. Am J Enol Vitic 48:291–302

Marzorati M, Alma A, Sacchi L, Pajoro M, Palermo S, Brusetti L et al (2006) A novel *Bacteroidetes* symbiont is localized in *Scaphoideus titanus*, the insect vector of Flavescence dorée in *Vitis vinifera*. Appl Environ Microbiol 72:1467–1475

Matus JT, Vega A, Loyola R, Serrano C, Cabrera S, Arce-Johnson P (2008) Phytoplasma and virus detection in commercial plantings of *Vitis vinifera* cv. Merlot exhibiting premature berry dehydration. Electron J Biotechnol 11:1–10

Mazzoni V, Anfora G, Ioriatti C, Lucchi A (2008) Role of winter host plants in vineyard colonization and phenology of *Zygina rhamni* (Hemiptera: Cicadellidae: Typhlocybinae). Ann Entomol Soc Am 101:1003–1009

Mazzoni V, Lucchi A, Cokl A, Presern J, Virant-Doberlet M (2009) Disruption of the reproductive behavior of *Scaphoideus titanus* by playback of vibrational signals. Entomol Exp Appl 133:174–185

McClure MS (1980) Role of wild host plants in the feeding, oviposition, and dispersal of *Scaphytopius acutus* (Homoptera: Cicadellidae) a vector of peach X-disease. Environ Entomol 9:283–292

Meignoz R, Boudon-Padieu E, Larrue J, Caudwell A (1992) Flavescence dorée de la vigne. Présence de MLO et effets cytopathogènes associés dans le liber de la vigne. J Phytopathol 134:1–9

Miller TA, Lauzon CR, Lampe D, Durvasula R, Matthews C (2006) Paratransgenesis applied to control insect-transmitter plant pathogens: the Pierce's disease case. In: Bourtzis K, Miller TA (eds) Insect symbiosis 2. Taylor and Francis/CRC Press, London/Boca Raton, pp 243–267

Moya-Raygoza G, Palomera-Avalos V, Chacon-Torres NM, Becerra-Chiron IM (2006) The parasitoid *Gonatopus bartletti* reduces presence of plant-pathogenic *Spiroplasma kunkelii* within the leafhopper vector *Dalbulus maidis*. Entomol Exp Appl 119:189–196

Musetti R, Marabottini R, Badiani M, Martini M, Sanità di Toppi L et al (2007) On the role of H_2O_2 in the recovery of grapevine (*Vitis vinifera* cv. Prosecco) from Flavescence dorée disease. Funct Plant Biol 34:750–758

Myers AL, Sutton TB, Abad JA, Kennedy GG (2007) Pierce's disease of grapevines: identification of the primary vectors in North Carolina. Phytopathology 97:1440–1450

Nielson MW (1968) The leafhopper vectors of phytopathogenic viruses (Homoptera, Cicadellidae) taxonomy, biology and virus transmission. Agric Res Serv – U S Dep Agric Tech Bull 1382:1–386

Nusillard B, Malausa JC, Giuge L, Millot P (2003) Assessment of a two year study of the natural enemy fauna of *Scaphoideus titanus* ball in its North American native area. IOBC/WPRS Bull 26:237–240

Olivier CY, Lowery DT, Stobbs LW, Vincent C, Galka B, Saguez J et al (2009) First report of aster yellow phytoplasmas ('*Candidatus* phytoplasma asteris') in Canadian grapevines. Plant Dis 93:669

Olsen KN, Cone WW, Wright LC (1998) Influence of temperature on grape leafhoppers in South Central Washington. Environ Entomol 27:401–405

Orenstein S, Zahavi T, Nestel D, Sharon R, Barkalifa M, Weintraub PG (2003) Spatial dispersion patterns of potential leafhopper and planthopper (Homoptera) vectors of phytoplasma in wine vineyards. Ann Appl Biol 142:341–348

Pearson RC, Pool RM, Gonsalves D, Goffinet DC (1985) Occurrence of flavescence doree-like symptoms on 'White Riesling' grapevines in New York, U.S.A. Phytopathol Mediterr 24:82–87

Purcell AH, Finlay AH (1979) Evidence for non-circulative transmission of Pierce's disease bacterium by sharpshooters. Phytopathology 69:393–395

Puterka GJ, Reinke M, Luvisi D, Ciomperlik MA, Bartels D et al (2003) Particle film, surround WP, effects on glassy-winged sharpshooter behavior and its utility as a barrier to sharpshooter infestations in grapes. Plant Health Prog Online. doi:10.1094/PHP-2003-0321-RS

Redak RA, Purcell AH, Lopes JRS, Blua MJ, Mizel RF III, Andersen PC (2004) The biology of xylem fluid-feeding insect vectors of *Xylella fastidiosa* and their relation to disease epidemiology. Annu Rev Entomol 49:243–270

Rousseau J (1997) Flavescence dorée: what can be done organically? Agric Biol 11(4):20–23

Schvester D, Carle P, Moutous G (1961) Sur la transmission de la flavescence dorée des vignes par une cicadelle. C R Acad Agric Fr 47:1021–1024

Seemuller E, Harries H (2010) Plant resistance. In: Weintraub PG, Jones P (eds) Phytoplasmas: genomes, plant hosts and vectors. CABI, Wallingford, pp 147–169

Sforza RD, Boudon-Padieu E (1998) Le principal vecteur de la maladie du Bois noir. Phytoma 510:33–37

Sforza R, Bourgoin T, Wilson SW, Boudon-Padieu E (1999) Field observations, laboratory rearing and descriptions of immatures of the planthopper *Hyalesthes obsoletus* (Hemiptera: Cixiidae). Eur J Entomol 96:409–418

Sharon R, Weintraub PG, Zahavi T (2003) Effect of rootstock on grapevine yellows – facts and explanations. In: Proceedings of the 14th ICVG conference, Locorotondo, Italy, pp 73–74, 12–17 Sept 2003

Sōgawa K (1982) The rice brown planthopper: feeding physiology and host plant interactions. Annu Rev Entomol 27:49–73

Steffek R, Reisenzein H, Zeisner N (2007) Analysis of the pest risk from grapevine Flavescence dorée phytoplasma to Austrian viticulture. EPPO bull 37:191–203

Taylor RAJ, Reling D (1986) Preferred wind direction of long distance leafhopper (*Empoasca fabae*) migrants and its relevance to the return migration of small insects. J Anim Ecol 55:1103–1114

Trichilo PJ, Wilson LT, Grimes DW (1990) Influence of irrigation management on the abundance of leafhoppers (Homoptera: Cicadellidae) on grapes. Environ Entomol 19:1803–1809

Tubajika KM, Civerolo EL, Puterka GJ, Hashim JM, Luvisi DA (2007) The effects of kaolin, harpin, and imidacloprid on development of Pierce's disease in grape. Crop Prot 26:92–99

Urban JN, Cryan JR (2007) Evolution of the planthoppers (Insecta: Hemiptera: Fulgoroidea). Mol Phylogenet Evol 42:556–572

Valk M, Stevenson AB (1994) Aster leafhoppers. In: Howard RJ, Garland JA, Seaman WL (eds) Diseases and pests of vegetables crops in Canada. The Canadian Phytopathological Society and the Entomological Society of Canada, Ottawa, pp 158–159

Van Helden M, Pain G, Pithon J (2007) Landscape characteristics influencing pest populations in viticulture. In: Lozzia GC, Lucchi A, Di Chiara SR, Tsolakis H (eds) Proceedings of the working group "Integrated protection in viticulture" meeting, Marsala, Italy, pp 369–373, 25–27 Oct 2007

Vidano C (1964) Scoperta in Italia dello *Scaphoideus littoralis* Ball cicalina Americana collegata alla Flavescence dorée della vite. Ital Agric 101:1031–1049

Waite H, Crocker J, Fletcher G, Wright P, DeLaine A (2001) Hot water treatment in commercial nursery practice – an overview. Aust Grapegrow Winemak 449a:39–43

Walker A, Tenscher A, Riaz S, Ramming D (2009) Breeding Pierce's disease resistant winegrapes. Foundation Plant Services, University of California, FPS Grape Program Newsletter, October 2009:8–12

Wayadande AC, Fletcher J (1995) Transmission of *Spiroplasma citri* lines and their ability to cross gut and salivary gland barriers within the leafhopper vector *Circulifer tenellus*. Phytopathology 85:1256–1259

Wayadande AC, Baker GR, Fletcher J (1997) Comparative ultrastructure of the salivary glands of two phytopathogen vectors, the beet leafhopper, *Circulifer tenellus* (Baker), and the corn leafhopper, *Dalbulus maidis* Delong and Wolcott (Homoptera: Cicadellidae). Int J Ins Morphol Embryol 26:113–120

Weintraub PG, Beanland L (2006) Insect vectors of phytoplasmas. Annu Rev Entomol 51:91–111

Weintraub PG, Jones P (2010) Phytoplasmas: genomes, plant hosts and vectors. CABI, Wallingford

Weintraub PG, Wilson MR (2010) Control of phytoplasma diseases and vectors. In: Weintraub PG, Jones P (eds) Phytoplasmas: genomes, plant hosts and vectors. CABI, Wallingford, pp 233–249

Weintraub PG, Pivonia S, Gera A (2008) Physical control of leafhoppers. J Econ Entomol 101:1337–1340

Weiss MR (2006) Defecation behavior and ecology of insects. Annu Rev Entomol 51:635–661

Wolf TK (2000) Grapevine yellows research in Virginia. Wines Vines, October 2000:28–35

Wolf RK, Prince JP, Davis RE (1994) Occurrence of grapevine yellows in Virginia vineyards. Plant Dis 78:203

Yigit A, Erckle L (1992) Studies on bio-ecology and control of grape leafhopper (*Arboridia* (=*Erythroneura*) *adanae* Dlab.) (Homoptera: Cicadellidae) in southern Anatolia region. Zirai Mücadele Arastirma Yilligi 22(23):25–28

Zahavi T, Peles S, Harari AR, Soroker V, Sharon R (2007) Push and pull strategy to reduce *Hyalesthes obsoletus* population in vineyards by *Vitex agnus castus* as trap plant. Bull Insectol 60:297–298

Zhou L, Hoy CW, Miller SA, Nault LR (2003) Marking methods and field experiments to estimate aster leafhopper (*Macrosteles quadrilineatus*) dispersal rates. Environ Entomol 32:1177–1186

Zimmerman R, Kondratieff B, Nelson E, Sclar C (1996) The life history of two species of grape leafhoppers on wine grapes in Western Colorado. J Kansas Entomol Soc 69:337–345

Chapter 12
Biology and Management of Mealybugs in Vineyards

Kent M. Daane, Rodrigo P.P. Almeida, Vaughn A. Bell, James T.S. Walker, Marcos Botton, Majid Fallahzadeh, M. Mani, Jose Luis Miano, René Sforza, Vaughn M. Walton, and Tania Zaviezo

12.1 Introduction

Economic losses resulting from vineyard mealybug infestations have increased dramatically during the past decade. In response, there has been a cosmopolitan effort to improve control strategies and better understand mealybug biology and ecology, as well as their role as vectors of plant pathogens. Mealybugs are named for the powdery secretions covering their bodies. The most important vineyard mealybugs belong to the subfamily Pseudococcinae (Hardy et al. 2008). Although numerous mealybug species are found in vineyards, this chapter will cover only those that have risen to the level of primary pest. These are the grape mealybug, *Pseudococcus maritimus* (Ehrhorn),

---

K.M. Daane (✉) • R.P.P. Almeida
Department of Environmental Science, Policy and Management, University of California, Berkeley, CA 94720-3114, USA
e-mail: daane@uckac.edu; rodrigoalmeida@berkeley.edu

V.A. Bell • J.T.S. Walker
New Zealand Institute for Plant & Food Research Limited, Private Bag 1401, Havelock North, 4157 New Zealand
e-mail: vaughn.bell@plantandfood.co.nz; jim.walker@plantandfood.co.nz

M. Botton
Embrapa Grape and Wine, Livramento st 515, PO Box 130, 95700-000 Bento Gonçalves, RS, Brazil
e-mail: marcos@cnpuv.embrapa.br

M. Fallahzadeh
Department of Entomology, Jahrom Branch, Islamic Azad University, Jahrom, Iran
e-mail: mfalahm@yahoo.com

M. Mani
Division of Entomology and Nematology, Indian Institute of Horticultural Research, Bangalore 560-089, India
e-mail: mmani1949@yahoo.com

N.J. Bostanian et al. (eds.), *Arthropod Management in Vineyards: Pests, Approaches, and Future Directions*, DOI 10.1007/978-94-007-4032-7_12,

obscure mealybug, *Pseudococcus viburni* (Signoret), longtailed mealybug, *Pseudococcus longispinus* (Targioni-Tozzetti), citrophilus mealybug, *Pseudococcus calceolariae* (Maskell), vine mealybug, *Planococcus ficus* (Signoret), citrus mealybug, *Planococcus citri* (Risso), pink hibiscus mealybug, *Maconellicoccus hirsutus* (Green), and the newly identified Gill's mealybug, *Ferrisia gilli* Gullan. Meanwhile in Brazil and India, *Dysmicoccus brevipes* (Cockerell) and *Xenococcus annandalei* Silvestri respectively, feed on vine roots. Collectively, these species will be referred to as the vineyard mealybugs, although their host range is diverse and many are pests of other agricultural crops and ornamental plants (McKenzie 1967; Ben-Dov 1995).

Outwardly, the vineyard mealybugs look similar. Mealybug females are wingless with an elongate-oval body (3–5 mm) that can be covered with wax secretions forming distinctive spine-like filaments. However, each species has distinct biological characteristics that result in different geographic ranges, host plant preferences, economic injury, and management strategies. This chapter presents a generalized description of their biology, damage, and life history, and summarizes the current information on cultural, biological, and chemical control practices. It provides brief descriptions of their regional significance and future control needs. For further reference, McKenzie (1967), Williams and Granara de Willink (1992), Ben-Dov (1995), and Hardy et al. (2008) provide reviews of Pseudococcidae taxonomy, geographic and/or host range and biology. Noyes and Hayat (1994) provide a review of some of the Anagyrini parasitoids attacking Pseudococcidae, and the ScaleNet (2011) website is an excellent reference tool.

12.2 Mealybug Biology and Development

12.2.1 Nomenclature and Geography

In order to provide even a brief description of the world's vineyard mealybugs some background on their nomenclature and geographic distribution is needed (Table 12.1).

J.L. Miano
INTA EEA Mendoza, San Martín 3853, Luján de Cuyo, Mendoza 5507, Argentina
e-mail: jlmiano@agrohuarpes.com

R. Sforza
USDA-ARS-European Biological Control Laboratory, Campus International de Baillarguet, CS90013 Montferrier sur Lez, 34988 St-Gely du Fesc, France
e-mail: rsforza@ars-ebcl.org

V.M. Walton
Department of Horticulture, Oregon State University, Corvallis, OR 97331-7304, USA
e-mail: waltonv@hort.oregonstate.edu

T. Zaviezo
Departamento de Fruticultura y Enología, Universidad Católica de Chile,
Casilla 302-22, Santiago, Chile
e-mail: tzaviezo@uc.cl

Table 12.1 The common vineyard mealybug species showing the pests geographic origin (by terrestrial ecozone) and present regional distribution, as well as common synonyms

Species (author and year)	Geographic origin	Current distribution	Common synonyms
Pseudococcus maritimus (Ehrhorn)	Nearctic	North America	*Dactylopius maritimus* Ehrhorn, *Pseudococcus bakeri* Essig, *P. omniverae* Hollinger
Pseudococcus viburni (Signoret)	Neotropic	Australia, Europe, New Zealand, North America (California), South Africa, South America	*Dactylopius indicus* Signoret, *D. viburni* Signoret, *D. affinis* Maskell, *Pseudococcus viburni* (Signoret), *Ps. affinis* (Maskell), *Ps. obscurus* Essig, *Ps. capensis* Brain, *Ps. nicotianae* Leonardi, *Ps. longispinus latipes* Green, *Ps. fathyi* Bodenheimer, *Ps. malacearum* Ferris, *Ps. affinis* (Maskell)
Pseudococcus longispinus (Targioni-Tozzetti)	Australasia	Australia, Europe, New Zealand, North America (California), South Africa, South America	*Coccus adonidum* L., *C. laurinus* Boisduval, *Dactylopius longispinus* Targioni Tozzetti, *D. adonidum* (L.), *D. adonidum* Auctorum, *D. hoyae* Signoret, *D. pteridis* Signoret, *D. longifilis* Comstock, *Oudablis lauri* Cockerell, *Pediculus coffeae* L., *Pseudococcus hoyae* (Signoret), *Ps. adonidum* (L.) *Ps. laurinus* (Boisduval), *Ps. adonidum* (Auctorum)
Pseudococcus calceolariae (Maskell)	Australasia	Australia, Europe, New Zealand, South America, South Africa, North America	*Pseudococcus citrophilus* Clausen, *Ps. fragilis* Brain, *Ps. gahani* Green
Planococcus citri (Risso)	Palearctic	Australia, Europe, New Zealand, North America, South Africa, South America	*Coccus tuliparum* Bouché, *C. citri* Boisduval, *Dactylopius alaterni* Signoret, *D. ceratoniae* Signoret, *D. citri* Signoret, *D. cyperi* Signoret, *D. robiniae* Signoret, *D. brevispinus* Targioni-Tozzetti, *D. destructor* Comstock, *D. secretus* Hempel, *Dorthesia citri* Risso, *Lecanium phyllococcus* Ashmead, *Phenacoccus spiriferus*, *Planococcoides cubanensis* Ezzat & McConnell, *Pl. citricus*, *Pl. cucurbitae* Ezzat & McConnell, *Pseudococcus citri*, Cockerell, *Ps.* (*citri*) *phenacocciformis* Brain
Planococcus ficus (Signoret)	Palearctic	North America (California and Mexico), South Africa, South America (Argentina), Europe (Italy), Middle East	*Coccus vitis* Lindinger, *Dactylopius ficus* Signoret, *D. vitis* Signoret, *D. subterraneus* Hempel, *Planococcus vitis* Ezzat & McConnell, *Pseudococcus citrioides* Ferris, *Ps. vitis* Bodenheimer

(continued)

Table 12.1 (continued)

Species (author and year)	Geographic origin	Current distribution	Common synonyms
Dysmicoccus brevipes (Cockerell)	Indo-Malaya	Australia, Africa, Asia, Middle East, South America (Brazil)	*Dactylopius brevipes* Cockerell, *D.* (*Pseudococcus*) *ananassae* Kuwana, *Dysmicoccus brevipes* (Cockerell), *Pseudococcus brevipes* (Cockerell), *Ps. missionum* Cockerell, *Ps. palauensis* Kanda, *Ps. cannae* Green, *Ps. longirostralis* James, *Ps. defluiteri* Betrem, *Ps. pseudobrevipes* Mamet
Ferrisia gilli (Gullan)	Nearctic	North America (California)	none
Maconellicoccus hirsutus (Green)	Indo-Malaya	Australia, Africa, Asia (India), Middle East, South America, Mexico, California	*Maconellicoccus perforatus* (De Lotto), *M. pasaniae* (Borchsenius), *Paracoccus pasaniae* Borchsenius, *Phenacoccus hirsutus* Green, *Ph. quarternus* Ramakrishna Ayyar, *Ph. quarternus* Shafee et al., *Pseudococcus hibisci* Hall, *Ps. glomeratus* Green, *Ps. crotolariae* Miller, *Ps. crotolariae* Yunus & Ho, *Spilococcus perforatus* De Lotto

Regions underlined indicate that the mealybug species is considered to be a primary pest

Historically, vineyard mealybug species were often misidentified, leading to confusion on their geographic distribution and economic importance. For example, many of the early North American specimens of mealybugs on grapes and pome fruit were described as *Ps. maritimus*, and yet, from the slides labeled as *Ps. maritimus* at the United States Museum of Natural History, there were at least 10 different species (Miller et al. 1984). It was particularly difficult to separate *Ps. maritimus* and *Ps. viburni* (Fig. 12.1) until the needed taxonomic descriptions of these closely related species were provided (Miller et al. 1984). Separation of *Pl. ficus* (Fig. 12.2) and *Pl. citri* is similarly difficult and can be made only through careful slide preparation to discern slight differences in multilocular pores and tubular ducts on adult females (Williams and Granara de Willink 1992). Demontis et al. (2007) and Cavalieri et al. (2008) provide a molecular separation of these species. Adult *Ps. calceolariae*, *M. hirsutus*, *F. gilli* are more easily distinguished. For example, *Ps. calceolariae* has distinctive dark stripes and short caudal filaments, *M. hirsutus* lacks lateral filaments, and *F. gilli* has glass-like rods (Fig. 12.3).

Complicating their proper identification is the fact that these pests have been often moved from their geographic origin such that many are now found in multiple regions (Table 12.1). The mealybug with the most limited range in vineyards is *F. gilli*, a Nearctic species that has been reported as vineyard pest only in California's Sierra foothills. This mealybug was only recently described, initially found infesting California pistachios (Gullan et al. 2003). It is included here as it could be misidentified

Fig. 12.1 Adult females of *Pseudococcus maritimus* (**a**) and *Ps. viburni* (**b**). These closely related species can only be discerned through slide preparation to view differences in multilocular pores and tubular ducts, or through the use of molecular techniques. A relatively reliable field tool is the color of the ostiolar fluid, extruded when the insect is prodded with a sharp object, which is *red* for *Ps. maritimus* and *clear* to *opaque* for *Ps. viburni*

Fig. 12.2 *Planococcus ficus* on the petiole of a grape berry (**a**) provides the classic view of the large (3–5 mm) adult female mealybug. However, a better portrayal of mealybug size and appearance in the field is provided by the infested grape leaf (**b**) that has more than 1,000 *Pl. ficus* of all developmental stages, but primarily second and third instars

as *Ferrisia malvastra* (McDaniel), found in other grape-growing regions of the world. *Dysmicoccus brevipes* and *X. annandalei* are reported only as important vineyard pests of Brazil and India, respectively. Although *Ps. maritimus* is well known, it was reported only from California vineyards until the 1950s. It is now known to occur in all North American vineyard regions from Canada to Mexico and from California to New York (Ben-Dov 1995; ScaleNet 2011). *Maconellicoccus hirsutus* is an Indo-Malaya native, and although it is now found in numerous regions, it is a primary vineyard pest only in India (Table 12.1). The other vineyard mealybug species are commonly found in more than one of the world's vineyards regions, although their pest status varies. For example, *Pl. ficus* and *Pl. citri* are found across a wide geographic range, but only in a few countries (primarily Spain, Italy and

Fig. 12.3 Adult female *Ferrisia gilli* with glass-like rods that accompany the production of live crawlers (small *yellow-orange* insects in the photo)

Brazil) *Pl. citri* is consistently cited as a vineyard pest (Cabaleiro and Segura 1997), whereas *Pl. ficus* is cited as a pest in Europe, the Middle East, northern Africa, South Africa, South America, California, and Mexico (Ben-Dov 1995; ScaleNet 2011). The transport of vine wood (both legal and illegal) and fruit is often suspected in the movement of mealybugs. However the wide host range of many of these species, which includes commonly used ornamental plant species (Ben-Dov 1995; ScaleNet 2011), makes border screening for the more ubiquitous mealybug species a daunting task.

12.2.2 Life History

There are slight variations among the species, but vineyard mealybugs generally have three larval instars for the female and four instars for the male (McKenzie 1967; Ben-Dov 1995; Wakgari and Giliomee 2005). The unsettled first instar, or crawler, moves quickly to find a feeding spot and is considered to be the dispersal stage. The first instar is about 0.6 mm long. Viewed from above, it is elongate-oval in shape, but from the side it is extremely flat. There are three molts, resulting consecutively in the second instar, third instar, and the 'immature' adult. Each of these stages resembles the previous except for an increasing size and amount of wax secretion. Females are unwinged and as they mature, become more sessile. Immature males are slightly longer and more slender than females. At the fifth instar, the male goes through a cocoon or prepupal stage and the emerged adult male is winged.

Fig. 12.4 Adult mealybug males are winged, as shown here for *Planococcus ficus*, next to an adult female producing an ovisac

12.2.3 Reproduction

The mature or gravid adult female begins to grow in size as the ovaries develop, ending at about 4–5 mm in length and far less dorso-ventrally flattened. The adult male is about 1.5 mm in length, with long wings, a brown colored body and two multi-segmented antennae that are about half the body length (Fig. 12.4). Sex determination of the vineyard mealybugs is unusual and worth noting as it impacts pest management programs. These mealybugs have the lecanoid type of the paternal genome elimination system, where both sexes develop from fertilized eggs (i.e., diploidy), but during early development of the male the paternal half is deactivated through heterochromatinization (Ross et al. 2010a). This system suggests females would produce a male-biased sex ratio when alone, and a more female-biased sex ratio when crowded with other females. However, in one study with *Pl. citri*, the opposite effect of crowding was observed, with a more male-biased sex ratio, suggesting that some mealybug species may selectively adjust their sex ratio (Ross et al. 2010b).

As suggested, mealybug reproduction can be quite variable. For vineyard mealybugs, mating is probably necessary (e.g. Zaviezo et al. 2010; Waterworth et al. 2011), although facultative parthenogenesis has been reported for *Pl. citri* (da Silva et al. 2010). To attract adult males, females emit a sex pheromone. For those species tested, females mate multiple times, and the number of matings affects egg production (Waterworth et al. 2011). Most vineyard mealybugs place their eggs in cotton-like ovisacs. *Pseudococcus longispinus*, *F. gilli*, *D. brevipes* and *Heliococcus bohemicus* Sulc (Bohemian mealybug), are the exceptions being ovoviviparous (depositing live first instars). The number of offspring produced per female varies depending on the species, environmental conditions, and food supply (Zaviezo et al. 2010). It has been reported ranging from about 50 to over 800.

12.2.4 Seasonal Development

Temperature is the driving force for mealybug development, although development times and temperature thresholds differ among species. For example, *Ps. maritimus* will have two generations in California's interior valleys (Geiger and Daane 2001), whereas *Pl. ficus* can have seven generations in the same region (Gutierrez et al. 2008) but is reported to have only three generations per year in Italy (Ben-Dov 1995). Similarly, *Pl. citri* in Brazil has six generations per year in the south, but up to 11 per year in the northeast where grapes are produced year round (two harvests per season). Other than *Ps. maritimus* and *H. bohemicus*, there does not appear to be winter dormancy for vineyard mealybugs.

There is also variation in seasonal feeding location and movement on the vine among and within species, depending on factors such as regional temperatures and vineyard management practices, as described for *Ps. maritimus* (Geiger and Daane 2001; Grasswitz and James 2008), *Pl. citri* (Cid et al. 2010), and *Pl. ficus* (Becerra et al. 2006). Here, a *Pl. ficus* infestation in an untreated table grape vineyard in California's Central Valley is used to exemplify the seasonal population dynamics (Daane et al. 2011). The mealybug population overwinters primarily under the bark of the trunk and cordon, with some of the population found underground on the roots, especially when tended by ants. There is no diapause. On warm days, development may occur during the winter months, with completion of the first generation almost entirely under the bark. From spring to summer, the *Pl. ficus* population follows the movement of plant resources from roots to shoots to leaves. Four to five generations are completed and population density can increase rapidly, although high summer temperatures, in excess of 40°C, may slow the growth of the population and increase mortality. As berries ripen and sugars develop, mealybugs move into the berry clusters, first attacking those near the vine cordon. The rapid population increase in summer is followed by an equally rapid decline after harvest, resulting from biological controls and abiotic mortality associated with high temperatures and vine senescence.

12.3 Mealybug Damage

12.3.1 Mealybug Feeding and Contamination

Mealybugs are phloem feeders that use long, slender mouthparts to suck out plant fluids (McKenzie 1967). Most of the vineyard mealybugs can feed on the vine's root, trunk, canes, leaves, or berry clusters. There are, however, differences in the amount of damage caused by each species. This is often related to those factors that determine population size (e.g., number of annual generations and female fecundity), preferred feeding locations, and temperature tolerances.

Fig. 12.5 Most mealybug species can feed on the vine roots, trunk, cane, leaves and fruit clusters. Severe infestations can result in defoliation, cluster infestation and rot, as shown for a *Planococcus ficus* infestation (**a**). Most mealybug populations remain at low levels with only a few berries in a cluster infested, as shown in (**b**) for *Pseudococcus maritimus*

As the mealybugs feed, they eliminate carbohydrate-rich honeydew, which can accumulate on the leaves and in the grape clusters, especially in late summer and early fall (Charles 1982). The mealybug 'flicks' honeydew away from its location, but it still accumulates on the vine. It has long been noted that honeydew serves as a substrate for the development of sooty mold fungi that can result in further vine damage. For table grape growers, any live or dead mealybugs and the honeydew or sooty molds will cause cosmetic damage to the grape cluster and reduce its marketability (Daane et al. 2011). In most raisin, juice, and wine grapes, the contamination from a small mealybug population, and the resultant honeydew droplets, will not cause economic damage. Although honeydew can be dissolved by light rain and will dry in warm temperatures, when mealybug populations are severe, honeydew can accumulate to form a hard, wax-like layer that covers the infested plant (Fig. 12.5). Feeding damage can result in defoliation and, after repeated annual infestations, cause vine death (Walton and Pringle 2004b).

12.3.2 Grapevine Leafroll Disease

In most of the world's wine grape regions, the transmission of viruses, rather than mealybug feeding or contamination, is the primary concern (Walton and Pringle 2004b; Charles et al. 2009; Bertin et al. 2010; Tsai et al. 2010). Grapevine leafroll disease (GLD) is caused by a complex of several viruses, collectively known as grapevine leafroll-associated viruses (GLRaVs). In cool-climate regions, the pathogen can be damaging to vines, crop, and wine quality. The most obvious GLD symptoms become apparent in the fall, when red cultivars display leaf reddening with green venation (Fig. 12.6). Symptoms are not as apparent in white cultivars where

Fig. 12.6 For many wine grape growers, grape leafroll disease is a greater concern than mealybug contamination. Many mealybug species have been shown to transmit the viruses that cause grape leafroll disease. The most apparent field symptoms are the *reddening* of leaves on *red* cultivars and the rolling of the leaf margin (**a**). The survival of mealybug on vine roots – even after the vine above has been pulled – is a major concern in the control of grape leafroll disease (**b**)

there is a slight leaf chlorosis. Both red and white cultivars develop the classic downward rolling of leaf margins and phloem disruption. GLRaV infections impact berry development and growth by delaying budbreak, flowering, and berry maturation, including changes in color, reduced sugar content, and increased acidity in fruit juice (Martelli et al. 2002; Charles et al. 2006).

Grapevine leafroll disease is associated with many distinct closteroviruses sequentially named GLRaV-1, -2 and so on; so far 10 species have been proposed (Martelli et al. 2002). Within this family of large single stranded RNA viruses, the majority causing GLD are ampeloviruses. GLRaV-2 belongs to the genus *Closterovirus*, and GLRaV-7 remains unassigned. GLRaV-3 is the predominant species in most vineyards with evidence of vector-driven disease spread (Cabaleiro and Segura 2006; Charles et al. 2009; Sharma et al. 2011) and reported yield losses of as much as 40% (Golino et al. 2002; Charles et al. 2006). All GLRaVs are graft-transmissible (Bertazzon et al. 2010) and this was initially assumed to be the main form of spread. However, researchers began to notice disease spread within vineyards that appeared to have a pattern of movement from a point source (Rosciglione and Castellano 1985; Habili et al. 1995). These spatial patterns implicated insect transmission, and have since been verified by monitoring the spread of infected vines over time (Cabaleiro et al. 2008; Charles et al. 2009).

In the 1980s, plant to plant transmission of GLRaV-3 by *Pl. ficus* was demonstrated (Engelbrecht and Kasdorf 1990). Since then, several species of mealybugs and soft scales have been shown to be GLRaV vectors, including *Ps. maritimus*, *Ps. viburni*, *Ps. longispinus*, *Ps. calceolariae*, *Pl. ficus*, *Pl. citri*, *H. bohemicus*, *Phenacoccus aceris* (Signoret) (apple mealybug), and *Pseudococcus comstocki* (Kuwana) (Comstock mealybug) (Rosciglione and Castellano 1985; Golino et al. 2002; Sforza et al. 2003; Cid et al. 2010). Additionally, GLRaVs can be transmitted by the soft scales *Pulvinaria vitis* (L.) (cottony vine scale) and *Parthenolecanium corni* (Bouché) (European fruit lecanium scale).

Most vector transmission studies focused on the identification of insect species capable of transmitting various GLRaV species, although recent studies have addressed transmission biology in more detail (Tsai et al. 2008, 2010). Importantly, transmission research has focused on GLRaV-3, which is the predominant species encountered in regions with disease spread. Although all mealybug and scale life stages may be capable of transmitting GLRaV-3, the smaller stages (e.g. crawlers or first instars) appear to be more efficient (Petersen and Charles 1997; Tsai et al. 2008). This is also the dispersal stage, with crawlers often being carried by the wind (Barrass et al. 1994) and other stages being moved on personnel, equipment, and infested nursery stock (e.g. Haviland et al. 2005). GLRaV-3 transmission by *Pl. ficus* occurs in a semi-persistent manner (Tsai et al. 2008), as would be expected for this genus and family of viruses. Acquisition and inoculation occur within 1 h of plant access period, although transmission efficiency increases proportionally with plant access time up to 24 h. The absence of an observable latent period required for transmission, together with the loss of vector infectivity over a period of days after acquisition, are hallmarks of semi-persistent transmission of plant viruses. Under laboratory conditions transmission efficiency of GLRaV-3 by *Pl. ficus* was ca. 10% per individual per day (Tsai et al. 2008). Although this value appears to be low when compared to other systems, the high fecundity of mealybugs places many potential vectors on each vine during each generation. Furthermore, the dispersal capability of minute first instar mealybugs is large, as previously shown in field studies in New Zealand.

Control of GLD is further hampered as both mealybug and virus can survive on the vine roots many years after the vine above ground has been pulled (Walton and Pringle 2004b; Bell et al. 2009). Generally, when removing diseased vines (roguing), all above-ground plant material is removed off-site and destroyed but the same is not always true of the roots. It is estimated that following vine removal, 70–80% of the roots may persist *in situ*, potentially for many years, although the quantity will vary according to factors like vine age, rootstock, and soil type (Bell et al. 2009). The retention of root debris following roguing is problematic as infected vine roots may sustain subterranean mealybug colonies (Bell et al. 2009), thereby leaving an unbroken link between virus and vector. Under these circumstances, South Africa and New Zealand managers argued that renewed disease pressure observed in some re-plant situations could be attributed to subterranean mealybug populations, feeding on and acquiring leafroll virus from residual vine roots, followed by dispersal to the roots of newly planted vines.

Although transmission of the various GLRaV species may follow the general trends observed with GLRaV-3 transmission by *Pl. ficus*, it should be noted that more research on the characterization of GLRaV transmission by various vector species is needed. Transmission studies aimed at identifying new vector species are essential to develop GLD management strategies, but yield little information on various aspects of transmission biology. Surprisingly, there is no evidence of virus-vector specificity in this system (Tsai et al. 2010). For example, different mealybug species transmit GLRaV-3, while *Pl. ficus* transmits at least five different GLRaV species. This finding has important epidemiological consequences: mealybug control may be necessary to limit disease spread, regardless of GLRaV (*Ampelovirus*) species.

12.3.3 Export Markets

Quarantine issues are a major concern for all vineyard mealybugs. As an example, molecular studies have shown that *Pl. ficus* in California probably originated from plant material in Israel and is thought to have been smuggled into the US on grape wood for commercial use. This pest eventually entered nursery material and was then spread within the state. Hot water dip and other procedures have been developed to clean nursery stock (Haviland et al. 2005; Liu et al. 2010). Still, the authors agree that movement of mealybug infested material across regional, provincial and state, and especially country borders is a serious concern.

12.4 Control Methods

12.4.1 Monitoring

There are no simple and effective methods to visually monitor vineyard mealybugs, and the process itself can be time-consuming and laborious. As exemplified for *Ps. maritimus*, the accuracy of monitoring plant material will depend on the mealybug population density, and the number of samples needed for an accurate count is often high because most mealybugs have a clumped distribution pattern, often being found on only a small percentage of the vines (Geiger and Daane 2001). The appropriate sampling programs will also vary throughout the season, depending largely on mealybug location as there are periods when much of the population is hidden (e.g. under bark) rather than exposed (e.g. on leaves). Also, as species have different numbers of annual generations and preferred feeding locations throughout the season, there is not a single sampling procedure appropriate for all vineyard mealybugs.

In most vineyards, signals of an infested vine can be used to aid the sampling program. First, ants are closely associated with mealybugs (Ripa and Rojas 1990; Addison and Samways 2000, Chap. 18) and their presence can help select vines for further sampling. Second, honeydew on the leaves can also be a good signal; a large population hidden under the bark will excrete enough honeydew that the infested trunk region will have a darker, wet appearance (Daane et al. 2011). Third, when some mealybug species numbers build, their feeding damage may cause leaves to turn yellow or brown and drop from the vine (Daane et al. 2011). Finally, at harvest time, berry clusters in direct contact with the spurs or trunk are more likely to be infested and by selecting these clusters a higher mealybug count can be made (Geiger and Daane 2001).

A faster sampling method is the use of sticky traps baited with sex pheromone to lure in and trap adult winged males. It has long been known that sexually mature female *Pl. citri* emit a sex pheromone to attract the winged adult males (Rotundo and Tremblay 1972). These pheromones can be synthesized and used in the field (Bierl-Leonhardt et al. 1981). Numerous sex pheromones have recently been identified, including for *Pl. ficus* (Hinkens et al. 2001), *M. hirsutus* (Zhang et al. 2004), *Ps. viburni* (Millar et al. 2005), *Ps. maritimus* (Figadère et al. 2007), *Ps. longispinus* (Zou and Millar 2009), and *Ps. calceolariae* (El-Sayed et al. 2010). They are being tested as management tools to detect mealybug populations. Researchers have shown that trap counts can even be used to predict berry damage (Walton et al. 2004). Some of these synthetic sex pheromones are commercially available. However, both conventional sampling and pheromone trapping have advantages and disadvantages and, for that reason, both methods should be used in combination.

12.4.2 Pesticides

Historically, pesticides have been a large part of vineyard mealybug control. Early programs included potassium cyanide, sodium cyanide, and sulfur fumigation (e.g., Essig 1914), which gave way to the chlorinated hydrocarbons (e.g., DDT) and organophosphates (e.g. parathion) from the 1940s to the 1990s (e.g., Frick 1952; Grimes and Cone 1985b). These materials were effective. For example, rates as low as 48 g a.i./ha of ethyl parathion provided *Ps. maritimus* control (Frick 1952). Eventually, however, most of these materials became less effective (Flaherty et al. 1982) or were ultimately banned from use because of concerns on non-target organisms.

Many organophosphates are still effectively used (Gonzalez et al. 2001; Walton and Pringle 2001; Sazo et al. 2008). Newer materials, with more novel modes of action, have also gained in popularity, including neonicotinoids, insect growth regulators, botanicals, and biosynthesis inhibitors (Daane et al. 2006b; Sunitha et al. 2009; Lo and Walker 2010). A major difference between the older and newer materials is the importance of coverage. As mentioned, a portion of the mealybug population is often under the bark, and for some species, on the vine roots. Many of the older foliar sprays did not effectively contact and kill mealybugs in these more protected locations. Some of the more novel materials have systemic properties, either applied through the irrigation system or as a foliar spray. For organic or sustainable farming programs, neem, light mineral oils, lime-sulfur, citrus products, and fatty acid soaps have been used. The few studies of these products have provided mixed results (Srinivas et al. 2007).

Another historical difference is that the earlier materials were often broad spectrum and killed more than just the targeted mealybugs. Flaherty et al. (1982) stated that 'extensive use of DDT and other synthetic insecticides used to control leafhoppers apparently disrupted natural control of grape mealybug [*Ps. maritimus*].' Other researchers have since discussed the impact of broad spectrum insecticides on mealybug natural enemies (e.g. Mani and Thontadarya 1988; Satyanarayana et al. 1991; Walton and Pringle 2001; Mgocheki and Addison 2009a). The cosmopolitan goal of managing vineyards with fewer broad spectrum pesticides, along with the development of resistance to common pesticides has fueled use and further research with the more novel insecticide materials.

Application timing is critical to control mealybugs with most insecticides. Exposed mealybugs are more easily killed than those under the bark, and the smaller stages are more susceptible than the larger mealybugs. This is especially true for insecticides with a short residual period. Much research, therefore, has been aimed at proper application timing and developing materials with better penetration into the protected habitats of mealybugs. For example, dormant season or early spring application takes advantage of the leafless vine, but mealybugs are in more protected locations. Applications with systemic insecticides near bloom are often used as the insecticide moves quickly in the vines to the leaves. After bloom, foliar materials are applied beneath the leaf canopy and aimed towards the grape clusters and interior canes. Late season applications can have issues with insecticide residues for both domestic and export market, because of complicated residue regulations. In addition, fresh market table grapes possess a dull haze or dust on the skin, termed 'bloom', and the use of some insecticides can remove the bloom and lower the crop value.

12.4.3 Cultural Control

A number of cultural controls are practiced and these vary greatly among regions. Few have been sufficiently evaluated. Many practices are specific to the table grape market. For example, the crop load on each vine is commonly thinned to increase berry size, and by thinning out grape clusters that come in direct contact with the trunk or cordon, the more susceptible clusters are also removed (Geiger and Daane 2001). Berry cluster manipulations are not always feasible for either raisin or wine grape production because of the trellising system used, the cost of thinning, and the need for optimal yield. Similarly, trellising systems for cane-pruned cultivars result in grape clusters that hang away from the trunk and cordons, and this reduces cluster infestation. Harvest date also impacts mealybug infestation levels, which can be higher in cultivars harvested later in the season because of greater exposure time to the later mealybug broods (Daane et al. 2011).

Mealybugs are found underneath the bark of the trunk, cordon, spurs, and canes. These locations provide some protection from insecticides, natural enemies, and environmental conditions. Stripping the bark exposes the mealybugs to these mortality factors. The infested bark should be destroyed rather than left in the row middles as the mealybugs can move back onto the vine. Common treatments after bark stripping include pesticides, as well as flaming to kill the mealybugs or banding the trunk with Stickum® to reduce movement of both mealybugs and ants from the trunk upwards to the clusters. While this effectively lowers mealybug density, it is labor intensive and too costly in many grape markets worldwide.

Cover crops have been used to improve soil health and lower pest densities by increasing natural enemy numbers or diversity. In vineyards, parasitoids that attack mealybugs could utilize floral nectaries found on some cover crop species as a food source to increase adult longevity. Generalist predators, such as the lacewings and some ladybeetle species, might also utilize these floral food resources as well as herbivores in the cover crop as alternate prey. However, many mealybug species can feed on ground vegetation. For example, *Pl. ficus* and *Ps. viburni* have been found on a number of common weeds such as *Malva parviflora* L. Therefore, the addition of a ground cover might also provide an alternate habitat for the mealybug. More work on the effect of ground covers on mealybugs and their natural enemies is warranted.

Overly vigorous vines can increase mealybug populations in two ways. First, excess nitrogen has been shown to increase the size of mealybug females and the number of eggs in each ovisac. Second, the increased foliage associated with overly vigorous vines provides better shelter for the mealybugs by reducing temperatures inside the vine leaf canopy, and may reduce the amount of applied foliar insecticide that reaches the mealybug. Controlling vine vigor is therefore a practice that can help improve mealybug control, in addition to being important for achieving viticultural goals.

12.4.4 Biological Control

Hundreds of natural enemies can attack mealybugs, making this brief review incomplete. A worldwide review of some of the earlier importation efforts is provided by Bartlett (1978) and Noyes and Hayat (1994). ScaleNet (2011) is also a good reference source. Here, the more common natural enemy groups are described, with specific mention of several key natural enemy species and programs (Figs. 12.7 and 12.8).

A number of predators contribute to mealybug control. Few specialize on mealybugs, whereas most are generalists that prey on any small, soft-bodied arthropods. For many of these natural enemies, there are no studies of their impact on mealybug populations. The most well known predator is the mealybug destroyer, *Cryptolaemus montrouzieri* Mulsant, which is native to Australia, but has been exported throughout the world. Both adults and larvae kill mealybugs. The larvae, to some extent, are mealybug mimics, possessing wax-like filaments similar to those of mealybugs. This 'camouflage' allows beetle larvae to forage without too much disturbance from mealybug-tending ants (Daane et al. 2007). One drawback is the poor tolerance of the predator to winter temperatures common in some vineyard regions (Smith and Armitage 1920). Surprisingly, there have been few studies that document the impact of *C. montrouzieri* on mealybug densities (but see Mani and Thontadarya 1989).

Other lady beetle species also attack mealybugs. Many beetle larvae in the subfamily Scymninae are covered with wax, similar to the mealybug, and are often mistakenly identified as *C. montrouzieri*. For example, these include species of *Hyperaspis*, *Nephus* (= *Scymnobius*), and *Scymnus*, which may be the most abundant mealybug predators in vineyards. However, because the taxonomic keys for these Scymninae beetles poorly differentiate among species, many of the observed beetles are seldom properly identified.

Fig. 12.7 Many parasitoid species attack mealybugs. The examples here are (**a**) a female *Anagyrus pseudococci* (ca. 2 mm) next to a vine mealybug 'mummy' showing the round parasitoid exit hole, (**b**) the smaller (ca. 1.3 mm) male *A. pseudococci*, which has a different color pattern and 'hairy' antennae, feeding on a drop of honeydew, (**c**) a female *Leptomastidea abnormis* 'host feeding' on a vine mealybug crawler, (**d**) *Leptomastix epona*, which was imported for obscure mealybug biological control in California but did not establish because of Argentine ant interference, (**e**) the small (ca. 1 mm) and fast-moving *Acerophagus flavidulus* closing in on a *Pseudococcus viburni*, and (**f**) *Coccidoxenoides perminutus* (ca. 1 mm) next to *Planococcus ficus* first instar

Migratory lady beetles, notably those in the subfamily Coccinellinae, are often attracted to large mealybug infestations and their honeydew. These include some of the large and recognizable species such as the convergent ladybeetle (*Hippodamia convergens* Guérin-Méneville) and the transverse lady beetle (*Coccinella transversoguttata* Falderman). More work is needed to document the effectiveness of the native lady beetles, found throughout the world's grape regions, as mealybug predators.

Lacewings have long been associated with mealybugs. For example, *Chrysoperla carnea* (Stephens) was first shown to suppress mealybugs (*Ps. maritimus*) in pears (Doutt and Hagen 1950). Lacewing larvae are effective predators of smaller mealybugs. They may have a difficult time feeding on eggs in the mealybug ovisac where waxy secretions provide some protection from the predator. Larger mealybugs excrete an ostiolar fluid that can act as a defensive mechanism. Native brown and green lacewing species are often overlooked while *C. carnea* has received more attention.

Cecidomyiid flies (i.e., predaceous midges) are another common mealybug predatory group (Abbas 1999). In most regions, little is known about their impact on mealybug population densities. However, Charles (1985) reported that *Diadiplosis koebelei* (Koebele) reduced *Ps. longispinus* in New Zealand vineyards by about 30%. Midges associated with mealybugs include *Dicrodiplosis californica* Felt in California (Geiger and Daane 2001), *D. koebelei* in New Zealand (Charles 1985), and a *Triommata coccidivora* Felt in India (Mani et al. 1987). The adult fly, which

Fig. 12.8 Common mealybug predators include lady beetles. Examples here are (**a**) an adult *Scymnus* sp. feeding on a grape mealybug, and (**b**) a large *Cryptolaemus montrouzieri* larva near the smaller obscure mealybug. The larvae of many of these lady beetle species have waxy filaments to mimic the mealybugs and reduce interference from mealybug-tending ants, (**c**) a cecidomyiid larva about to feed on *Pseudococcus maritimus*, and (**d**) a third instar green lacewing (*Chrysoperla carnea*) larva attacking a *Ps. maritimus* and prompting the mealybug to secrete a ball of red ostiolar fluid in defense

is not predatory, deposits its eggs in or near the mealybug ovisac and the maggot-like larvae feed, primarily, on mealybug eggs and small larvae. The fly larvae typically pupate in the ground.

Most successful biological control programs rely primarily on encyrtid parasitoids that are mealybug specialists, some attacking only a few specific mealybug species (Noyes and Hayat 1994). These parasitoids are typically internal koinobionts, but can be either solitary or gregarious and preferentially attack varying host stages. Parasitoids have been credited with some level of control for vineyard mealybugs throughout the world. For example, *Anagyrus pseudococci* (Girault), as a parasitoid of *Pl. citri* and *Pl. ficus*, is one of the most well-studied (Blumberg et al. 1995; Islam and Copland 2000; Daane et al. 2004) and widely distributed natural enemies (e.g., Israel (Berlinger 1977), Europe (Duso 1989), South Africa (Walton and Pringle 2004b), and elsewhere).

In some cases, parasitoid performance can be linked to geographic strains of the targeted mealybug. In New Zealand, for example, *Ps. viburni* was brought under exceptional control by release of the parasitoid *Acerophagus maculipennis* Mercet (Charles et al. 2010), whereas in Chile *Ps. viburni* is controlled by *Acerophagus flavidulus* (Brèthes) (Ripa and Rojas 1990). The biology of *A. maculipennis* (Sandanayaka et al. 2009) and *A. flavidulus* (Karamaouna and Copland 2000, 2009) have been studied. Nevertheless, it is still unclear how these species exhibit a level of host discrimination that may differentiate between geographic strains of *Ps. viburni*. This intriguing level of discrimination, combined with the geographic location of these parasitoid species, has been used to assess the origin of *Ps. viburni* (Charles 2011).

Some parasitoid species are attracted to the mealybug's sex pheromone (Walton et al. 2006), which may act as a kairomone (Franco et al. 2008). For example, the parasitoid *A. pseudococci* was caught in *Pl. ficus* pheromone-baited traps (Millar et al. 2002). It was later observed that parasitism levels of *Pl. ficus* were higher in vineyards with mating disruption (Walton et al. 2006). Ongoing studies are screening the attractiveness of different parasitoid species to mealybug sex pheromones, to test the hypothesis that some parasitoid species spend more time searching for mealybugs in vineyards where a mating disruption program is implemented, thereby increasing parasitism rates.

Ants have long been associated with outbreaks of honeydew-producing homopterans. The mutualistic association has clear benefits for the ants, which are provided with a carbohydrate food source, and in return, ant-tending has been credited with protecting homopterans from natural enemies. Not surprisingly, ants have been shown to disrupt mealybug biological control in vineyards from South Africa (Mgocheki and Addison 2009b) to North America (Daane et al. 2007). Ant species vary in dominance in different vineyard regions (Addison and Samways 2000; Cooper et al. 2008). The Argentine ant, *Linepithema humile* (Mayr), is one of the world's most damaging invasive insects and it is now common in many vineyards in association with mealybugs and soft scale pests (Addison and Samways, Chap. 18).

12.4.5 Mating Disruption

Mating disruption was first attempted against *Pl. ficus* in North America (Walton et al. 2006), and is currently gaining in popularity. However, prior to this work, researchers in Europe and Israel investigated attract and kill for adult male *Pl. citri* in citrus. But that initial work found that the extent of male reduction was not enough to decrease fruit infestation (Franco et al. 2009). It is likely that future mealybug control programs will rely more heavily on novel control strategies using semiochemicals, especially if the price of synthetic sex pheromones for mealybugs can be reduced.

12.5 Mealybugs in Some Major Grape-Producing Areas

12.5.1 North America

Pseudococcus maritimus is the primary North American mealybug pest in vineyards. It is found from California to Canada, and from Washington to New York. Insecticides are generally not needed to control *Ps. maritimus*. Parasitoids have long been credited with *Ps. maritimus* control in North America, while early records indicate that *Zarhopalus corvinus* (Girault) was the dominant parasitoid and, in combination with *Anagyrus yuccae* (Coquillet), *Acerophagus notativentris* (Girault), *Anagyrus clauseni* Timberlake, and *Pseudleptomastix squammulata* Girault, provided up to 80% parasitism (Clausen 1924). Later surveys have reported *A. notativentris* and *Acerophagus angelicus* (Howard) to be the dominant parasitoids (Grimes and Cone 1985a; Grasswitz and Burts 1995; Geiger and Daane 2001).

For most of North America, *Ps. maritimus* is the only mealybug of concern. The occasional outbreak of *Ps. maritimus* generally results from pesticide usage removing the natural enemies, or outbreaks associated with ant populations. California, however, presents a more unique situation as most of the other vineyard mealybug species, discussed herein, can be found in the state. A review from the least to the most important vineyard mealybugs – other than *Ps. maritimus* – would begin with *Ps. calceolariae*, which was first recorded in California in 1913 as a citrus pest in southern California. A classic biological control program was initiated with natural enemies imported from Australia, including the first introduction of *C. montrouzieri* in 1916 (Smith and Armitage 1920). In the 1920s the importation of the encyrtids *Coccophagus gurneyi* Compere and *Tetracnemoidea brevicornis* (Girault) (formerly *Tetracnemus pretiosus* Timberlake) was credited with reducing *Ps. calceolariae* densities to 'almost negligible numbers' (Compere and Smith 1932).

Maconellicoccus hirsutus, the primary mealybug pest in India, is found in southern California, near the desert table grape region in the Coachella Valley. However, this mealybug is not a pest in California vineyards because of a successful biological control program, which was initiated for the Caribbean in 1994 and later extended to Mexico and southern California. The parasitoids *Anagyrus kamali* Moursi, *Gyranusoidea indica* Shafee, Alam & Agarwal, and *Allotropa* sp. nr. *mecrida* (Walker) are credited with reducing *M. hirsutus* densities to non-economic levels throughout the state (Roltsch et al. 2006) and it is not currently found in vineyards.

Pseudococcus longispinus was first reported as a citrus pest in California and, to help control this invasive pest, parasitoids were imported in the 1920s, including *Tetracnemoidea sydneyensis* (Timberlake) (from Australia), *Anagyrus fusciventris* (Girault) (from Hawaii), and *Tetracnemoidea peregrina* (Compere) (from Argentina) (Bartlett 1978). DeBach (1949) suggested that parasitoids helped suppress *Ps. longispinus* in citrus, but that predators, especially *C. montrouzieri*, were more important. Currently, *Ps. longispinus* infests a small number of vineyards in California's coastal region. Recent surveys found *T. sydneyensis*, *T. peregrina*, *A. angelicus*, *A. pseudococci*, *Leptomastidea abnormis* (Girault), *Leptomastix dactylopii* Howard, and *Coccidoxenoides perminutus* Girault attacking this mealybug (Daane et al. 2008a).

Little is known about *F. gilli* as this species was only described in 2003, initially found infesting pistachios in the San Joaquin Valley (Gullan et al. 2003). Nevertheless, it became the primary vineyard pest in some of California's Sierra Foothill appellations. Similar to *Ps. maritimus*, *F. gilli* has two generations per year, overwintering under the bark, and moving onto the leaves and berry clusters during the summer. Some parasitoid species have been recorded, with a key parasitoid being *Acerophagus* sp. nr. *meritorius* Gahan that was most likely present in California as a parasitoid of the closely related *F. virgata*.

Pseudococcus viburni has long been in California, but only became a key vineyard pest when the wine grape industry expanded into the Central Coast region. Nevertheless, the range of this pest seems to be increasing. Prior to 1993, there were no effective parasitoids of *Ps. viburni* in California and for this reason, *Acerophagus flavidulus* (Brèthes) and *Leptomastix epona* (Walker) were imported from Chile (Daane et al. 2008a). Both *A. flavidulus* and *L. epona* were initially recovered. However, foraging ants diminished their impact (Daane et al. 2007). Insecticides are currently used for most *Ps. viburni* populations, especially when *Pl. ficus* is also found.

Planococcus ficus is currently the most damaging vineyard mealybug in California as well as in Mexico. *Planococcus ficus* appears capable of surviving across a wide geographic range, from desert table grapes to cool coastal wine grapes (Daane et al. 2007), with from 3 to 10 generations per year, depending on the temperature. To control this pest, parasitoids have been imported from Spain, Israel, and South Africa, and they include *A. pseudococci*, *L. abnormis*, *C. perminutus* and *L. dactylopii* (Daane et al. 2008b). Although these natural enemies provide some suppression, biological traits of *Pl. ficus* limit their effectiveness (Daane et al. 2004; Gutierrez et al. 2008). Mating disruption has shown some promise (Walton et al. 2006) and is being used on a larger scale each year. Nevertheless, insecticides are the primary control tool for *Pl. ficus*. Currently, most North American insecticide programs are based on the use of one or more of the following insecticides: imidacloprid (systemic – near bloom time), buprofezin (foliar – late spring or early summer), acetamiprid (late spring to harvest), clothianidin (foliar or systemic – from late spring to harvest), spirotetramat (late spring to early summer, or post-harvest), and chlorpyrifos (delayed dormant or post-harvest).

For North America, much of the future mealybug research concerns GLDs, such as determining the required treatment thresholds for mealybugs in order to reduce GLRaV spread. Connected to this is the development of better monitoring programs, using synthetic sex pheromones to determine the abundance and species of mealybugs. Better ant controls are also needed (Daane et al. 2007; Tollerup et al. 2007). Researchers have investigated the use of ant baits to deliver small but lethal amounts of toxicant to the ant colony by exploiting their social behavior to distribute food via trophallaxis, thereby delivering the toxicant to the nest population to provide season-long control (Tollerup et al. 2004; Daane et al. 2006a; Nelson and Daane 2007;

Cooper et al. 2008). In contrast, broad spectrum insecticide sprays targeted at ants may kill foraging ants, but unlike baits they have little effect on nests, allowing population resurgence.

12.5.2 South America

Grape production has distinct management practices and mealybug pest problems through South America. Here, mealybugs in Chile, Argentina and Brazil will be discussed as examples of the dynamics of South American mealybug problems and controls.

In Chile, grape is one of the oldest and most economically important crops, with ca. 180,000 ha, of which about one-third are destined for table grape production (42% of the total fruit exports). Mealybugs are the main phytosanitary problem for Chilean table grapes because of their quarantine importance for many markets. They have been responsible for up to 70% of table grape rejections by inspectors prior to export. In contrast, the economic impact of mealybugs in wine grapes is not well understood, although populations have increased over the years and recent work has demonstrated the potential negative impact of mealybugs. The questionable issue is the presence of GLD in Chilean vineyards (Herrera and Madariaga 2001). To some extent, GLDs are not considered as important in Chilean grape production as in other vineyard regions because the vines are not grafted (own-rooted vines), which are thought to be more tolerant to GLD than modern rootstocks.

Several mealybug species have been associated with grapes in Chile (Artigas 1994), but by far the most common is *Ps. viburni*, with *Ps. longispinus*, *Ps. calceolariae* and *Pl. citri* being rare (Gonzalez 2003; Ripa and Luppichini 2010), despite being common on other subtropical fruit crops such as citrus and avocados (Ripa and Larral 2008). Two other species have also been mentioned, *Ps. maritimus* and *Pl. ficus*, but the literature is contradictory in this regard (Artigas 1994; Gonzalez 2003; Gonzalez and Volosky 2005), and presently they are believed to be misidentifications. Earlier records of *Ps. maritimus* might correspond to a new species, which is in the process of being formally described.

Vineyard mealybug control in Chile has been mostly accomplished through applications of organophosphate insecticides, and more recently neonicotinoids and insect growth regulators (Gonzalez et al. 2001; Sazo et al. 2008; Salazar et al. 2010). Additionally, as organic wine grape production has increased, the use of augmentative biological control has increased accordingly, including the release of the endemic parasitoid *A. flavidulus*, predators like *C. montrouzieri* and *Sympherobius maculipennis* Kimmins, and entomopathogens such as the soil-inhabiting fungus *Metarhizium anisopliae* (Metschnikoff) Sorokin (Ripa and Larral 2008; Ripa and Luppichini 2010; Salazar et al. 2010).

In Argentina, viticulture began with the initial Spanish colonization in the sixteenth century. Currently, there are about 228,575 ha in grape production, with about 93% in wine grapes. Mendoza is the most important grape-growing province, containing approximately 70% of Argentina's grape production which is mostly dedicated to wine grapes, followed by San Juan province with about 22% with 11,800 ha of table and raisin varieties.

Historically, Argentina had relatively few grape pests because of its hot and dry climate. Mealybug pest problems began in 2001 when *Pl. ficus* was first found. This invasive pest soon developed to damaging populations, initially in the table grape, and later in wine grape regions. Currently, *Pl. ficus* is distributed throughout most of Argentina's grape production valleys. Practices such as mechanical harvesting have hastened its movement among vineyards and regions. *Planococcus ficus* is not the only mealybug found in Argentina vineyards (Cordo et al. 2004), but it is the only one reported to cause significant economic damage. *Planococcus ficus* has six generations annually, with the first generation beginning in early spring (September to October). The mealybug directly infests the grape clusters, beginning with the third generation in midsummer (December) and building throughout the season, especially when tended by ants. Ants are associated with *Pl. ficus* spread, and in the Mendoza region, the ant species in the genera *Dorymyrmex*, *Linepithema*, *Pheidole*, *Solenopsis*, *Camponotus*, and *Brachymyrmex* have been observed tending this pest (Cucchi and Becerra 2009). The widespread distribution of *Pl. ficus* also presents the danger of further spread of GLRaVs in Argentina (de Borbon et al. 2004).

To develop improved control programs, research is now clarifying the extent of GLRaVs present in Argentina and their natural dispersion by mealybugs, including the use of epidemiological models. Initially, control programs relied on insecticide applications of neonicotinoid (e.g., imidacloprid) and organophosphate products (e.g., dimethoate and methyl pirimiphos). Although these pesticides are still used, current research has investigated semiochemical (mating disruption) and tetramic acid based pesticides (e.g., spirotetramat) as alternate control tools. Natural enemies, such as the lady beetle *Hyperaspis lanatii* González & Gordon, the lacewing *Chrysoperla asoralis* Banks, and the parasitoids *Anagyrus* sp., *Leptomastix* sp., *Leucopis* sp., have been found associated with *Pl. ficus* (Cucchi and Becerra 2009).

In Brazil, viticulture is a relatively new industry, with about 82,000 ha, primarily in the southern states (Paraná, Santa Catarina, São Paulo and Rio Grande do Sul) near Argentina and Uruguay, and in the eastern states of Bahia and Pernambuco. The vines are grown for table and wine grapes, with Rio Grande do Sul producing about 60% of the juice and wine grapes, whereas in São Paulo grapes are grown primarily for the table grape market, including the export market.

Mealybugs are a recent concern for Brazilian growers mainly due to direct infestation of table grape clusters. Although there is growing awareness of mealybugs as vectors of GLRaVs, their role in Brazil is still not well understood (Fajardo et al. 2003). *Planococcus citri* is the most abundant vineyard mealybug species (Morandi Filho et al. 2009), whereas *Pl. ficus* is rarely reported (Foldi and Kozar 2006). Surprisingly, the root-infesting mealybug *D. brevipes* is second most in importance, and unlike other root-infesting mealybug species, it is also found above ground and will infest the berry clusters in Brazil. Other species of mealybugs associated with

Brazilian grapes are *Ps. viburni*, *Ps. maritimus*, and *Planococcus minor* (Maskell). However, they are considered of secondary importance.

One particular situation of Brazilian viticulture is the use of *Vitis labrusca* L. cultivars Niagara, Isabel and Ives for juice, wine, and table grapes, representing about 50% of all grape cultivars. The importance of these *V. labrusca* cultivars is that, in many regions, they are grown next to *Vitis vinifera* L., but as *V. labrusca* can host but not show GLD symptoms, they may provide an undetected refuge for these pathogens. Currently, researchers in Brazil are working to improve wine quality where the presence and spread of GLRaV is considered a primary issue for replanting vineyards and the future development of the wine industry. Because *D. brevipes* is a root mealybug, this species may be an issue (if it is also a vector of a GLRaV) for replanting *V. labrusca* with *V. vinifera*.

Mealybug management is based on chemical treatments, primarily with neonicotinoid insecticides (e.g. imidacloprid and thiamethoxam). These are typically applied as a soil drench directed to the grape roots. Many Brazilian researchers suggest that the previous use of broad spectrum pesticides to control other vineyards pests (e.g. South American fruit fly, thrips) have destroyed much of the natural enemy population that attacks *D. brevipes* and other vineyard pests. Future research will investigate improving natural controls and monitoring programs, as well as testing novel insecticides such as spirotetramat. Because mealybugs were an often overlooked pest group in Brazilian vineyard management, another goal is to survey and correctly identify mealybug pests and to extend information on mealybugs as vectors of plant pathogens.

12.5.3 Europe

Modern studies of mealybugs in European vineyards began in the 1980s with the examination of *Planococcus* spp. in the transmission of GLRaV (Rosciglione and Castellano 1985) and the serological characterization of GLRaVs. Because grape cultivation in Europe is dominated by wine rather than table grapes, GLD is the main concern with mealybugs. Besides the outward GLD symptoms mentioned earlier, leafroll damage is considered to be the reduction in sugar content and an increase in acidity in the berries. Nevertheless, it is commonly accepted by European growers that GLD is not as severe as grapevine yellows (induced by phytoplasmas and transmitted by leafhoppers and planthoppers) or grapevine fanleaf disease (induced by viruses and transmitted by nematodes). In Europe, four mealybug species are known to develop on grapes: *Pl. ficus*, *Pl. citri*, *H. bohemicus*, and *Ph. aceris* (Sforza et al. 2003; Bertin et al. 2010; Cid et al. 2010), along with four coccid scales and a diaspid scale. For example, in vineyards of France (Champagne, Burgundy, Alsace), four hemipterans are sympatric in leafroll-infected vineyards, namely the mealybugs *H. bohemicus*, *Pl. ficus*, and *Ph. aceris* and the coccids *Pa. corni* and *Pu. vitis* (Sforza et al. 2003).

The biology of *Pl. ficus* and *Pl. citri* is poorly known even though they are natives of Eurasia and polyphagous in European agroecosystems. These pseudococcids were only recently considered as economically important pests. *Planococcus ficus* is present throughout the Mediterranean Basin, and often sympatric and misidentified, with the closely related *Pl. citri*, as discussed previously. Both mealybugs are reported as vectors of GLRaV-3,-5 in several European countries (Cabaleiro and Segura 1997). *Planococcus ficus* may be the only European GLRaV vector with multiple generations. There have been no concerted control programs for either *Pl. ficus* or *Pl. citri*. Survey studies have identified natural enemies and quantified their abundance during the growing season (Sforza et al. 2003), in order to improve the understanding of resident biological control agents. Currently, natural regulation of *Pl. ficus* and *Pl. citri* is primarily provided by the activity of resident *A. pseudococci*.

Heliococcus bohemicus became a primary vineyard pest in the last two decades in Hungary, Switzerland, Italy, France, and Germany, but it is only reported as a GLRaV vector on grape in France (GLRaV-1) and Italy (GLRaV-1 and GVA) (Kozar et al. 1994; Sforza et al. 2003; Zorloni et al. 2006). This mealybug is univoltine in France, and bivoltine in Italy. A natural enemy survey showed parasitism levels of at least 35%, attributed to the encyrtids *Leptomastidea bifasciata* (Mayr) and *Anagyrus szodensis* Erdös, with activity from spring through summer.

Phenacoccus aceris is common on vines as well as some tree species (e.g., oak, apple, chestnut). This is a univoltine species, with a high fecundity rate that is reported to range from 800 to 3,600 eggs per female. It is found throughout Eurasia, where it is reported as a virus vector on grapes, as well as 'little cherry disease' on cherry (Kosztarab and Kozar 1988). In French vineyards, it transmits GLRaV-1, GLRaV-3, and GVA, and GVB from grape to grape (Sforza et al. 2003). The encyrtid *Anagyrus schoenherri* (Westwood) has been reported attacking second instar *Ph. aceris* (Sforza et al. 2003) and releases of *C. montrouzieri* have shown promise.

12.5.4 New Zealand

Three introduced mealybug species, *Ps. calceolariae*, *Ps. longispinus* and *Ps. viburni*, have been primary pests in New Zealand vineyards (Charles 1993). A more recent survey revealed that only *Ps. calceolariae* and *Ps. longispinus* were commonly encountered, with *Ps. viburni* now regarded as an insignificant component of the mealybug fauna, resulting from a successful biological control program (Charles et al. 2010). Damage from these pests includes berry contamination with insects, honeydew and sooty molds (Charles 1982). However, as New Zealand production is primarily for wine grapes, the mealybugs are more recognized as vectors of GLRaVs (Petersen and Charles 1997). Indeed, mealybugs and GLD are considered the primary destructive pests and disease affecting vines (Charles et al. 2006).

In the North Island of New Zealand, *Ps. calceolariae* and *Ps. longispinus* have three generations per year (Charles 1981) and it is likely that the same number of generations occur in Marlborough in the South Island, the major wine grape region. However, other aspects of the biology of these pests remain poorly understood. For example, *Ps. calceolariae* is frequently found on vine roots but the proportion of the population on roots at any point in time and its relative mobility in this environment remain unknown. Subterranean mealybugs hamper monitoring, may limit the effectiveness of contact insecticides and some natural enemies, and may also reduce the effectiveness of programs to remove GLRaV-infected vines, which is commonly practiced in New Zealand wine regions (Bell et al. 2009). Given what is known of *Ps. calceolariae*, it is conceivable that their survival on remnant roots could, in part, explain the relatively rapid reappearance of GLDs observed in replanted vineyards. Research is now underway to assess the likelihood of this mechanism perpetuating a renewed incidence of leafroll virus in re-plants.

Grapes in New Zealand are primarily grown for wine production, and as mentioned, the goal of mealybug control is not so much on preventing crop damage from mealybug infestations, but in managing the incidence and spread of GLRaVs (Charles et al. 2009). Consequently, tolerance for mealybugs is very low, especially where they co-exist with leafroll virus. The use of insecticides for mealybug control is largely limited to North Island vineyards, and includes an organophosphate (e.g., prothiofos) near budbreak and two in-season applications of an insect growth regulator (e.g., buprofezin). In 2008, the average number of buprofezin applications per block was 0.31, whereas applications of prothiofos (not endorsed under the Sustainable Winegrowing Program, SWNZ) averaged just 0.15 per block. Research into other insecticides, including systemic products, is underway. Recently, the label claim of one such active ingredient, imidacloprid, was extended to include vines but its use in New Zealand is restricted to that of a soil drench that can only be applied to non-cropping vines infected with leafroll virus and about to be removed (Lo and Walker 2010). This strategy attempts to reduce the incidence of viruliferous mealybugs on the remnant roots of rogued vines.

In New Zealand, three important issues are the likely focus of future vineyard pest management research: (1) the development of efficient mealybug monitoring systems, (2) the determination of the role of biological control in regulating mealybug populations, and (3) control measures based on treatment thresholds and the use of novel insecticides. First, the recent identification of the sex pheromone for *Ps. calceolariae* and *Ps. longispinus* (Millar et al. 2009; El-Sayed et al. 2010) will enable researchers to develop monitoring programs, better study pest phenology, and detect new mealybug incursions. Sex pheromones may also offer the potential to control these mealybug pests through mating disruption and/or male 'lure and kill'. Second, the potential for biological control of *Ps. calceolariae* and *Ps. longispinus* in New Zealand has not been fully explored, despite a good understanding of the species composition and regional distribution of many mealybug natural enemies (Charles 1981, 1985, 1993; Charles et al. 2010). Still, the current changes in pesticide use patterns, particularly the virtual elimination of mid- to late season organophosphate applications, may improve the vineyard ecosystem for natural enemy use (Charles et al. 2010). Third, the long-term challenge facing the wine sector is the development and implementation of a leafroll virus program. New Zealand operates a high-health plant program to ensure that vineyards receive virus-free plant material but has only recently increased the focus on roguing virus-infected vines and mealybug control. This situation not only demands a better understanding of mealybug ecology and cost-effective control measures but also an appreciation of the social and economic challenges confronting communities of growers that now need to share strategies when implementing area-wide virus-elimination programs.

12.5.5 India

Severe mealybug outbreaks have been reported in India's vineyards, adversely affecting grape production by as much as 90% in extreme cases in the state of Andhra Pradesh. Amongst the eight mealybug species that have been reported on vines in India, *M. hirsutus*, *Pl. citri*, *Nipaecoccus viridis* (Newstead) and the root mealybug, *X. annandalei*, are the primary mealybug pests. Cultivars harvested in late fall suffer heavily from mealybug damage. The increasing mealybug problem in recent years may be due to frequent and indiscriminate use of insecticides to control other pests, which may disrupt natural enemies responsible for the suppression of mealybugs. Although the root mealybug will not be covered here in detail, root damage by this pest reduces vine vigor and yield, and shortens fruit-bearing canes (Rajagopal et al. 1997).

Maconellicoccus hirsutus is the most important of the vineyard mealybugs in peninsular India, with severe infestations leading to berry and shoot damage. The mealybug occurs on the vine throughout the year (Mani and Thontadarya 1987c). After harvest, the mealybug population is confined to vegetative parts, where it overwinters. In spring, the vines are given a 'foundation pruning' (usually in April–May), and *M. hirsutus* remains on the leaves, stem, and trunk from this period until harvest. From mid- to late summer, the population density is typically low until late fall when the vines are given a 'berry pruning'. The mealybug population density starts increasing from mid-December onwards and by January (midwinter – but temperatures in India are, of course, different from most grape growing regions, as are the grape cultural practices) the *M. hirsutus* population migrates from the trunk, cordons, and shoots to developing berries. The pest population build-up coincides with high temperatures (30–40°C), low humidity (less than 40%) and berry development. Peak population is reached before harvest in spring (March–April). An early harvested crop usually reduces mealybug damage as compared to a late harvested crop. Also, heavy rains and cool temperatures of less than 20°C can result in a temporary reduction in the *M. hirsutus* population, often encountered in winter and rainy seasons.

For biological controls, six parasitoids and seven predators have been associated with *M. hirsutus*. The parasitoids are *Anagyrus dactylopii* (Howard), *Allotropa* sp. nr. *japonica* Ashmead, *Anagyrus mirzai* Agarwal & Alam, *Alamella flava* Agarwal, *Leptopilinia* sp., and *Chartocerus* sp. nr. *walkeri* Hayat. The predators are *Scymnus gratiosus* Wiese, *Scymnus coccivora* Ayyar, *C. montrouzieri*, *Chrysopa* sp., *Spalgis epius* Westwood, *Cacoxenus perspicax* (Knab), and *Triommata coccidivora*. Among these, *A. dactylopii* and *S. coccivora* are the most important, causing up to 70% parasitism (Mani et al. 1987). Studies have investigated the biology of the more important natural enemies. For example, *A. japonica* can be reared on 15–20 day old *M. hirsutus* (Mani and Krishnamoorthy 1989) and larva of *S. coccivora* consumed 308 eggs or 62 nymphs (Mani and Thontadarya 1987a). The lady beetle *C. montrouzieri* showed the potential to consume about 1,000 eggs or 300-500 mealybug nymphs (Mani and Thontadarya 1987b). The augmentative release of this beetle showed promise against *M. hirsutus* in field trials (Mani and Thontadarya 1989). Studies also investigated the impact of various pesticides on these natural enemies. For example, application of dichlorvos, diazinon, phosalone, fish oil rosin soap, and the commonly used fungicides proved to be safe to *A. dactylopii* (Mani and Thontadarya 1988) and could be integrated with the release of *C. montrouzieri*.

Prevention is better than cure. Cultural, mechanical, biological, and chemical methods of control have to be integrated to reduce the mealybug populations and reduce berry damage. Cultural practices include: (1) the collection and destruction of the mealybug from infested clusters at harvest time (March–April), (2) the collection and destruction of all the pruned material from mealybug infested vines during the foundation pruning (April–May), (3) bark stripping – or the removal and destruction of loose bark (April–May), (4) a similar removal of weeds and other alternate host plants that harbor mealybugs in and around the vineyards (season-long), and (5) removal of ant colonies that tend the mealybugs.

Insecticide practices to manage mealybugs include the following: (1) drenching ant colonies with chlorpyrifos or malathion dust (April–May), (2) treating the trunk and cordons with dichlorvos (April–May), (3) systemic application of imidacloprid applied to basins in the soil around the trunk or through drip irrigation system (April–May), (4) foliar applications of buprofezin and/or methomyl (about 30 days of soil drenching, or 30–60 days before harvest), (5) foliar sprays of dichlorvos or azadirachtin (3–15 days before harvest), and (6) releasing of *C. montrouzieri* (at 5,000/ha from August–September) or foliar sprays of a mixture of *Verticillium lecanii* (Zimmerman) / *Beauveria bassiana* (Balsamo) Vuillemin at 15-day interval in the rainy season (July–August). These steps may also be repeated after the second harvest (October–November).

12.5.6 The Middle East

Planococcus ficus is the primary vineyard pest of the Middle East, and has been reported as a pest in Iran (Williams and Moghaddam 2000), Iraq, Israel, Lebanon, Libya, Egypt (Ben-Dov 1995), Syria, Tunisia (Mahfoudhi and Dhouibi 2009), and Turkey (Kaydan et al. 2005). For example, *Pl. ficus* is found in many vineyard and fig production areas and has become a serious vineyard pest in southern Iran

(Williams and Moghaddam 2000; Fallahzadeh et al. 2009). However, the pest distribution and pest status is uneven across the Middle East and, for example, *Pl. ficus* has never been reported to cause damage in northern Iran vineyards.

In southern Iran, *Pl. ficus* has five generations, with population density increasing rapidly from spring (May) into summer, and then declining after harvest (August to September). After the fifth generation, all developmental stages of *Pl. ficus* can be found overwintering on roots (November to March). Along with the change in population density is the expected change in feeding location. In spring, mealybugs are primarily found on the trunk and canes, while in summer they are primarily found on leaves, new canes, and berries. However, a portion of the population is always found in protected locations (Fallahzadeh et al. 2009).

Seven primary, two primary/secondary, three secondary parasitoid species, as well as two coccinellids, and four other predator species are associated with *Pl. ficus* in southern Iran (Fallahzadeh et al. 2011). The primary parasitoids are *A. pseudococci*, *L. dactylopii*, *A. dactylopii*, *A. mirzai*, *Anagyrus agraensis* Saraswat, *Leptomastix flava* Mercet, and *Chartocerus kurdjumovi* (Nikol'skaya). The primary/secondary parasitoids are *Prochiloneurus bolivari* Mercet and *Pachyneuron muscarum* (L.). The secondary parasitoids are *Marietta picta* (André), *Aprostocetus trjapitzini* (Kostjukov), and *Baryscapus sugonjaevi* (Kostjukov), and these attack either the *Anagyrus* or *Leptomastix* species.

In other Middle East regions, the encyrtid parasitoids *L. dactylopii*, *L. abnormis*, *Clausenia josefi* Rosen, and *Neoplatycerus* sp. nr. *palestinensis* (Rivnay) were found attacking *Pl. ficus* in Egyptian vineyards. In Tunisia, both *Pl. citri* and *Pl. ficus* were found in vineyard regions, where parasitoids were more frequently recorded than predators as natural enemies. *Anagyrus pseudococci* had a parasitism rate of 80.3%, followed by *L. abnormis* (12.1%), *C. perminutus* (4.5%), and *L. dactylopii*, whereas only two coccinellids (*Rhyzobius lophanthae* (Blaisdell) and *Scymnus* sp.) were associated with these mealybugs (Mahfoudhi and Dhouibi 2009). In Israel, both *Pl. citri* and *Pl. ficus* occur, with the former being primarily a citrus pests and the latter being more of a vineyard pest. Natural enemies attacking *Pl. ficus* include *C. josefi* and *A. pseudococci* (Berlinger 1977). More recently, the use of *A. pseudococci* against *Pl. ficus*, as well as the use of semiochemicals for monitoring *Pl. ficus*, has been undertaken in Israel (Franco et al. 2003; Zada et al. 2003).

Other mealybugs and scale insects have been reported from Middle East vineyards. The mealybug *Chorizococcus viticola* Kaydan & Kozár was collected on vineyards from southern Iran and a related species, *Chorizococcus shaferi* (Hollinger), found on grapes is a presumed invasive species from North America (Fallahzadeh et al. 2010). *Chorizococcus viticola* can reach damaging levels, and in some parts of Iran it is the most damaging vineyard pest, where it can reach high densities by midsummer (July), especially on berry clusters. The damage caused by this pest has increased in recent years in Beyza, Kavar and Akbar Abad (Fallahzadeh et al. 2010). Two encyrtid parasitoids, *Gyranusoidea iranica* Japoshvili & Fallahzadeh and *Anagyrus matritensis* (Mercet) and the lady beetle predator *Nephus bipunctatus* (Kugelann), were recorded as natural enemies of *C. viticola* (Fallahzadeh and Japoshvili 2010;

Fallahzadeh et al. 2010). Other mealybugs reported from Middle East vineyards include *M. hirsutus*, *N. viridis*, *Pl. citri* and *Ps. maritimus*, but these mealybugs are not regarded as important vineyard pests.

12.5.7 South Africa

Planococcus ficus is the key economic mealybug species occurring in vineyards in South Africa. *Planococcus ficus* was initially identified in the Western Cape Province as *Pl. citri* (Joubert 1943) after this pest was accidentally introduced to the area. Other pseudococcid species have since been recorded from vines in South Africa and include *Ps. longispinus* and *F. malvastra*. The most recent records of mealybugs as well as their distribution in South African vineyards can be found in Walton et al. (2009). As with many other regions, the primary concern of *Pl. ficus* is the transmission of GLRaVs.

In South Africa, the influence of temperature on the development of *Pl. ficus* was reported by Walton and Pringle (2005), who estimated up to six annual generations of *Pl. ficus*. Seasonal development showed an upward migration of the population on the trunk from spring or early summer, with populations starting to develop on new growth and continuing until near harvest, reaching peak population densities in mid- to late summer. Mealybugs found in the vine canopy after harvest formed the nuclei of winter colonies. Winter population levels of *Pl. ficus* were low and consisted mainly of non-ovipositing adult females. The most recent advance in *Pl. ficus* management is the development of pheromone monitoring for South African vineyards, which can aid with treatment decisions (Walton et al. 2004).

Planococcus ficus populations are attacked by a range of natural enemies (Walton and Pringle 2004a, b). These include, in descending order of abundance, *Anagyrus* spp., *C. perminutus*, and *L. dactylopii* for parasitoids, and *Nephus bineavatus* (Mulsant), *Nephus angustus* (Casey) and *Nephus quadrivittatus* (Mulsant) for predatory beetles. Biological control is severely hampered by the presence of ants, such as *L. humile*, *Formica perpilosa* Wheeler, and *Crematogaster peringueyi* Emery as they provide biological refuges for the mealybugs (Addison and Samways 2000; Mgocheki and Addison 2009b). Management of ant colonies has led to marked increases of parasitism and ultimately biological control of these pests (Mgocheki and Addison 2010).

Chemical control of *Pl. ficus* is based on two treatments of organophosphates applied 2 weeks apart, just before bud burst. An additional supplementary treatment of a chemical with a short residual period is sometimes applied prior to harvest. The use of insect growth regulators and systemic neonicotinoids has increased and these are currently being used as in-season pest control options. Mating disruption by use of pheromone impregnated dispensers for *Pl. ficus* (Walton et al. 2006) is being investigated as an alternative in high value grape production units. Because ants impact mealybug densities and damage, chemical control measures for ants using directed sprays or chemical barriers have also been developed (Addison 2002).

12.6 Conclusion

Mealybugs may be the most universally important vineyard insect pest, causing crop damage through their presence, as well as the accumulation of honeydew and sooty molds. They also reduce vine vigor through repeated annual infestations and vector grapevine leafroll associated viruses. These innocuous-looking insects can be found in most of the world's vineyard regions. Although the mealybug species and their level of damage often vary, this review of vineyard mealybugs in Europe, the Middle East, North America, New Zealand, South Africa, South America, and India reveals remarkable similarity in pest issues and control strategies. For the most part, biological controls are a key component of mealybug pest suppression measures. In most regions there is still a reliance on insecticides when mealybug densities become too high, and vineyard managers worldwide have moved along similar lines of insecticide materials, with most regions now using organophosphates more sparingly and developing new programs based on neonicotinoids, insect growth regulators, and/or tetramic acid derivatives. Cultural practices can be used to enhance both biological controls and insecticide measures, but appear to be relatively labor intensive, and therefore, too costly in some regions. Future control measures will focus on novel methods to monitor mealybugs, using synthetic sex pheromones that may even find commercial use in mating disruption programs. For most wine grape regions, there is a need to better understand grapevine leafroll disease and the role of mealybugs and other scale insects in the dispersion of this important plant pathogen. It is towards this goal that grape pest researchers have joined a cosmopolitan effort towards the study and control of mealybugs.

References

Abbas MST (1999) Studies on *Dicrodiplosis manihoti* Harris (Diptera, Cecidomyiidae), a common predator of mealybugs. J Pest Sci 72:133–134

Addison P (2002) Chemical stem barriers for the control of ants (Hymenoptera: Formicidae) in vineyards. S Afr J Enol Vitic 23:1–8

Addison P, Samways MJ (2000) A survey of ants (Hymenoptera: Formicidae) that forage in vineyards in the Western Cape Province, South Africa. Afr Entomol 8:251–260

Artigas JN (1994) Economic entomology: insects of agriculture, forestry, medical and veterinary, vol 1. Ediciones Universidad de Concepción, Concepción, Chile (in Spanish)

Barrass IC, Jerie P, Ward SA (1994) Aerial dispersal of first- and second-instar longtailed mealybug, *Pseudococcus longispinus* (Targioni Tozzetti) (Pseudococcidae: Hemiptera). Aust J Exp Agric 34:1205–1208

Bartlett BR (1978) Pseudococcidae. In: Clausen CP (ed) Introduced parasites and predators of arthropod pests and weeds: a world review, Agricultural handbook no. 480. USDA, Washington, D.C, pp 137–170

Becerra V, Gonzalez M, Herrera ME, Miano JL (2006) Population dynamics of vine mealybug *Planococcus ficus* sign. (Hemiptera: Pseudococcidae) in vineyards. Rev Fac Cien Agrar Univ Nac Cuyo 38:1–6 (in Spanish)

Bell VA, Bonfiglioli RGE, Walker JTS, Lo PL, Mackay JF, McGregor SE (2009) Grapevine leafroll associated virus 3 persistence in *Vitis vinifera* remnant roots. J Plant Pathol 91:527–533

Ben-Dov Y (1995) A systematic catalogue of the mealybugs of the world (Insecta: Homoptera: Coccoidea: Pseudococcidae, and Putoidae) with data on geographical distribution, host plants, biology, and economic importance. Intercept Ltd., Hampshire

Berlinger MJ (1977) The Mediterranean vine mealybug and its natural enemies in southern Israel. Phytoparasitica 5:3–14

Bertazzon N, Borgo M, Vanin S, Angelini E (2010) Genetic variability and pathological properties of grapevine leafroll-associated virus 2 isolates. Eur J Plant Pathol 127:185–197

Bertin S, Cavalieri V, Graziano C, Bosco D (2010) Survey of mealybug (Hemiptera: Pseudococcidae) vectors of *Ampelovirus* and *Vitivirus* in vineyards of northwestern Italy. Phytoparasitica 38:401–409

Bierl-Leonhardt BA, Moreno DS, Schwarz M, Fargerlund J, Plimmer JR (1981) Isolation, identification and synthesis of the sex pheromone of the citrus mealybug, *Planococcus citri* (Risso). Tetrahedron Lett 22:389–392

Blumberg D, Klein M, Mendel Z (1995) Response by encapsulation of four mealybug species (Homoptera: Pseudococcidae) to parasitization by *Anagyrus pseudococci*. Phytoparasitica 23:157–163

Cabaleiro C, Segura A (1997) Field transmission of grapevine leafroll associated virus 3 (GLRaV-3) by the mealybug *Planococcus citri*. Plant Dis 81:283–287

Cabaleiro C, Segura A (2006) Temporal analysis of grapevine leafroll associated virus 3 epidemics. Eur J Plant Pathol 114:441–446

Cabaleiro C, Couceiro C, Pereira S, Cid M, Barrasa M, Segura A (2008) Spatial analysis of epidemics of grapevine leafroll associated virus-3. Eur J Plant Pathol 121:121–130

Cavalieri V, Mazzeo G, Garzia GT, Buonocore E, Russo A (2008) Identification of *Planococcus ficus* and *Planococcus citri* (Hemiptera: Pseudococcidae) by PCR-RFLP of COI gene. Zootaxa 1816:65–68

Charles JG (1981) Distribution and life history of the long-tailed mealybug, *Pseudococcus longispinus* (Homoptera: Pseudococcidae) in Auckland New Zealand vineyards. N Z J Zool 8:285–294

Charles JG (1982) Economic damage and preliminary economic thresholds for mealybugs (*Pseudococcus longispinus* T-T.) in Auckland vineyards. N Z J Agric Res 25:415–420

Charles JG (1985) *Diadiplosis koebelei* new record (Diptera: Cecidomyiidae) a predator of *Pseudococcus longispinus* (Homoptera: Pseudococcidae) from New Zealand. N Z J Zool 12:331–334

Charles JG (1993) A survey of mealybugs and their natural enemies in horticultural crops in North Island, New Zealand, with implications for biological control. Biocontrol Sci Technol 3:405–418

Charles JG (2011) Using parasitoids to infer a native range for the obscure mealybug, *Pseudococcus viburni*, in South America. BioControl 56:155–161

Charles JG, Cohen D, Walker JTS, Forgie SA, Bell VA, Breen KC (2006) A review of the ecology of grapevine leafroll associated virus type 3 (GLRaV-3). N Z Plant Prot 59:330–337

Charles JG, Froud KJ, van den Brink R, Allan DJ (2009) Mealybugs and the spread of grapevine leafroll-associated virus 3 (GLRaV-3) in a New Zealand vineyard. Australas Plant Pathol 38:576–583

Charles JG, Bell VA, Lo PL, Cole LM, Chhagan A (2010) Mealybugs (Hemiptera: Pseudococcidae) and their natural enemies in New Zealand vineyards from 1993–2009. N Z Entomol 33:84–91

Cid M, Pereira S, Cabaleiro C, Segura A (2010) Citrus mealybug (Hemiptera: Pseudococcidae) movement and population dynamics in an arbor-trained vineyard. J Econ Entomol 103:619–630

Clausen CP (1924) The parasites of *Pseudococcus maritimus* (Ehrhorn) in California (Hymenoptera, Chalcidoidea). Part II. biological studies and life histories. Univ Calif Publ Entomol 3:253–288

Compere H, Smith HS (1932) The control of the citrophilus mealybug, *Pseudococcus gahani*, by Australian parasites. Hilgardia 6:585–618

Cooper ML, Daane KM, Nelson EH, Varela LG, Battany MC, Tsutsui ND, Rust MK (2008) Liquid baits control Argentine ants sustainably in coastal vineyards. Calif Agric 62:177–183

Cordo H, Logarzo G, Braun K, Di Iorio O (2004) Catalog of phytophagus insects of Argentina and their associated plants. Argentina Entomological Society, Buenos Aires

Cucchi NJA, Becerra V (2009) Phytosanitary treatment manual for temperate crops under irrigation. Instituto Nacional de Tecnología Agropecuaria, Mendoza (in Spanish)

da Silva EB, Mendel Z, Franco JC (2010) Can facultative parthenogenesis occur in biparental mealybug species? Phytoparasitica 38:19–21

Daane KM, Malakar-Kuenen RD, Walton VM (2004) Temperature-dependent development of *Anagyrus pseudococci* (Hymenoptera: Encyrtidae) as a parasitoid of the vine mealybug, *Planococcus ficus* (Homoptera: Pseudococcidae). Biol Control 31:123–132

Daane KM, Sime KR, Hogg BN, Bianchi ML, Cooper ML, Rust MK, Klotz JH (2006a) Effects of liquid insecticide baits on Argentine ants in California's coastal vineyards. Crop Prot 25:592–603

Daane KM, Bentley WJ, Walton VM, Malakar-Kuenen R, Millar JG, Ingels CA et al (2006b) New controls investigated for vine mealybug. Calif Agric 60:31–38

Daane KM, Sime KR, Fallon J, Cooper ML (2007) Impacts of Argentine ants on mealybugs and their natural enemies in California's coastal vineyards. Ecol Entomol 32:583–596

Daane KM, Cooper ML, Triapitsyn SV, Andrews JW, Ripa R (2008a) Parasitoids of obscure mealybug, *Pseudococcus viburni* (Hem: Pseudococcidae) in California: establishment of *Pseudaphycus flavidulus* (Hym: Encyrtidae) and discussion of related parasitoid species. Biocontrol Sci Technol 18:43–57

Daane KM, Cooper ML, Triapitsyn SV, Walton VM, Yokota GY, Haviland DR et al (2008b) Vineyard managers and researchers seek sustainable solutions for mealybugs, a changing pest complex. Calif Agric 62:167–176

Daane KM, Bentley WJ, Smith RJ, Haviland DR, Weber E, Gispert C et al (2011) Vine mealybug. In: Bettiga L, Bentley WJ (eds) University of California grape pest management manual. University of California Press, Oakland, pp 125–135

de Borbon CM, Gracia O, Talquenca GSG (2004) Mealybugs and grapevine leafroll-associated virus 3 in vineyards of Mendoza, Argentina. Am J Enol Vitic 55:283–285

DeBach P (1949) Population studies of the long-tailed mealybug and its natural enemies on citrus trees in southern California, 1946. Ecology 30:14–25

Demontis MA, Ortu S, Cocco A, Lentini A, Migheli Q (2007) Diagnostic markers for *Planococcus ficus* (Signoret) and *Planococcus citri* (Risso) by random amplification of polymorphic DNA-polymerase chain reaction and species-specific mitochondrial DNA primers. J Appl Entomol 131:59–64

Doutt RL, Hagen KS (1950) Biological control measures applied against *Pseudococcus maritimus* on pears. J Econ Entomol 43:94–96

Duso C (1989) Bioecological study on *Planococcus ficus* (Sign.) in Veneto. Boll Lab Entomol Agrar Filippo Silvestri 46:3–20

El-Sayed AM, Unelius CR, Twidle A, Mitchell V, Manning LA, Cole L et al (2010) Chrysanthemyl 2-acetoxy-3-methylbutanoate: the sex pheromone of the citrophilous mealybug, *Pseudococcus calceolariae*. Tetrahedron Lett 51:1075–1078

Engelbrecht DJ, Kasdorf GGF (1990) Transmission of grapevine leafroll disease and associated closteroviruses by the vine mealybug, *Planococcus ficus*. Phytophylactica 22:341–346

Essig EO (1914) The mealybugs of California. Mon Bull Calif State Comm Hortic 3:18–143

Fajardo TVM, Kuhn GB, Nickel O (2003) Viral diseases (in Portuguese). In: Fajardo TVM (ed) Grape processing: pant. Brasília: Embrapa Information Technology, vol 35. Fruits of Brazil, Assessoria de Comunicação Social, Brasília, pp 45–62

Fallahzadeh M, Japoshvili G (2010) Checklist of Iranian encyrtids (Hymenoptera: Chalcidoidea) with descriptions of new species. J Insect Sci 10:1–26

Fallahzadeh M, Saghaei N, Ostovan H (2009) Seasonal abundance of *Planococcus ficus* (Hemiptera, Pseudococcidae) in Jahrom vineyards, Fars Province-IranFars Province-Iran. Plant Prot J (Iran) 1:263–276 (in Farsi)

Fallahzadeh M, Kaydan MB, Kozar F (2010) Description of a new species of *Chorizococcus* (Hemiptera: Coccoidea: Pseudococcidae) infesting *Vitis vinifera* in Iran. Turk J Entomol 34:157–163

Fallahzadeh M, Japoshvili G, Saghaei N, Daane KM (2011) Natural enemies of *Planococcus ficus* (Hemiptera: Pseudococcidae) in Fars Province vineyards, Iran. Biocontrol Sci Technol 21:427–433

Figadère BA, McElfresh JS, Borchardt D, Daane KM, Bentley W, Millar JG (2007) Trans-alpha-Necrodyl isobutyrate, the sex pheromone of the grape mealybug, *Pseudococcus maritimus*. Tetrahedron Lett 48:8434–8437

Flaherty DL, Peacock WL, Bettiga L, Leavitt GM (1982) Chemicals losing effect against grape mealybug. Calif Agric 36:15–16

Foldi I, Kozar F (2006) New species of *Cataenococcus* and *Puto* from Brazil and Venezuela, with data on others species (Hemiptera: Coccidea). Nouv Rev Entomol 22:305–312

Franco JC, Gross S, da Silva EB, Suma P, Russo A, Mendel Z (2003) Is mass-trapping a feasible management tactic of the citrus mealybug in citrus orchards? An Inst Super Agron 49:353–367

Franco JC, Silva EB, Cortegano E, Campos L, Branco M, Zada A, Mendel Z (2008) Kairomonal response of the parasitoid *Anagyrus* spec. nov. near *pseudococci* to the sex pheromone of the vine mealybug. Entomol Exp Appl 126:122–130

Franco JC, Zada A, Mendel Z (2009) Novel approaches for the management of mealybug pests. In: Ishaaya I, Horowitz AR (eds) Biorational control of arthropod pests: application and resistance management programs. Springer, New York, NY, pp 233–278

Frick KE (1952) The value of some organic phosphate insecticides in control of grape mealybug. J Econ Entomol 45:340–341

Geiger CA, Daane KM (2001) Seasonal movement and distribution of the grape mealybug (Homoptera: Pseudococcidae): developing a sampling program for San Joaquin Valley vineyards. J Econ Entomol 94:291–301

Golino DA, Sim ST, Gill R, Rowhani A (2002) California mealybugs can spread grapevine leafroll disease. Calif Agric 56:196–201

Gonzalez RH (2003) Mealybugs of agricultural and quarantine importance in fruit orchards in Chile (Hem: Pseudococcidae). Rev Frutic 24:5–17

Gonzalez RH, Volosky C (2005) Mealybugs and fruit moth: quarantine problems affecting fresh fruit exports. Rev Frutic 25:41–62

Gonzalez RH, Poblete JG, Barria PG (2001) The tree fruit mealybug in Chile, *Pseudococcus viburni* (Signoret), (Homoptera: Pseudococcidae). Rev Frutic 22:17–26

Grasswitz TR, Burts EC (1995) Effect of native natural enemies on the population dynamics of the grape mealybug, *Pseudococcus maritimus* (Hom: Pseudococcidae), in the apple and pear orchards. Entomophaga 40:105–117

Grasswitz TR, James DG (2008) Movement of grape mealybug, *Pseudococcus maritimus*, on and between host plants. Entomol Exp Appl 129:268–275

Grimes EW, Cone WW (1985a) Life history, sex attraction, mating, and natural enemies of the grape mealybug, *Pseudococcus maritimus* (Homoptera: Pseudococcidae). Ann Entomol Soc Am 78:554–558

Grimes EW, Cone WW (1985b) Control of the grape mealybug, *Pseudococcus maritimus*, (Hom: Pseudococcidae), on Concord grape in Washington. J Entomol Soc B C 82:3–6

Gullan PJ, Downie DA, Steffan SA (2003) A new pest species of the mealybug genus *Ferrisia* Fullaway (Hemiptera: Pseudococcidae) from the United States. Ann Entomol Soc Am 96:723–737

Gutierrez AP, Daane KM, Ponti L, Walton VM, Ellis CK (2008) Prospective evaluation of the biological control of vine mealybug: refuge effects and climate. J Appl Ecol 45:524–536

Habili N, Fazeli C, Ewart A, Hamilton R, Cirami R, Saldarelli P et al (1995) Natural spread and molecular analysis of grapevine leafroll-associated virus 3 in Australia. Plant Dis 85:1418–1422

Hardy NB, Gullan PJ, Hodgson CJ (2008) A classification of mealybugs (Hemiptera: Pseudococcidae) based on integrated molecular and morphological data. Syst Entomol 33:51–71

Haviland DR, Bentley WJ, Daane KM (2005) Hot-water treatments for control of *Planococcus ficus* (Homoptera: Pseudococcidae) on dormant grape cuttings. J Econ Entomol 98:1109–1115

Herrera MG, Madariaga VM (2001) Presence and incidence of grapevine viruses in the central zone of Chile. Agric Tech 61:393–400 (in Spanish)

Hinkens DM, McElfresh JS, Millar JG (2001) Identification and synthesis of the sex pheromone of the vine mealybug, *Planococcus ficus*. Tetrahedron Lett 42:1619–1621

Islam KS, Copland MJW (2000) Influence of egg load and oviposition time interval on the host discrimination and offspring survival of *Anagyrus pseudococci* (Hymenoptera: Encyrtidae), a solitary endoparasitoid of citrus mealybug, *Planococcus citri* (Hemiptera: Pseudococcidae). Bull Entomol Res 90:69–75

Joubert CJ (1943) Mealybugs on vines. Bull Dep Agric S Afr 243:1–20

Karamaouna F, Copland MJW (2000) Host suitability, quality and host size preference of *Leptomastix epona* and *Pseudaphycus flavidulus*, two endoparasitoids of the mealybug *Pseudococcus viburni*, and host size effect on parasitoid sex ratio and clutch size. Entomol Exp Appl 96:149–158

Karamaouna F, Copland MJ (2009) Fitness and life history parameters of *Leptomastix epona* and *Pseudaphycus flavidulus*, two parasitoids of the obscure mealybug *Pseudococcus viburni*. BioControl 54:65–76

Kaydan MB, Kilincer N, Kozar F (2005) Studies on Pseudococcidae (Hemiptera Coccoidea) fauna of urban ecosystem of Ankara Province, Turkey. Boll Zool Agrar Bachic 37:85–95

Kosztarab M, Kozar F (1988) Scale insects of central Europe. Springer, Netherlands

Kozár F, Guignard E, Bachmann F, Mani E, Hippe C (1994) The scale insect and whitefly of Switzerland (Homoptera: Coccoidea and Aleyrodoidea). Bull Soc Entomol Suisse 67: 151–161

Liu Y-B, Bettiga LJ, Daane KM (2010) Ultralow oxygen treatment for control of *Planococcus ficus* (Hemiptera: Pseudococcidae) on grape benchgrafts. J Econ Entomol 103:272–276

Lo PL, Walker JTS (2010) Good results from a soil-applied insecticide against mealybugs. N Z Winegrower 14:125–127

Mahfoudhi N, Dhouibi MH (2009) Survey of mealybugs (Hemiptera: Pseudococcidae) and their natural enemies in Tunisian vineyards. Afr Entomol 17:154–160

Mani M, Krishnamoorthy A (1989) Life cycle, host stage suitability and pesticide susceptibility of the grape mealybug parasitoid, *Allotropa japonica* sp. n. J Biol Control 3:7–8

Mani M, Thontadarya TS (1987a) Biological studies on the grape mealybug predator *Scymnus coccivora* (Ayyar). J Biol Control 1:89–92

Mani M, Thontadarya TS (1987b) Development and feeding potential of the coccinellid, *Cryptolaemus montrouzieri* Muls. on grape mealybug, *Maconellicoccus hirsutus* (Green). J Biol Control 1:19–22

Mani M, Thontadarya TS (1987c) Population dynamics of the mealybug *Maconellicoccus hirsutus* (Green) and its natural enemies in the grapevine ecosystem. J Biol Control 1:93–97

Mani M, Thontadarya TS (1988) Studies on the safety of different pesticides to the grape mealybug natural enemies, *Anagyrus dactylopii* (How.) and *Scymnus coccivora* Ayyar. Indian J Plant Prot 16:205–210

Mani M, Thontadarya TS (1989) Field evaluation of *Cryptolaemus montrouzieri* Muls. in the suppression of *Maconellicoccus hirsutus* green on grapevine. J Biol Control 2:14–16

Mani M, Thontadarya TS, Singh SP (1987) Record of natural enemies on the grape mealybug, *Maconellicoccus hirsutus* (Green). Curr Sci (India) 56:624–625

Martelli GP, Agranovsky AA, Bar-Joseph M, Boscia D, Candresse T, Coutts RHA et al (2002) The family Closteroviridae revised. Arch Virol 147:2039–2044

McKenzie HL (1967) Mealybugs of California with taxonomy, biology and control of North American species. University of California Press, Berkeley

Mgocheki N, Addison P (2009a) Effect of contact pesticides on vine mealybug parasitoids, *Anagyrus* sp. near *pseudococci* (Girault) and *Coccidoxenoides perminutus* (Timberlake) (Hymenoptera: Encyrtidae). S Afr J Enol Vitic 30:110–116

Mgocheki N, Addison P (2009b) Interference of ants (Hymenoptera: Formicidae) with biological control of the vine mealybug *Planococcus ficus* (Signoret) (Hemiptera: Pseudococcidae). Biol Control 49:180–185

Mgocheki N, Addison P (2010) Spatial distribution of ants (Hymenoptera: Formicidae), vine mealybugs and mealybug parasitoids in vineyards. J Appl Entomol 134:285–295

Millar JG, Daane KM, McElfresh JS, Moreira JA, Malakar-Kuenen R, Guillén M et al (2002) Development and optimization of methods for using sex pheromone for monitoring the mealybug *Planococcus ficus* (Homoptera: Pseudococcidae) in California vineyards. J Econ Entomol 95:706–714

Millar JG, Midland SL, McElfresh JS, Daane KM (2005) (2,3,4,4-tetramethylcyclopentyl)methyl acetate, a sex pheromone from the obscure mealybug: first example of a new structural class of monoterpenes. J Chem Ecol 31:2999–3005

Millar JG, Moreira JA, McElfresh JS, Daane KM, Freund AS (2009) Sex pheromone of the long-tailed mealybug: a new class of monoterpene structure. Org Lett 11:2683–2685

Miller DR, Gill RT, Williams DJ (1984) Taxonomic analysis of *Pseudococcus affinis* (Maskell), a senior synonym of *Pseudococcus obscurus* Essig, and a comparison with *Pseudococcus maritimus* (Ehrhorn) (Homoptera: Coccoidea: Pseudococcidae). Proc Entomol Soc Wash 86:703–713

Morandi Filho WJ, Grützmacher AD, Botton M, Bertin A (2009) Chemical control of mealybugs, *Planococcus citri* (Risso) (Hemiptera: Pseudococcidae) in vineyards of different ages. Arq Inst Biol 76:427–435

Nelson EH, Daane KM (2007) Improving liquid bait programs for Argentine ant control: Bait station density. Environ Entomol 36:1475–1484

Noyes JS, Hayat MS (1994) Oriental mealybug parasitoids of the Anagyrini (Hymenoptera: Encyrtidae). CAB International Press, Wallingford

Petersen CL, Charles JG (1997) Transmission of grapevine leafroll-associated closteroviruses by *Pseudococcus longispinus* and *P. calceolariae*. Plant Pathol 46:509–515

Rajagopal BK, Viraktamath CA, Gowda VN (1997) Incidence of ant associated mealybug *Xenococcus annandalei* (Homoptera: Pseudococcidae) on grapes in south India. Entomon 22:165–166

Ripa SR, Larral P (2008) Avocado and citrus pest management. Colección Libros INIA N° 23. Instituto de Investigaciones Agropecuarias, Chile. (in Spanish)

Ripa SR, Luppichini P (2010) Vineyard pest management. Colección Libros INIA N°26. Instituto de Investigaciones Agropecuarias, Chile. (in Spanish)

Ripa SR, Rojas PS (1990) Management and biological control of the white vine mealybug. Rev Frutic 11:82–87

Roltsch WJ, Meyerdirk DE, Warkentin R, Andress ER, Carrera K (2006) Classical biological control of the pink hibiscus mealybug, *Maconellicoccus hirsutus* (Green), in Southern California. Biol Control 37:155–166

Rosciglione B, Castellano MA (1985) Further evidence that mealybugs can transmit grapevine virus A (GVA) to herbaceous hosts. Phytopathol Mediterr 24:186–188

Ross L, Pen I, Shuker DM (2010a) Genomic conflict in scale insects: the causes and consequences of bizarre genetic systems. Biol Rev 85:807–828

Ross L, Langenhof MBW, Pen I, Beukeboom LW, West SA, Shuker DM (2010b) Sex allocation in a species with paternal genome elimination: the roles of crowding and female age in the mealybug *Planococcus citri*. Evol Ecol Res 12:89–104

Rotundo G, Tremblay E (1972) Studies on a sexual pheromone of *Planococcus citri* (Risso) (Homoptera, Coccoidea). I Boll Lab Entomol Agrar Filippo Silvestri Portici 30:217–230

Salazar A, Gerding M, Luppichini P, Ripa R, Larraín P, Zaviezo T, Larral P (2010) Biology, management and control of mealybugs. Bol. INIA 204. Instituto de Investigaciones Agropecuarias, Chile. (in Spanish)

Sandanayaka WRM, Charles JG, Allan DJ (2009) Aspects of the reproductive biology of *Pseudaphycus maculipennis* (Hym: Encyrtidae), a parasitoid of obscure mealybug, *Pseudococcus viburni* (Hem: Pseudococcidae). Biol Control 48:30–35

Satyanarayana J, Murthy MS, Srinivasa N (1991) Impact of pesticides on grapevine mealybug predator, *Cryptolaemus montrouzieri* Mulsant. 1. Residual toxicity and safety of pesticidal sprays to predatory grubs. Indian J Entomol 53:587–592

Sazo L, Araya JE, Cerda JDL (2008) Effect of a siliconate coadjuvant and insecticides in the control of mealybug of grapevines, *Pseudococcus viburni* (Hemiptera: Pseudococcidae). Cien Invest Agrar 35:177–184

ScaleNet (2011) ScaleNet. http://www.sel.barc.usda.gov/scalenet/scalenet.htm

Sforza R, Boudon-Padieu E, Greif C (2003) New mealybug species vectoring grapevine leafroll-associated viruses-1 and-3 (GLRaV-1 and-3). Eur J Plant Pathol 109:975–981

Sharma AM, Wang J, Duffy S, Zhang S, Wong MK, Rashed A et al (2011) Occurrence of grapevine leafroll-associated virus complex in Napa Valley. PLoS One 6(10): e26227

Smith HS, Armitage HM (1920) Biological control of mealybugs in California. Mon Bull Calif Dep Agric 9:104–158

Srinivas T, Prasad KS, Shekhar MA, Manjunath D (2007) Evaluation on neem based formulations vis-a-vis dichlorvos against *Meconellicoccus hirsutus,* Uttar Pradesh. J Zool 27:13–20

Sunitha ND, Jagginavar SB, Biradar AP (2009) Bioefficacy botanicals and newer insecticides against grape vine mealybug, *Maconellicoccus hirsutus* (Green). Karnataka J Agric Sci 22:710–711

Tollerup KE, Rust MK, Dorschner KW, Phillips PA, Klotz JH (2004) Low-toxicity baits control ants in citrus orchards and grape vineyards. Calif Agric 58:213–217

Tollerup K, Rust MK, Klotz JH (2007) *Formica perpilosa*, an emerging pest in vineyards. J Agric Urban Entomol 24:147–158

Tsai CW, Chau J, Fernandez L, Bosco D, Daane KM, Almeida RPP (2008) Transmission of grapevine leafroll-associated virus 3 by the vine mealybug (*Planococcus ficus*). Phytopathology 98:1093–1098

Tsai C, Rowhani A, Golino DA, Daane KM, Almeida RPP (2010) Mealybug transmission of grapevine leafroll viruses: an analysis of virus-vector specificity. Phytopathology 100:830–834

Wakgari WM, Giliomee JH (2005) Description of adult and immature females of six mealybug species (Hemiptera: Pseudococcidae) found on citrus in South Africa. Afr Entomol 13:281–332

Walton VM, Pringle KL (2001) Effects of pesticides and fungicides used on grapevines on the mealybug predatory beetle *Nephus 'boschianus'* (Coccinellidae, Scymnini). S Afr J Enol Vitic 22:107–110

Walton VM, Pringle KL (2004a) A survey of mealybugs and associated natural enemies in vineyards in the Western Cape Province, South Africa. S Afr J Enol Vitic 25:23–25

Walton VM, Pringle KL (2004b) Vine mealybug, *Planococcus ficus* (Signoret) (Hemiptera: Pseudococcidae), a key pest in South African vineyards. A review. S Afr J Enol Vitic 25:54–62

Walton VM, Pringle KL (2005) Developmental biology of vine mealybug, *Planococcus ficus* (Signoret) (Homoptera: Pseudococcidae), and its parasitoid *Coccidoxenoides perminutus* (Timberlake) (Hymenoptera: Encyrtidae). Afr Entomol 13:143–147

Walton VM, Daane KM, Pringle KL (2004) Monitoring *Planococcus ficus* in South African vineyards with sex pheromone-baited traps. Crop Prot 23:1089–1096

Walton VM, Daane KM, Bentley WJ, Millar JG, Larsen TE, Malakar-Kuenen R (2006) Pheromone-based mating disruption of *Planococcus ficus* (Hemiptera: Pseudococcidae) in California vineyards. J Econ Entomol 99:1280–1290

Walton VM, Kruger K, Saccaggi DL, Millar IM (2009) A survey of scale insects (Sternorryncha: Coccoidea) occurring on table grapes in South Africa. J Insect Sci 9:47

Waterworth RA, Wright IM, Millar JG (2011) Reproductive biology of three cosmopolitan mealybug (Hemiptera: Pseudococcidae) species, *Pseudococcus longispinus*, *Pseudococcus viburni*, and *Planococcus ficus*. Ann Entomol Soc Am 104:249–260

Williams DJ, Granara de Willink MC (1992) Mealybugs of Central and South America. CAB International, Wallingford
Williams DJ, Moghaddam M (2000) Mealybug species of the genus *Planococcus* Ferris in Iran (Homoptera: Coccoidea: Pseudococcidae) with a discussion of *Planococcus vovae* (Nasonov). J Entomol Soc Iran 18(1/2):32–43 (in Farsi)
Zada A, Dunkelblum E, Assael F, Harel M, Cojocaru M, Mendel Z (2003) Sex pheromone of the vine mealybug, *Planococcus ficus*, in Israel: occurrence of a second component in a mass-reared population. J Chem Ecol 29:977–988
Zaviezo T, Cadena E, Flores MF, Bergmann J (2010) Influence of different plants substrates on development and reproduction for laboratory rearing of *Pseudococcus calceolariae* (Maskell) (Hemiptera: Pseudococcidae). Cien Invest Agrar 37:31–37
Zhang AJ, Amalin D, Shirali S, Serrano MS, Franqui RA, Oliver JE et al (2004) Sex pheromone of the pink hibiscus mealybug, *Maconellicoccus hirsutus*, contains an unusual cyclobutanoid monoterpene. Proc Natl Acad Sci U S A 101:9601–9606
Zorloni A, Prati S, Bianco PA, Belli G (2006) Transmission of grapevine virus A and grapevine leafroll-associated virus 3 by *Heliococcus bohemicus*. J Plant Pathol 88:325–328
Zou Y, Millar JG (2009) Synthesis of the pheromone of the longtailed mealybug, a sterically congested, irregular monoterpenoid. J Org Chem 74:7207–7209

Chapter 13
Leaf-Eating Lepidoptera in North American Vineyards

Walter J. Bentley and Richard L. Coviello

13.1 Introduction

Leaf eating Lepidoptera commonly found in North American vineyards include four species. Two of these, the grape leaffolder, *Desmia funeralis* (Hübner) (Pyralidae) and the western grapeleaf skeletonizer, *Harrisina brillians* Barnes & McDunnough (Zygaenidae), are considered major pests. A fifth species, the omnivorous leafroller, *Platynota stultana*, Walsingham (Tortricidae), can be found webbing leaves but primary damage is through direct feeding on flowers and berries that results in secondary infection of bunch rot organisms. Damage to leaves by this insect is of no economic importance. Omnivorous leafroller is a major grape pest of western United States. Less common defoliating lepidopterans include the achemon sphinx moth, *Eumorpha achemon* (Drury) (Sphingidae) and the whitelined sphinx, *Hyles lineata* (F.) (Sphingidae).

Damage caused by leaf feeding Lepidoptera has generally decreased within the last decade. This is primarily because of better methods of detecting their presence and the use of reduced-risk and non-disruptive insecticides. These products allow the survival of parasitoids and predators so that biological control can be incorporated into a truly integrated program of arthropod management.

W.J. Bentley (✉)
Division of Agriculture and Natural Resources, Kearney Agricultural Research and Extension Center, University of California, Parlier, CA 93648, USA
e-mail: walt@uckac.edu

R.L. Coviello
Division of Agriculture and Natural Resources, University of California, Fresno, CA 93702, USA
e-mail: rlcoviello@earthlink.net

N.J. Bostanian et al. (eds.), *Arthropod Management in Vineyards: Pests, Approaches, and Future Directions*, DOI 10.1007/978-94-007-4032-7_13,

13.2 Grape Leaffolder

Prior to the invasive pests such as the vine mealybug, *Planococcus ficus* (Signoret) (Hemiptera: Pseudococcidae) and the glassy winged sharpshooter, *Homalodisca vitripennis* (Germar) (Hemiptera Cicadellidae), grape leaffolder was ranked as the fourth arthropod causing damage to vines in California (Jensen et al. 1975). The grape leaffolder is found throughout the United States, eastern Canada and northern Mexico (Jensen et al. 1975; Mead and Webb 2008). It is thought to be native to the eastern United States (Strauss 1916; Jensen et al. 1992). In 1855, Glover described injury on grapes caused by this insect in the District of Columbia, South Carolina and Georgia (Strauss 1916). Although all cultivars are attacked, American grape cultivars (*Vitis* sp.) are preferred over *Vitis vinifera* L.

13.2.1 Description

The moth is predominantly black. Two white spots are present on the forewings. The hind wings of the female also have two smaller white spots whereas the hind wings of the male have a single irregular white spot (Fig. 13.1). There is a slight

Fig. 13.1 Female *Desmia funeralis* (Photo J. Dibble)

Fig. 13.2 *Desmia funeralis* egg (Photo F. Jensen)

white fringe on both pairs of wings. The wingspan is approximately 25 mm. The abdomen has two bright white bands across the back (Strauss 1916; Jensen et al. 1975, 1992; Mead and Webb 2008). The antenna of the female is moniliform and straight, whereas the male antenna is elbowed. Moths fly in the evening and throughout the night.

Eggs are approximately 0.88 mm long, elliptical and flat in shape (Fig. 13.2). The chorion has small hexagonal reticulations (Strauss 1916; Jensen et al. 1975; Mead and Webb 2008). They are laid singly, usually on the bottom side of the leaves along the juncture of a leaf vein and the leaf surface. If the leaves are pubescent, they are deposited on the upper side. Multiple eggs are laid when moths are abundant, they may be laid adjacent to each other. Eggs are difficult to detect and are best seen when the leaf is directly exposed to the sun.

Larvae (Fig. 13.3) range in size from 1.6 mm at hatch to 22 mm at full growth. At hatching, the larva is translucent cream to brown (Strauss 1916; Jensen et al. 1975, 1992). As it feeds on leaf tissue the abdomen will appear green. There are five larval instars. The first two instars have no distinctive sclerotization but characteristically feed in groups between two touching leaves. The third instar larva can be field identified by the presence of a black spot above the second pair of true legs on the thorax. The fourth instar possesses two spots and the fifth instar has three spots above the legs and a spot above the anus. The later larval stages can be found in rolled leaves. When the leaf rolls are handled, larvae actively wriggle backwards and drop from the leaf to the ground on a silken thread.

Fig. 13.3 *Desmia funeralis* larvae (Photo F. Jensen)

The pupal stage is brown and is found in leaf rolls or, infrequently, in the soil beneath the vine. It averages 13 mm in length (Strauss 1916; Jensen et al. 1975, 1992).

13.2.2 Damage

Leaf feeding within rolled and tied leaves results in various levels of defoliation. The amount of damage is dependent upon larval population and time of year. In the process of leaf rolling, larvae spin a silk thread that anchors them to the leaf margin. It then repeats this behavior with a thread attached towards the middle of the leaf. This behavior is continued 200–300 times (Jensen et al. 1975, 1992). The leaf gradually curls as the filaments contract. More strands are later placed along the margin that aid in forming a tight tube that is open at both ends (Fig. 13.4). Leaf rolling occurs at night. Feeding by older larvae occurs within these rolls and only the leaf edge, within the roll, is fed upon. Usually one roll is found per leaf but, where larvae are abundant, more than one is common (Jensen et al. 1975). Generally each roll contains a single larva. At least two leaf rolls are required for larger larvae to complete development. Feeding by young larvae is characterized by the removal of all but the epidermal layer of the leaf, where two leaves touch. They never eat through the leaf and the remaining epidermal cells are colorless, and when dry, turn brown.

First brood damage is usually minimal. Second brood damage can result in up to 50% defoliation and third brood feeding can completely defoliate vineyards

Fig. 13.4 Leaf rolled by *Desmia funeralis* larvae (Photo F. Jensen)

Fig. 13.5 Vineyard defoliation due to *Desmia funeralis* larval feeding (Photo F. Jensen)

(Fig. 13.5) (Jensen et al. 1992). When vine defoliation occurs, larvae will feed on grape clusters. In areas where temperatures in the fall are quite warm, a partial fourth generation is possible. Unless biological control is disturbed, because of the application of broad spectrum insecticides aimed at other pests, severe defoliation

Table 13.1 Average moth flight periods and time required for completion of various developmental stages of the grape leaffolder in Fresno, CA (Barnes 1944)

Brood	Moth flight period	Time required for egg hatch (days)	Total larval time (weeks)	Total pupal time (days)	Time from egg to adult (weeks)
First	April 2–May 24	10–17	3–4	10–14	6.5–7.5
Second	June 15–July 15	4–5	2–3	7–11	4–5
Third	August 3–September 5	4–5	3–5	Overwinter	

does not occur. Kliewer (1970) studied the effects of defoliation on 'Thompson' yield and quality. Their studies showed that more than 20% leaf loss within 1 month after berry set stage would impact yield. Two months after berry set, the same level of defoliation reduced yield and total sugars. On table grapes, even low levels of defoliation can impact grape quality through sun burning of clusters (Kliewer and Lider 1968; May et al. 1969).

13.2.3 Life History

Grape leaffolder overwinters as a pupa in old leaf remnants on the soil of vineyards (Strauss 1916; Jensen et al. 1975, 1992). Adult emergence begins just after bud break of cv. Thompson seedless in California. Usually moths are first detected in early April and continue to fly into June (averaging 53 days) in California. The average moth flight periods in Fresno, CA, over a 4-year period are presented in Table 13.1. In Florida, moths can be found as early as February. Three generations are common in most areas of North America. The three broods are very distinct.

AliNiazee (1974) reared grape leaffolder larvae in the laboratory at 23.9 ± 1°C. Average development time for the first, second, third, fourth and fifth larval instars was 3.5, 4.0, 4.4, 4.3, and 6.1 days, respectively. Moths live approximately 9 days under field conditions and females deposit most of their eggs on the second to the fifth day. Females can lay on average 200 eggs (AliNiazee 1974).

Adults fly primarily at night. They do not readily fly during windy periods but are often seen near windbreaks and structures. Earliest egg deposition tends to be on suckers near the ground (Jensen et al. 1992), and the earliest evidence of feeding will be found in such areas.

13.2.4 Monitoring and Economic Thresholds

Damage due to grape leaffolder is quite variable from year to year. However, vineyards with a history of infestation should be monitored every year. Grape leaffolder tends to reinfest the same areas of vineyards. These areas should be noted and visited regularly for monitoring. Also, vigorous, densely foliated vines tend to be preferred over weak and stressed vines (Jensen et al. 1992).

Evidence of a sex pheromone for grape leaffolder was first demonstrated in 1972. Virgin female moths were isolated and shown to attract male moths in both the laboratory and the field (AliNiazee and Stafford 1972a). Later work determined the pheromone to be a blend of (*Z*, *Z*)-11,13-hexadecadienal, 11-hexadecynal, and (*Z*)-11-hexadecenal (Millar et al. 2002). In field tests, male moths were trapped equally well in traps baited with a concentration range of pheromone from 0.2 to 6 mg. Lures effectively trapped males for 5 weeks. Prior to the monitoring with virgin females or pheromones, malt syrup liquid baits were used to identify moth activity (Barnes 1944). Moths can also be caught in bait traps containing terpinyl acetate or with blacklight traps (AliNiazee and Stafford 1972b).

Because leaf damage from larval feeding is easily seen, few scouts rely on pheromone traps to monitor grape leaffolder. In only the most extreme situations do first generation larvae reach levels that impact grape yield or quality adversely and this allows farmers to evaluate larval populations during that generation. Searching for early larval feeding on leaves in the upper grape canopy is the most common way to monitor larval abundance (Jensen et al. 1992). Young larvae feeding in groups can easily be recognized at least 1 week before leaf rolling. Even if these signs are missed, some leaf rolls begin to appear before the bulk of the generation rolls the leaves. These first rolls appear on the upper part of the vine where they are easily detected by examining for small larvae. Concentrate searches in areas with a history of infestation. It is important to recognize feeding early if treatments are to be made. Once roll forming is established, the larvae are well protected from insecticides inside the leaf roll.

There is no current valid economic threshold for grape leaffolder. First brood treatments are not recommended because examining larval rolls made by the first brood can give an indication of parasitism, which is quite common. In the absence of parasitism, each brood will increase its abundance two- to five-fold (Jensen et al. 1975). Where increased leaf damage is seen in the second brood, and there is no evidence of parasitism, spray applications should be considered, particularly on fresh market grapes. Both wine and raisin grapes can withstand more feeding than fresh market grapes.

13.2.5 Natural Control

Strauss (1916) made the initial recovery of parasitoids. He reported six species of Hymenoptera and three species of Diptera attacking grape leaffolder in Virginia. The most common and effective parasite is *Bracon* (Formerely *Microbracon*) *cushmani* (Musebeck) (Donohoe and Barnes 1934). Two other hymenoptera are less commonly found and have little impact: *Brachymeria ovata* (Say), a pupal parasitoid, and *Coccygomimus sanguinipes* (Cresson). The tachinid parasitoids include *Nemorilla pyste* (Walker) and *Erynnia tortricis* (Coquillett). Both are parasitoids of

Fig. 13.6 *Desmia funeralis* larval parasitism by *Bracon cushmani* (Photo F. Jensen)

mid to late larval instars of grape leaffolder. Females of both species deposit eggs on the thorax and head of the leaffolder. Upon hatching, larvae burrow into the host. *Trichogramma* sp. will also parasitize grape leaffolder eggs but parasitism levels are quite low. In 1968, searches for parasitoids and predators resulted in the establishment of three parasitic wasps in Tulare County California (Doutt et al. 1969). These are *Macrocentrus nuperus* Cresson, *Apanteles canarsiae* Ashmead, and *Pardiaulomella ibseni* Girault.

The most effective biological control is provided by *B. cushmani* that develops on third, fourth and fifth instar larvae. The female stings and paralyzes the larval host and lays one or more eggs upon it. In minimally managed vineyards, parasitism usually ranges from 30 to 40%, even as high as 78% (Jensen et al. 1975). Several parasitoid larvae may feed externally on a grape leaffolder larva, slowly consuming it (Fig. 13.6). The sclerotized portions of the host often remain inside the leaffold along with the parasitoid pupae and host excrement (Jensen et al. 1992). The thinner integument of the host epidermis is attached to its head and appears like a thin flag.

13.2.6 Management

Grape leaffolder is easily managed with non-disruptive insecticides when applications are timed correctly. However, sprays must be applied prior to substantial leaf rolling by older instar larvae to obtain maximal control. Wettable powder and dust formulations of *Bacillus thuringiensis* Berliner have provided control equal to that of carbaryl (Jensen 1969; AliNiazee and Jensen 1972). Reduced-risk insecticides such as spinosad and methoxyfenozide showed exceptional efficacy and relatively long residual activity (Bentley et al. 2009). Cryolite, a stomach poison, is effective and non-disruptive to natural enemies but it is often restricted by wineries because of residues. As methoxyfenozide does not interfere with biological control, it is a key factor in establishing long term control of the grape leaffolder. Spinosad has moderate impact on parasitoids and predators. Broadly toxic insecticides, such as permethrin and carbaryl, should be avoided.

13.3 Western Grapeleaf Skeletonizer

There are two species of grapeleaf skeletonizer found in North America and their damages are identical. Both feed on Virginia creeper, *Parthenocissus quinquefolia* (L.) Planchon, wild grape (*Vitis* sp.), cultivated grape (*Vitis vinifera*) and redbud (*Cercis* sp.). The earliest identified was *Harissina americana* (Guérin), with the common name of grapeleaf skeletonizer (Jones 1909). Guérin-Méneville described it as *Agloape americana* Boisduval but the listed dates are questionable (Jones 1909). In 1839, Harris described the species as *Procris americana* and noted that it is probably the same species that Guérin-Méneville incorrectly placed in the genus *Agloape* (Jones 1909). Harris found *P. americana* to be damaging vines in Chapel Hill, North Carolina. Early literature reports its presence in Canada, New England, New York, New Jersey, Washington D.C., Georgia, Florida, Ohio, Missouri, and Arizona.

The Arizona records of 1893 probably refer to western grapeleaf skeletonizer *H. brillians*. This species is a pest of southwestern commercial vineyards in California, Arizona, Colorado, Utah, Nevada, New Mexico, and the states of Sonora, Chihuahua, Coahuila, and Aquascalientes in Mexico (Stern et al. 1983). Although not an annual problem, it is a serious vineyard pest in western North America. The following information deals with the western grapeleaf skeletonizer. It was first reported in California near San Diego in 1941 (Lange 1944). It was first found in California's San Joaquin Valley on October 7, 1974 in Visalia. Since then, it has established itself throughout California.

13.3.1 Description

The western grapeleaf skeletonizer moth is black throughout. When seen under bright sunlight it appears to have a bluish to green blue metallic iridescence (Fig. 13.7) (Jones 1909; Lange 1944; Robinson 1950; Stern et al. 1992). Maximum size of the adult is 14 mm long with a wing expanse of 31 mm. When larval food is limited, usually due to population abundance late in the year, the adult may be half

Fig. 13.7 Female *Harrisina brillians* (Photo F. Jensen)

these measurements. Male and female moths are similar, with the exception of lateral posterior tufts on the abdomen of the male. Both sexes have bi-pectinate antennae. An orange collar is present on the pronotum in few specimens. The moths fly during the day and predominantly in the early morning.

Eggs are laid in groups ranging from 12 to more than 300 with an average of 155 (Lange 1944; Robinson 1950) (Fig. 13.8). They are shaped as a capsule and are 0.60–0.65 mm long and 0.40–0.45 mm wide (Allen et al. 1974). They have a fine surface reticulation and are cream to pale yellow in color. Eggs are usually laid on the leaf undersurface and multiple moths will oviposit on the same leaf.

Larvae, when fully grown, are approximately 12 mm long with a slightly tapering body at each end (Lange 1944; Robinson 1950; Allen et al. 1974) (Fig. 13.9). The head is black. The body is yellow below with two transverse purple bands. The first band covers the last thoracic and the first two abdominal segments. The rear band covers the sixth and seventh abdominal segments. There are also transverse bands of setae on the thorax and abdominal segments 1, 2, 4, 6, 7, 8, and 9. These setae are irritating to humans, causing a dermal rash. Young larvae are predominately cream-colored in the first instar and, after the second molt, gradually develop coloration. The young larvae are marked with scattered black setae over the body. Full coloration and setal patterns appear in the fourth instar.

The pupa is reddish brown and somewhat flattened. It averages 7–10 mm long and 3–4 mm wide. The abdominal segments 2–7 possess transverse bands of spicules and spiracles are evident on abdominal segments 1–8 (Robinson 1950). The silken cocoon is convex and somewhat flattened. Initially it is white and weathers to a dirty white coloration. It is 14–17 mm long and 5–9 mm wide. Cocoons are found

Fig. 13.8 Eggs of *Harrisina brillians* (Photo F. Jensen)

Fig. 13.9 Late instar larvae of *Harrisina brillians* defoliating grape

Fig. 13.10 Characteristic early instar larval feeding of *Harrisina brillians* (Photo F. Jensen)

under the bark or between the trunk and grape stakes grouped tightly together (Allen et al. 1974). The winter is spent in the pupal stage.

13.3.2 Damage

The feeding by young larvae is very characteristic in that the first three instars, and the early fourth, feed only on the outer epidermis of the underside of the leaf leaving a thin layer of translucent tissue that appears white in contrast to the remainder of the green leaf. The multiple larvae feed adjacent to each other in a line (Fig. 13.10). Such feeding behavior is easily seen and can be used to evaluate the potential for vine defoliation. In the late fourth and fifth instar, the larvae disperse to feeding individually and will remove all but the leaf veins and petiole (Fig. 13.9). When populations are abundant, complete defoliation occurs resulting in crop sunburn. Premature summer defoliation results in vine regrowth during the fall that reduces the crop the following spring. Initial infestations in 1943 resulted in crop loss of 90% in some vineyards (Lange 1944). Western grapeleaf skeletonizer does not damage berries until leaves have been completely consumed. Their feeding on berries can allow disease organisms such as *Botrytis* to enter the fruit. In addition to crop damage, even moderate populations of grapeleaf skeletonizer will affect workers picking the crop. Severe dermatitis due to urticating setae on the larvae results from larval contact with the skin.

Table 13.2 Development stages of western grapeleaf skeletonizer in Tulare County, California commercial vineyards (Stern et al. 1992)

Generation	Adults	Eggs	Larvae
First	Late April to mid-May	Late April to Late May	Early May to late June
Second	Late June to early July	Late June to mid-July	July to mid-August
Third	Late August to early September	Late August to mid-September	September and October

13.3.3 Life History

There are three distinct generations of western grapeleaf skeletonizer and, in warm years, a partial fourth in California's San Joaquin Valley (Roltsch et al. 1992). The number of generations varies in other states, but there is always a minimum of two (Table 13.2) (Stern et al. 1992). The first and second generations are quite distinct, with no overlapping of stages. The second and third generations are also distinct, but there is a slight overlap of adults and larvae and, to a lesser extent, eggs.

The winter is spent as a pupa beneath vine bark or at the base of vines near the soil surface. In central California, moths begin emergence in late March or early April but, in some areas, not until early May. Flight activity predominates in the early morning hours, and less during the late hours of the day. Mating occurs quickly, within the first 3 days of emergence (Robinson 1950). Eggs are laid within the first 4 days of mating on the abaxial leaf surfaces and most often, in the more shaded areas of the canopy. Eggs of the first spring generation require 12–16 days to hatch whereas those of the summer generation need 10 days (Robinson 1950). Eggs laid by the emerging moths of the overwintering generation are usually deposited on sucker growth or leaves low in the canopy while those of subsequent generations are deposited on leaves throughout the vine. The moths are short lived, surviving for approximately 1 week in the spring and less than a week in the summer. During this time they feed on nectar from blossoms of various weeds and, possibly, flowering citrus.

Feeding of young larvae is first seen on leaves in the lower canopy and characteristically results in the removal of the lower epidermis (Fig. 13.10). As larvae reach the fourth and fifth instars they form a coalesced mass, molt, and then migrate out to feed singularly. The last two instars consume the interveinal leaf tissue. Progressively more damage occurs with each generation unless biological or chemical control is accomplished.

Mature larvae migrate to crevices under the bark, between the grape stake and vine, or drop to the soil surface to pupate (Lange 1944; Stern et al. 1983, 1992). Some pupae from the second generation will enter diapause and will emerge the following year. All pupae from the third generation enter diapause. Although almost all pupae transform into adults the following year, some have remained in diapause for up to 2 years (Lange 1944).

A phenology model has been developed for western grapeleaf skeletonizer (Roltsch et al. 1990). The lower threshold for development is 9.0°C and the upper threshold is 28.2°C. Using a single sine method of calculation with a horizontal cutoff, average generation development time is 808 degree-days (DD). Average development time for eggs is 145 DD, for the five larval instars is 385 DD, for pupae is 278 DD. Field validation of the phenology model has shown it to accurately predict generations and life stages of the pest.

13.3.4 Monitoring and Economic Thresholds

Monitoring western grapeleaf skeletonizer can be done by observing the development of larvae, and their characteristic feeding damage to the vine, or through the use of pheromone trapping. Initial pheromone studies identified four compounds as components of the sex pheromone emitted by the female (Myerson et al. 1982). Only 2-butyl-(Z)-7-tetradecenoate effected trapping of males (Myerson et al. 1982; Soderstrom et al. 1985). Furthermore, the attractant of the male is due to the S-(+) enantiomer (Soderstrom et al. 1985). Studies later identified that the optimum dosage required in a rubber septum to be 100 μg of the S-(+)-2-butyl-(Z)-7-tetradecenoate. Still, virgin female baited traps were slightly more effective (Curtis et al. 1989). The synthesis of the western grapeleaf skeletonizer sex pheromone has greatly aided in development and evaluation of its phenology model and the clear identification of three distinct generations in the San Joaquin Valley (Roltsch et al. 1991, 1992).

Although pheromone trapping of western grapeleaf skeletonizer has been quite useful, most grape growers rely on finding the presence of young larvae to identify the potential for damage and the period of susceptibility to insecticides. The first larval generation is particularly important to evaluate in that the grape canopy is still not fully developed and spray coverage can be maximized with non-disruptive insecticides. Monitoring of the first generation should focus on vines at the end of rows or along the border rows (Stern et al. 1992). Dispersal from these sites later establishes the moth throughout the vineyard. Monitoring locations for infestation can be easily estimated based on where moths are seen flying during the early morning hours. Moth activity, based on pheromone trap catches, peaks at 0800 hours (Carr et al. 1992). Because eggs are clustered on leaves closest to the base of the vine and that the eggs are easily seen, vines can be marked and searched systematically for hatch and larval feeding. Currently, there is no validated economic threshold for initiating treatment of western grapeleaf skeletonizer. Because the potential for defoliation and field worker injury is high, low populations are usually treated. However, the past history of infestation and the presence of biological control are often taken into consideration in deciding upon the need for treatment. First generation damage is minimal and an insecticide application decision is best made during the second larval generation, after evaluating biological control of the first generation.

13.3.5 Natural Control

Western grapeleaf skeletonizer was first found in California in 1941 and an intensive quarantine program was established in 1942 by the California Department of Agriculture. The quarantine prohibited the movement of plant material out of San Diego County (Lange 1944). Also, fruit containers were required to be fumigated with methyl bromide. In 1952, a mandatory eradication program to spray vineyards and wild grapes was instituted (Clausen 1961). The program was halted in 1955 when moth abundance had dropped dramatically. The insect was still able to persist in southern California, however. In conjunction with the initial quarantine, a biological control program was initiated by the California Department of Agriculture and the Agricultural Commissioner of San Diego County in 1950. This program was continued even though the eradication attempts were stopped (Clausen 1961). An intensive search for natural enemies was made in San Diego County California and throughout the southwest US (Clausen 1961). At the time, there were no native parasitoids of western grapeleaf skeletonizer in California and none recorded in its native range in Arizona, New Mexico or Mexico (Smith and Langston 1953). They conducted the searches in Arizona and the Mexican states of Sonora and Chihuahua. The searches resulted in little success until reports of sporadic infestations at Emery Park, Arizona, were investigated. The parasites *Ametedoria misella* (Wulp) (= *Sturmia harrisinae* Coquillette) (Tachinidae) and *Apanteles harrisinae* Muesebeck (Braconidae) were recovered in moderate numbers. Also discovered in the process of attempting to rear parasitoids for release was a granulosis virus that attacked skeletonizer larvae, causing approximately 98% mortality in the rearing facilities (Steinhaus and Hughes 1952; Smith et al. 1956). A bacteria, believed to be *Bacillus cereus* F. & F. (Steinhaus and Hughes 1952; Clausen 1961), was also recovered from skeletonizer colonies.

Continued exploration in Chihuahua, Mexico, where both *H. brillians* and another moth, *Malthaca* sp. (Zyganidae), were present, resulted in the recovery of *A. harrisinae, Hatcheller* sp. and two species of Tachinidae, but in very small numbers (Clausen 1961). Searches in Missouri, New York, and Florida, where *H. americana* is prevalent, resulted in no parasitoid detections. In Florida, the parasitoid *Pelecystoma harrisinae* (Ashmead) (Braconidae) was recovered from larvae of *Acoloithus* sp. (Zyganidae). Other parasitoïds found, included *Phorocera sp.* (Tachinidae), *A. misella, A. harrisinae*, and *Haltichella* sp. (Chalcididae). Fifteen primary parasitoids were eventually identified and six were propagated and released for use against the pest (Clausen 1961). Only *A. misella, A. harrisinae* and the granulosis virus were sufficiently effective to justify release against western grapeleaf skeletonizer.

Apanteles harrisinae attacks early larval stages of western grapeleaf skeletonizer (Clausen 1961). It was first released in 1951 at Los Coches Creek, California, primarily on wild grapes. It became established after these releases with parasitism rates ranging from 20 to 86% in 1951 and 1952. It is a gregarious internal parasite of western grapeleaf skeletonizer (Clausen 1961). It is active in the early morning or early evening hours. The wasp inserts three to nine eggs within the abdomen of late first (predominant), second and third instar larvae. The adult parasitoid lives for approximately 4 days. Development from egg to pupa is completed in approximately 18 days, while the cocoon stage lasts 13–34 days (Clausen 1961). The larvae of *A. harrisinae* will emerge immediately after pupation of the host. Up to 28 wasps may emerge from a single parasitized larva. The small white cocoons will surround the body of the host and appear as a white cottony mass. The sex ratio in the field is 1:1. It overwinters as a mature larva within the host cocoon.

Ametedoria misella (Wulp), a solitary internal parasite, oviposits externally, primarily in fourth and fifth instar *H. brillians*. Once in the host body, the larva will briefly move in the body and then settles in one of the silk glands. It remains there until the host becomes a pupa. The parasitic fly then leaves the silk gland and cuts into the pupal integument at the prothoracic spiracle (Clausen 1961). The mature larva then emerges between the thorax and abdomen and the pupal stage is formed where the puparium often remains attached to the dead host. The time spent as a pupa ranges from 9 to 15 days. The parasite overwinters as a first instar larva within the host pupa. It begins emergence during April and May in California. The parasitic fly is slightly larger than a housefly and possesses the typical large bristles on the abdomen. A minimum of two generations occur each year.

By far, the most effective biological control of western grapeleaf skeletonizer has been due to activity of *Harrisina brillians* granulosis virus (HbGV). It was first found in field-collected specimens to be used for parasitoid rearing in the insectary at Fort Huachuca, Arizona (Clausen 1961). The presence of the virus in California is attributed to the introduction of contaminated larva used in the production of parasitoids. The first field infections (1951) were in San Diego County where the dissemination was aided by the release of contaminated parasitoids of western grapeleaf skeletonizer (Steinhaus and Hughes 1952).

The principal site of infection by the virus is the midgut epithelium and that can result in almost complete destruction of the tissue (Federici and Stern 1990; Stern and Federici 1990). The virus replicates in the gut cells of both larvae and adults. Larvae develop diarrhea within 4 days after infection and the feces remain infectious, thus serving to infect other larvae. In the 1970s, when western grapeleaf skeletonizer was found in the San Joaquin Valley, the virus was apparently absent from field populations. A single vial of lyophilized powder of diseased larvae, prepared in 1951 by Steinhaus and held at –20°C at the Department of Entomological Sciences at University of California at Berkeley, served as the source of reintroduction (Federici and Stern 1990). This source was shown to be infective in the laboratory. A double antibody sandwich enzyme-linked immunosorben assay (ELISA) is available to detect the presence of HbGV (Stark et al. 1999).

Visual evidence of larval infection includes random deposition of eggs that fail to hatch, young larvae wandering singly on a leaf instead of feeding in rows, and abnormal development and coloration of larvae. There is also a distinctive appearance of leaves whereby only small opaque patches of leaf tissue, where larvae have fed, are present and the absence of live larvae. Larger larvae will either drop to the ground or remain attached to leaves by their prolegs, hanging upside down. Eventually, the dead insects darken to black (Stern et al. 1992).

Experiments demonstrated that both parasitoids *A. misella* and *A. harrisinae* are capable of transmitting the virus with rates of infection of 38.5 and 25% respectively (Smith et al. 1956). *Harrisina brillians* moths emerging from overwintering pupae tested positive for HbGV (Stark et al. 1999). As the virus and two parasitoids have become established in California, the incidence of pest outbreaks has declined (Stern et al. 1992).

13.3.6 Management

With the reintroduction of HbGV into native moth populations, particularly where grapes grow wild along riparian areas, both the incidence and severity of western grapeleaf skeletonizer has dropped dramatically within the last 20 years. However, sporadic outbreaks still occur, and the use of insecticides is sometimes necessary particularly in fresh market grapes. During the extended eradication program in California, insecticides such as cryolite and *B. thuringiensis* were widely and effectively used to manage the pest (Hall 1955). Subsequent experimentation resulted in similar results with *B. thuringiensis* var. *kurstaki* (Btk) and cryolite providing over 90% control with a single spray (Stern et al. 1983). Both these insecticides and the insect growth regulator methoxyfenozide and spinosad have proven to be effective controls (Bentley et al. 2009).

There are no validated economic thresholds for larval damage or moth abundance in vineyards. Sampling is done to detect the presence of larvae and the initiation of feeding. The first generation is most easily seen with minimal damage being done. Also, the presence of the granulosis virus HbGV can be evaluated based on the feeding pattern of small larvae. If the virus is detected, most grape growers will not apply insecticides. However, if no evidence of viral infection is found an insecticide will be applied. Because insecticides such as methoxyfenozide, Btk, cryolite, and abamectin are not toxic to parasites and predators, a truly integrated program can be utilized to manage western grapeleaf skeletonizer in vineyards.

13.4 Omnivorous Leafroller

The omnivorous leafroller (OLR), *P. stultana,* is thought to be native to northern Mexico. It was first described in Sonora, Mexico, in 1884 (Powell and Opler 2009). It has often been found in shipments of fruits and peppers from Mexico to the United States. It was first collected in southern California in 1898 on ornamentals. The first record of it infesting grape was on July 19, 1962 in Fresno County (Nakata et al. 1974). By 1968, it was known throughout the San Joaquin Valley and into the Napa Valley. Omnivorous leafroller has, as its name implies, an extremely wide host range. The larvae feed on weeds, fruit trees, field crops and vegetable crops. A short list of crops attacked includes alfalfa, apple, apricot, avocado, bushberries, celery,

Fig. 13.11 Adult *Platynota stultana* (Photo J. Dibble)

citrus, cotton, eggplant, lettuce, melons, peach pepper, plum, sorghum, sugar beet, strawberry, tomato, and walnut (Atkins et al. 1957). It has since spread to Florida, Illinois, Massachusetts, Michigan, Texas, Virginia, and Washington (Atkins et al. 1957). Although it is termed a leafroller, it does not truly roll leaves but folds leaf tissue together. The insect does not diapause and, in colder climates, it is primarily a greenhouse pest.

13.4.1 Description

Omnivorous leafroller wings are distinctively bell shaped (Fig. 13.11). The forewings are brown at the base and bright orange on the distal half. Males are approximately 4–6 mm in length and possess a broad costal fold on the forewing (Powell and Opler 2009). Females are 6.5–9 mm in length and predominantly reddish brown in coloration. Both sexes possess labial palpi that protrude forward and appear as a long snout. Eggs are laid in masses that average 97 eggs (Gilligan and Epstein 2009) (Fig. 13.12). They are usually deposited on smooth upper plant surfaces. The individual eggs are elliptical in shape and overlap like fish scales. They are light green in color. They are 0.9–1 mm by 0.6–0.7 mm in size. As the eggs age, coloration becomes light tan. In most cases a female will deposit eggs in a single group in one night. The eggs are fastened to each other and the mass is covered by a cement like secretion of the colleterial glands (Nelson 1936). Interestingly, the eggs from a single egg mass will hatch at the same time (Atkins et al. 1957).

Fig. 13.12 *Playtnota stultana* egg cluster (Photo J. Dibble)

Larvae complete 5–6 instars. At hatching they are approximately 1.6 mm in length, and at maturity, 13 mm (Atkins et al. 1957). Upon hatching the larvae spin a network of silken filaments that allow for emergence from the egg chorion. These filaments appear necessary for movement on plant surfaces. The early stage larvae are cream colored and with light brown head and thoracic shield. The larvae are heliotropic. Later stages are green to brown in color with dark mahogany head and thoracic shield. These larger larvae possess a dorsal longitudinal darkened stripe from the first to last abdominal segment. A single, oval pinacula is present on each abdominal segment on each side of the midline. The pinaculae appear cream colored and are distinctive of larvae. The mature OLR larva is shown in Fig. 13.13. If disturbed, the larvae will wiggle vigorously and drop from a small silken thread. Prior to pupation, the larvae become inactive and the body shortens and a brown pupal case is formed within the feeding nest.

13.4.2 Damage

Minor leaf damage results from larval feeding in grape. The primary damage is due to feeding on the grape cluster after veraison. Larvae can be found in clusters during vine bloom (Fig. 13.14) but damage from such early cluster feeding is minimal, affecting only few berries. Larval feeding on soft berries more often results in bunch rot (*Botrytis cinerea* Persoon ex. Fries), as they mature and carbohydrates build.

Fig. 13.13 Mature *Platynota stultana* larva (Photo F. Jensen)

Fig. 13.14 *Platynota stultana* larval nest in grape cluster (Photo F. Jensen)

The development of *Botrytis* in the cluster is the most important damage resulting from OLR feeding. It is a severe pathogen in stored grape. Infected berries will appear water soaked and, within few days, the berry skin will slip. Eventually a gray mold develops and moves throughout the cluster.

Although the primary damage is done when the berries have softened, chemical control is directed at larvae infesting grape clusters during bloom. Spray coverage is maximized with this timing (Coviello et al. 1992).

13.4.3 Life History

Omnivorous leafroller is quiescent (not diapausing) during the colder winter months and unable to survive prolonged periods of freezing. This condition limits its range in North America. The overwintering stages are the third through fifth instar larvae (Atkins et al. 1957). They are predominantly in mummified berries on or at the base of the vine (Nakata et al. 1974; Coviello et al. 1992). However, in southern California and Arizona, weeds from Chenopodiaceae and Leguminosae often serve as hosts. Mortality during the winter is high due to larval exposure. Birds are a major source of predation in winter (Nakata et al. 1974).

Studies utilizing virgin female OLR moths in the San Joaquin Valley demonstrated the presence of a sex pheromone (AliNiazee and Stafford 1972a). This pheromone was later synthesized and identified by Hill and Roelofs (1975) as (*E*)-11-tetradecenyl acetate ((*E*)-11-14:Ac) and (*Z*)-11-tetradecenyl acetate ((*Z*)-11-14 Ac) in the ratio of 12:88. Additionally (*E*)-11-tetradecen-1-ol ((*E*)-11-14:OH) and (*Z*)-11-tetradecen-1-ol ((*Z*)-11-14:OH) were found in abdominal extracts of the female in the same *Z*:*E* ratio as the acetates. There was also a small amount of tetradecyl acetate (14:Ac) and tetradecan-1-ol (14:OH) present. The most frequent ratio of acetates to alcohols was 2:1. The proportion of *E*:*Z* 11-14-Ac found to be most attractive to males was 91–96% (Hill and Roelofs 1975). Also, the addition of the 11–14:OH to 11–14:Ac's synergized the attraction of males when mixed at 0.2–2%.

The use of the synthesized OLR pheromone has allowed for monitoring male activity throughout the year. Male flight in the San Joaquin Valley begins in late February to early March (Baker et al. 1975, 1978). In coastal areas and southern California, males can be trapped even during the winter months. Mating occurs during the evening hours after sunset when moths are most active (AliNiazee and Stafford 1972c; Coviello et al. 1992). Egg laying begins on the second or third day after moth emergence. Four to six generations are reported in central California (Nakata et al. 1974; Baker et al. 1975; Coviello et al. 1992). The completion of a single generation, including a preoviposition period, requires 698 DD (>8.7°C and <30.6°C) (Bentley et al. 2009). Assuming 698 DD for completion of a single generation and beginning calculations on February 15 and ending on December 31, only four generations appear possible in the San Joaquin Valley. This is based on a 30-year temperature average at Parlier, California. During extremely warm years, there may be a partial fifth generation.

Egg deposition occurs on smooth upper leaf surfaces and newly formed clusters. They are usually deposited at sunset or at dawn (Atkins et al. 1957). Eclosion occurs within 5–11 days, depending upon temperature (Nakata et al. 1974).

First generation larvae coincide with bloom of grape. The young caterpillars move upward on the plant and will feed, singly on leaves or in a flower bud. If touched, they immediately wriggle backwards and will drop and remain suspended on a silken thread made from larval spinnerets (Atkins et al. 1957). Larvae are generally

Fig. 13.15 *Eumorpha achemon* larva defoliating a vine (Photo F. Jensen)

solitary but whenever abundant, several will be found within webbed nests in clusters. In some cases, larvae may create channels into vine tendrils, stems, or petioles. These plant parts will die and the tissue will be incorporated into the larval nests. Webbing in flower clusters is a good indicator of OLR presence. Here, they feed creating cavities in the cap stem and the feeding site becomes susceptible to botrytis bunch rot as the berries ripen. Although larval feeding in any of the three generations can lead to severe berry damage, the second and third generation of OLR during and after veraison are the most problematic. The final generation of larvae occurs after harvest of most cultivars and it is not damaging. The fourth generation pupates within the vineyard or on weeds or other crop hosts such as alfalfa outside the vineyard. Pupation during the summer months is completed in 5 days.

13.4.4 Monitoring and Economic Thresholds

Monitoring is accomplished with pheromone baited sticky traps for moths and by examining a selected number of clusters for larval infestation. Pheromone traps are used primarily to establish a biological fix for timing insecticide sprays (Bentley et al. 2009).

Pheromone trap catches of moths allow for identifying the beginning of each flight of males and are not correlated with harvest infestation. Traps should be in place by January 1 in California coastal areas and in southern California and Arizona (Bentley et al. 2009). In California's San Joaquin Valley, trap placement should be done by February 15 (Bentley et al. 2009). Male capture is useful in determining insecticide application timing and also for estimating where in the vineyard larvae may be most abundant. Traps should be monitored twice per week until consistent moth flight is determined. Once consistent flight is determined traps are monitored once per week. A minimum of three traps per vineyard should be used to determine the biofix.

Sampling for larvae begins generally just before grape bloom. This is carried out by monitoring 200 flower clusters to detect larval presence. Ten clusters are sampled on each of 20 vines randomly chosen throughout the vineyard (Shaw et al. 1983; Bentley et al. 2009). Vineyards larger than 20 ha may require more intensive sampling. If any larvae are found at bloom, an insecticide should be applied. After bloom, insecticide application is necessary if two or more clusters are infested (Shaw et al. 1983; Bentley et al. 2009).

13.4.5 Natural Control

Because OLR does not diapause, freezing winter temperatures limit its distribution in North America and most likely can result in significant mortality in those areas where periodic winter freezes occur. However, where it is well established on the west coast, Arizona and Mexico, the action of parasitoids provide the greatest level of biological control. There are 10 species of parasitic Hymenoptera and two species of parasitic Diptera that are commonly found (Coviello et al. 1992). These include the following: *Goniozus platynotae* Ashmead (Bethylidae), *Trichogramma* sp. (Trichogrammatidae), *Apanteles* sp. (Braconidae), *Macrocentrus ancylivorus* Rohwer (Braconidae), *Cremastus platynotae* Cushman (Ichneumonidae), *Diadegma compressus* Cresson (Icheumonidae), *Elachertus proteoteratis* (Howard) (Eulophidae), *Spilochalcis* sp. (Chalcididae), *Erynnia tortricis* (Coquillett) (Tachinidae), and *Nemorilla pyste* (Walker) (Tachinidae).

Predators in vineyards can also account for low levels of OLR mortality. These include *Chrysoperla carnea* (Stephens), *Orius tristicolor* (White), *Nabis sp.*, and theridiid spiders (Coviello et al. 1992). Overall, biological control in vineyards is considered minor, accounting for 10% mortality at its highest level (Coviello et al. 1992).

13.4.6 Management

Management of OLR includes cultural and chemical methods. Additionally, the use of reduced-risk insecticides and mating disruption will allow for maximum biological control activity.

13.4.6.1 Sanitation

Vineyard sanitation involves winter removal of unharvested clusters from vines, destroying weeds and vine debris. It is an important component of reducing within vineyard populations of OLR. The widely used technique of row plowing, termed French plowing, has been largely supplanted with herbicides to manage weeds. While this removes weeds that harbor larvae, it does not bury grounded unharvested clusters or vine debris and this appears to have led to increased incidence of OLR (Nakata et al. 1974). Where sources of OLR are from outside the vineyard, plowing and cluster destruction may have minimal impact.

13.4.6.2 Early Harvest

Early harvest of grapes is not always possible but, in the case of raisin and certain table grape cultivars, this can be used to avoid the build up of late season OLR populations and subsequent rot problems. Whenever possible, harvesting prior to September 1 reduces the impact of the third generation larvae (Coviello et al. 1992).

13.4.6.3 Mating Disruption

Field experimentation has documented the effectiveness of mating disruption as a management tool for OLR (Shorey et al. 1995, 1996). Male moths find females by following a plume of sex pheromone released by them. Efficacy in disruption was noted when evaporators, dispensing 0.9 mg/ha/day of pheromone were spaced at 20 m apart and this was equivalent to those spaced at 5.5 m. Successful disruption declined when dispensers were spaced 100 m apart (Shorey et al. 1996). The large concentration of synthetic pheromone emitted by thousands of dispensers in a vineyard saturates the air to the extent that males cannot follow the individual plume to a single female. Experience has shown that such disruption is best accomplished in large plantings or plantings that are isolated from other sources of OLR. Pheromone dispensers should be in place for a minimum of two generations for optimum efficacy.

Dispensers are available in several types. These include plastic twist-ties, laminar membranes, aerosol dispensers, paraffin emulsions, and microencapsulated sprayable material. They are placed in the vineyard at the beginning of the flight, as indicated by pheromone trap catch of male moths (Bentley et al. 2009).

13.4.6.4 Chemical Control

Selective insecticides have played an important role in the management of leaf feeding lepidopterans in general and, particularly, the OLR (Bentley et al. 2009). Insecticides such as Btk, methoxyfenozide, spinosad, and spinetoram are considered reduced-risk to farm workers and the environment, and are as effective as the older and more broadly toxic insecticides. These products have different modes of action, slowing the development of insecticide resistance, and do not result in secondary pest outbreaks. They have largely resulted in the elimination of need for the more broadly toxic organophosphate, carbamate, pyrethroid insecticides and toxicity issues associated with their use. Spring treatments are the most effective compared to sprays applied later in the growing season primarily due to spray coverage issues.

For insecticide management, sprays are timed by using the OLR phenology model and based on the accumulation of DD. To utilize DD timing, OLR pheromone traps should be placed in the vineyard by January (coastal and southern California) and mid-February (central California) to determine when moths of the overwintering generation start to fly. Consistent trapping of the moths signals the biofix, i.e., start calculating DD. The best reduction of OLR and its damage occurs when treatments are timed within the range of 371–482 DD (8.89°C lower threshold and 30.6°C upper threshold) (Bentley et al. 2009). The treatment timing is effective for the first and second flight of omnivorous leafroller. For the third flight the range of timing is 260–371 DD due to the overlapping of generations. Once veraison has occurred, penetration of the insecticide into the cluster is reduced and may not provide the pest control equivalent to that of an earlier generation.

13.5 Achemon and Whitelined Sphinx Moths

Achemon sphinx moth, *E. achemon,* and whitelined sphinx, *H. lineata,* are periodic and bivoltine pests that can defoliate grapes. Severe defoliation results from feeding of the second generation larvae of both species. They are common throughout the United States and Mexico. Quayle (1907) reported that either of these pests at 150 larvae per vine had caused complete defoliation of a vineyard.

13.5.1 Description

The adult achemon sphinx is a beautiful hawk moth that begins emergence from the overwintering pupal stage in late April in California. Its wingspan ranges from 75 to 100 mm and is marked with a mixture of colors. The base color is tan and there are distinct dark brown spots on the forewings and the base of the wings on each side of the thorax (Essig 1958; Bentley 1992). The hind wings are deep pink with a brown border and several scattered dark spots toward the back edge of the wings. They fly at night and can often be seen hovering near petunias, evening primrose, and rhododendrons.

The whitelined sphinx is slightly smaller than the achemon sphinx with a wingspan ranging from 60 to 90 mm. It is also striking in appearance with a base gray to brown color highlighted by white lined veins and a broad white band running from the tip to the base in the middle of the forewings (Essig 1958; Bentley 1992). The hind wings are darker, with a rose colored band across the middle of the wing. There are also a series of white tufts of scales on the first four or five abdominal segments.

Adults emerge in April from overwintering pupae in the soil. In the evening, they visit a wide range of flowers including petunias, honeysuckle, and columbines.

The eggs of both species are round, smooth, and are laid singly on leaves, usually on the upper leaf surface. The diameter of the eggs for both species is on average 1.5 mm (Quayle 1907; Essig 1958).

The larval stage of each species has unique characteristics. Achemon sphinx larvae (Fig. 13.15) average 75 mm in length at maturity (Essig 1958) and range from green, when young, to green or reddish when mature. There is a series of 6–8 white or pale yellow oblique bars crossing the spiracles on the abdomen. After the penultimate molt, the larva loses the long black anal horn (Essig 1958). Whitelined sphinx larvae reach approximately 70 mm in length and have two color phases. The predominant color is green with black lines running longitudinally along the body (Essig 1958). The head and anal horn is yellow. The less common form is predominantly black with yellow lines along the body. Both color phases can be found in populations. The whitelined sphinx moth attacks a wide range of host plants including beet, chickweed and knotweed (Essig 1958).

Pupation of both species occurs in the soil. The chrysalis is mahogany brown, with a smooth surface and pointed at the anal end. The outline of the wings and mouthparts can be seen along the chrysalis. Pupae of the achemon sphinx range from 45 to 50 mm and those of the whitelined sphinx from 30 to 35 mm.

13.5.2 Damage

Feeding by newly emerged larvae results in numerous round holes in the center of the leaf. Feeding can cause complete defoliation of the vines and other hosts. A mature achemon sphinx larva will consume nine grape leaves in a 24-h period (Stafford and Doutt 1974). Damage caused by the first larval generation is localized to a few plants but the second generation, when abundant, can spread over larger sections causing severe defoliation. Adult moths commonly pollinate rangeland plants and such areas can be a source of this pest.

13.5.3 Life History

Both moth species overwinter in the soil as pupae at 5–15 cm deep under the soil surface (Essig 1958; Stafford and Doutt 1974). Where tillage is common, the source of moths is commonly outside the vineyard, particularly for whitelined sphinx. Where there is minimal tillage moths originate mostly from within the vineyard.

In California, the achemon sphinx begins its emergence in early May and completes development by early July (Stafford and Doutt 1974; Bentley 1992). The second generation is found in July and it can result in severe vine damage during August. In some years, a third generation may occur in warmer grape growing areas.

Emergence of the whitelined sphinx also occurs in May, and development of the first generation is complete by mid-July (Stafford and Doutt 1974). As with the achemon sphinx, a second generation will also develop.

13.5.4 *Monitoring and Economic Thresholds*

No formal monitoring method has been developed for sphinx moths. Because the larvae are so large and feeding quite obvious, simple observation while monitoring for other pests has been adequate. In general, if non-parasitized larvae are found in July, an insecticide should be applied.

13.5.5 *Natural Control*

Although not well documented, biological control does play an important part in limiting the damage by sphinx moths. Tachinid flies have been observed to parasitize larvae. However, they have not been identified to species. Often a fly egg will be found on the head or thorax of the larvae. Upon hatching, the maggot will burrow into the sphinx larva and feed internally. Areas bordered by native habitats often result in skunks feeding on wintering pupa (Stafford and Doutt 1974).

13.5.6 *Management*

Early larval instars of both species are quite easily controlled. *Bacillus thuringiensis* var. *kurstaki* is the primary larvicide recommended because it is non-disruptive, very effective and inexpensive. Application of an insect growth regulator insecticide (methoxyfenozide) against the smaller stage larvae can be highly effective.

13.6 Conclusion

Leaf-feeding Lepidoptera in grapes can be managed with a truly integrated pest management (IPM) approach in North America. The number of selective and non-disrupting insecticides, with a wide range of modes of action, allows for maximum utilization of biological control. The major obstacle to a comprehensive IPM program is the lack of accurate, easily used sampling methods and action thresholds, in part due to the focus on direct-feeding pests that limit fruit quality. Given the infrequent occurrence of all but omnivorous leafroller and grape leaffolder in California vineyards, this information is likely not to be developed in the immediate future.

References

AliNiazee MT (1974) Contribution to the bionomics of the grape leaffolder, *Desmia funeralis* (Hübner): a laboratory study with field observations. Pan-Pac Entomol 50:269–278

AliNiazee MT, Jensen FL (1972) Microbial control of the grape leaffolder with different formulations of *Bacillus thuringiensis*. J Econ Entomol 66:157–158

AliNiazee MT, Stafford EM (1972a) Virgin female traps aid control survey for omnivorous leaf roller in San Joaquin valley vineyards. Calif Agric 26:5–6

AliNiazee MT, Stafford EM (1972b) Seasonal flight patterns of the omnivorous leafroller and grape leaffolder in central California vineyards as determined by blacklight traps. Environ Entomol 1:65–68

AliNiazee MT, Stafford EM (1972c) Sex pheromone studies with the omnivorous leafroller, *Platynota stultata* (Lepidoptera: Tortricidae): effect of various environmental factors on attraction of males to the traps baited with virgin females. Ann Entomol Soc Am 65:958–961

Allen RP, Kane EA, Paddock EL (1974) Grapeleaf skeletonizer and western grapeleaf skeletonizer. State Calif Dep Food Agric Detect Manual:1–2

Atkins EL Jr, Frost MH Jr, Anderson LD, Deal AS (1957) The "omnivorous leaf roller", *Platynota stultana* Wlshm., on cotton in California: nomenclature, life history, and bionomics (Lepidoptera:Tortricidae). Ann Entomol Soc Am 50:251–259

Baker JL, Hill AS, Carde RT, Kurokawa A, Roelofs WL (1975) Sex pheromone field trapping of the omnivorous leafroller, *Platynota stultana*. Environ Entomol 4:90–92

Baker JL, Hill AS, Roelofs WL (1978) Seasonal variations in pheromone trap catches of male omnivorous leafroller, *Platynota stultana*. Environ Entomol 7:399–401

Barnes DF (1944) Notes on the life history and other factors affecting control of the grape leaf folder. U S Dep Agric E-616:1–8

Bentley WJ (1992) Sphinx moths. In: Flaherty DL, Christensen LP, Lanini WT, Marios JJ, Philips PA, Wilson LT (eds) Grape pest management, vol 3343, 2nd edn. University of California, Division of Agriculture and Natural Resources, Oakland, pp 235–236

Bentley WJ, Varela LG, Zalom FG, Smith RJ, Purcell AH, Phillips PA et al (2009) Insects and mites. In: Ohlendorf B, O'Neil MJ (eds) UC IPM pest management guidelines: grape, vol 3448. University of California, Agriculture and Natural Resources, Oakland, pp 12–69

Carr WC Jr, Roltsch WJ, Mayse MA (1992) Diurnal and generational flight activity of the western grapeleaf skeletonizer (Lepidoptera: Zygaenidae): comparison of monitoring methods. Environ Entomol 21:112–116

Clausen K (1961) Biological control of western grape leaf skeletonizer (*Harrisina brillians* B. and McD.) in California. Hilgardia 31:613–638

Coviello R, Hirshefelt DJ, Barnett WW (1992) Omnivorous leafroller. In: Flaherty DL, Christensen LP, Lanini WT, Marios JJ, Philips PA, Wilson LT (eds) Grape pest management, vol 3343, 2nd edn. University of California, Division of Agriculture and Natural Resources, Oakland, pp 166–173

Curtis CE, Landolt PJ, Heath RR, Murphy R (1989) Attraction of western grapeleaf skeletonizer males (Lepidoptera: Zygaenidae) to S-(+)-2-butyl-(Z)-7-tetradecenoate. J Econ Entomol 82:454–457

Donohoe HC, Barnes DF (1934) *Microbracon cushmani* Mues. attacking *Desmia funeralis* Hbn. in the San Joaquin valley, California. J Econ Entomol 27:859

Doutt RL, Nakata J, Skinner FE (1969) Parasites for control of grapeleaf folder. Calif Agric 23:4

Essig EO (1958) Insects and mites of western North America. Macmillan Company, New York, NY

Federici BA, Stern VM (1990) Replication and occlusion of a granulosis virus in larval and adult midgut epithelium of the western grapeleaf skeletonizer, *Harrisina brillians*. J Invertebr Pathol 56:410–414

Gilligan TM, Epstein ME (2009) LBAM ID: tools for diagnosing light brown apple moth and related western U.S. leafrollers (Tortricidae: Archipini). Colo State Univ, Calif Dep Food

Agric, and U. S. Dep Agric Aphis, PPQ. Factsheet: *Platynota stultana*. http://keys.luciddcentral.org/keys/v3/LBAM/Platynota_stultana.html
Hall IM (1955) The use of *Bacillus thuringiensis* Berliner to control the western grapeleaf skeletonizer. Dept. Bio. Control, Univ. Calif. Exp. Stn, Riverside
Hill AS, Roelofs WL (1975) Sex pheromone components of the omnivorous leafroller mother, *Platynota stultana*. J Chem Ecol 1:91–99
Jensen FL (1969) Microbial insecticides for control of grape leaffolder. Calif Agric 23:5–6
Jensen F, Nakata J, Flaherty D (1975) Grape leaffolder. Grape pest management in the southern San Joaquin valley. San Joaquin Valley Agriculture Research and Extension Center Parlier, CA
Jensen FL, Hirschfelt DJ, Flaherty DL (1992) Grape leaffolder. In: Flaherty DL, Christensen LP, Lanini WT, Marios JJ, Philips PA, Wilson LT (eds) Grape pest management, vol 3343, 2nd edn. University of California, Division of Agriculture and Natural Resources, Oakland, pp 133–139
Jones PR (1909) The grape-leaf Skeletonizer (*Harrisina americana* Guérin-Méneville). U S Dep Agric B E Bull 68(Part VIII):77–90
Kliewer WM (1970) Effect of time and severity of defoliation on growth and compositon of Thompson seedless grapes. Am J Enol Vitic 21:37–47
Kliewer WM, Lider LA (1968) Influence of cluster exposure to the sun on composition of Thompson seedless fruit. Am J Enol Vitic 19:175–184
Lange WH Jr (1944) The western grape leaf skeletonizer *Harrisina brillians* in California. Bull Dep Agric 33(2):98–104
May P, Shaulis NJ, Antcliff AJ (1969) The effect of controlled defoliation in the Sultana vine. Am J Enol Vitic 20:237–250
Mead FW, Webb SE (2008) Grape leaffolder, *Desmia funeralis* (Hübner) (Instecta: Lepidoptera: Crambidae). Featured creatures. Entomol Nematol Univ Fla Coop Ext Ser, EENY-192
Millar J, Mclefresh JS, De Assis Marques F (2002) Unusual acetylenic sex pheromone of grape leaffolder (Lepidoptera:Pyralidae). J Econ Entomol 95:692–698
Myerson J, Hadden WF, Soderstrom EO (1982) Sec-butyl-(Z)-7-tetradecenoate. A novel sex pheromone component from the western grapeleaf skeletonizer, *Harrisina brillians*. Tetrahedron Lett 23:2757–2760
Nakata J, Doutt RL, Lynn CD, Flaherty DL, Jensen F (1974) Omnivorous leaf roller: grape pest management in the southern San Joaquin valley. San Joaquin Valley Agriculture Research and Extension Center, Oakland
Nelson RH (1936) Observations on the life history of *Platynota stultana* Wlshm. on greenhouse rose. J Econ Entomol 29:306–312
Powell JA, Opler PA (2009) Moths of western North America. University of California Press, Berkeley/Los Angeles
Quayle HJ (1907) Insects injurious to the vine in California. Coll Agric, Univ Calif Agric Exp Stn Bull 192:140
Robinson DW (1950) Description, life history and habits of the western grape skeletonizer, *Harrisina brillians* B. & McD. Bull Dep Agric State Calif 39:149–151
Roltsch WJ, Carr WC Jr, Mayse MA, Dodds RS (1992) Western grapeleaf skeletonizer pheromone trap catch seasonal patterns in central California. Southwest Entomol 17:223–231
Roltsch WJ, Mayse MA, Clausen K (1990) Temperature-dependent development under constant and fluctuating temperatures: comparison of linear versus nonlinear methods for modeling development of western grapeleaf skeletonizer (Lepidoptera: Zygaenidae). Environ Entomol 19:1689–1697
Roltsch WJ, Carr WC Jr., Mayse MA (1991) Seasonal patterns of western grapeleaf skeletonizer (Lepidoptera: Zygaenidae) pheromone trap catch. Res Bull Vitic Enol Res Center Calif State Univ:1–6
Shaw PB, Kido H, Flaherty DL, Barnett WW, Andris HL (1983) Spatial distribution of infestations of *Platynota stultana* (Lepidoptera: Tortricidae) in California vineyards and a plan for sequential sampling. Environ Entomol 12:60–65

Shorey HH, Sisk CB, Gerber RG (1995) Disruption of pheromone communication in *Platynota stultana* (Lepidoptera: Tortricidae) in grape vineyards. Environ Entomol 24:1270–1274

Shorey HH, Sisk CB, Gerber RG (1996) Widely separated pheromone release sites for disruption of sex pheromone communication in two species of Lepidoptera. Environ Entomol 25:446–451

Smith OJ, Langston RL (1953) Continuous laboratory propagation of western grape leaf skeletonizer and parasites by prevention of diapause. J Econ Entomol 46:477–484

Smith OJ, Hughes KM, Dunn PH, Hall IM (1956) A granulosis virus disease of the western grape-leaf skeletonizer and its transmission. Can Entomol 88:507–515

Soderstrom EL, Brandl DG, Myerson J, Buttery RG, Mackey BE (1985) Sex pheromone for attracting western grapeleaf skeletonizer *Harrisina brillians* (Lepidoptera Zygaenidae). J Econ Entomol 78:799–801

Stafford EM, Doutt RL (1974) Insect grape pests on northern California. Calif Agric Exp Stn Ext Serv Circ 566:74

Stark DM, Mills NJ, Purcell AH (1999) Interactions between the parasitoid *Ametadoria misella* (Diptera: Tachinidae) and the granulovirus of *Harrisina brillians* (Lepidoptera: Zygaenidae). Biol Control 14:146–151

Steinhaus EA, Hughes KM (1952) A granulosis virus of western grapeleaf skeletonizer. J Econ Entomol 45:744–745

Stern VM, Federici BA (1990) Granulosis virus: biological control for western grapeleaf skeletonizer. Calif Agric 44:21–22

Stern VM, Flaherty DL, Peacock WL (1983) Control of the western grapeleaf skeletonizer (Lepidoptera:Zygaenidae), a new grape pest in the San Joaquin valley, California. J Econ Entomol 76:192–195

Stern VM, Peacock WL, Flaherty DL (1992) Western grapeleaf skeletonizer. In: Flaherty DL, Christensen LP, Lanini WT, Marios JJ, Philips PA, Wilson LT (eds) Grape pest management, vol 3343, 2nd edn. University of California, Division of Agriculture and Natural Resource, Oakland, pp 214–221

Strauss JF (1916) The grape leaf-folder. U S Dep Agric Bull 419:1–16

Chapter 14
Grape Berry Moths in Western European Vineyards and Their Recent Movement into the New World

Claudio Ioriatti, Andrea Lucchi, and Lucia G. Varela

14.1 Introduction

This chapter addresses the topic of pest biology and management for cluster-feeding Lepidoptera of European origin, from the viewpoint of applied entomologists working in Europe and North America. We describe the main biological, morphological, and behavioral features of the Palaearctic moths harmful to grapes. Five species of Lepidoptera feed on grape clusters in Europe. *Lobesia botrana* (Denis & Schiffermüller) and *Eupoecilia ambiguella* (Hübner) are key pests that require specific control measures. *Argyrotaenia ljungiana* (Thunberg), *Cryptoblabes gnidiella* (Millière) and *Ephestia parasitella unicolorella* Staudinger are occasionally harmful to vineyards. As the five species are similar in size and occur almost in the same ecological niche, morphological features allowing species identification are provided in Table 14.1. For comparison, a description of the North American grape berry moth *Paralobesia viteana* (Clemens) is provided by Isaacs et al. (Chap. 15).

C. Ioriatti (✉)
Centre for Technology Transfer, FEM-IASMA, 38010 San Michele all'Adige, TN, Italy
e-mail: claudio.ioriatti@iasma.it

A. Lucchi
Department of CDSL, Section of Agricultural Entomology, University of Pisa,
Via del Borghetto 80, 56124 Pisa, Italy
e-mail: alucchi@agr.unipi.it

L.G. Varela
UC Cooperative Extension, Division of Agriculture and Natural Resources,
University of California, 133 Aviation Blvd., Suite 109,
Santa Rosa, CA 95403-2894, USA
e-mail: lgvarela@ucdavis.edu

N.J. Bostanian et al. (eds.), *Arthropod Management in Vineyards: Pests, Approaches, and Future Directions*, DOI 10.1007/978-94-007-4032-7_14,

Table 14.1 Main morphological features of the principal grape cluster-feeding moth pests in European vineyards

Species	Family, sub family	Adult	Egg	First instar larva	Fifth instar larva	Pupa
Lobesia botrana	Tortricidae, Olethreutinae	Wingspan 11–13 mm. Forewings tan-cream, mottled with gray-blue, brown, and black blotches. Hindwings gray with a fringed border. Wings held in a bell shape over the abdomen at rest	0.6 × 0.7 mm Slightly elliptical, yellow straw at laying, gradually turns to transparent light gray with bright iridescent reflections	1 mm Creamy white with a black head	10 mm Tan to yellowish brown head and prothoracic shield, rear edge of the prothoracic shield has a darker brown to black border. White tubercles at the base of the body hairs. Anal comb has 5–6 dark brown teeth	4–6 mm Slim with the cranial and the caudal end rounded. Cremaster with 8 hooked bristles. Female pupae generally more stocky and larger than males
Eupoecilia ambiguella	Tortricidae, Tortricinae	Wingspan 12–15 mm Forewings yellow-brown with dark brown band. Hind wings slate gray fringed	0.6 × 0.8 mm Light yellow when first laid later becoming spotted with bright orange	1 mm Creamy white with brown head	12 mm Reddish to brownish yellow with bristles over its whole body. Head capsules and prothoracic plate and thoracic legs are black. Anal comb with 6–7 teeth	5–7 mm Reddish brown. Cremaster with 16 hooked bristles. Female pupae generally more stocky and larger than males
Cryptoblabes gnidiella	Pyralidae, Phycitinae	Wingspan 15 mm Forewings dark gray, punctuated by tiny black spots, veiled in white and dotted with reddish scales characterized by indistinct lighter bands. Hind wings shiny white and streaked with terminal gray lines. Horn-shaped process on the third male antennomerous	0.7 × 0.4 mm Subcircolar shape with one slightly flatter pole. White at laying, it gradually assumes a yellowish color	1 mm Light yellow with brown head	11 mm Yellow to light brown body with two narrow longitudinal darker bands on the dorsal side. Brown-reddish head with small black areas at the base of blackish bristles	7 mm Brown-reddish which will turn gradually toward the dark red. Very sharp creamy abdomen, distally displays oblique hooks

Ephestia parasitella unicolorella	Pyralidae, Phycitinae	Wingspan 15–18 mm Fore wings light brown with transverse reddish brown and barely perceptible strips	Subcircur shape, white at laying	White-pinkish with dark brown head	10 mm Light yellow body with shades of pink and numerous bristles with elongated blackish hair tubercles. Reddish head	7–8 mm Reddish-brown
Argyrotaenia ljungiana	Tortricidae, Tortricinae	Wingspan 12–16 mm Forewing silver-white, strigulated with gray; markings dark reddish brown, sprinkled with black; margins of basal and median fasciae irregular. Hindwing gray	Batches of 40–50, yellow, turning brown during development		18 mm Clear green body. Light green or yellowish green head. Anal comb with 6–8 prongs	Pale brown; in a silken cocoon in spun leaves, or overwintering in a cocoon in debris on the ground

14.2 *Lobesia botrana* and *Eupoecilia ambiguella*

14.2.1 *Taxonomy and Occurrence*

The European grapevine moth (EGVM) *L. botrana* was described in 1775 by Denis and Schiffermüller as *Tortrix*, and subsequently as *Eudemis* and *Polychrosis*. Currently, as *Lobesia* Guenée 1845, it is in the family Tortricidae, subfamily Olethreutinae, tribe Olethreutini (Razowski 1995). *Lobesia botrana* is historically present in Europe, Asia, and Africa (CAB 1974). Although widespread in all grapevine-growing areas, it is economically important mostly in southern Europe. In southern France, in central and southern Spain, Portugal, Greece, Italy, and the islands of the Mediterranean Basin, *L. botrana* is the only moth species to have an important impact on grapevine production (Thiéry 2005). Recently, it has expanded its geographic range and was found in Chile in 2008, California in 2009, and Argentina in 2010 (Gonzales 2010; Varela et al. 2010).

The European grape berry moth (EGBM), *E. ambiguella*, is considered a significant insect pest in many grapevine-growing areas where it can cause considerable damage to grapes. It is found from Britain to Japan, from the Mediterranean Basin to the Scandinavian countries (farther north than grapevine-growing regions) (CAB 1986). First described by Hübner in 1796 as *Tinea ambiguella*, it was subsequently included in the genera *Cochylis* and *Clysia* and lastly, according to Razowski (1995), in the genus *Eupoecilia.* Known as Einbindiger Traubenwickler, Polilla de la vid, Cochylis de la vigne and Tignola della vite, it was recognized as the major grape berry pest in Europe until the 1920s. More recently, and in many areas, it has been gradually replaced as a major pest by *L. botrana.* The shift started in the Mediterranean Basin and is now extending to Switzerland, Austria and southern Germany where populations of *L. botrana* and *E. ambiguella* overlap.

14.2.2 *Host Plants*

Lobesia botrana larvae feed on grapevine flowers and berries and on a number of other plants growing in warm-dry environments, such as one finds in most Mediterranean countries. Its host range includes about 40 species belonging to 27 different families (Coscollá 1997). The spurge flax *Daphne gnidium* L. is hypothesized to be its original host plant (Bovey 1966; Lucchi and Santini 2011; Tasin et al. 2011). However EGVM is frequently associated with other hosts in habitats where suitable host plants occur. These hosts include for example olive tree inflorescence, Virginia creeper, jujube, rosemary, evergreen clematis, dogwood, ivy, currant (Bovey 1966; Coscollá 1997; Stavridis and Savopoulou-Soultani 1998; Thiéry 2005). *Eupoecilia ambiguella* is also polyphagous and shares several host plants with EGVM. Even though mugwort, *Artemisia vulgaris* L., is sometimes reported as its native host plant, grapevine is now accepted as its original host.

Other cultivated and wild host plants include black and red currant, lemon citrus, Virginia creeper, blackthorn, smooth bedstraw, wayfaring tree, highbush cranberry, laureltinus, privet, ash, maple, dogwood, and many others (Solinas 1962). Since grape berry moths are mostly restricted to grapevines in grapevine-growing areas, the nutritional and ecological polyphagy of these two tortricid species needs to be researched.

14.2.3 Life History

Eupoecilia ambiguella and *L. botrana* have a similar biology, but slightly different climatic preferences: *L. botrana* prefers warm and dry conditions and *E. ambiguella* prefers cool and humid climates (Bournier 1977). *Lobesia botrana* and *E. ambiguella* are typical multivoltine species with facultative diapause. In northern Europe and in the Mediterranean Basin, EGVM has 2–4 generations per year on *Vitis vinifera* L., depending on latitude and microclimates (Roditakis and Karandinos 2001; Harari et al. 2007). Two to three generations per year are the rule in Germany, Switzerland, Austria, and northern France. It has three generations in the warmer climates of southern France, Spain, Portugal, Greece, and Italy, where the species can sometimes give rise to a fourth flight and a fourth generation, partial or complete. In Israel, Egypt, and Crete some EGVM populations do not undergo diapause and spend the winter in the larval stage, continuing to feed on unharvested clusters or on alternate hosts. EGBM usually has only two generations per year. A third generation has been frequently observed in France and Italy (Marcelin 1985; Varner and Mattedi 2004).

Lobesia botrana moths are crepuscular insects flying at canopy levels at twilight (60–80 lux). Males are more active than females. The flight of virgin females is interrupted by several stationary phases, in which they release a pheromone, the main component of which is (*E,Z*)-7,9-dodecadienyl acetate (Roelofs et al. 1973). *Eupoecilia ambiguella* moths are active from twilight to dawn, resting motionless during daylight. They feed, call, and mate during the night and early morning. Females lay eggs in the afternoon and evening (Bovey 1966). The main component of the pheromone is (*Z*)-9-dodecenyl acetate (Arn et al. 1976; Saglio et al. 1977).

Adults of *L. botrana* (Fig. 14.1a) feed from emergence to sexual maturity. Mating occurs about 24 h after emergence and oviposition starts 3 days thereafter. Mating lasts from a few minutes to 2 h. Males mate several times with different females, and females have a rare tendency toward polyandry. Males are much less attracted to mated females than to virgins. On average a female lays from 50 to 80 eggs (Fig. 14.1b, c), most of which are laid in the first week of life. The average life span of the adult moth is approximately 15–20 days, usually shorter for males than females. The first generation of both species develops on flowers (anthophagous) and the other generations on berries (carpophagous) (Figs. 14.1d and 14.2b–d).

Single eggs are laid on bracts, caps and stems of the flower clusters in spring and on the berries in summer. First generation larvae feed on several pre-bloom flowers

Fig. 14.1 *Lobesia botrana* (**a**) adult, (**b**) egg, (**c**) egg in the black-head stage, (**d**) larva, (**e**) pupa

and web them together with silk. Webbing gradually thickens to form the so-called glomerulus or nest (Fig. 14.2a).

In hot weather, *E. ambiguella* larvae sometimes bore into the rachis and peduncle, seeking moisture. Within the glomerulus, EGBM larvae construct a silk case to which they retreat during the hottest hours of the day. In these nests, the hidden larvae feed on flowers whose remnants are used to increase the size of the case. First generation grape berry moths eventually pupate either inside the glomerulus or on the underside of a leaf (Figs. 14.1e and 14.3d). Their pupal stage lasts about 2 weeks.

The emerging moths (Fig. 14.3a) lay single eggs on the berries (Fig. 14.3b). EGBM hardly ever lays eggs on the rachis or on peduncles. After hatching, the

Fig. 14.2 *Lobesia botrana* (**a**) larval nest, (**b, c**) injured berries, (**d**) grape cluster infected by gray mold and sour rot

Fig. 14.3 *Eupoecilia ambiguella* (**a**) adult, (**b**) egg, (**c**) larva, (**d**) pupa

larvae (Fig. 14.3c) feed on the berries and may cause severe damage. When disturbed they drop down on a silk thread. European grape berry moth larvae are sluggish while EGVM larvae are highly mobile. Feeding on berries promotes infection by gray mold (*Botrytis cinerea* Persoon ex Fries), which leads to even greater injury than that caused by the insect itself. Non-diapausing larvae pupate preferably within the cluster or on the leaves. Photoperiod induces diapause that lasts for months, but mature EGBM larvae stay for several months in a prepupal stage and low temperatures are necessary to advance to the pupal stage (Fig. 14.3d). The diapausing larvae build their cocoons mainly under exfoliating bark, in crevices and cracks of the trunk and cordons.

14.2.4 Economic Importance and Control

14.2.4.1 Grape Berry Moth Damage

The two species are quite similar in behavior and damage. Sometimes they are present at the same time in the same vineyard but with a different degree of population density. Until the end of the nineteenth century, *E. ambiguella* was widespread in all the southern European regions. From the beginning of the twentieth century, *L. botrana* gradually became the predominant species south of the Alps.

Grape berry moth infestation levels depend on the growth characteristics of the cultivars, the agronomic practices, the climatic conditions, and the number of generations per year. The anthophagous generation does not generally cause yield reduction. The following carpophagous generations are the most destructive due to larval feeding on green and ripe berries, which results in yield reduction. The presence of larvae, webbing and rotten berries, leads to downgrading of table grapes. Moreover, secondary infections of gray mold, *B. cinerea*, develop rapidly on damaged berries causing bunch rot which substantially degrades wine quality. Black aspergilli's rot, *Aspergillus niger* and *A. carbonarius*, producers of ochratoxin A (Cozzi et al. 2006), are also often related to larval feeding activity. Because of these undesirable economic effects, *L. botrana* and *E. ambiguella* must be managed to keep their damage at an acceptable level.

14.2.4.2 Population Density and Risk Assessment

Time of the first appearance of adults and hatching of the first eggs can be forecasted by predictive models based on temperature requirements of individual instars and critical conditions for oviposition (Moravie et al. 2006). Unfortunately, forecast models based on DD are empirical and their robustness is strongly dependent on the environment in which they have been validated. Alternative forecasting techniques are currently under development, such as the evaluation of larval age distribution during the previous generation in order to predict the distribution of female emergence (Delbac et al. 2010).

Trapping females with food-baited traps is a valuable tool to predict the onset of oviposition, an event used to properly time insecticide treatments (Thiéry et al. 2006). However, baited-trap maintenance and monitoring are very time-consuming chores for growers and consultants. Pheromone traps are easier to use. They are a sensitive tool to monitor flight of males. They can be useful to time an ovicidal treatment, and to properly schedule scouting activities in the vineyard.

Forecasting models and moth trapping alone do not provide sufficient population density information and need to be supplemented with appropriate field scouting of eggs and young larvae (Shahini et al. 2010). Based on the resulting infestation assessment (% of injured clusters, number of nests per inflorescence, number of eggs and larvae per cluster, number of injured berries per cluster), insecticides are applied according to action thresholds (AT). The action thresholds vary widely depending on the generation, susceptibility of the cultivar to the subsequent infection by *B. cinerea*, as well as whether berries are produced for table grape, raisins or wine production.

First generation larvae may not necessarily need to be controlled by chemical treatments, especially if they develop on cultivars with abundant blossom not subject to intense shedding. Between flowering and harvest, damage to inflorescences is compensated by the increase in weight of uninjured berries in the majority of cultivars. That explains the lack of a defined injury threshold for the anthophagous generation (Moschos 2005).

Chemical control for the first generation is exclusively applied when the pest population density is particularly high or if it exceeds an AT of more than 50% infested inflorescences (Bagnoli et al. 2009). For the following generations, the suggested AT ranges from 5% to 15% of infested (eggs or young larvae) clusters respectively for compact and loosely-bunched cultivars, according to their susceptibility to rot.

Knowledge of the spatial distribution of the population is important for the development of efficient sampling programs, that allow a more accurate estimate of the damage and AT.

14.2.4.3 Natural Enemies and Biological Control

The cohort of *L. botrana* and *E. ambiguella* natural enemies varies considerably in time and space due to insect physiology, activity and ecological niche of individual species. Fungi of the genera *Spicaria*, *Beauveria*, *Paecilomyces*, *Aspergillus*, *Cephalosporium*, *Cladosporium*, *Penicillium*, *Citromyces*, *Verticillium* and *Stemphylium* can infect a large percentage of overwintering pupae. The bacteria *Bacillus thuringiensis* Berliner var. *kurstaki* (Btk) and var. *aizawai* are effectively and extensively used against EGVM and EGBM, both in conventional and organic vineyards (Scalco et al. 1997; Vidal 1997; Keil and Schruft 1998; Shahini et al. 2010). Arthropods associated with grape berry moths include predators such as spiders (Clubionidae, Theridiidae, Tomisidae, Linyphiidae, Salticidae), mites (Thrombididae) and insects belonging to Dermaptera, Hemiptera, Neuroptera, Diptera and Coleoptera (Solinas 1962; Coscollá 1997). Among insect parasitoids, species associated with EGVM in

Europe belong to the Hymenoptera (Ichneumonidae, Braconidae, Chalcididae, Pteromalidae, Eulophidae, Elasmidae, Trichogrammatidae) and Diptera (Tachinidae). As with most natural enemies including parasitoids, the natural control achieved by each species varies greatly in time and space. Typically, the frequency of egg and larval parasitism is high in the first two generations and decreases drastically in the overwintering generation, which is mainly affected by larval-pupal and pupal parasitoids.

Extensive scientific efforts to develop biological control as an effective solution for practical use in the field are still needed. Egg parasitoids of the genus *Trichogramma* have been mass-released in an inundative strategy with mixed results (Castaneda-Samayoa et al. 1993; Hommay et al. 2002; Ibrahim 2004). The pteromalids *Dibrachys affinis* Masi and *D. cavus* (Walker) are gregarious generalist larval-pupal parasitoids of Lepidoptera, Diptera and Hymenoptera that can be readily reared in the laboratory. However, due to lack of host specificity and because they are also hyperparasites, they are not good candidates for release. The most frequent and efficient species in European vineyards is the larval parasitoid *Campoplex capitator* Aubert (Ichneumonidae). It is regarded as the best candidate for EGVM biological control, but to date, releases have not taken place because of the difficulties associated with artificially mass-rearing the species (Thiéry and Xuéreb 2004).

14.2.4.4 Chemical Control

Most insecticides applied in the past against grape berry moths have been gradually replaced by more selective and less toxic products. New neurotoxic insecticides (spinosyns and oxadiazines), chitin synthesis inhibitors, compounds accelerating molting, microbial insecticides, and more recently some avermectins and anthranilic diamides, have been introduced in current integrated control strategies. Nevertheless, the organophosphates chlorpyriphos and methyl chlorpyriphos are still largely used in European vineyards. Control with insecticides that are larvicidal with some ovicidal activity gives remarkable flexibility on application timing. The efficacy of these products depends on the optimal treatment of the most susceptible stages, so prediction of life cycle events is critical for each moth species. Because of increasing accuracy of the forecasting tools, a single insecticide application to control the second generation of either species is usually effective in most grapevine districts in Europe. More treatments are needed in the southern regions to control *L. botrana*. In terms of selectivity, *B. thuringiensis* has undoubtedly the highest ecological value, but its use is still limited due to its short persistence. Successful application timing can be achieved with adequate population monitoring with pheromone traps and egg field scouting.

14.2.4.5 Pheromone-Mediated Control Strategies

The use of pheromones for control of grape berry moths has increased in vineyards due to the high selectivity and low environmental impact. Mating disruption

Table 14.2 European vineyards treated with pheromone mating disruption for management of grape berry moth pests during 2010 (IBMA 2011) in relation to the total vineyard surface of each country (OIV 2007)

Country	Total vineyard surface (ha)	Vineyard treated with MD (ha)	% treated
Germany	102,000	70,000	68.6
France	867,000	20,000	2.3
Italy	847,000	16,500	1.9
Spain	1,169,000	14,500	1.2
Switzerland	14,800	7,000	47.3
Austria	49,900	2,400	4.8
Czech Republic	17,700	1,300	7.3
Portugal	248,000	1,200	0.5
Hungary	75,000	300	0.4
Slovakia	17,600	100	0.6
Cyprus	15,300	100	0.7
Total	3,423,300	133,400	3.9

(MD) with hand-applied dispensers is the most well-studied and widely used pheromone-mediated control technique against grape berry moths in the European grapevine-growing regions (Stockel et al. 1992; Neumann et al. 1993; Charmillot and Pasquier 2000). Currently it is applied on approximately 140,000 ha in European vineyards, i.e. about 3–4% of the total grapevine-growing area in the European Union (Table 14.2). Recent MD area-wide applications have also been conducted in Chile and California where EGVM was accidentally introduced (Witzgall et al. 2010; Ioriatti et al. 2011).

The most common hand-applied dispensers available on the market for grape berry moths are Shin-Etsu twist-ties ropes (Isonet L®, Lplus®, LE® in Europe; Isomate® in the US), the BASF twin ampoules (RAK1+2®, RAK 2®) and, for EGVM only, the Suterra® membranes. The active ingredients in these dispensers are the main pheromone components, (*E,Z*)-7,9-dodecenyl acetate and (*Z*)-9-dodecenyl acetate for EGVM and EGBM, respectively.

Five hundred dispensers per hectare (the number of dispensers may vary depending on manufacturer) must be deployed in the vineyards before the onset of the first seasonal flight, because late deployment will likely cause control failures. Dispensers must be evenly distributed in the vineyard, and should be attached to vine shoots to ensure protection by foliage from direct exposure to sun and high temperatures. Twice as many dispensers must be hung along the vineyard edges. Border effects are obviously much reduced when MD is applied in area wide projects as in certain growing regions of Germany, France, Switzerland, northern Italy, and Spain (Kast 2001; Ioriatti et al. 2008).

Depending on the vineyard layout, the time to attach the dispensers on the vines may vary between 1.5 and 3 h/ha. The surface area of vineyards in Europe under pheromone-mediated (MD) control of grape berry moths is still limited, despite intensive research and substantial experience with practical applications during the last two decades. This is because of socio-cultural and economical conditions existing in the different vine growing areas where interest in innovative methods is often low. Increasing quality standards for wine and table grapes, with respect to pesticide residues, are creating new opportunities for extensive adoption of MD in IPM programs. However, high costs of MD (about 110 €/ha for EGVM and 150 €/ha for both insects) have hampered the diffusion of this method to date. Cost reduction must be considered for a wider adoption of MD in European vineyards.

Novel pheromone application systems to control Lepidoptera pests such as autoconfusion, lure and kill, aerosol puffers, microencapsulated sprayables, and nanofibers may represent future opportunity for grape berry moth control (Underwood et al. 2002; Charmillot et al. 2005; Nansen et al. 2007; Anfora et al. 2008; Hein et al. 2011). New investments in fundamental research are critical for an effective improvement in semiochemical applications. The research should address the reproductive, physiological, and behavioral mechanisms by which the pheromone affects the target insects, as well as explain how volatile compounds are involved in tritrophic interactions.

14.2.4.6 *Lobesia botrana* as an Invasive Species in the Americas

It was first detected in Chile in April of 2008. Grapes are grown from the region of Atacama in the north to the region of Araucanía in the south. In surveys conducted in the growing season of 2008–2009, moths were detected in all grapevine-growing regions. Low levels of catches were detected in the 2010–2011 season in grapevine-growing areas from Atacama to Araucanía. However, large urban areas remain as moth reservoirs. The current control strategy is to attempt eradication of the pest and major efforts are concentrated towards vineyards and urban areas surrounding grapevine-growing areas. After three seasons of control with insecticide and MD, moth catches in monitoring traps have decreased significantly and the number of foci per region has also decreased considerably.

In the United States, the first report of this pest was in September 2009 in the Napa Valley, California. Surveys conducted in 2010 show that the highest infestation is in Napa County, with a few moths detected, and a few foci in nine other counties of California. No detection has been made in any other US state, despite trap-based surveillance in many of the primary grape production regions. The strategy in California is also eradication and in the first year of control populations decreased dramatically from the first to the third generation.

It is unclear how EGVM was first introduced into Chile or California. At low population levels the damage caused by the larvae is inconspicuous. By the time the first infestations are detected, the spread may be extensive due primarily to movement of grapevines with undetected infestations and movement of unsanitized machinery.

In both Chile and California the primary host for EGVM is grapevine flower clusters and berries. In extensive surveys in Chile it has only been detected in plums next to an infested vineyard. In California, it has been detected in low numbers only in olive flowers adjacent to vineyards. To date, surveys conducted in riparian vegetation in infested areas of California have not detected larvae in wild grapes.

Chile began eradication programs 2 years before California. In the first year, control measures were applied to vineyards to a radius of 5,000 m from a detection site. This was reduced to 1,000 m the following year. The recommendations are to make two insecticide applications for the first generation, one for the second and one for the third generation, plus the application of pheromone dispensers for mating disruption. In urban districts surrounding grapevine-growing areas, homeowners have a choice of destroying the fruit or accepting insecticide treatments.

In California, EGVM has three generations a year. In the first year control measures were applied to a radius of 1,000 m from a detection site and it was reduced to 500 m the following year. With the goal of eradicating the pest, one insecticide application is targeted for each of the first and second generations, when the larvae are most exposed. To further suppress populations the use of MD dispensers (the product registered as Isomate®-EGVM in the US) is highly encouraged. Control in urban areas has been limited to those counties with low trap catches. Homeowners mostly adopted fruit removal as a management technique.

Given that the first flight and egg laying period is very extended, if the application for this generation is done before egg laying or too early in the egg laying period, a second application may be needed to cover the prolonged egg hatch. Furthermore, at this time, the flower cluster is rapidly expanding, decreasing the surface covered by an insecticide. Thus, it is best to wait and time the control of the first generation when the highest proportion of larvae is about to emerge from eggs. The timing for this event can be determined by following the male moth flight with pheromone traps and monitoring egg development. If the insecticide used has some ovicidal properties the recommendation is to make the application when the heads of the larvae are visible in 20% of the eggs. When eggs are too few to monitor, treatment is applied shortly after peak flight. Insecticides registered for organic production are strictly larvicidal and are applied at egg hatch. Due to the short residue of organic materials, two or more applications are recommended starting at egg hatch and weekly for as long as larvae are detected forming glomeruli in the flower cluster.

The insecticide timing for the second generation depends on whether the insecticide has some ovicidal properties or if it is strictly larvicidal. If it is ovicidal, the applications can start a few days after the first males of the second flight are caught in a trap. For larvicidal insecticides (conventional or organic), the applications can start 10–14 days after the first moths of the second flight are caught, if eggs are too few to monitor. The second generation is substantially shorter than the first, lasting approximately 4 weeks. This makes timing for control of the second generation easier to predict. If treatments are timed appropriately for the first and second generations, treatment of the third generation should not be necessary. Treatments for the third generation are limited in their efficacy in California because of overlap in generations, the difficulty in penetrating a closed cluster and the short period between egg hatch and the larvae penetrating the berry.

The strategy of targeting control measures towards the first and second generation supplemented with mating disruption has proven extremely successful at drastically reducing populations. This is probably aided by the fact that all control measures are taking place in an area wide manner since all growers are strongly encouraged to participate. So far, alternate hosts do not appear to contribute significantly to population levels. In California, chlorpyriphos is not registered for seasonal use in vineyards, and the insecticides most used to control *L. botrana* are insect growth regulators, diamides and Btk, and to a lesser extent, spinosyns and avermectin. A major concern of the program was to avoid disruption of natural control of native pseudococcids pests. This was achieved by using selective insecticides.

A major challenge is to achieve complete control in urban areas. Another challenge is to determine when a population has truly been eradicated. Delimitation of infestations is done using pheromone traps. Pheromone traps are effective; however the male moth does not fly more than 100 m, with an average distance <50 m (Roehrich and Carles 1981). This entails having a very high density of traps with no catches during several generations. The risk of a false negative is significant since it will be tempting to declare the pest eradicated when in reality populations are breeding at undetectable levels.

In February 2010, EGVM was also detected in the major grape production region of the Province of Mendoza in Argentina. As of 2011, the first year of control is underway.

14.3 *Cryptoblabes gnidiella*

14.3.1 Taxonomy and Occurrence

Among Pyralidae Phycitinae, the honeydew moth (HM), *C. gnidiella* (Fig. 14.4a–e), is the most frequent and harmful species on grapes in the Mediterranean Basin. Described for the first time by Millière in 1867 as *Ephestia gnidiella*, it was then reported by Briosi as *Albinia wockiana* in 1877. The current classification is due to Hartig (1939), who redescribed the species from specimens collected in central Italy. Widespread throughout the Mediterranean region, HM is reported from Malaysia, New Zealand, Hawaii, some African and Asian countries, and many tropical and subtropical regions of North and South America.

14.3.2 Host Plants

Cryptoblabes gnidiella is a polyphagous species associated with about 60 different host plants belonging to 30 families. These include *Actinidia deliciosa* (Chevalier), *Citrus* spp., *Daphne* spp., *Daucus carota* L., *Diospyros kaki* L., *Eriobotrya japonica* (Thunberg) Lindley, *Gossypium herbaceum* L., *Malus* spp., *Persea Americana* Miller, *Prunus* spp., *Pyrus* spp., *Ricinus communis* L., *Tamarix* spp., and *Vitis* spp. (Zocchi 1971; Yehuda et al. 1991–1992; Sing and Sing 1997). Very frequently HM shares the host plant with other insects, either Lepidoptera (e.g., the European grapevine moth *L. botrana*) or Hemiptera (aphids and pseudococcids) which

Fig. 14.4 *Cryptoblabes gnidiella* (**a**) adult, (**b**) young larva, (**c**) mature larva, (**d**) pupa, (**e**) infested clusters

produce honeydew of which HM larvae are active consumers, hence the common name "Honeydew moth".

14.3.3 Life History

In the Mediterranean Basin, *C. gnidiella* has 3–4 generations per year depending on latitude, with a first flight in May–June, a second in July, a third in August–September and a fourth in October–November, with possible overlapping generations on late harvest grape varieties (Bagnoli and Lucchi 2001). In Israel the species can have up to seven generations on grapes and citrus (Avidov and Harpaz 1969). In the grapevine-growing areas of northeast Brazil, where climatic conditions allow two annual crops, HM can have as many as nine generations per year (Bisotto-de-Oliveira et al. 2007). It overwinters as a larva and pupation takes place inside a silken cocoon. The sex pheromone of *Cryptoblabes gnidiella* is a mixture of quaternary aldehydes (Bjostad et al. 1981).

14.3.4 Economic Importance and Control

The economic importance of *C. gnidiella* varies greatly according to geographical areas. Though it mainly occurs in coastal areas characterized by heavy infestation of

L. botrana and *Planococcus* spp., it is also able to infest healthy pre-veraison grapes, feeding on cluster stems. During ripening, larvae may feed superficially on berry skins. Regardless of the feeding damage, since it is highly gregarious the number of larvae determines the level of damage caused (Lucchi et al. 2011). Because of the highly aggregated larval distribution, affected clusters are fully compromised by the inevitable and rapid development of rots enhanced by the presence of drosophilids and by nitidulids. In Israel and Brazil, the species is considered a key pest of vineyards (Harari et al. 2007; Bisotto-De-Oliveira et al. 2007). Protection of grapes from *C. gnidiella* infestation is achieved with effective control of *L. botrana*. If well timed, it eliminates the need for a specific spray against phycitin larvae. Moreover, Btk can be usefully employed in case of asynchronous outbreaks.

14.4 *Ephestia parasitella unicolorella*

Larvae of *E. parasitella unicolorella* (Pyralidae: Phycitinae) (Fig. 14.5a–f) are found within the cluster after veraison as a secondary pest on wilted or dried berries and very often hidden within them. Winter is spent in the larval stage, in a thin cocoon spun by the mature larva on woody structures of the vine or on support poles. It is not yet clear where this insect resides outside the vineyard, especially during the spring, nor the number of generations that the species has in Europe. The economic importance of *E. parasitella unicolorella* is negligible. Deseo (1980) reports that the young larva feeds on the rachis and petiole of the bunch, whereas the older larva can penetrate and develop on a single berry, feeding on the pulp. At harvest mature larvae are frequently found in the most internal parts of the bunch associated with or inside rotten or dried berries, almost motionless and folded in a C shape. Xuéreb et al. (2003) advised to destroy the unharvested clusters to avoid the further development of the species in the vineyard. In a recent review the name of *Ephestia unicolorella woodiella* Richard & Thomson has been proposed for this species (Huertas Dionisio 2007).

14.5 *Argyrotaenia ljungiana*

Argyrotaenia ljungiana (Fig. 14.6a–d) (syn. *A. pulchellana* Haworth) is present throughout the Palaearctic region with the exception of Japan. Females deposit their eggs in batches of 40–50 eggs, usually on the upper surface of the leaves. Larvae feed primarily on leaves of host plants. Pupation occurs in a silken cocoon inside webbed leaves. In the Mediterranean region, the species has three generations per year, the first generation occurring in April and May, the second from the end of June to July and the third in August–September. *Argyrotaenia ljungiana* overwinters in the pupal stage, inside a cocoon in debris on the ground. A highly polyphagous species, it feeds on many wild and cultivated plants, including grapevine and apple.

Fig. 14.5 *Ephestia parasitella unicolorella* (**a**) adult, (**b, c, d**) larva, (**e**) overwintering larva, (**f**) pupa

Occasionally, *A. ljungiana* can give rise to important outbreaks in vineyards, where it feeds on inflorescences and berry clusters. The harmfulness of this tortricid on grapes has been described in Italy (Varner et al. 2001), France (Marcelin 1985), Bulgaria (Kharizanov 1976), and Hungary (Voigt 1972). On the berries it may cause superficial but extensive excavations, which are different from those caused by the other two tortricids, but it can also provide entry points to rots when feeding on ripening grapes. Sometimes the larvae can deeply abrade the cluster rachis causing desiccation of the grapes.

Fig. 14.6 *Argyrotaenia ljungiana* (**a**) adult, (**b**) larva, (**c, d**) damage on grapes

14.6 Conclusion

Of the five Lepidoptera species feeding on clusters in European vineyards, EGVM and EGBM have the highest adverse economic impact. The other three species described here, two Pyralidae and one Tortricidae, are occasional or secondary pests. EGVM has increased its geographic range in the twentieth century throughout Europe and the Middle East, invading the Americas early in the twenty-first century. In the Mediterranean region south of the Alps, EGBM is being replaced by EGVM as the major lepidopteran pest. Where EGVM is established, insecticide control for the first generation is not practiced given that, in most varieties, damage to the inflorescence has no impact on yield. Insecticidal control is targeted primarily at the second generation larvae. In recent years, more selective insecticides have been introduced, with some having ovicidal activity. In Europe, the area-wide approach based on the use of pheromones to control EGVM and EGBM represents an important development in a more environmentally acceptable control of these insects. In regions were eradication is being pursued a program combining mating disruption and insecticides targeted for first and second generation larvae has achieved a drastic reduction in population levels.

Acknowledgements We are much indebted to Bruno Bagnoli and Vittorio Veronelli for productive discussions on grape moths. Photo acknowledgments: Bruno Bagnoli: Figs. 14.1b, c; 14.3c; 14.4b–d; 14.5b, c. Paolo Giannotti: Figs. 14.1a, e; 14.3a, d. Mauro Varner: Fig. 14.6b, c. Elena Pozzolini: Fig. 14.6d.

References

Anfora G, Baldessari M, De Cristofaro A, Germinara GS, Ioriatti C, Reggiori F et al (2008) Control of *Lobesia botrana* (Lepidoptera: Tortricidae) by biodegradable Ecodian sex pheromone dispensers. J Econ Entomol 101:444–450

Arn H, Rauscher S, Buser HR, Roelofs WL (1976) Sex pheromone of *Eupoecilia ambiguella*: cis-9-dodecenyl acetate as a major component. Z Naturforsch Teil C Biochem Biophys Biol Virol 31:499–503

Avidov Z, Harpaz I (1969) Plant pests of Israel. Israel University Press, Jerusalem

Bagnoli B, Lucchi A (2001) Bionomics of *Cryptoblabes gnidiella* (Millière) (Pyralidae Phycitinae) in Tuscan vineyards. IOBC/WPRS Bull 24(7):79–83

Bagnoli B, Lucchi A, Pollini A (2009) Contro le tignole della vite si interviene sulla seconda generazione. L'Informatore Agrario 22:9–11

Bisotto-de-Oliveira R, Redaelli LR, Sant'Ana J, Cover C, Botton M (2007) Ocorrência de *Cryptoblabes gnidiella* (Millière) (Lepidoptera: Pyralidae) relacionada à fenologia da videira em Bento Goncalves. Neotrop Entomol 36:555–559

Bjostad LB, Gurevitz E, Gothilf S, Roelofs WL (1981) Sex attractant for the honeydew moth, *Cryptoblabes gnidiella*. Phytoparasitica 9:95–99

Bournier A (1977) Grape insects. Annu Rev Entomol 22:355–376

Bovey P (1966) Super-famille des Tortricoidea. In: Balachowsky AS (ed) Entomologie appliquée à l'agriculture. Lépidoptères. Édition Masson et Cie, Paris, pp 859–887

CAB (1974) Distribution maps of pests, map 70. CAB International, Wallingford

CAB (1986) Distribution maps of pests, map 76. CAB International, Wallingford

Castaneda-Samayoa O, Holst H, Ohnesorge B (1993) Evaluation of some *Trichogramma* species with respect to biological control of *Eupoecilia ambiguella* Hb. and *Lobesia botrana* Schiff. (Lep., Tortricidae). Z Pflanzenk Pflanzen 100:599–610

Charmillot PJ, Pasquier D (2000) Lutte par confusion contre les vers de la grappe: succès et problèmes rencontrés. IOBC/WPRS Bull 23(4):145–147

Charmillot PJ, Degan T, Pasquier D, Briand F (2005) Nouveaux procédés à base de phéromones pour lutter contre les vers de la grappe: essais préliminaires en 2004. Rev Suisse Vitic Arboric Hortic 5:283–288

Coscollá R (1997) La polilla del racimo de la vid (*Lobesia botrana* Den. y Schiff.). Generalitat Valenciana, Conselleria de Agricultura, Pesca y Alimentación, Valencia

Cozzi G, Pascale A, Perrone G, Visconti A, Logrieco A (2006) Effect of *Lobesia botrana* damages on black aspergilli rot and ochratoxin A content in grapes. Int J Food Microbiol 111:S88–S92

Delbac L, Lecharpentier P, Thiéry D (2010) Larval instars determination for the European grapevine moth (Lepidoptera: Tortricidae) based on the frequency distribution of head-capsule widths. Crop Prot 29:623–630

Deseo KV (1980) Due fitofagi di secondaria importanza nei vigneti emiliani: *Euzophera bigella* Zell. ed *Ephestia parasitella* ssp. *unicolorella* Staud. (Lepidoptera, Pyralidae). Inf fitopatol 30:7–9

Gonzales M (2010) *Lobesia botrana*: polilla de la uva. Rev Enol 2:2–5

Harari AR, Zahavi T, Gordon D, Anshelevich L, Harel M, Ovadia S et al (2007) Pest management programmes in vineyards using male mating disruption. Pest Manag Sci 63:769–775

Hartig F (1939) Contributo alla conoscenza della fauna lepidotterologica dell'Italia centrale. Mem Soc Entomol Ital 18:186–198

Hein DF, Breuer M, Hummel HE, Greiner A, Wendorff JH, Hellmann C et al (2011) Electrospun nanofibers as novel carriers of insect pheromones: communication disruption strategy against the tortricid moths *Eupoecilia ambiguella* and *Lobesia botrana* in vineyards. IOBC/WPRS Bull 67:183–187

Hommay G, Gertz C, Kienlen JC, Pizzol J, Chavigny P (2002) Comparison between the control efficacy of *Trichogramma evanescens* Westwood (Hymenoptera: Trichogrammatidae) and two *Trichogramma cacoeciae* marchal strains against grapevine moth (*Lobesia botrana* Den. & Schiff.) depending on their release density. Biocontrol Sci Technol 12:569–581

Huertas Dionisio M (2007) Estados inmaturos de lepidoptera (XXX). Tres especies del género *Ephestia* Guenée, 1845 en huelva, España (Lepidoptera: Pyralidae, Phycitinae). Sociedad Hispano-Luso-Americana de Lepidopterologia, Madrid, España, SHILAP Rev. Lepidopterol 35:381–399

IBMA (2011) International biocontrol manufacturers association. http://www.ibma-global.org

Ibrahim R (2004) Biological control of grape berry moths *Eupoecilia ambiguella* Hb. and *Lobesia botrana* Schiff. (Lepidoptera: Tortricidae) by using egg parasitoids of the genus *Trichogramma*. Ph.D. dissertation, Institute of Phytopathology and Applied Zoology, Justus Liebig University of Giessen, Giessen, Germany

Ioriatti C, Anfora G, Tasin M, De Cristofaro A, Witzgall P, Lucchi A (2011) Chemical ecology and management of *Lobesia botrana* (Lepidoptera; Tortricidae). J Econ Entomol 104(4):1125–1137

Ioriatti C, Lucchi A, Bagnoli B (2008) Grape area wide pest management in Italy. In: Koul O, Cuperus GW, Elliott N (eds) Areawide pest management: theory and implementation. CABI International, Wallingford, pp 208–225

Kast WK (2001) Twelve years of practical experience using mating disruption against *Eupoecilia ambiguella* and *Lobesia botrana* in vineyards of the Wuerttemberg region, Germany. IOBC/WPRS Bull 24(2):71–73

Keil S, Schruft G (1998) Effectiveness of *Bacillus thuringiensis* on the grape vine and grape berry moth (*Lobesia botrana* and *Eupoecilia ambiguella*). Conference Information: Integrated control in viticulture. In: Proceedings of the meeting at Gödöllo, Hungary, 4–6 Mar 1997 IOBC/WPRS Bull 21(2):63–65

Kharizanov A (1976) *Argyrotaenia pulchellana* Haw. A new pest of vine. Rastitelna Zashchita 24:32

Lucchi A, Botton M, Bagnoli B (2011) Tignola rigata su vite da tenere sotto controllo. L'informatore agrario 31:65–69

Lucchi A, Santini L (2011) Life history of *Lobesia botrana* on *Daphne gnidium* in a natural park of Tuscany. IOBC/WPRS Bull 67:197–202

Marcelin H (1985) Control of grape tortricids. Phytoma 370:29–32

Moravie MA, Davison AC, Pasquier D, Charmillot PJ (2006) Bayesian forecasting of grape moth emergence. Ecol Model 197:478–489

Moschos T (2005) Yield loss quantification and economic injury level estimation for the carpophagous generations of the European grapevine moth *Lobesia botrana* Den. et Schiff. (Lepidoptera: Tortricidae). Int J Pest Manag 52:141–147

Nansen C, MacDonald KM, Rogers CD, Thomas M, Poppy GM, Baxter IH (2007) Effects of sex pheromone in electrostatic powder on mating behaviour by *Lobesia botrana* males. J Appl Entomol 131:303–310

Neumann U, Schmid A, Ioriatti C, Varner M, Castillo R, Lucas A et al (1993) La technique par confusion contre les "vers" de la grappe en Europe aujourd'hui. Phytoma 456:15–17

OIV (2007) Organisation internationale de la vigne et du vin. http://www.oiv.int/oiv/cms/index

Razowski J (1995) The catalogue of the species of Tortricidae (Lepidoptera): part IV: Palaearctic Olethreutinae Microcorsini, Bactrini, Endotheniini and Olethreutini. Acta Zool Cracov 32:285–324

Roditakis NE, Karandinos MG (2001) Effects of photoperiod and temperature on pupal diapause induction of grape berry moth *Lobesia botrana*. Physiol Entomol 26:329–340

Roehrich R, Carles JP (1981) Observations sur les déplacements de l'Eudemis, *Lobesia botrana*. Boll Zool Agr Bach II 16:10–11

Roelofs WL, Kochansky J, Cardé R, Arn H, Rauscher S (1973) Sex attractant of the grape vine moth, *Lobesia botrana*. Mitt Schweiz Entomol Ges 46:72–73

Saglio P, Descoins C, Gallois M, Lettéré M, Jaouen D, Mercier J (1977) Etude de la phéromone sexuelle de la cochylis de la vigne *Eupoecilia* (*Clysia*) *ambiguella*, HB Lépidoptère Tortricoidea, Cochylidae. Ann Zool Ecol Anim 9:553–562

Scalco A, Charmillot PJ, Pasquier D, Antonin P (1997) Comparaison de produits a base de *Bacillus thuringiensis* dans la lutte contre les vers de la grappe: du laboratoire au vignoble. Rev Suisse Vitic Arboric Hortic 29:345–350

Shahini S, Kullaj E, Cakalli A, Lazarevska S, Pfeiffer DG, Gumeni F (2010) Population dynamics and biological control of European grapevine moth (*Lobesia botrana*: Lepidoptera: Tortricidae) in Albania using different strains of *Bacillus thuringiensis*. Int J Pest Manag 56:281–286

Sing YP, Sing DK (1997) Host plants, extent of damage and seasonal abundance of earhead caterpillar *Cryptoblabes gnidiella* (Millière). Adv Agric Res India 7:133–137

Solinas M (1962) Studio morfo-biologico sulla *Clysiana ambiguella* Hb. Ann Facoltà Agraria Univ Cattolica Piacenza 4:327–361

Stavridis DG, Savopoulou-Soultani M (1998) Larval performance on and oviposition preference for known and potential hosts by *Lobesia botrana* (Lepidoptera: Tortricidae). Eur J Entomol 95:55–63

Stockel JP, Schmitz V, Lecharpentier P, Roehrich R, Neumann U, Torres-Vila M (1992) Three years experience in the control of the grape moth *Lobesia botrana* (Den. and Schiff.) using mating disruption in a Bordeaux vineyard. IOBC/WPRS Bull 15:117–120

Tasin M, Lucchi A, Ioriatti C, Mraihi M De Cristofaro A, Boger Z, Anfora G (2011) Oviposition response of the moth *Lobesia botrana* to sensory cues from a host plant. Chemical Sense 36(7):633–639

Thiéry D (2005) Vers de la grappe. Les Connaître pour s'en Protéger. Vigne & Vin, Bordeaux

Thiéry D, Xuéreb A (2004) Vers une lutte biologique contre Eudémis (*Lobesia botrana*). In: Proceedings, Mondiaviti, 1–2 Dec 2004, Bordeaux, France, pp 47–52

Thiéry D, Rétaud P, Dumas-Lattaque L, Féru R, Xuéreb A, Bourriau F (2006) Trapping *Lobesia botrana* females with apple juice: a valuable tool to predict oviposition? IOBC/WPRS Bull 29(11):235–238

Underwood K, Howse P, Loughlin D (2002) Potential applications of ExoSect® Auto-confusion™. In: Book of abstracts of the IOBC/WPRS Working group meeting "Pheromones and other semiochemicals in integrated production", Erice, Italy, 22–27 Sept 2002

Varela LG, Smith RJ, Cooper ML, Hoenisch RW (2010) European grapevine moth, *Lobesia botrana*, in Napa Valley vineyards. Practical Winery & Vineyard. March/April:1–5

Varner M, Mattedi L (2004) Le tignole nella Piana Rotaliana. L'Informatore Agrario 26:63–69

Varner M, Mattedi L, Lucin R (2001) Mating disruption in Trentino viticulture: 10 years experience in "Cantine Mezzacorona". IOBC/WPRS Bull 24(7):143–150

Vidal G (1997) Les bio-insecticides. Le *Bacillus thuringiensis*, comment s'y retrouver? Pro Agric Vitic 114:299–302

Voigt E (1972) Damage caused by *Argyrotaenia pulchellana* Haw. to grape vines in Hungary. Pflanzenschutzberichte 43:13–23

Witzgall P, Kirsch P, Cork A (2010) Sex pheromones and their impact on pest management. J Chem Ecol 36:80–100

Xuéreb A, Mautrait E, Laguerre M, Thiéry D (2003) Une pyrale polyphage pouvant causer des dégâts au vignoble. Phytoma 559:30–32

Yehuda SB, Wysoki M, Rosen D (1991–1992) Phenology of the honeydew moth, *Cryptoblabes gnidiella* (Millière) (Lepidoptera: Pyralidae), on avocado in Israel. Isr J Entomol 25–26:149–160

Zocchi R (1971) Contributo alla conoscenza dell'entomofauna delle tamerici in Italia. Redia 52:31–129

Chapter 15
Biology and Management of Grape Berry Moth in North American Vineyard Ecosystems

Rufus Isaacs, Luis A.F. Teixeira, Paul E. Jenkins, Natalia Botero Neerdaels, Greg M. Loeb, and Michael C. Saunders

15.1 Introduction

The grape berry moth, *Paralobesia viteana* (Clemens), is one of the most widespread and damaging insect pest of grapes in eastern North America. It was renamed from *Endopiza viteana* Clemens (Brown 2006). Larvae (Fig. 15.1c) of this pest bore into berries causing direct injury, reducing yield, and opening berries to opportunistic pathogens (Fig. 15.1d). Where this pest reaches high populations, berries may not be harvestable due to contamination by larvae or diseases that reduce fruit quality, forcing grape growers to leave heavily-infested regions of vineyards unharvested. In the past 50 years, prevention of damage and infestation by grape berry moth has been achieved primarily by the use of broad-spectrum insecticides, but

R. Isaacs (✉) • P.E. Jenkins
Department of Entomology, Michigan State University,
East Lansing, MI 48824, USA
e-mail: isaacsr@msu.edu; jenki132@msu.edu

L.A.F. Teixeira
Product Support and Renewal, DuPont Crop Protection, Wilmington, DE, USA
e-mail: luis.a.teixeira@usa.dupont.com

N. Botero Neerdaels
Driscoll's Strawberry Associates Inc., 151 Silliman Rd., Watsonville, CA 95076, USA
e-mail: natalia.neerdaels@driscolls.com

G.M. Loeb
Department of Entomology, New York State Agricultural Experiment Station,
Cornell University, Geneva, NY 14456, USA
e-mail: gme1@cornell.edu

M.C. Saunders
Department of Entomology, Pennsylvania State University, University Park, PA 16802, USA
e-mail: mcs5@psu.edu

N.J. Bostanian et al. (eds.), *Arthropod Management in Vineyards: Pests, Approaches, and Future Directions*, DOI 10.1007/978-94-007-4032-7_15,

Fig. 15.1 Grape berry moth, *Paralobesia viteana*, the primary insect pest of grapes in eastern North America. (**a**) Adult moth, (**b**) Frass and webbing of first generation larval injury, also showing premature coloration of red grape berries, (**c**) Mature larva on developing berry, (**d**) Entry hole of larva from late season infestation with associated berry splitting and early disease development

increased restrictions on these chemicals in food crops and the risk of resistance to insecticides continues to stimulate the search for alternative control methods. This review includes the current status of knowledge about the biology of *P. viteana* and management strategies for its control. An earlier review of this pest and its management is provided by Dennehy et al. (1990a). We also refer readers to Ioriatti et al. (Chap. 14) for comparison with European species of berry-infesting Lepidoptera. In this chapter, we highlight future research opportunities that may improve the sustainability of vineyard integrated pest management programs while reducing crop damage from *P. viteana*.

15.2 Biology

15.2.1 Taxonomy

For many years, the North American species of grape berry moth was considered the same species as the European vine moth, *Lobesia botrana* (Denis & Schiffermüller). This was a consequence of misidentifications by early taxonomists and reports that it fed on a wide range of host plants, similar to the European species (Johnson and Hammar 1912). To further complicate the issue, *P. viteana* had historically been placed in several other genera, including *Polychrosis* and *Lobesia*, and was described under the synonym of *Penthina vitivorana* by Packard in 1869 (Brown 2006). There is consensus today, however, that these three taxonomic designations were incorrect. Early in the twentieth century, Slingerland (1904) demonstrated that the grape berry moth was an exclusively American species, adopting the genus by Clemens of *Endopiza*, and pointing out differences in habits that permitted separation of *E. viteana* from the European species. Unlike *L. botrana*, larvae of the North American species feed principally on native and cultivated grapes (*Vitis* spp., Vitaceae), whereas *L. botrana* is highly polyphagous (Slingerland 1904; Thiéry and Moreau 2005). Additionally, pupation and overwintering by the North American species occur in the leaf litter rather than on posts or grape canes as is the case for *L. botrana.* Though this species has been named previously in research articles as *P. viteana* (Taschenberg and Roelofs 1977), there has been no consistent convention used in the literature until recently when detailed clarification of the taxonomy of this species was provided by Brown (2006). When searching for literature on this insect over the past 50 years, it is therefore advisable to seek articles using both *Endopiza viteana* and *Paralobesia viteana.*

15.2.2 Geographic Distribution

Paralobesia viteana is distributed across eastern North America, from Ontario and New England in the north to Florida in the south and to Texas in the west, coinciding with the range of its ancestral host, wild grapevines (Isely 1917). Johnson and Hammar (1912) listed the following states in decreasing importance where *P. viteana* had greatest effect on grape production: Ohio, New York, Pennsylvania, Indiana, Illinois, Michigan, Missouri, New Jersey, Virginia, Maryland, West Virginia, Iowa, Delaware, and Arkansas. These states were the main regions of grape production, while Massachusetts, Connecticut, Kentucky, Kansas, Texas, Nebraska, and Wisconsin were less affected because of their limited grape industries. As grape production has spread into new regions of eastern North America, this pest has quickly colonized new vineyards, including Quebec (Bostanian et al. 2003).

15.2.3 Host-Plant Relationships

Until recently *P. viteana* was thought to be active on only a single host plant genus, colonizing wild and cultivated *Vitis* spp. grapes (Slingerland 1904). Recent studies by Saunders and Timer (unpubl. data) have demonstrated oviposition and development of *P. viteana* on berry-producing weeds and additional crops, indicating further potential for non-vineyard habitats to support this species. This suggests that *P. viteana* has some flexibility in its oviposition behavior, and this may be a recent development or it may reflect an adaptive strategy to spread risk across host plants while retaining the highest fidelity to *Vitis* spp.

More than 50 species of wild grape occur in North America (Moore 1991; Fergusson-Kolmes and Dennehy 1993) and these are commonly found in stands of young woods, perturbed habitats, and on the borders of mature forests (Morano and Walker 1995). Galet (1979) lists 10 principal American species, with the most common being *Vitis riparia* Michaux, called riverbank grape or Bermuda vine; *Vitis aestivalis* Michaux, the common blue grape or winter grape; and *Vitis labrusca* L., known as the fox grape, or northern muscadine. Wood lots containing wild grapevines frequently border vineyards throughout the range of *P. viteana*. When clusters from these wild grapevines were sampled in Michigan, infestation by *P. viteana* was on average 84.9% in deciduous woods adjacent to vineyards (Botero-Garcés and Isaacs 2004a), similar to results from New York State where 50–80% of wild grape berries have been reported to be infested (Dennehy et al. 1990a). Populations in these non-crop habitats likely provide a source of insects to infest nearby vineyards.

Host plant volatiles are also important in the chemical ecology of grape berry moth. In a wind tunnel, females are attracted to volatiles released from flowers, unripe and ripe berries, mature leaves and shoot tips of grapevines (Cha et al. 2008a). Eleven plant-released volatiles that elicited responses from antennae of *P. viteana* were identified by Cha et al. (2008b). Surprisingly, none of these compounds are specific to grape, indicating that this moth uses a specific blend of common compounds to find its host plant. Rubber septa loaded with these 11 compounds or a 7-compound subset induced upwind flights of females at the same level as freshly cut grape shoots (Cha et al. 2008b) (Fig. 15.2). More recent experiments have demonstrated a high level of plasticity in the response of the moth to blends of these volatile components (Cha et al. 2011). Moreover, panel traps baited with either the 7 or 11 component blends captured significantly more female moths than traps baited with hexane only in a commercial vineyard (Cha et al. 2008a) (Fig. 15.3). This indicated the potential for using host plant-based lures for monitoring the activity of female moths in the field, and thereby improving timing of crop protectant applications and other pest management operations. In the long term, a better understanding of the role of plant volatiles may make it possible to change volatile profiles of grapes to decrease host finding or host acceptance by grape berry moth and/or develop attractants to pull grape berry moth away from vineyards.

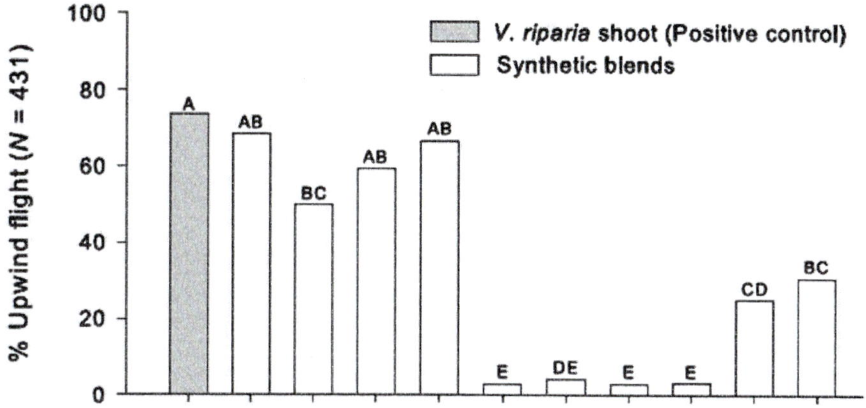

Treatments	Shoot	10	11	12	13	14	15	16	17	18	19
(*Z*)-3-hexenyl acetate		•	•	•							
(*E*) & (*Z*)-Linalool oxide		•	•	•	•		•	•	•	•	•
Nonanal		•	•	•	•	•		•	•	•	•
Linalool		•									
tDMNT**		•	•	•	•	•	•		•	•	•
Methyl salicylate		•	•	•							
Decanal		•	•	•	•	•	•	•		•	•
β-caryophyllene		•		•	•	•	•	•	•	•	
Germacrene-D		•		•	•	•	•	•	•		
α-farnesene		•									

Fig. 15.2 Upwind flight response (%) of female *Paralobesia viteana* in a flight tunnel to wild grape, *Vitis riparia*, shoots and various synthetic volatiles released by grape shoots (From Cha et al. 2008a)

15.2.4 Life Cycle

Paralobesia viteana overwinters as pupae in leaf litter, both inside and outside vineyards. Moths emerge from late April to June in northern regions (Luciani 1987; Tobin et al. 2002). Emergence of moths in spring usually spans 6 weeks. It is determined largely by genetic factors and temperature. Some overwintering individuals emerge as early as April, while others may delay emergence until July or August, making generations difficult to distinguish (Tobin et al. 2002). The first males begin flight activity a few days before females are present (Tobin et al. 2002). After emergence, females release pheromones to attract males and mating occurs within few days. Females then oviposit on developing buds, florets, and berries (Clark and Dennehy 1988). Eggs hatch in 3–5 days depending on temperature and there are four larval instars (Luciani 1987). The number of generations varies with geographical location based on temperature accumulation, from two to three generations reported

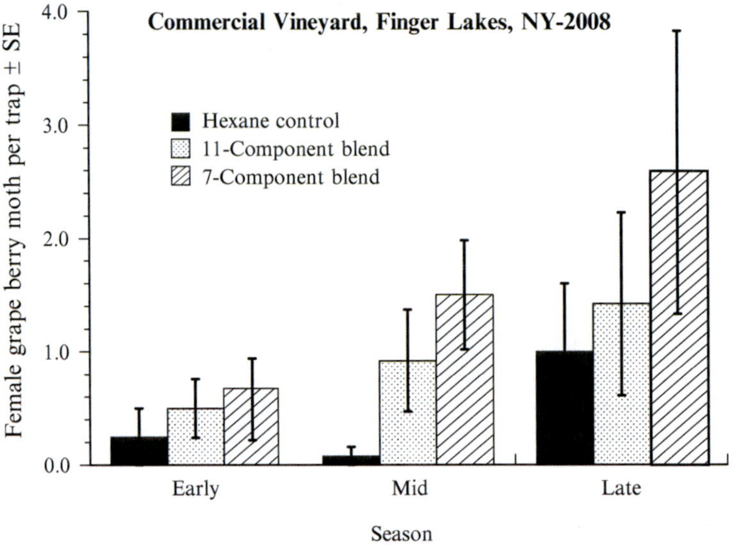

Fig. 15.3 Number of female *P. viteana* captured per panel trap baited with either synthetic host plant lure (an 11-component or 7-component blends) or hexane control at three times during the field season at a commercial vineyard in the Finger Lakes region of New York State

in the Lake Erie region (Ingerson 1920; Gleissner and Worthley 1941) and central New York State (Hoffman et al. 1992), and up to four generations in southern Missouri and Arkansas (Biever and Hostetter 1989). Projections of different climate change scenarios indicate increasing mean surface temperatures on the earth will lead to additional grape berry moth generations, with associated increases in the risk of crop infestation prior to harvest (Tobin et al. 2008; Chen et al. 2011).

The adult moth varies in length from 4 to 6 mm, and is brown with slate-gray patches on its dorsal wing surface, shaped like a saddle (Fig. 15.1a). The adult lifespan of *P. viteana* ranges from 4 to 23 days (Johnson and Hammar 1912), and average male and female longevity at 23°C is 18.5 days. Females of the first flight are active during grapevine bloom, thereby having access to flowers and small berries where eggs are laid singly. The eggs are circular, flattened, but slightly convex, 0.8 mm in diameter, and translucent. Females lay an average of 33 eggs in their lifetime (Luciani 1987), and most egg laying occurs 1 h before and after dark (Clark and Dennehy 1988). The black head capsule of larvae can be observed through the chorion from the third day of egg development. After 4 days, a 1 mm long larva crawls out from a slit it cuts in the chorion using its mandibles. During the first generation, larvae crawl on stems and produce webbing in the developing cluster. Larvae are greenish-yellow when small, becoming dark purple in color at maturity. After feeding within protective webbing or in small berries (Fig. 15.1b), fourth instar larvae leave the clusters to pupate. They do this by cutting a flap in a leaf and webbing it over themselves to spin their cocoon (Johnson and Hammar 1912).

The pupa is blue-green soon after the molt, but turns dark brown until adult emergence (Luciani 1987). Given that a small proportion of the grape flowers are set into berries and most of the egg laying by *P. viteana* occurs later in the season, the economic impact of feeding by first generation larvae of *P. viteana* is minimal. Indeed, Dennehy et al. (1990a) found no effect of early-season *P. viteana* control on infestation levels at harvest.

During the second flight, females lay eggs individually on the small developing berries. Upon hatching, first instar larvae direct themselves toward the stem or to a point of berry contact where they enter the berry, taking 73 min on average to enter the berry (Isaacs and VanderWerp, unpubl. data). Once inside, larvae feed under the berry skin, causing discoloration and premature ripening of the area which increases as larvae continue feeding. Larvae may also exit one berry and enter adjacent berries to feed before pupation. Full-grown larvae (Fig. 15.1c) emerge from the berry in the fourth instar and move to nearby leaves to pupate.

Diapause is induced in eggs and neonate larvae exposed to photoperiods shorter than the critical day length of 14–15 h that triggers development to stop at the pupal stage (Nagarkatti et al. 2001). In the Lake Erie region of Pennsylvania, diapause was induced in an increasing proportion of individuals from 2 July to 13 August, when all individuals entered diapause. Laboratory studies found that 12.8°C temperatures and short days also led to strong diapause induction (Nagarkatti et al. 2001). During long warm summers egg laying has been observed into October in New York and Michigan, suggesting some temperature-based plasticity in the level of diapause induction. Pupae remain in diapause for a variable duration, with emergence in spring ranging over a 6 week period (Tobin et al. 2001). This creates a significant challenge for management of *P. viteana* because of the difficulty of timing insecticide applications accurately to target peak egg laying.

The emergence phenology of this species is characterized by a high degree of variability in time and in space, and many authors have explored the factors controlling emergence timing of *P. viteana.* Spatial variation of vineyard infestation in northern regions may also be driven by low winter temperatures. Pupae die after a single exposure to −24°C, and vineyards with higher infestation are less likely to experience this low winter temperature than less infested ones (Dennehy et al. 1990b). High levels of snow accumulation in winter also favor pupal survival (Dennehy et al. 1990b; Martinson et al. 1991) due to thermal protection. Phenological studies by several authors have attributed this variation to differences in soil types (Pfeiffer et al. 1992), degree-day (DD) accumulation (Hoffman et al. 1992; Tobin et al. 2001), and moth races (Tobin et al. 2003). Coincidence between key phenological stages of grapevine development and events in *P. viteana* phenology suggest that DD models may predict the development of this insect despite the broad emergence period described above. The DD requirement for development from egg to adult was estimated to be 423.9 DD (>8.41°C) (Tobin et al. 2001). Emergence in the field from overwintering pupae started at 148 DD accumulated since 1 January (Tobin et al. 2002).

Recent detailed observations of clusters through the season have revealed the temporal pattern of oviposition by *P. viteana*, helping to explain high levels of

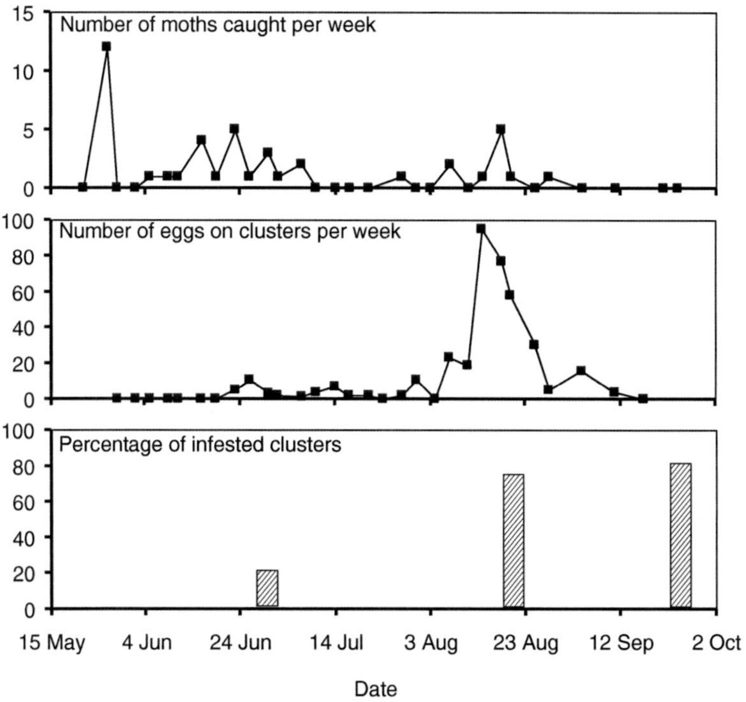

Fig. 15.4 Phenology of *Paralobesia viteana* in high risk commercial *Vitis labrusca* grape vineyards in SW Michigan, US. *Top*: weekly male moth captures in traps baited with sex pheromone; *Middle*: number of fresh eggs laid on clusters each week; *Bottom*: % clusters infested by *P. viteana*

infestation sometimes experienced at harvest time. In southwest Michigan vineyards, oviposition starts in early June, coinciding with early grape bloom. It drops slightly after bloom but continues at a lower level until mid- to late July, intensifying in August close to veraison, and ending in September often before harvest (Teixeira et al. 2009) (Fig. 15.4). In vineyards with high populations, there are no periods without oviposition that would indicate discrete generations, consistent with the earlier findings of Tobin et al. (2002) indicating wide variability in moth emergence from overwintering pupae.

An alternative, though time-consuming, approach for tracking *P. viteana* generations is to trap female *P. viteana* with either malaise or light traps. Data from these trap catches have laid the framework for validation of the Tobin et al. (2001) degree-day model. More data are needed to substantiate the model including local variations, but the peaks of emergence indicated by Malaise and blacklight trap capture data coincide with the peaks generated by using Tobin's model to estimate generational peaks by degree-day calculations. They represent a significant advance over the information provided by the sex pheromone trap catches of males (Fig. 15.5).

Finding a reliable and practical starting point (or biofix) for initiating a phenology model is essential to successfully using the degree-day model to predict given

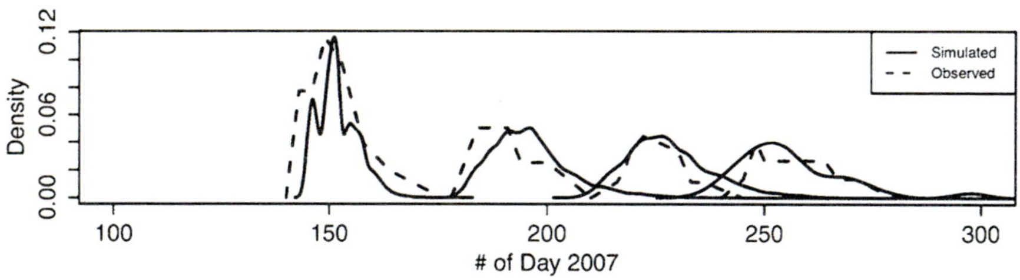

Fig. 15.5 Phenology of female *Paralobesia viteana* in Malaise and light traps for 2007 in northeast, PA, US (*dashed line*) overlain by temperature-dependent model predictions of within-season phenology (*solid line*) (From Chen et al. 2011)

developmental events. Trapping adult female moths, either with light traps, Malaise traps, or panel traps baited with synthetic host volatiles, can provide fairly clear information on female flight activity. Peak captures of females in the spring flight were good predictors of flight activity in subsequent generations. However, these trapping techniques are not currently suitable for widespread adoption by vineyard managers. Further studies are needed to develop more practical methods for detecting the emergence of female *P. viteana*. Until then, traps baited with the sex pheromone which catch male moths will remain the primary method of detecting moth activity.

15.2.5 Reproductive Biology

The presence of a sex pheromone and identification of its main component as (*Z*)-9-dodecenyl acetate (*Z*9-12Ac) was reported by Roelofs et al. (1971). Witzgall et al. (2000) extracted pheromone glands from female grape berry moths and found that each contained approximately 1.2 ng of pheromone on the second to third day of life. Apart from the main component, eight other compounds were identified from pheromone glands: (*Z*) and (*E*)-11 tetradecenyl acetate, (*Z*)-9-dodecyl acetate, dodecanol, (*Z*)-9-dodecenol, tetradecyl acetate, tetradecanol and hexadecanol. Some of these were synergists to the primary pheromone, some had no effect, while (*Z*)-9-dodecenol reduced moth capture in traps (Witzgall et al. 2000). Although no studies have reported the daily timing of pheromone release in grape berry moth, females attain maximum pheromone titres on the second and third days of calling (Witzgall et al. 2000). Mating occurs within the first 3 days of adult life for both sexes and most egg laying occurs on the fourth to seventh day of adult life.

While there have been significant advances towards elucidation of the chemical ecology of grape berry moth, the reproductive biology of this species remains somewhat unclear. For example, it is not known whether *P. viteana* mates once or multiple times and mating behavior has not been studied in detail. Mating on the clusters of vines is expected, as host plants are often used by host-specializing insects to locate mates (Thornhill and Alcock 1983; Landolt and Phillips 1997). The attraction of moths to volatiles from vegetative grapevine tissues described above suggests that moths locate grape patches using this general cue, followed by more local searching for clusters, and subsequently for a mate. Both male and female moths are attracted to grape host volatiles in flight tunnel trials (Cha et al. 2008b) indicating a potential role of the host plant in mate location. The details of this and subsequent courtship behaviors remain to be described and may provide important insights into mechanisms for disruption of the *P. viteana* life cycle.

15.2.6 Distribution in Vineyard Ecosystems

The distribution of *P. viteana* is closely linked to the host plants on which it lays eggs and feeds as larvae, as predicted for specialist phytophages (Schowalter 2000). Most grape agroecosystems in eastern North America consist of vineyards bordered by woodlots, riparian areas unsuited for agriculture, windbreaks made of tree rows, fallow and cultivated fields, or grasses. Surveys of wild grapevine clusters invariably find eggs and larvae of *P. viteana* (Seaman et al. 1990; Nagarkatti et al. 2002a). They are usually most abundant on clusters in the borders of adjacent vineyards (Hoffman and Dennehy 1989; Botero-Garcés and Isaacs 2003). This damage at vineyard borders often results in growers making insecticide applications that are restricted to these areas.

More *P. viteana* are typically caught in traps placed at the edge of vineyards rather than in traps placed inside vineyards (Biever and Hostetter 1989; Hoffman and Dennehy 1989; Trimble et al. 1991; Botero-Garcés and Isaacs 2003). Hoffman and Dennehy (1989) trapped grape berry moths along a transect through different habitats, showing that a greater proportion of moths were caught in deciduous woods at the beginning of the season compared with near harvest when more moths were caught inside the vineyards than anywhere else. Observations in unmanaged vineyards indicate a more even distribution, suggesting that in-vineyard pest control is partially responsible for the observed distribution. Vertical and horizontal sampling of the crop and non-crop habitats within Michigan grape farms (Botero-Garcés and Isaacs 2003) revealed that in adjacent woodlands captures of male *P. viteana* in pheromone traps increased up to 10 m in height but were restricted to the vine canopy within vineyards. It seems that this pest is closely linked in space to the distribution of its host plant.

15.2.7 Dispersal Behavior

Understanding the potential of *P. viteana* moths to move could help elucidate how this species is able to adapt to the changes in abundance of wild and cultivated host plants, and would help with the design of management programs. While some tortricid fruit pests have high flight capacity (Gu and Danthanarayana 1990; Schumacher et al. 1997), *P. viteana* is considered a weak flyer. After releasing fluorescent dust-marked moths in a mature vineyard, Botero-Garcés and Isaacs (2004b) found that male moths were recaptured up to 58 m from the release site while female moths were recaptured 41 m from the release site. Although the proportion of female moths recaptured was low, when they were released in woods and recaptured in adjacent vineyards, the average maximum displacement of females was greater (79 m) than when moths were released in the vineyard habitat (19 m). This suggests that moths will fly greater distances when host resources are scarce. Considering the evolutionary history of this species, the ability to fly between patches of wild grape would help maintain populations in years when spring frost damage or falling trees temporarily prevent fruiting of vines. Movement of this species at the landscape scale has not been studied, but this may provide insights into the importance of large- and small-scale movements for colonization of vineyards. Furthermore, there is little knowledge of how abiotic factors and reproductive status influence flight behavior in this species, although moths are reluctant to fly in the flight tunnel with wind speeds >0.66 m/s (Cha et al. 2008a).

15.2.8 Natural Enemies

A survey of natural enemies of *P. viteana* in Pennsylvania suggested that *Trichogramma minutum* Riley was the only native egg parasitoid with potential for controlling *P. viteana*. However, natural parasitism by *T. minutum* was not dependable since it was found more often on wild *Vitis* spp. in wooded habitats compared with cultivated grapes (Nagarkatti et al. 2002a). Recent availability of highly selective insecticides for *P. viteana* control may provide an opportunity for these egg parasitoids to contribute more to pest population reduction. Parasitoid wasps that attack *P. viteana* have also been studied in New York and Michigan vineyards. Seaman et al. (1990) found three dominant hymenopteran parasitoids (*Trichogramma pretiosum* Riley, *Glypta mutica* Cushman, and *Apanteles polychrosidis* Viereck) on *P. viteana* in three different habitats: wild grapes, organically managed commercial vineyards, and conventionally managed commercial vineyards. Parasitism by the egg parasitoid *T. pretiosum* was greater than other natural enemies (4.5–20.2%), with the highest parasitism occurring in wild habitats, likely due to greater survival in wooded habitats rather than the more open and pesticide treated vineyards. The larval parasitoids *G. mutica* and *A. polychrosidis* caused lower levels of mortality (0.01–6.4% and 0–11.5%, respectively). Combined, the three species caused 12–40% mortality of *P. viteana* (Seaman et al. 1990). In Arkansas vineyards, Williamson and Johnson (2005) found from 3% to 49% parasitism, with braconid and ichneumonid species represented. In Michigan juice grape vineyards, a complex of parasitoids was reared from larvae in infested berries, including the *G. mutica* and *A. polychrosidus* reported previously, but this community also included *Enytus obliteratus* (Cresson) and a *Sinophorus* sp. which was the most common species found (Jenkins 2006).

Generalist natural enemies including lady beetles, lacewings, syrphids, ground beetles, and spiders are common in vineyard ecosystems (Costello and Daane 1999; Williamson and Johnson 2005; Jenkins 2006) but the role of this complex in suppressing *P. viteana* populations has not been studied in detail.

15.3 Management

15.3.1 Pest Status

Grape berry moth evolved in the wooded ecosystems of North America in association with wild grape populations. With the advent of commercial vineyards in the New World planted with European and adapted North American cultivars, *P. viteana* populations had the opportunity to exploit this abundant food source. By the early 1860s, more than 2,428 ha of grapes were grown east of the Mississippi River, and 40 years later the area had increased to 97,128 ha (Hendrickson 1913). Coincident with the increase in grape production, problems with grape berry moth infestation became more significant and the object of study by applied entomologists. Grape berry moth was first reported as an injurious pest in 1869, because of yield losses estimated to be up to 50% in Ohio, Missouri, and southern Illinois (Johnson and Hammar 1912). Isely (1917) stated that the pest status was unclear since there were few economic estimates of grape loss, and there were only erratic patterns of infestation within farms and around grape growing regions.

Management of this species in the 1920s included the few available insecticides and cultural methods based on behavioral and phenological studies of the pest. Pettit (1933) recommended applications of Bordeaux mixture and arsenical poison sprays to protect grape clusters. Cultural control measures included sanitation of vineyards by removal or destruction of trash and leaf litter to reduce pupal densities (Smythe 1913). Pettit (1933) pointed out that neither neglected vineyards nor wild grapes ought to be allowed near vineyards. This prompted the removal of abandoned vineyards by the Michigan Department of Agriculture with the help of growers (Ball and Lovitt 1968). This eradication program removed over 3,000 ha of vineyard in 12 years (Ball and Lovitt 1969, 1971), but there is no record of a resulting decline in pest pressure. Recent studies to examine the effect of removing wild grapevines from woods did not find any consistent reduction in *P. viteana* infestation in adjacent vineyards (Jenkins and Isaacs 2007).

15.3.2 Monitoring and Phenology Prediction

Since the identification of the major component of the sex pheromone in the 1970s, pheromone traps have been used to understand the phenology of *P. viteana*, with the goal of predicting optimal timing of insecticide applications. Pheromone-baited traps may be useful to time management practices against the first generation of grape berry moth, but not subsequent generations (Dennehy et al. 1990a). These researchers noted that moth captures did not correlate with the timing or abundance of cluster infestation in vineyards, making the traps of limited practical use. Instead, they developed a Grape Berry Moth Risk Assessment Protocol (GBMRAP) (Martinson et al. 1991) that guided growers to assign risk to different vineyards based on the history of infestation, winter snow cover, and proximity to woods. Based on this initial risk ranking, if vineyards were low or intermediate then the need to protect clusters was further refined by sampling in July, with a threshold of 6% infested clusters, and again in August in high risk vineyards if infestation exceeded 15% infested clusters. This approach was adopted widely in New York and Pennsylvania vineyards, leading to significant reduction in insecticide use (Dennehy et al. 1990b). Despite this success, the GBMRAP has provided less effective control in recent years, perhaps because of changes in *P. viteana* phenology or because of restrictions on some highly effective insecticides with long residual control (e.g. azinphos-methyl) that helped ensure the success of this approach. Moreover, the GBMRAP was primarily developed for grapes grown for juice (primarily *Labrusca*-based cultivars). Lower economic injury levels should probably be used for wine varieties because of their greater economic value and also because they tend to be more prone to fruit rots.

Detailed studies on the phenology of *P. viteana* (Nagarkatti et al. 2001; Tobin et al. 2001, 2002, 2003) have provided some important insights to the developmental parameters of this species. In particular, elucidation of the base temperature for development (8.41°C) (Tobin et al. 2003), coupled with calculation of the number of DD for completion of a generation, now provide the foundation for a DD model to predict the start of egg laying by the economically important second and third generations. In juice grape vineyards across the northern tier of the eastern United States, entomologists are currently testing a model that uses the timing of wild grape bloom in the spring as the biofix, coinciding with egg laying by the first generation, and egg laying by subsequent generations predicted to start 450 and 900 growing DD (base 8.41°C). In Michigan, based on the timing of wild grape bloom, online weather data are available with an automatically generated table that guides growers to the egglaying of the first and subsequent generations (Enviroweather 2011). A similar system has been developed for New York growers (NEWA 2011). In recent tests to validate this model, use of these timings for cluster protection has provided improved control of *P. viteana* compared with standard management programs (Fig. 15.6).

15.3.3 *Insecticidal Control*

Before the 1980s, 40–70% of New York vineyards were treated with insecticides, which were mainly prophylactic applications owing to the low cost of chemical

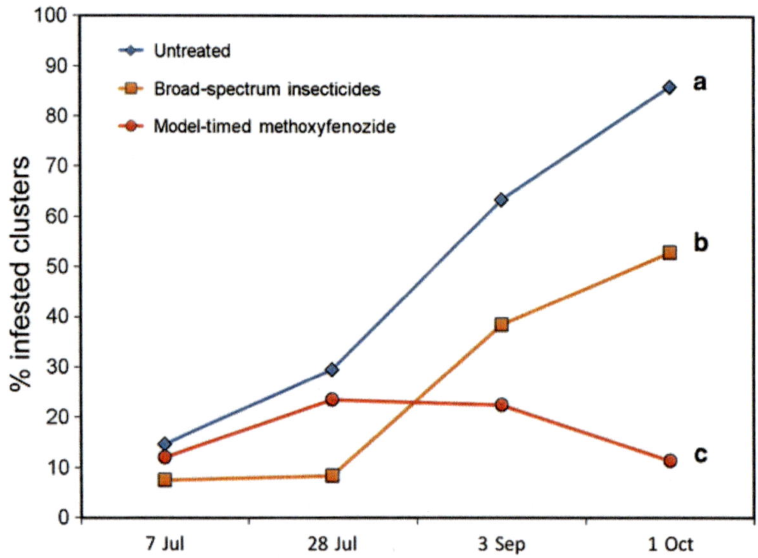

Fig. 15.6 Infestation by *Paralobesia viteana* on vines that were either untreated with insecticide (*blue*), treated with a standard broad-spectrum insecticide program that included a pyrethroid, a carbamate, and an organophosphate insecticide applied for generations 1, 2, and 3, respectively (*orange*), or treated with methoxyfenozide for generations 2 and 3 and timed using a degree-day model (*red*). Pre-harvest collections of berries infested with *P. viteana* larvae from vines in each treatment revealed 94%, 44%, and 14% survival to pupation, respectively. Pre-harvest infestation on October 1 with different letters are significantly different ($P<0.05$)

products and relatively high prices paid for grapes (Dennehy et al. 1990a). These insecticides, such as methyl parathion and azinphosmethyl, were also highly active with long residual activity (Dennehy et al. 1990a; Nagarkatti et al. 2002b). However, implementation of the Food Quality Protection Act of 1996 brought about the loss of registrations for some of the most important insecticides. In addition, resistance to carbaryl has been detected recently in populations of *P. viteana* in New York and Pennsylvania (Nagarkatti et al. 2002b).

The increasing availability of reduced-risk insecticides for use by grape growers against *P. viteana* provides an opportunity for improved control in some cases (Isaacs et al. 2005), particularly if these insecticides are applied with good coverage of berry clusters (Wise et al. 2010). Some new selective insecticides such as methoxyfenozide, an insect growth regulator, may also allow for conservation of natural enemies of *P. viteana*. However, these chemicals have had relatively little adoption until recently and evidence for conservation is lacking in eastern US vineyards. Jenkins and Isaacs (2007) sampled vineyards receiving broad-spectrum insecticide programs compared with others receiving the selective insect growth regulator methoxyfenozide for *P. viteana* control. Lower infestation by *P. viteana* was found in the selective program, but there was no consistent increase in parasitism. The lower levels of survival by *P. viteana* larvae in vineyards managed with reduced-risk insecticides in this study may in part be due to improved timing of insecticide applications. Insecticide applications for this study were based on weekly scouting information, whereas the conventional vineyards were sprayed in response to regional recommendations or the grower's standard spray timing (Jenkins and Isaacs 2007).

Expanding on this research, studies have combined methoxyfenozide which provides long residual activity against *P. viteana* (Isaacs et al. 2005) with application time based on degree-day accumulation. By targeting the start of increased egg laying in the mid to late period of the season and thereby treating a high proportion of the eggs laid during the year, control equivalent to a broad-spectrum insecticide program was achieved (Teixeira et al. 2009). More recent studies have employed applications based on degree-day timings to improve alignment with egg laying of both the second and third generations. This builds on the earlier studies of *P. viteana* phenology by Tobin et al. (2003). Because egg laying by the first generation of *P. viteana* coincides with bloom of the wild grape ancestral host, *V. riparia*, applications of methoxyfenozide have been tested at 450 and 900 growing DD after wild grape bloom for improving control compared to standard calendar-based programs for *P. viteana* control. The lower infestation and lower survival of *P. viteana* in grapes treated using this program (Fig. 15.6), coupled with the lower cost, provide compelling reasons for adoption of methoxyfenozide in vineyards infested by this pest. Our recent continuation of this research through expansion into commercial settings has included spatially focused applications of methoxyfenozide within the canopy and within vineyards. By making small adjustments to nozzle settings and by considering the regions of farms that contain infestation levels worth controlling, vineyard managers can improve control and reduce costs.

The phenology prediction system described above employs wild grape bloom as a biofix because this occurs after spring temperature fluctuations and because it coincides with egg laying by the first generation. An alternative approach is to use the start of male flight, detected using pheromone traps, to set biofix. This approach was recommended in Arkansas vineyards to time insecticide applications within an IPM program (Lewis and Johnson 1999). Using this system, first generation larvae are controlled at vineyard borders at 400–600 DD (base 10°C) after first moth capture, followed by re-application at 1,250 and 2,250 DD after biofix if additional moths are caught in traps or if the infestation of clusters increases by 1%.

15.3.4 Pheromone Mating Disruption

The application of sex attractant pheromones to crops for disruption of mating and reducing pest infestation has been successful in many fruit systems (Cardé and Minks 1995). This is an environmentally-safe alternative to insecticides because the materials are innocuous to non-target organisms and have very low human toxicity. Research on the use of grape berry moth sex pheromone to prevent grape infestation has been ongoing for several decades, starting in 1971 with the identification of (Z)-9 dodecenyl acetate as the major component of the pheromone (Roelofs et al. 1971). Taschenberg et al. (1974a, b) used this compound in mass trapping and mating disruption for control of grape berry moth. Their encouraging results with mating disruption were followed by other trials evaluating different formulations of the pheromone for release in vineyards, such as microcapsules and hollow fibers (Taschenberg and Roelofs 1977). Further trials were conducted after improvements in pheromone synthesis and release technology with the introduction of polyethylene tube dispensers, showing economically acceptable control of *P. viteana* in New York, and Ontario but also occasional insufficient control (Trimble et al. 1991; Trimble 1993). Recent trials using a commercial product with similar polyethylene tube technology (Isomate®-GBM) in vineyards in Michigan, New York and Pennsylvania found no significant improvement in control when twist-ties were added over the top of an insecticide-based control program (Isaacs et al. 2012). A different formulation of microencapsulated pheromone for mating disruption (3M Sprayable Pheromone®), supplemented with insecticide sprays when necessary, was found to provide acceptable control (Trimble et al. 2003; Trimble 2007). A major advantage of this formulation was that it could be applied using an airblast sprayer, but this product is no longer being manufactured.

Another product for mating disruption that has recently been evaluated is SPLAT-GBM® (ISCA Technologies, Riverside, CA), a viscous wax matrix containing 3% of pheromone. When applied manually in 1 ml drops to vineyards at a rate of 400 drops/ha, SPLAT-GBM® caused a 27% reduction in infestation compared to nearby vineyards not treated with pheromone (Jenkins and Isaacs 2008). The development of a mechanical applicator allowed larger-scale testing of SPLAT-GBM® application to vineyards. The applicator was designed to treat two grape rows at a time and was mounted in the bed of a two-wheel drive 'Gator' ATV (John Deere, Moline, IL). The system consisted of a tank holding SPLAT-GBM® connected to an air tank reservoir pressurized by a 12 V portable air compressor. The air tank was custom-fitted with a regulator to control air pressure outflow. The holding tank was connected through spray hoses to two solenoid valves controlled by a programmable relay timer. The solenoids allowed precise control over the size of the drops and the rate of application. When driven at 16 km/h in the vineyard, the applicator enabled application of SPLAT-GBM® to vineyards at a rate of 4 ha/h, approximately 10-fold faster than manual application. Two applications of SPLAT-GBM® at 2.5 kg/ha in 0.8 g drops for a total density of 3,089 drops/ha resulted in a 50% reduction in cluster damage. Comparison of different rates and timings of this new formulation provides guidance for optimal deployment strategies (Teixeira et al. 2010). Current field trials are aiming to integrate spatially-specific applications of this tactic with insecticide tools for reduced cost programs. Further progress in the development of mating disruption control tactics for grape berry moth is hindered by a lack of data on the behavioral mechanisms underlying mating disruption. This knowledge can be gathered with behavioral observations of moths reacting to mating disruption products (Stelinski et al. 2004), or by conducting field trials to compare observed patterns of moth catch to the pattern expected when disruption is caused by a specific behavior (Miller et al. 2006).

15.3.5 IPM Program Implementation

There have been some major advances toward developing sustainable and cost-effective IPM programs for *P. viteana* in recent years. However, losses to this pest continue to challenge producers by reducing yield. In severe cases it causes rejections of grape loads by processors and wineries from heavily-infested clusters (Fig. 15.1d). Implementation of IPM programs can help reduce crop infestation and minimize the risk that *P. viteana* and other insects reach economic injury levels. Growers in New York saved $46.28/ha in insecticide costs by scouting for *P. viteana* and leafhoppers (Snyder et al. 1992). In Michigan regular vineyard scouting led to an average saving of one insecticide spray (S. Van Timmeren and R. Isaacs, unpubl. data).

Vineyard IPM programs have recently been promoted within an overall emphasis on food safety and as part of sustainability programs that guide environmental, social, and economic responsibility (Baldwin 2009). Thus, management of this key pest is increasingly considered within broader whole-farm sustainability programs, such as Vine Balance in New York and the Grape*A*Syst program in Michigan. These programs aim to advance sustainability goals, but they are also being used to help growers, wineries and juice cooperatives show the public and food distributors that they are making measurable progress towards environmental goals. Continued investment in agricultural extension programs that explain and demonstrate new pest management approaches will be needed to ensure widespread adoption of these tactics to minimize the impact of *P. viteana* on viticulture in this region.

15.4 Conclusion

Grape berry moth has been a pest of cultivated grapes in North America since the early commercialization of this crop in the New World. Despite this long association, there remain some key aspects of the biology of this species that are not well understood. Details of the mating biology remain unclear, with the courtship behaviors, location of mating, and relative role of sex pheromones and plant volatiles as areas of potential future exploration. In biological control, parasitoid associations with *P. viteana* larvae have been elucidated in some regions but are unreported for much of its geographic range, and the role of generalist predators in regulating populations is an area for potentially fruitful research. New developments in linking weather conditions with pest development, and delivery of this information online for growers to use, are expected to reduce harvest losses to *P. viteana*. Coupling these tools for improved timing of crop protection products with new selective insecticides such as insect growth regulators and effective mating disruption approaches should provide the combined benefits of reduced crop damage, increased biological control, reduced worker risk, and improved environmental quality in vineyards. Testing and documenting these potential benefits remains a goal of grape IPM programs in eastern North America, and we expect that this will increasingly need to be considered at the farm-wide scale to take into account the movement of pests and natural enemies through agricultural landscapes.

Acknowledgements Our thanks to Keith Mason, Steve Van Timmeren, Elly Maxwell, Mark VanderWerp, Tim Weigle, Steve Hesler, Jody Timer and Andy Muza, for their collaboration on grape berry moth research. We also thank the many grape growers in Michigan, New York, and Pennsylvania for access to their farms and for their willingness to collaborate on applied research projects. Funding support is gratefully acknowledged from USDA-CSREES Crops at Risk, Viticulture Consortium East, and North Central IPM Programs; Michigan State University's Project GREEEN; National Grape Cooperative; and the Michigan Grape and Wine Industry Council.

References

Baldwin CJ (2009) Sustainability in the food industry. Wiley-Blackwell, Ames

Ball BD, Lovitt DF (1968) Plant pest control programs: orchard and vineyard inspection and removal. Michigan Department of Agriculture, Lansing

Ball BD, Lovitt DF (1969) Plant pest control programs: neglected orchard and vineyard removal. Michigan Department of Agriculture, Lansing

Ball BD, Lovitt DF (1971) Neglected orchard and vineyard removal. Michigan Department of Agriculture. Pest Control Programs, Lansing

Biever KD, Hostetter DL (1989) Phenology and pheromone trap monitoring of the grape berry moth, *Endopiza viteana* Clemens (Lepidoptera: Tortricidae) in Missouri. J Entomol Sci 24:472–481

Bostanian NJ, Vincent C, Goulet H, Lesage L, Lasnier J, Bellemare J, Mauffette Y (2003) The arthropod fauna of Quebec vineyards with particular reference to phytophagous arthropods. J Econ Entomol 96:1221–1229

Botero-Garcés N, Isaacs R (2003) Distribution of grape berry moth, *Endopiza viteana* (Lepidoptera: Tortricidae), in natural and cultivated habitats. Environ Entomol 32:1187–1195

Botero-Garcés N, Isaacs R (2004a) Influence of uncultivated habitats and native host plants on cluster infestation by grape berry moth, *Endopiza viteana* Clemens (Lepidoptera: Tortricidae), in Michigan vineyards. Environ Entomol 33:310–319

Botero-Garcés N, Isaacs R (2004b) Movement of the grape berry moth, *Endopiza viteana*: displacement distance and direction. Physiol Entomol 29:443–452

Brown JW (2006) Scientific names and pest species in Tortricidae (Lepidoptera) frequently cited erroneously in the entomological literature. Am Entomol 52:182–189

Cardé RT, Minks AK (1995) Control of moth pests by mating disruption. Annu Rev Entomol 40:559–585

Cha DH, Nojima S, Hesler SP, Zhang A, Linn CE Jr, Roelofs WL, Loeb GM (2008a) Identification and field evaluation of grape shoot volatiles attractive to female grape berry moth (*Paralobesia viteana*). J Chem Ecol 34:1180–1189

Cha DH, Hesler SP, Moser CL, Nojima S, Linn CE, Roelofs WL, Loeb GM (2008b) Flight tunnel responses of female grape berry moth (*Paralobesia viteana*) to host plants. J Chem Ecol 34:622–627

Cha DH, Linn CE Jr, Teal PEA, Zhang A, Roelofs WL, Loeb GM (2011) Eavesdropping on plant volatiles by a specialist moth: significance of ratio and concentration. PLoS One 6:e17033. doi:10.1371/journal.pone.0017033

Chen S, Fleischer SJ, Tobin PC, Saunders MC (2011) Projecting insect voltinism under high and low greenhouse gas emission conditions. Environ Entomol 40:505–515

Clark LG, Dennehy TJ (1988) Oviposition behavior of grape berry moth. Entomol Exp Appl 47:223–230

Costello MJ, Daane KM (1999) Abundance of spiders and insect predators on grapes in central California. J Arachnol 27:531–538

Dennehy TJ, Hoffman CJ, Nyrop JP, Saunders MC (1990a) Development of low-spray, biological and pheromone approaches for control of grape berry moth, *Endopiza viteana* Clemens, in the eastern United States. In: Bostanian NJ, Wilson L, Dennehy TJ (eds) Monitoring and integrated management of arthropod pests of small fruit crops. Intercept Ltd., Andover, pp 261–282

Dennehy TJ, Hoffman CJ, Kamas JS (1990b) Grape berry moth risk assessment reduces insecticide use 70 percent. Chautauqua County Agricultural News 72. Cornell Cooperative Extension, Geneva, NY

Enviroweather (2011) Enviroweather – weather–based pest, natural resources, and production management tools. http//www.enviroweather.msu.edu

Fergusson-Kolmes LA, Dennehy TJ (1993) Differences in host utilization by populations of North American grape phylloxera (Homoptera: Phylloxeridae). J Econ Entomol 86:1502–1511

Galet P (1979) A practical ampelography. University Press, Cambridge

Gleissner BD, Worthley HN (1941) Evidence for a third brood of the grape berry moth, *Polychrosis viteana* Clemens, in the great lakes region. J Econ Entomol 34:426–431

Gu H, Danthanarayana W (1990) Age-related flight and reproductive performance of the light brown apple moth, *Epiphyas postvittana*. Entomol Exp Appl 54:109–115

Hendrickson AH (1913) Forty-second annual report of the secretary of the state horticultural society of Michigan for the year 1912, Lansing, MI

Hoffman CJ, Dennehy TJ (1989) Phenology, movement, and within-field distribution of the grape berry moth, *Endopiza viteana* (Clemens) (Lepidoptera: Tortricidae), in New York vineyards. Can Entomol 121:325–335

Hoffman CJ, Dennehy TJ, Nyrop JP (1992) Phenology, monitoring, and control decision components of the grape berry moth (Lepidoptera: Tortricidae) risk assessment program in New York. J Econ Entomol 85:2218–2227

Ingerson HG (1920) Life history of the grape berry moth in northern Ohio. U S Dep Agric Bull 911:1–38

Isaacs R, Mason KS, Maxwell E (2005) Stage-specific control of grape berry moth, *Endopiza viteana* (Clemens) (Lepidoptera: Tortricidae), by selective and broad-spectrum insecticides. J Econ Entomol 98:415–422

Isaacs R, Mason KS, Teixeira LAF, Loeb G, Hesler S, Weigle T, Muza A, Timer J, Saunders M (2012) Comparison of three dispenser distribution patterns for pheromone mating disruption of *Paralobesia viteana* in vineyards. J Econ Entomol (in press)

Isely D (1917) Control of the grape-berry moth in the Erie-Chautauqua grape belt. Washington, D.C. U S Dep Agric Bull 550:1–44

Jenkins PE (2006) Control of the grape berry moth, *Paralobesia viteana*, using reduced-risk insecticides, cultural controls, and conservation of natural enemies. M.S. thesis, Michigan State University, East Lansing

Jenkins PE, Isaacs R (2007) Reduced-risk insecticides for control of grape berry moth (Lepidoptera: Tortricidae) and conservation of natural enemies. J Econ Entomol 100:855–865

Jenkins PE, Isaacs R (2008) Mating disruption of *Paralobesia viteana* (Lepidoptera: Tortricidae) in vineyards using pheromone deployed in SPLAT-GBM™ wax droplets. J Chem Ecol 34:1089–1095

Johnson F, Hammar AG (1912) The grape berry moth. Washington, D.C. U S Dep Agric Bull 116:15–71

Landolt PJ, Phillips TW (1997) Host plant influences on sex pheromone behavior of phytophagous insects. Annu Rev Entomol 42:371–391

Lewis BA, Johnson DT (1999) Grape berry moth management program. Proc Okla Ark Hortic Ind Show 18:56–60

Luciani MA (1987) The biology of the grape berry moth, *Endopiza Viteana*, Clemens (Lepidoptera: Tortricidae) in southern Ontario. Ph.D. dissertation, University of Guelph, Ontario

Martinson TE, Hoffman CJ, Dennehy TJ, Kamas JS, Weigle T (1991) Risk assessment of grape berry moth and guidelines for the management of the eastern grape leafhopper. N Y Food Life Sci Bull 138:1–10

Miller JR, Gut LJ, de Lame FM, Stelinski LL (2006) Differentiation of competitive vs. non-competitive mechanisms mediating disruption of moth sexual communication by point sources of sex pheromone (Part 2): case studies. J Chem Ecol 32:2115–2143

Moore MO (1991) Classification and systematics of eastern North American *Vitis* L. (Vitaceae) north of Mexico. Sida Contrib Bot 14:339–367

Morano LD, Walker MA (1995) Soils and plant communities associated with three *Vitis* species. Am Midl Nat 134:254–263

Nagarkatti S, Tobin PC, Saunders MC (2001) Diapause induction in the grape berry moth (Lepidoptera: Tortricidae). Environ Entomol 30:540–544

Nagarkatti S, Muza AJ, Saunders MC, Tobin PC (2002a) Role of the egg parasitoid *Trichogramma minutum* in biological control of the grape berry moth, *Endopiza viteana*. BioControl 47:373–385

Nagarkatti S, Tobin PC, Muza AJ, Saunders MC (2002b) Carbaryl resistance in populations of grape berry moth (Lepidoptera: Tortricidae) in New York and Pennsylvania. J Econ Entomol 95:1027–1032

NEWA (2011) Network for environment and weather applications, New York State Integrated Pest Management Program. http://www.newa.cornell.edu

Pettit RH (1933) The principal grape insects of Michigan. Agric Exp Stn Mich State Coll Agric Appl Sci Spec Bull 239:3–18

Pfeiffer DG, Boucher TJ, Lachance MW, Killian JC (1992) Entomological research in Virginia (USA) vineyards. In: Bostanian NJ, Wilson LT, Dennehy T (eds) Monitoring and integrated management of arthropod pests of small fruit crops. Intercept Ltd., Andover, pp 45–61

Roelofs WL, Tette JP, Taschenberg EF, Comeau A (1971) Sex pheromone of the grape berry moth: identification by classical and electroantennogram methods, and field tests. J Insect Sci 17:2235–2243

Schowalter TD (2000) Insect ecology: an ecosystem approach. Academic, San Diego

Schumacher P, Weyeneth A, Weber D, Dorn S (1997) Long flights in *Cydia pomonella* L. (Lepidoptera: Tortricidae) measured by a flight mill: influence of sex, mated status and age. Physiol Entomol 22:149–160

Seaman AJ, Nyrop JP, Dennehy TJ (1990) Egg and larval parasitism of the grape berry moth (Lepidoptera: Tortricidae) in three grape habitats in New York. Environ Entomol 19:764–770

Slingerland MV (1904) The grape berry moth. Ithaca, N.Y. Cornell Univ Bull 223:41–80

Smythe RA (1913) Forty-second annual report of the secretary of the state horticultural society of Michigan for the year 1912, Lansing, MI

Snyder DP, Weigle TH, White GB (1992) Economics of integrated pest management practices for insects in grape production. Cornell University. Coll Agric Life Sci 92:1–9

Stelinski LL, Pelz KS, Liburd OE (2004) Field observations quantifying attraction of the parasitic wasp, *Diachasma alloeum* (Hymenoptera: Braconidae) to blueberry fruit infested by the blueberry maggot fly, *Rhagoletis mendax* (Diptera: Tephritidae). Fla Entomol 87:124–129

Taschenberg EF, Roelofs WL (1977) Mating disruption of the grape berry moth, *Paralobesia viteana*, with pheromone released from hollow fibers. Environ Entomol 6:761–763

Taschenberg EF, Cardé RT, Hill A, Tette JP, Roelofs WL (1974a) Sex pheromone trapping of the grape berry moth. Environ Entomol 3:192–194

Taschenberg EF, Cardé RT, Roelofs WL (1974b) Sex pheromone mass trapping and mating disruption for control of redbanded leafroller and grape berry moths in vineyards. Environ Entomol 3:239–242

Teixeira LAF, Mason KS, Isaacs R (2009) Control of grape berry moth (Lepidoptera: Tortricidae) in relation to oviposition phenology. J Econ Entomol 102:692–698

Teixeira LAF, Mason K, Mafra-Neto A, Isaacs R (2010) Mechanically-applied wax matrix (SPLAT-GBM) for mating disruption of grape berry moth (Lepidoptera: Tortricidae). Crop Prot 29:1514–1520

Thiéry D, Moreau J (2005) Relative performance of European grapevine moth (*Lobesia botrana*) on grapes and other hosts. Oecologia 143:548–557

Thornhill R, Alcock J (1983) The evolution of insect mating systems. Harvard University Press, Boston

Tobin PC, Nagarkatti S, Saunders MC (2001) Modeling development in grape berry moth (Lepidoptera: Tortricidae). Environ Entomol 30:692–699

Tobin PC, Nagarkatti S, Saunders MC (2002) Diapause maintenance and termination in grape berry moth (Lepidoptera: Tortricidae). Environ Entomol 31:708–713

Tobin PC, Nagarkatti S, Saunders MC (2003) Phenology of grape berry moth (Lepidoptera: Tortricidae) in cultivated grape at selected geographic locations. Environ Entomol 32:340–346

Tobin PC, Nagarkatti S, Loeb G, Saunders MC (2008) Historical and projected interactions between climate change and insect voltinism in a multivoltine species. Glob Change Biol 14:951–957

Trimble RM (1993) Efficacy of mating disruption for controlling the grape berry moth, *Endopiza viteana* (Clemens) (Lepidoptera: Tortricidae), a case study over three consecutive growing seasons. Can Entomol 125:1–9

Trimble RM (2007) Comparison of efficacy of pheromone dispensing technologies for controlling the grape berry moth (Lepidoptera: Tortricidae) by mating disruption. Environ Entomol 100:1815–1820

Trimble RM, Pree DJ, Vickers PM, Ker KW (1991) Potential of mating disruption using sex-pheromone for controlling the grape berry moth, *Endopiza viteana* (Clemens) (Lepidoptera: Tortricidae), in Niagara peninsula, Ontario vineyards. Can Entomol 123:451–460

Trimble RM, Vickers PM, Nielsen KE, Barinshteyn G (2003) Sprayable pheromone for controlling the North American grape berry moth by mating disruption. Agric For Entomol 5:263–268

Williamson JR, Johnson DT (2005) Effects of grape berry moth management practices and landscape on arthropod diversity in grape vineyards in the southern United States. HortTechnology 15:232–238

Wise JC, Jenkins PE, Schilder AM, Vandervoort C, Isaacs R (2010) Sprayer type and water volume influence pesticide deposition and control of insect pests and diseases in juice grapes. Crop Prot 29:378–385

Witzgall P, Bengtsson M, Trimble RM (2000) Sex pheromone of grape berry moth (Lepidoptera: Tortricidae). Environ Entomol 29:433–436

Chapter 16
Grape Root Borer

J. Christopher Bergh

16.1 Introduction

Over a century ago, a statement by the observant entomologist F. E. Brooks from West Virginia, the US, captured one of the fundamental issues underlying the infestation of vineyards in the eastern US by larvae of the grape borer, *Vitacea polistiformis* (Harris): 'So inconspicuous is the insect itself, and its manner of working, that a vineyard may be suffering greatly from its attacks and yet those who have the care of the vines remain entirely ignorant of the cause of the trouble' (Brooks 1907). Seven decades later, Dutcher and All (1979a) reiterated this ongoing problem, stating that 'Due to the cryptic nature of the grape root borer larva and the chronic, yet pernicious impact of larval feeding on vine vigor, infestations are often not noticed until severe damage has occurred'. Despite long-standing recognition of the potential threat from grape root borer (reviewed in Brooks 1907) and a considerable body of published research and observations on aspects of its biology, behavior and management, the deleterious effects from this troublesome pest continue to plague many eastern US vineyards. This is likely due to a combination of factors, not least of which is the insidious nature of the development of a grape root borer infestation. Detection of, and early intervention against infestation requires experience and education, that many growers or crop consultants do not possess and a continuing level of diligence that many have not applied. Furthermore, the expansion of the eastern US wine grape industry since the 1970s was primarily with European grape (*Vitis vinifera* L.) varieties and their hybrids. Many of these were grafted to rootstocks developed from crosses of American grape species considered susceptible to

J.C. Bergh (✉)
Alson H. Smith, Jr. Agricultural Research and Extension Center,
Virginia Polytechnic Institute and State University, 595 Laurel Grove Road,
Winchester, VA 22602, USA
e-mail: cbergh@vt.edu

N.J. Bostanian et al. (eds.), *Arthropod Management in Vineyards: Pests, Approaches, and Future Directions*, DOI 10.1007/978-94-007-4032-7_16,

grape root borer. Such factors would be expected to increase the incidence and severity of problems from the pest, especially as vineyards age. Perhaps most importantly however, the development and delivery of monitoring and management tactics for grape root borer based on modern IPM principles have not kept pace with those now considered routine for managing many other lepidopteran pests of agricultural crops. A truly integrated strategy for managing grape root borer is not presently available. This, I believe, is due to some fundamental knowledge gaps concerning the ecology and behavior of the grape root borer that have impeded the development of the best management practices.

There are several review articles on grape root borer (e.g. Dutcher and All 1979b; Olien et al. 1993), including an annotated bibliography of the earlier literature (Williams and Snow 1991). This chapter will provide only a brief overview of its biology and pest status. A review of management recommendations for grape root borer will be followed by a discussion of the current status of research on entomopathogenic nematodes and mating disruption as potential control options. Finally, I will identify the aspects of its biology, ecology and behavior that are poorly understood but important to the development and/or optimization of new or existing monitoring and management tactics, and will highlight research avenues and grower education activities that are relevant to these goals.

16.2 Grape Root Borer Biology and Pest Status

Endemic to the eastern United States, Snow et al. (1991) reported grape root borer captures in sex pheromone traps from Michigan to Florida and west to Missouri. Considered oligophagous on members of the Vitaceae (Brooks 1907; All et al. 1987), infestations in commercial vineyards are thought to originate from moths that developed on wild grapevines. Unlike the females of many other clearwing moths, which tend to oviposit on or near specific plant parts serving as larval food, grape root borer is much less discriminating in its deposition of eggs. Brooks (1907) observed oviposition on vine wood and foliage but most commonly on non-host plants growing in vine rows. This relatively unselective oviposition behavior may be due primarily to the subterranean location of larval food. Upon hatching, larvae must burrow down through the soil to find, and as their common name implies, bore into grape roots. During their development, grape root borer larvae grow considerably (Fig. 16.1). Their average length and width increase from 2.4 to 29 mm and from 0.4 to 6 mm, respectively (Bambara and Neunzig 1977); their feeding channels, which are typically packed densely with reddish frass (Fig. 16.2), increase in circumference as they mine roots of increasing diameter from distal locations toward the vine crown (Dutcher and All 1978a) (Fig. 16.3). Upon the cessation of feeding, larvae leave the roots, move up through the soil and construct cocoons beneath the soil surface, within which they pupate. In commercial vineyards, adult moths emerge from the soil in the morning, leaving a relatively large, amber colored pupal exuvia protruding from the soil or lying on it (Fig. 16.4). Teneral moths typically walk to vines

Fig. 16.1 Late instar grape root borer larva in grape root

Fig. 16.2 Excavated grape root showing the dense, *reddish* frass that packs feeding channels

and sit on the lower trunk for some period prior to moving up into the canopy (Fig. 16.5), where females call and mating occurs in the afternoon (Dutcher and All 1978b). Mated females deposit the majority of their eggs during the first 2 days after mating, laying 354 eggs on average, from which larvae hatch in about 14 days (Dutcher and All 1978b). J. R. Meyer (pers. comm.) found that the developmental duration of larvae on potted vines varied from 1 year in Georgia to up to 3 years in Ohio, with intermediate durations in North Carolina.

The use of sex pheromone traps for grape root borer has revealed substantial variation among geographic regions in the seasonal period of its flight activity. In the northern and central portions of its range, captures typically begin in late June

Fig. 16.3 Grape root showing grape root borer feeding channel from which frass has been removed

Fig. 16.4 Grape root borer pupal exuvia

or early July, peak between mid-July and mid-August and end by early September (Snow et al. 1991), whereas in Florida, initial and peak captures have ranged from May through early August and from mid-August through early October, respectively (Snow et al. 1991; Webb et al. 1992; Weihman and Liburd 2007). The protracted flight of grape root borer in Florida, which may extend into December

Fig. 16.5 Teneral grape root borer female expanding her wings at the base of a grape vine

(Webb et al. 1992), has been attributed to its 1-year generation time in that area (J. R. Meyer, pers. comm.). Although pheromone traps deployed in commercial plantings have been useful to detect the presence and relative abundance of grape root borer, the relationship between captures in traps and the infestation status of vineyards remains unclear (see Sect. 16.5.4.3). Pheromone trap-based surveys of grape root borer in Virginia (Bergh et al. 2005) and Florida (Weihman and Liburd 2007) showed that the pest is widely distributed among commercial vineyards, posing a risk to many plantings.

While grape root borer is considered a significant pest in many States in the eastern US, it has been especially problematic in the southeastern portion of its range. Among eight states in this region that developed Crop Profiles for wine grapes between 1999 and 2008, grape root borer was identified as the most serious pest in North Carolina and a major pest in Pennsylvania, Ohio, Tennessee, and Arkansas (USDA 2011). Grape root borer was identified as a pest of concern in the North Central Region Grape Industry's 2007 Pest Management Strategic Plan (USDA 2011).

Damage from larval feeding on roots can be expressed in a number of ways. Infested vines can show a gradual reduction in overall vigor and productivity that can manifest as reduced shoot growth, smaller leaves, fewer bunches, smaller berries and vine death (Dutcher and All 1979a). However, since some of these symptoms are commonly associated with pathological conditions, such as systemic viral or fungal infections, or over-cropping of vines and other forms of vineyard mismanagement, they are not reliable indicators of grape root borer damage and the lack of distinct symptoms further hampers effective management.

16.3 Management Options for Grape Root Borer

Compared with many other insect pests, management options for grape root borer remain extremely limited, and while some alternative tactics show promise for the future (see Sect. 16.4), recommendations for its control have not changed significantly in many years. Given that the developmental duration of larval grape root borer exceeds 1 year in much of its range, infested vines may harbor larvae from overlapping generations and in these areas any management approach must be implemented for at least two consecutive years.

16.3.1 Cultural Tactics

16.3.1.1 Soil Mounds and Synthetic Barriers

Brooks (1907) suggested that grape root borer might be managed through cultivation of the soil around the base of vines after pupation, thereby preventing adult emergence by burying pupae or exposing them to surface conditions. Sarai (1970) used a hoe pulled by a tractor to create a ridge of soil in vine rows in early July and reported that the number of moths collected from the treated plot was reduced by 85%. Sarai (1970) also compared the effect of pupal burial depth on adult emergence in the laboratory and concluded from the two studies that a soil ridge at least 19 cm high and 60 cm wide would provide effective control. All et al. (1985) constructed 0.3 m high × 1.0 m wide soil mounds in June in replicated plots in a Georgia vineyard and reported an 83% reduction in the number of pupal exuviae recovered from them, compared with non-mounded vines, 65 days later. Sarai (1970) noted the importance of constructing mounds between the period of pupation and adult emergence, since larvae leaving roots after mounding would pupate near the mound surface. For this reason, Webb and Mortensen (1990) stated that mounding would likely not be effective in Florida, where the annual period of grape root borer emergence is prolonged. To ensure the effectiveness of this tactic in areas where larvae feed for more than 1 year, mounds must be removed at the end of the season and then re-built at the beginning of the following season. This also prevents the deleterious effects of scion rooting in grafted vines and the growth of potentially susceptible roots into mounds (All et al. 1987). Although apparently an effective tactic, soil mounding is not widely practiced by grape growers, requiring time and energy, precluding the use of cover crops in vine rows and likely increasing soil erosion on sloped terrain. Furthermore, soil cultivation around the base of vines may not be practical for all grape species or rootstocks, since some (e.g., muscadines, Olien 1990) produce shallow roots that may be susceptible to damage.

In an approach related to soil mounding, plastic or synthetic woven barriers were installed in vine rows (Attwood and Wylie 1963) and around the base of potted vines (Yonce 1995) to prevent larval penetration of the soil and/or adult emergence. Although the results from these studies suggested that synthetic barriers may have some utility, this tactic has not been embraced by growers, likely due to the cost of purchasing, installing and maintaining them and their impermanence.

16.3.1.2 Weed Control

Since first instar grape root borer larvae are quite susceptible to desiccation (Sarai 1972), weed control in vine rows has been thought to help reduce grape root borer infestations by creating hotter and drier surface and soil conditions (Olien et al. 1993). However, Townsend (1991) subjected vines to treatments including bare soil strips, bark or hay mulch and grass/weed cover with and without irrigation and reported no significant differences among treatments in the total number of pupal exuviae collected over 5 years. In Virginia, J. C. Bergh (unpubl. data) has collected numerous pupal exuvia from vineyard blocks in which wide weed-free strips have been maintained for many years. Annual variations in rainfall and temperature during the period of peak oviposition and larval eclosion may have a greater impact on the survival and establishment of young larvae than weed control. Weed management cannot be considered a stand-alone management solution for grape root borer.

16.3.1.3 Wild Vine Removal

Many eastern US vineyards are in close proximity to native forests where wild grape is common and often prolific. The removal of these potential hosts to reduce pest pressure has been attempted by some growers. However, while intuitively appealing as a cultural control strategy, the effectiveness of this laborious process has not been demonstrated experimentally and its practicality likely varies because of differences among vineyards in terrain, restrictions associated with land ownership or the time at which this occurs relative to when vineyards are established. Furthermore, the relative suitability of different native *Vitis* spp. as hosts for grape root borer is unknown, adding further uncertainty to the potential effectiveness of wild vine removal.

16.3.1.4 Host Plant Resistance

There are no rootstocks or cultivars that have been unequivocally demonstrated to provide protection from grape root borer based on antibiosis or antixenosis resistance mechanisms. *Post hoc* measurements of root damage, larval survivorship or numbers of pupal exuvia recovered from a range of cultivars or rootstocks in established vineyards as well as evaluations of larval survival and root damage in potted vines (Alderz and Hopkins 1981; Webb and Mortensen 1990) have shown some relative differences in susceptibility. Although Webb and Mortensen (1990) concluded that further evaluation of potential resistance in the Florida leatherleaf grape, *Vitis shuttleworthii* House, was warranted, the last two decades have not produced new research on potential sources of resistance to grape root borer (P. Cousins, pers. comm.).

16.3.2 Biological Control

Various authors have identified predators, parasites and pathogens associated with the different life stages of grape root borer (reviewed in Olien et al. 1993). Dutcher and All (1978c) showed that predation on cohorts of sentinel eggs in an insecticide-treated and an untreated vineyard was 11.6% and 61.7%, respectively. They concluded that the insecticide program had reduced predator populations, although the identity and abundance of potential predators was not determined. Aside from augmentative releases of certain nematodes (see Sect. 16.4.1), natural enemies of grape root borer are not generally considered to contribute substantially to the suppression of infestations, despite the fact that current insecticide regimens for grape pest management rely on products that are much less disruptive to natural enemy populations than those used in the past.

16.3.3 Chemical Control

Chlorpyrifos remains the only insecticide registered for grape root borer control, applied as a soil drench around the base of vines to create a toxic barrier to burrowing neonates. For optimal efficacy, All et al. (1987) recommended that vine rows should be free of weeds and other impediments to thorough spray coverage. All et al. (1985) reported at least 4 weeks of residual chlorpyrifos activity against neonates, although soil residues were relatively ineffective against pupae and adults and are not considered to affect larvae established on roots. However, soil application of chlorpyrifos is rarely included in standard vineyard insecticide programs. Rather, it is most often used in response to recognition of an established infestation, and in those instances, I am aware of applications that were likely relatively ineffective due to poor spray coverage and/or inadequate spray volume. Chlorpyrifos applications are timed to span the peak of larval eclosion, although its 35-day preharvest interval precludes its utility for grape cultivars harvested early or in states such as Florida, where peak adult emergence, oviposition and hatch occur in late summer and into fall. Finally, some growers are philosophically averse to this approach, believing that its disruptive effects on soil biodiversity have detrimental impacts on general vine health and, ultimately, berry (and wine) quality. Although the registration of chlorpyrifos does not appear to be threatened in the near term, the current regulatory environment in the United States creates uncertainty about the long-term availability of this organophosphate insecticide for viticulture. Consequently, as eastern US vineyards increase in number and age, the development and delivery of alternative and sustainable management tactics and strategies for grape root borer is an increasingly pressing objective.

16.4 Research on Alternative Management Options for Grape Root Borer

Published efforts to develop and apply advanced pest management technologies to grape root borer span three decades and have focused primarily on entomopathogenic nematodes and mating disruption, both the subject of ongoing research.

16.4.1 Entomopathogenic Nematodes

All et al. (1981) and Saunders and All (1985) showed that the entomopathogenic nematode, *Steinernema carpocapsae* (Weiser), was present in soil from Georgia (US) vineyards and parasitized first instar grape root borer larvae in laboratory assays or small experimental plots in the greenhouse and field. However, All et al. (1981) reported that relatively low numbers of late instar larvae and pupae collected from vineyards were infected by *S. carpocapsae* and that inoculative releases of the nematode in vineyards were ineffective. An explanation for the failure of these attempts followed the discovery that infective juveniles of *S. carpocapsae* use an ambush host-finding strategy (Campbell and Gaugler 1993) and have limited mobility in soil. Compared with ambusher species and strains, entomopathogenic nematodes that use cruising behavior are considered better adapted to locate and parasitize root boring larvae (Kaya and Gaugler 1993).

In laboratory bioassays, Williams et al. (2002) compared the virulence of 17 species and strains of *Heterorhabditis* and *Steinernema* nematodes to larval grape root borer, including representatives of both ambush and cruise foraging strategists. All but one species caused some degree of larval mortality, with highest infection rates from *H. bacteriophora* strain GPS11 (92%) and *H. zealandica* strain X1 (86%), both cruiser nematodes. Focusing on these two strains, Williams et al. (2010) compared the effect of nematode application timing and number of applications on established grape root borer populations in commercial vineyards, based on weekly collections of pupal exuviae from treated and untreated plots. The *H. zealandica* strain was applied to plots in May or June or in both months in three consecutive years at two vineyards in Ohio and in the third or fourth week of May or in both weeks for two consecutive years at one site in Georgia, US. Analysis combining the results from each 2- or 3-year study showed that all treatments significantly reduced the number of pupal exuviae collected, compared with untreated plots, and there were no significant differences based on the timing or number of applications. Across all sites, overall control from *H. zealandica* ranged from 55% to 81%. The persistence of *H. zealandica* was evaluated by exposing wax moth, *Galleria mellonella* (L.), larvae to soil samples taken at several post-application intervals from the treated and untreated plots which indicated that *H. zealandica* persisted for only several weeks following application. The *H. bacteriophora* strain was applied in May or September or in both months (same growing season) at two vineyards in Ohio. Pupal exuviae were collected from one vineyard over two consecutive years while post-treatment soil samples were taken from both vineyards to assess the persistence of this strain. As with *H. zealandica*, combined data from both years showed that all treatments significantly reduced the number of pupal exuviae collected, compared with the control plots, and that there were no significant differences according to the number or timing of applications. At the treated vineyard in Ohio, control from *H. bacteriophora* ranged from 69% to 92% among the treatments. The persistence of *H. bacteriophora* in soil samples from both Ohio vineyards commonly extended over at least 12 months post treatment. Williams et al. (2010) concluded that the substantial difference in the persistence of the two nematodes evaluated did not appear to be due to seasonal soil moisture levels and suggested that the indigenous *H. bacteriophora* was better adapted to environmental conditions prevailing at the study sites than the non-indigenous *H. zealandica*.

While *H. bacteriophora* is commercially available in the US and can be used against grape root borer, most growers and many researchers and extension agents have no experience with this tactic. Consequently, considerable education about the technical aspects of nematode release, including quality control/nematode viability assessments and application techniques will be critical for widespread adoption of this approach. This objective could be achieved via demonstration trials of nematode efficacy in commercial vineyards across the geographic range of grape root borer, providing experience and a level of trust that growers and their advisors will require.

16.4.2 Mating Disruption

The primary component of the grape root borer sex pheromone, (*E,Z*)-2,13-octadecadienyl acetate ((*E,Z*)-2,13-ODDA), was identified by Schwarz et al. (1983) and followed by demonstration of greatly increased attractiveness by the addition of 1% (Z,Z)-3,13-ODDA (Snow et al. 1987). Disruption of pheromone-mediated communication by grape root borer was first reported by Johnson et al. (1981). Although males were not captured in traps baited with lures containing the minor component, (Z,Z)-3,13-ODDA, their behavioral response to virgin female grape root borers in cages surrounded by dispensers containing this compound was reduced.

Johnson et al. (1986) deployed laminated dispensers containing (*Z,Z*)-3,13-ODDA in several vineyards in Arkansas and North Carolina that showed no pre-treatment differences in exuviae counts between disrupted and non-disrupted plots at each site. In the season during which dispensers were deployed, male response to caged females was much reduced in treated plots and mating of sentinel virgin females was eliminated. In the second season after treatment, significantly fewer exuviae were collected from treated plots than from the controls. Subsequently, Johnson et al. (1991) showed that either component of the grape root borer sex pheromone formulated in rope dispensers significantly reduced the number of exuviae collected from seven of eight commercial vineyards, compared with untreated blocks in the same locale, although the main component, (*E,Z*)-2,13-ODDA, appeared to be more effective at preventing male response to traps.

Webb (1991) deployed rope dispensers containing the binary pheromone blend in a Florida vineyard and used a vineyard at a second location as the untreated control. In the two seasons during which dispensers were deployed, male moth captures in traps at the treated vineyard were eliminated and fewer mated female moths were collected from the treated than the untreated site. Since an average of only six pupal exuviae per year were collected from 100 vines in the untreated vineyard in two consecutive years, and despite continuing captures in pheromone traps, Webb (1991) was unable to evaluate and compare treatment effects on infestation levels.

Recent studies evaluated the effects of rope dispensers containing a blend of (*E,Z*)-2,13-ODDA and (*E,Z*)-3,13-ODDA, registered as Isonet® Z (Shin-Etsu Corp., Tokyo, Japan) for leopard moth, *Zeuzera pyrina* (L.), mating disruption in Europe. In Florida, Weihman and Liburd (2006) reported that 635 ropes/ha of this formulation eliminated captures of male grape root borer in pheromone traps. Pfeiffer et al. (2010) treated vineyard blocks with 247 ropes/ha and reported that pupal exuviae counts after 2 years of mating disruption were significantly lower than in the corresponding controls at two of three sites in Virginia.

Johnson et al. (1991) noted the tendency for grape root borer mating disruption to be less effective in heavily infested vineyards. They remarked that, to achieve >85% infestation reductions, the grape root borer population density should be <1.4 pupal exuviae per vine, and the entire vineyard along with adjacent vineyard blocks should be treated with the pheromone ties. This would prevent adjacent vineyards from serving as a reservoir for mated females. The immigration of mated females from wild grapevines is an ongoing concern for any grape root borer management program, even though their flight distance has yet to be studied.

At present, an application for registration of a mating disruption formulation for grape root borer is in preparation and it is anticipated that the product will be commercially available within 2 years. While this would be a valuable addition to the control tactics legally available, its commercial success will ultimately be driven by grower demand and adoption. The US registration of a mating disruption formulation for grape berry moth, *Paralobesia viteana* (Clemens), was recently cancelled by the registrant due to insufficient demand. It continues to be widely used in Canada.

16.5 Knowledge Gaps and Suggestions for Future Research

Grape growers do not yet have adequate resources to effectively mitigate and manage the risk from grape root borer in a manner considered sustainable and that is applicable throughout the pest's range and across all grape varieties. In this section, I will identify aspects of grape root borer biology about which we have an incomplete understanding but that impinge on our ability to develop and deliver an integrated management approach and suggest avenues of research that should enhance the achievement of this goal.

16.5.1 Risk Factors

In combination, the perennial effects of a number of horticultural, cultural and environmental factors likely influence the susceptibility of individual vineyards or vineyard blocks to attack and infestation by grape root borer. Vineyard age is thought to be directly related to the probability of root borer infestation and although young vines can be attacked (All et al. 1987), this relationship has not been examined systematically. Vineyard proximity to forest containing native hosts of grape root borer has long been considered another primary risk, but the susceptibility/suitability of each of the many native *Vitis* species in the eastern US (Massey 1945; Moore 1991) and their relative contributions to local populations is unknown. Expansion of the eastern US grape industry has primarily involved the production of *V. vinifera*. Although cultivars of this species are assumed to be equally susceptible to attack, horticultural differences in the rate at which different cultivar/rootstock combinations produce or replace roots or the depth of root growth may influence their ability to tolerate a given level of grape root borer infestation. Differences in the type, texture and compaction of vineyard soils and related water retention capacity may differentially influence root growth. Furthermore, larval survivorship prior to their establishment on roots may be related to soil type and texture. We are conducting an intensive 5-year assessment of the abundance and distribution of pupal exuviae in a large number of vineyard blocks in northern and central Virginia that differ in many of the factors identified above. Geospatial and principal components analyses will be used to compare infestation distributions and to identify the underlying contributions of each putative risk factor to differences in the extent of infestation among vineyards.

16.5.2 Behavioral Manipulation of Adult Male and Female Grape Root Borer

Results from grape root borer mating disruption studies suggest considerable plasticity in the pheromone component or blend of components that can be used to disrupt male response to virgin females or pheromone lures (Johnson et al. 1986,

1991; Webb 1991; Pearson and Meyer 1996; Weihman and Liburd 2006) or to reduce infestation levels (Johnson et al. 1991; Pfeiffer et al. 2010). This raises interesting questions about the behavioral and/or physiological mechanisms underlying their effects. Further research on the efficacy and behavioral response of male moths to these compounds and blends is warranted, especially the blend registered for leopard moth mating disruption that is considered a potential commercial product for grape root borer mating disruption. The behavioral and electrophysiological responses shown by female grape root borer to their pheromone components (Pearson 1992; Pearson and Meyer 1996; Pearson and Schal 1999) may also influence the relative effectiveness of different mating disruption formulations.

Assuming that immigration of mated female moths into commercial vineyards from wild grapevines implies a directed response to specific qualities of the host, determination and comparison of their behavioral and electrophysiological responses to olfactory and visual stimuli from grape plants may reveal factors that guide their location of vineyards and that may be amenable to manipulation.

16.5.3 Food-Finding by Grape Root Borer Larvae

Dutcher and All (1978c) reported that the grape root borer shows a type III survivorship curve, with highest mortality of eggs and especially first instar larvae. The infestation of vines is ultimately a function of the success with which the tiny neonate larvae find and establish on grape roots, but virtually nothing is known about their subterranean behavior. Controlled studies addressing the movement of neonates in soil and the effect of soil type on their movement and survival may yield important insights. Interestingly, some vineyard soil samples collected by Dutcher and All (1978a) contained small, medium or large larvae in distinct tunnels coated with a violet oral exudate, leading them to suggest that the larvae were migrating between roots. While it has been suggested that grape root borer larvae locate food by chance (Brooks 1907; All et al. 1987), their oligophagous habit on Vitaceae and the likelihood of a highly co-evolved relationship with grape raise the intriguing alternative hypothesis that food-finding is guided by cues from the host. Larval perception of compounds associated with grape roots may enable host recognition and acceptance upon contact or may potentially facilitate their orientation to roots. Further plausible support for this hypothesis arises when one considers the potential complexity of the larva's subterranean environment in its native habitat, in which roots of wild grape are spaced apart in the soil matrix and likely co-mingle with roots of non-host species.

Bergh et al. (2011) showed that newly hatched grape root borer larvae respond to alcohol-based extracts of grape roots applied to filter paper discs in small Petri dishes. Larvae were recorded significantly more often in quadrants containing discs treated with extracts of roots from several grape rootstocks and native *Vitis* spp. than in quadrants with no disc or with an alcohol-treated disc (Fig. 16.6). There was no apparent response of larvae to discs treated with an extract of roots from the non-host, apple. Moreover, when extracts from different Vitaceae root sources were

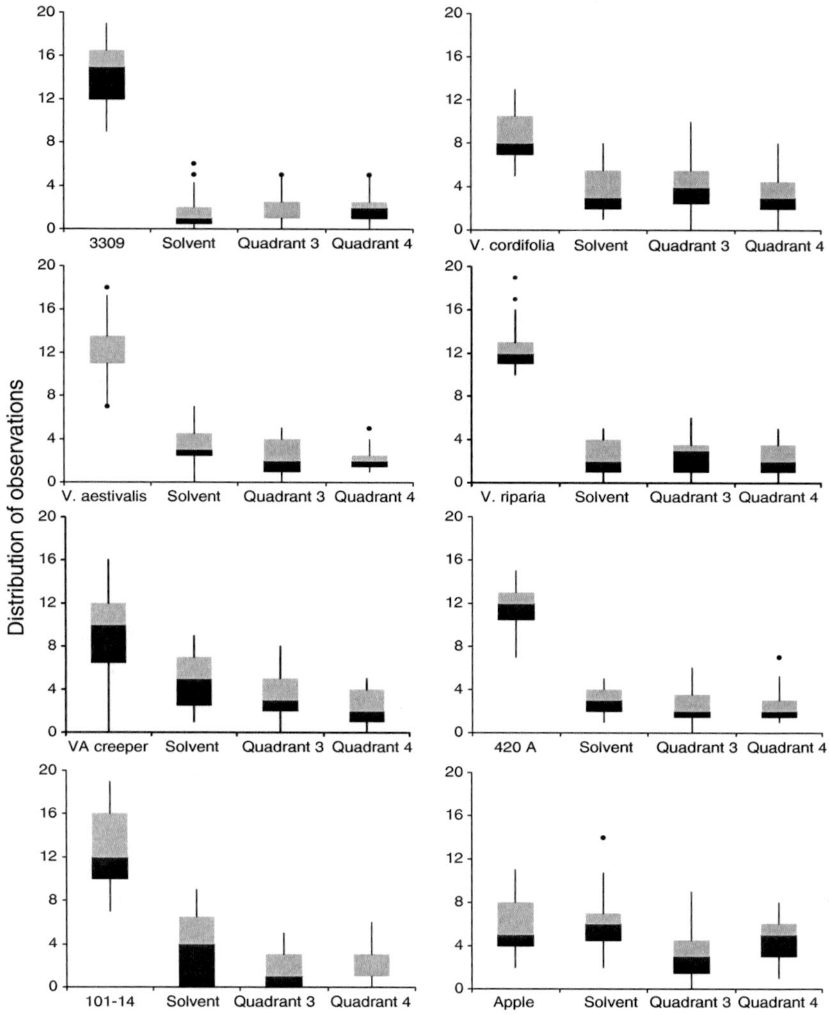

Fig. 16.6 The distribution of grape root borer neonates in Petri dish bioassays in which filter paper discs treated with alcohol-based extracts of roots from host and non-host plants were presented with solvent-treated discs

presented in pairs, using the 3309 rootstock as the standard, the response to the 420-A rootstock (*V. berlandieri* x *V. riparia*) and *V. riparia* 'Gloire' was significantly stronger than to others (Bergh et al. 2011). These findings raise questions about whether root compounds influence the interactions between grape root borer larvae and their host, and whether they may be related to differences in susceptibility or suitability among *Vitis* spp. and rootstocks. An intriguing concept is whether larval food-finding can be disrupted or prolonged by exposing larvae to synthetic sources of behaviorally active root compounds, hypothetically increasing their mortality by increasing searching time and energy, and time exposed to natural enemies in the soil.

The novel research on the behavioral responses of corn rootworm to stimuli from corn roots, and the manipulation of these responses, provides a compelling precedent for this possibility (Bernklau et al. 2004; Bernklau and Bjostad 2008, 2009).

16.5.4 Sampling, Monitoring and Management Decisions

Accurate assessment of grape root borer populations in vineyards remains challenging. Growers often become aware of an infestation when larvae or feeding damage are found on the roots of dead or weakened vines that have been pulled from the ground. The crude nature of vine removal provides only a rough indication of current infestation status, since some of the root system is usually left in the soil and portions of roots with feeding damage are especially weak.

16.5.4.1 Monitoring Using Acoustic Emissions Detection

Acoustic emissions detection may offer novel opportunities for researchers to address certain aspects of grape root borer biology, especially related to infestation distributions, but would likely not be a practical approach for growers or crop consultants. This species seems a probable candidate for application of this technology, because late instar larvae typically complete their feeding near the base of vines (Dutcher and All 1978b), usually on the crown itself or near the origin of main lateral roots, and the acoustic emissions from their chewing should be readily detectable using available instruments (Mankin et al. 2009; Mankin and Moore 2010).

16.5.4.2 Monitoring Pupal Exuviae

The only non-destructive means by which to confirm the infestation of individual vines or to compare the status of different vineyards is via sampling pupal exuviae, which is a laborious task but much less difficult and more accurate when an area around the vine base is kept free of vegetation during the period of adult emergence (Johnson et al. 1991; Webb 1991). Dutcher and All (1978b) found that 90% of larvae pupated within a 35-cm radius from the base of the trunk. Exuviae are often found lying on the soil and are subject to being blown away by orchard machinery. Consequently, sampling at weekly or biweekly intervals has been appropriately employed in some studies comparing management tactics (Johnson et al. 1991; Webb 1991; Williams et al. 2010). However, such intense sampling is impractical for most growers. The spatial distribution of grape root borer pupal exuviae within individual vineyard blocks has not been determined, precluding the development of a sampling protocol based on probabilities of detection. A standardized sampling scheme based on the distributions of exuvia would enable efficient, accurate and comparable evaluations by researchers assessing the effects of control options and would assist growers with management decisions.

16.5.4.3 Monitoring Using Pheromone Traps

Unfortunately, captures of male grape root borer in pheromone traps cannot yet be related directly to the infestation status of vineyard blocks. Although the active space of a pheromone lure is unknown, male grape root borers are strong, swift fliers and may respond to baited traps over considerable distances. Bergh (2006) placed traps in spatially isolated apple orchards and vineyards that were adjacent to forest with wild grapevines and showed that total captures during the period of peak flight activity in Virginia were not significantly different between apple orchards and vineyards. Consequently, captures by traps placed in vineyards often reflect populations from within and outside the planting.

Trap design has an important effect on the number of male grape root borer captured. Weihman and Liburd (2007) reported that green, bucket style traps captured significantly more grape root borers than wing style sticky traps in Florida vineyards. Since male grape root borer moths are relatively large and can quickly 'saturate' the sticky liner of wing and delta style pheromone traps, their use in vineyards with moderate to high pest pressure requires frequent servicing to maintain effectiveness and efficiency. In the Florida study, wing trap liners were replaced at 4–6 week intervals and Weihman (2005) reported that captures decreased over successive weeks when liners were not replaced but exceeded those in bucket traps during the period immediately following their replacement. My comparisons of delta traps and bucket traps over several years have consistently shown that when the liners of either delta or wing style sticky traps are replaced at regular intervals, these trap types capture numerically or significantly more grape root borer than bucket traps (Fig. 16.7). Of the two styles of sticky trap, delta traps proved to be somewhat more effective and certainly easier to use, due to their readily removable liner. While any of the different pheromone traps available will adequately indicate the annual onset, peak and cessation of grape root borer flight, thereby assisting certain management decisions, the interpretation of grape root borer captures in relation to the size of local populations will be relative to the style of trap used and the frequency of servicing. Weihman and Liburd (2007) noted that bucket traps are much better suited to grower needs, because their effectiveness is not known to be compromised by the number of moths captured and so require much less servicing than sticky traps. However, for some purposes such as mass trapping, bucket traps are likely not sufficiently efficient or effective.

16.5.5 Mass Trapping

In theory, mass trapping to manage grape root borer is compelling, since this species appears to meet most of the criteria identified by El-Sayed et al. (2006) as being important to the success of this approach. Grape root borer population densities are much lower than of most moth pests and it is univoltine throughout its range. Compared with many agroecosystems, eastern US vineyards are relatively small in size and males respond rapidly and persistently to pheromone-baited traps deployed within them. However, mass trapping requires an optimally effective

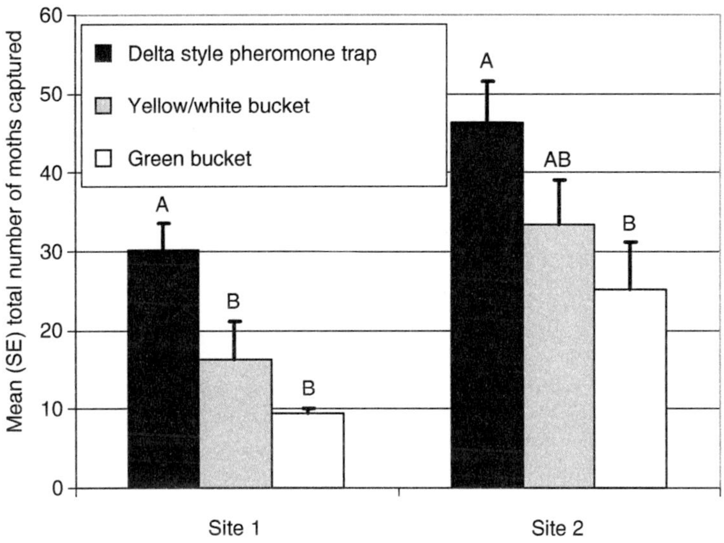

Fig. 16.7 Male grape root borer captures in delta traps, *green* bucket traps and *yellow/white* bucket traps at two commercial vineyards in Virginia, US, 2010. Traps were deployed from trellis wires at an elevation of about 1.5 m, with 15 and 25 m between traps within rows at Sites 1 and 2, respectively. Each trap treatment was randomly assigned to a position within each of four rows per site, with three buffer rows between trap rows. Traps were deployed on 14 July and rotated among positions within each row at weekly intervals for 5 weeks. Moth captures were recorded weekly, and total captures per trap were compared using analysis of variance and least significant difference means separation test

means by which to remove as many males as quickly as possible from the breeding population (El-Sayed et al. 2006), and as discussed previously, none of the traps currently available are ideally suited to this, each having unique drawbacks. We are using trapping studies and behavioral analyses of moth responses to traps to determine the features of traps that influence their relative effectiveness for capturing male grape root borer. Ultimately, the development of a maximally effective and efficient non-saturating trap would optimize evaluations of mass trapping as a potential management tactic for grape root borer.

16.6 Conclusion

In conclusion, there appear to be numerous opportunities to expand and improve monitoring and management options for incorporation into a multifaceted approach to grape root borer control. Some seem to offer promise in the near term while others will require considerable effort and time to develop or are, as yet, purely speculative. In the meantime, the grape research and extension community should actively and continuously seek to raise awareness among growers of the potential risk that grape root borer represents, especially given that the recent expansion of the eastern US wine grape industry has been driven largely by growers from non-agricultural backgrounds. These growers often require comprehensive and ongoing education about all facets of grape production, and in the absence of expert advice, may often overlook the growth of grape root borer infestations and so repeat the mistakes of the past. To help avoid this problem, Bergh (2006) advocated the deployment of grape root borer pheromone traps as soon as new plantings are initiated, and especially in those that are not in close proximity to older, established blocks. Trapping data from such vineyards should provide important baseline information on moth pressure and the relative risk of attack. It is important to encourage growers to maintain accurate, annual records of seasonal captures in such vineyards, as these may indicate changes due to the building up of infestations or intervention practices. However, our ability to address many of the questions about grape root borer biology and management that growers may pose, continues to be hindered by knowledge gaps in several areas. Much of the excellent foundational research on grape root borer biology and management conducted during the 1980s and 1990s focused on the pest in either 'Concord' or 'Muscadine' vineyards and while some of this work undoubtedly translates directly to *V. vinifera*, other results and conclusions may prove to be more or less broadly applicable. For example, Dutcher and All (1979a) reported that a single mature larva feeding at the base of 'Concord' vines can cause substantial girdling and significant yield reduction and calculated an economic injury level of 0.074 larvae (or pupal exuvia) per vine, or 73 larvae (or pupal exuviae) per ha. Their recommendation for intervention as soon as grape root borer is detected in a vineyard was based on a specific set of horticultural and environmental conditions and may prove to be conservative under different sets of conditions. Regional differences in the seasonal activity of grape root borer, its population density and the factors that may influence their populations also affect management recommendations. The research avenues that I have proposed are intended to promote a renewed consideration of the unresolved issues surrounding the behavior and ecology of this important pest and to stimulate the development of creative and sustainable solutions for its management.

Acknowledgments I offer sincere thanks to J. R. Meyer, North Carolina State University, and A. Zhang, USDA ARS Beltsville, Maryland, for their assistance with various aspects of my grape root borer research. Thanks also to T. K. Wolf for his valuable insights into viticulture, to the grape growers in Virginia who allowed us access to their vineyards, to my technician, J. P. Engelman, and to the several summer student interns who assisted us. I gratefully acknowledge the Virginia Wine Board, the Viticulture Consortium-East, and the Virginia Agricultural Council for their support of my research.

References

Adlerz WC, Hopkins DL (1981) Grape insects and diseases in Florida. Proc Fla State Hortic Soc 94:331–336

All JN, Saunders MC, Dutcher JD, Javid AM (1981) Susceptibility of grape root borer larvae, *Vitacea polistiformis* (Lepidoptera: Sesiidae) to *Neoaplectana carpocapsae* (Nematoda: Rhabditida): and potential of host kairomones for enhancement of nematode activity in grape vineyards. Misc Publ Entomol Soc Am 12:9–14

All JN, Dutcher JD, Saunders MC, Brady UE (1985) Prevention strategies for grape root borer (Lepidoptera: Sesiidae) infestations in Concord grape vineyards. J Econ Entomol 78:666–670
All JN, Dutcher JD, Saunders MC (1987) Control program for the grape root borer in grape vineyards of the eastern United States. Down Earth 43:10–12
Attwood VG, Wylie WD (1963) Grape root borer threatens vineyards. Arkansas Farm Res 12:6
Bambara SB, Nuenzig HH (1977) Descriptions of immature stages of the grape root borer, *Vitacea polistiformis* (Lepidoptera: Sesiidae). Ann Entomol Soc Am 70:871–875
Bergh JC (2006) Trapping grape root borer (Lepidoptera: Sesiidae) in vineyard and non-vineyard habitats in Virginia. J Entomol Sci 41:253–256
Bergh JC, Pfeiffer DG, Love KP (2005) Survey of grape root borer, *Vitacea polistiformis* (Harris), using pheromone traps in Virginia vineyards. J Entomol Sci 40:337–342
Bergh JC, Zhang A, Meyer JR, Kim D (2011) Response of grape root borer (Lepidoptera: Sesiidae) neonates to root extracts from vitaceae species and rootstocks. Environ Entomol 40:880–888
Bernklau EJ, Bjostad LB (2008) Identification of feeding stimulants in corn roots for western corn rootworm larvae (Coleoptera: Chrysomelidae). J Econ Entomol 101:341–351
Bernklau EJ, Bjostad LB (2009) Localized search cues in corn roots for western corn rootworm (Coleoptera: Chrysomelidae) larvae. J Econ Entomol 102:558–562
Bernklau EJ, Fromm EA, Bjostad LB (2004) Disruption of host location of western corn rootworm larvae (Coleoptera: Chrysomelidae) with carbon dioxide. J Econ Entomol 97:330–339
Brooks FE (1907) The grape vine root borer. W Va Agric Exp Stn Bull 110:19–30
Campbell JF, Gaugler R (1993) Nictation behavior and its ecological implications in the host search strategies of entomopathogenic nematodes (Heterorhabditidae and Steinernematidae). Behavior 126:155–169
Dutcher JD, All JN (1978a) Models of the distribution of subterranean stages of *Vitacea polistiformis* in Concord grape vineyards. Environ Entomol 7:461–465
Dutcher JD, All JN (1978b) Reproductive behavior of *Vitacea polistiformis* Harris. J Ga Entomol Soc 12:55–58
Dutcher JD, All JN (1978c) Survivorship of the grape root borer in commercial grape vineyards with contrasting cultural practices. Environ Entomol 7:461–465
Dutcher JD, All JN (1979a) Damage impact of larval feeding by the grape root borer in a commercial Concord grape vineyard. J Econ Entomol 72:159–161
Dutcher JD, All JN (1979b) Biology and control of grape root borer in Concord grape vineyards. Ga Agric Exp Stn Res Bull 232:1–15
El-Sayed AM, Suckling DM, Wearing CH, Byers JA (2006) Potential of mass-trapping for long-term pest management and eradication of invasive species. J Econ Entomol 99:1550–1564
Johnson DT, Mayes RL, Gray PA (1981) Status of grape root borer, *Vitacea polistiformis* (Lepidoptera: Sesiidae) management and feasibility of control by disruption of mating communication. Misc Publ Entomol Soc Am 12:1–7
Johnson DT, Meyer JR, Mayes RL (1986) Evaluation of Hercon laminated dispensers baited with Z, Z-3,13-octadecadien-1-ol acetate for suppression of the grape root borer, Vitacea polistiformis (Harris) (Lepidoptera: Sesiidae), populations in grapes. J Entomol Sci 21:231–236
Johnson DT, Lewis BA, Snow JW (1991) Control of grape root borer (Lepidoptera: Sesiidae) by mating disruption with two synthetic sex pheromone compounds. Environ Entomol 20: 930–934
Kaya HK, Gaugler R (1993) Entomopathogenic nematodes. Annu Rev Entomol 38:181–206
Mankin RW, Moore A (2010) Acoustic detection of *Oryctes rhinoceros* (Coleoptera: Scarabaeidae: Dynastinae) and *Nasutitermes luzonicus* (Isoptera: Termitidae) in palm trees in urban Guam. J Econ Entomol 103:1135–1143
Mankin RW, Samson PR, Chandler KJ (2009) Acoustic detection of melolonthine larvae in Australian sugarcane. J Econ Entomol 102:1523–1535
Massey AB (1945) Native grapes, vol 38. Virginia Polytechnic Institute, Blacksburg
Moore MO (1991) Classification and systematics of eastern North American *Vitis* L. (Vitaceae) north of Mexico. SIDA 14:339–367
Olien WC (1990) The muscadine grape: botany, viticulture, history, and current industry. Hortscience 25:732–739

Olien WC, Smith BJ, Hegwood CP Jr (1993) Grape root borer: a review of the life cycle and strategies for integrated control. Hortscience 28:1154–1156

Pearson GA (1992) Pheromone effects on mating success and female behavior in the grape root borer. Ph.D. dissertation, North Carolina State University, Raleigh, NC

Pearson GA, Meyer JR (1996) Female grape root borer (Lepidoptera: Sesiidae) mating success under synthetic sesiid sex pheromone treatment. J Entomol Sci 31:323–330

Pearson GA, Schal C (1999) Electroantennogram responses of both sexes of grape root borer (Lepidoptera: Sesiidae) to synthetic female sex pheromone. Environ Entomol 28:943–946

Pfeiffer DG, Luab CA, Jordan TA, Wallingford AK, Cassell M (2010) Control of grape root borer using mating disruption – 2009. In: Proceedings of the 85th Cumberland-Shenandoah fruit workers conference, Winchester, VA, 19–20 Nov 2009, pp 35–36

Sarai DS (1970) Effect of burial of grape root borer pupae on adult emergence. J Econ Entomol 62:1507–1508

Sarai DS (1972) Seasonal history and effect of soil moisture on mortality of newly hatched larvae of the grape root borer in southern Missouri. J Econ Entomol 65:182–184

Saunders MC, All JN (1985) Association of entomophilic rhabditoid nematode populations with natural control of first-instar larvae of the grape root borer, *Vitacea polistiformis*, in Concord grape vineyards. J Invertebr Pathol 45:147–151

Schwarz M, Klun JA, Leonhardt BA, Johnson DT (1983) (E,Z)-2,12-octadecadien-1-ol acetate. A new pheromone structure for sesiid moths. Tetrahedron Lett 24:1007–1010

Snow WJ, Schwarz M, Klun JA (1987) The attraction of the grape root borer, vitacea polistiformis (Harris) (Lepidoptera: Sesiidae) to (E, Z)-2,13 octadecadienyl acetate and the effects of related isomers on attraction. J Entomol Sci 22:371–374

Snow WJ, Johnson DT, Meyer JR (1991) The seasonal occurrence of the grape root borer (Lepidoptera: Sesiidae) in the eastern United States. J Entomol Sci 26:157–168

Townsend HG (1991) The effect of drip irrigation and ground cover on the grape root borer. Annual progress report to the grape and wine development program, Missouri Department of Agriculture, Jefferson City, MO, pp 60–65

USDA (2011) National Science Foundation Center for IPM, Raleigh, NC. http://www.ipmcenters.org/whatis.cfm

Webb SE (1991) Management of grape root borer in Florida with a pheromone. Proc Fla State Hortic Soc 104:3–5

Webb SE, Mortensen JA (1990) Evaluation of bunch grape rootstocks and muscadine varieties for resistance to grape root borer. Proc Fla State Hortic Soc 103:310–313

Webb SE, Sprenkel RK, Sharp JL (1992) Seasonal flight activity of grape root borer (Lepidoptera: Sesiidae) in Florida. J Econ Entomol 85:2161–2169

Weihman SW (2005) Monitoring and control tactics for grape root borer *Vitacea polistiformis* Harris (Lepidoptera: Sesiidae) in Florida vineyards. M.S. thesis, University of Florida, Gainesville, FL

Weihman SW, Liburd OE (2006) Mating disruption and attract-and-kill as reduced-risk strategies for control of grape root borer, *Vitacea polistiformis* (Lepidoptera: Sesiidae) in Florida vineyards. Fla Entomol 89:245–250

Weihman SW, Liburd OE (2007) Seasonal distribution and evaluation of two trap types for monitoring grape root borer, *Vitacea polistiformis* (Lepidoptera: Sesiidae) in Florida vineyards. Fla Entomol 90:480–487

Williams RN, Snow JW (1991) Annotated bibliography of the grape root borer (Lepidoptera: Sesiidae). Fla Entomol 74:320–334

Williams RN, Fickle DS, Grewal PS, Meyer JR (2002) Assessing the potential of entomopathogenic nematodes to control the grape root borer, *Vitacea polistiformis* (Lepidoptera: Sesiidae) through laboratory and greenhouse bioassays. Biocontrol Sci Technol 12:35–42

Williams RN, Fickle DS, Grewal PS, Dutcher J (2010) Field efficacy against the grape root borer, *Vitacea polistiformis* (Lepidoptera: Sesiidae) and persistence of *Heterorhabditis zealandica* and *H. bacteriophora* (Nematoda: Heterorhabditidae) in vineyards. Biol Control 53:86–91

Yonce CE (1995) Physical barriers as a means to reduce grape root borer, *Vitacea polistiformis* (Harris) infestations in vineyards and in greenhouse muscadine plants. J Entomol Sci 30:237–242

Chapter 17
Japanese Beetle and Other Coleoptera Feeding on Grapevines in Eastern North America

Douglas G. Pfeiffer

17.1 Introduction

17.1.1 Scope

The Japanese beetle, *Popillia japonica* Newman, and several other coleopteran foliavores, i.e. green June beetle, *Cotinus nitida* (L.), rose chafer, *Macrodactylus subspinosus* (F.), grape rootworm, *Fidia viticida* Walsh, and grape flea beetle, *Altica chalybea* Illiger cause conspicuous foliar injury. A number of these coleopterans also cause injury to other vine parts, i.e., primary buds, berries, or roots. Although such injury may be much more economically important than the leaf injury, we will not address them here because the leaf injury arouses the greatest amount of concern. Japanese beetle is the main species to be discussed. In an earlier study (Pfeiffer et al. 1990), it was found to be the target of most insecticide sprays in Virginia vineyards, owing to its conspicuous leaf injury. Although its populations fluctuate considerably from year to year, Japanese beetle remains an important pest to be addressed by grape pest management programs in eastern North America. The thrust of this chapter is on beetles that, in at least one life stage, inflict defoliation injury on grapevines. There is one additional coleopteran that may cause economic losses, the multicolored Asian lady beetle, *Harmonia axyridis* (Pallas). Injury that may be inflicted to fruit by this otherwise beneficial predator is discussed by Pfeiffer et al. (Chap. 19).

D.G. Pfeiffer (✉)
Department of Entomology, Virginia Tech., 205C Price Hall, Blacksburg, VA 24061-0319, USA
e-mail: dgpfeiff@vt.edu

N.J. Bostanian et al. (eds.), *Arthropod Management in Vineyards: Pests, Approaches, and Future Directions*, DOI 10.1007/978-94-007-4032-7_17,

17.1.2 The Grapevine Leaf System and Fruit Ripening

Grapevines are sometimes described as factories that transform sunlight to sugars. The process of photosynthesis produces non-structural carbohydrates, primarily sugars, that are important constituents of grape berries whether for wine production or for consumption at the table. Net photosynthesis (P_n) decreases with increasing Leaf Area Loss (LAL). This not only reduces P_n for the whole leaf, resulting from reduced photosynthetic surface, but it can also decrease P_n from the remaining leaf area. This decline becomes especially steep after 20% LAL (Boucher et al. 1987). The impact of reduced leaf area is therefore greater than expected from simple loss of leaf area. Many studies use mechanical removal of leaf area to simulate the impact of insect-induced defoliation. Care should be taken to simulate the actual injury as closely as possible. Simple removal of intact leaves (breaking the petiole from a shoot) is not a substitute for defoliation (Boucher 1986). On a broader ecological basis, evaluation of feeding impact by considering only leaf tissue removal can lead to an oversimplification of the effects of injury. This is because beetle feeding induces the production and release of plant volatiles that attract more beetles to the vine leading to further feeding, an effect not shown by simple mechanical leaf area removal.

During the development of grapevines and berry clusters, there are changes in leaf vulnerability and the source-sink relationships of sugars in the vine. Early in the season, leaves are thin and delicate. Later in the season, leaves are tougher and can tolerate more insect feeding. At that stage, they have already made important photosynthetic contributions to the vine. However, another factor mitigates the greater importance of late season photosynthesis. In the early part of the season, the main sinks for sugar production are shoots, as they are still in their growth phase. Sugar accumulation in berries is minimal at this time. After veraison, shoot growth slows down and berries become the sink for most of the sugar production. Furthermore, for red cultivars, berries begin to develop their characteristic color. Consequently, defoliation after veraison may have a greater impact on ultimate berry quality at harvest.

17.2 Japanese Beetle

17.2.1 Appearance

The adult Japanese beetle (JB), *P. japonica* (Scarabaeidae), is shiny green, with copper-colored elytra, with tufts of white setae arranged along the sides of the abdomen (Fig. 17.1). Male and female beetles are differentiated by an apical tibial spur on the front pair of legs that is pointed in the male and rounded in the female. The larvae are C-shaped white grubs with three pairs of legs on the thoracic segments and they are found in the root zone of grasses. The pattern of setae found on the

Fig. 17.1 Adult Japanese beetles with associated feeding injury to grape foliage (Photo by Rufus Isaacs)

raster (the underside of the last abdominal segment) is important in the identification of white grubs. In Japanese beetle larvae this pattern is typified by a V-shaped arrangement of setae, opening toward the hind end.

17.2.2 Biology

The biology of JB has been reviewed by Fleming (1972), and Potter and Held (2002). It is a native of Japan, but probably not mainland Asia (Potter and Held 2002). In the past, it had been of limited importance in Japan because of restricted habitats, and the presence of natural enemies. Recently, increased outbreaks have been associated with increases in grassy areas (Ando 1986). Other species of *Popillia* are also present in Asia (e.g. *Popillia lewisi* Arrow, *Popillia uchidai* Niijimi & Kinoshita, and *Popillia indigonacea* Motschulsky). Japanese beetle was first found in New Jersey in 1916, but judging by its numbers it had probably been already present for about 5 years (Fleming 1968). It has now spread across all states east of the Rocky Mountains, except Florida. Climatic conditions in many parts of the world are suitable for its establishment (Potter and Held 2002). Moreover, because of human modifications of the environment through irrigation, potential areas of spread may be even greater. The adults appear in late June to early July (or early June in the southern parts of its range) and begin to feed on a wide range of plants. Males first appear slightly before females, and beetle populations peak in

July, continuing into September. The eggs are laid in a series of ovipositional bouts, between which females return to host plants for additional feeding and mating. There is a single generation annually, but in the northern part of its range, some individuals require a second year to complete their development.

Japanese beetle is highly polyphagous, feeding on more than 300 species representing 79 different plant families, with Vitaceae among the most preferred (Potter and Held 2002). Some host plants favor higher JB reproduction than others (Ladd 1987), and the suitability of host plants appears to be dictated by secondary chemistry rather than quantitative traits, such as digestibility-reducing materials (Keathley and Potter 2008). Feeding causes the release of volatile plant compounds, which lead to the attraction of even more beetles (Loughrin et al. 1996). Usually considered as sun-loving insects, they will nevertheless spend part of their time in the shade. High levels of light and temperatures are known to enhance activity (Moore and Cole 1921) and the tendency to alight and fly toward lures (Heath et al. 2001). Companion plants have little effect (Held et al. 2003). On some hosts, beetles prefer to feed also on flowers (Held and Potter 2004), especially if the flowers are in an elevated position, but grape flowers are not very attractive. Japanese beetles prefer to lay eggs in soils with high moisture content (Allsopp et al. 1992b).

A collection of grape cultivars was rated in terms of vulnerability of JB attack by Langford and Cory (1948). Their rating system is as follows: **Group 1 –** Preferred cultivars (Injury very severe): Cabernet Sauvignon, Pinot Chardonnay, Baco No. 1, Delaware, Seibel 128, Seibel 1000, Seibel 1xx, Seibel 2xx, Seibel 2056, Seibel 6339, Seibel 5409, Seibel, 9110, Seibel 5279, Couderc 13, Couderc 4401, Bertile-Seyve 2862, Seyve-Villard 12309, Norton, Cynthiana (Norton and Cynthiana are both *Vitis aestivalis* Michaux, and generally considered synonymous), America, Bell, Brilliant, Manito, Rommel, Wine King, N.Y. 10839, N.Y. 1407, N.Y. 11456, N.Y. 13920, N.Y. 20159; **Group 2 –** Attractive cultivars (injury severe): Catawba, Delicatessen, Cloeta; **Group 3 –** Cultivars frequently attacked (injury moderate): Westfield (close to 'Concord') Lona, Diamond; **Group 4 –** Unattractive cultivars (injury light and occasional): Champanel. Most cultivars used in wine production fall into Group 1 (preferred).

An evaluation of JB preference for grape cultivars was carried out by Gu and Pomper (2008). A point system was assigned based on the % of damaged leaves per vine, as well as of leaf area loss. Cultivars with >70% incidence of injury were generally European or French hybrids, whereas those with <70% injury were either American cultivars, or hybrids with some *Vitis labrusca* L. parentage. The cultivars Marquis, Reliance, Catawba, Concord Seedless, Concord, Edelweiss, and Einset showed promise for arthropod management with reduced insecticide use. It is noteworthy that Catawba was nevertheless in Group 2 (attractive) of Langford and Cory (1948).

Beetles cause a skeletonizing type of injury to grape leaves, although they may eat completely through the leaves on some cultivars. Berries are rarely attacked by Japanese beetle. However, when this happens, injury can be exacerbated by secondary feeding by the green June beetle. Adults form dense aggregations on selected leaves, generally feed at the top of the canopy, a typical behavior on their host plants. Leaf injury is thus greatest in the upper parts of the canopy. This stratification toward upper parts of the canopy is due to visual orientation, not some nutritional factor (Rowe and Potter 1996).

Through their feeding, JB predispose vines to further infestation (Iwabuchi and Takahashi 1983). Feeding by JB induces the release of volatile compounds that are attractive to other beetles of both sexes, and mechanically injured leaves do not show this response (Loughrin et al. 1995).

Pairs of beetles are often seen together on plant tissues. Females are often first mated as they emerge from the ground (Fleming 1972). Copulation lasts about 2 min, but males may remain mounted for an additional 2 h (Barrows and Gordh 1978). Occasionally males contest for females, whereby the occupying male usually wins, unless the intruding male is significantly larger (Kruse and Switzer 2007). Eggs are fertilized by sperm from the most recent mating (Ladd 1966). During the period of adult activity, females will make repeated trips to the soil to lay eggs, and 1–4 eggs are laid at a time. Females prefer to oviposit in moist grassy areas. The eggs hatch in about 2 weeks, and then the larvae feed on grass roots until the onset of cold weather, when they descend several centimeters deep in the soil. In southern parts of the range, or when winters are mild, larvae may not leave the root zone. In spring, the larvae return to the root zone to resume feeding until they begin to pupate in May.

There are reports that JB infestations are most severe in the Mid-Atlantic States, where there are large acreages of larval habitats (pastures) adjacent to vineyards, the preferred adult food. This combination is very favorable for the growth of JB populations (Régnière et al. 1983). In Massachusetts, adults are active from mid-July to mid-August, peaking in late July. In a Massachusetts study, the majority of adults were not reproductively mature until late August, and eggs were recovered from turf in early September (Vittum 1986). In this northern part of the range, at least a portion of the population requires 2 years to complete its development. Infestations vary widely from year to year. This is partly influenced by seasonal rainfall patterns, because the eggs are susceptible to desiccation in dry soil. Rainfall should be at least 250 mm and distributed uniformly over the summer for good survival (Fleming 1970, 1972).

17.2.3 Impact of Japanese Beetle Feeding

A survey of grape grower practices in Virginia in the mid-1980s revealed that most of the insecticide sprays in Virginia vineyards were targeted against JB, because of its conspicuous feeding injury (Pfeiffer et al. 1990). Consequently, a study was initiated to determine the effects of this feeding on berry yield and quality (Boucher and Pfeiffer 1989). Four feeding treatments were compared on the French hybrid 'Seyval Blanc' in the upper Shenandoah Valley: a natural unprotected plot, a controlled plot where beetle feeding was prevented, and two caged plots where high numbers of beetles were contained on vines (1) from the beginning of beetle activity to veraison, and (2) from veraison to harvest. The natural infestation did not result in any significant reduction in berry quality, yield or vine growth, despite greater leaf area loss than in the control (6.5% versus 3% leaf area loss, respectively). Intensive feeding by JB after veraison caused the most severe effects on berry quality. These vines had 11% leaf area loss when averaged over the whole vine (initial visual impact of feeding may be misleading because feeding is more intense on the upper leaves). This loss occurred in less than one half the time relative to natural feeding, about 3 weeks compared with 6 weeks, respectively.

Although established vines can tolerate injury caused by JB feeding, young vines can be totally defoliated and should be protected more rigorously, especially when grown in tubes (i.e. plastic cylinders often placed around newly planted vine trunks, intended to provide protection). Hence, vines are most vulnerable when young, and also after veraison once they are mature.

A Michigan study compared the effects of feeding by rose chafer and JB on berry quality of *V. labrusca* 'Niagara' vines (Mercader and Isaacs 2003). These beetles attack vineyards in early and mid-season, respectively. Feeding around bloom by rose chafer resulted in a loss of less than 1% LAL. Feeding during veraison by JB resulted in about 7% LAL, somewhat lower than the levels determined by Boucher and Pfeiffer (1989). These levels of feeding caused no differences in vine growth parameters. That study also included an artificial leaf area removal experiment, removing up to 30% of each fully expanded leaf at either bloom or veraison. While this level of defoliation caused reduced trunk diameters measured at veraison, there were no differences among treatments by the time of leaf abscission in the fall. Berry parameters were not evaluated in that study. Young vines were able to tolerate levels of feeding exceeding those imposed by population levels used in the study.

A study conducted in Kentucky showed that there are important cultivar-specific differences in sensitivity of vines to JB feeding (Hammons et al. 2010). The study compared six cultivars: two American cultivars (*Vitis labrusca* L. 'Concord', *Vitis aestivalis* Michaux 'Norton'), two European cultivars (*Vitis vinifera* L. 'Cabernet Franc', 'Cabernet Sauvignon'), and two French-American hybrids (*V. vinifera* × *Vitis riparia* Michaux 'Chambourcin' and *Vitis* sp., interspecific hybrid 'Frontenac'). The percent defoliation levels noted for the 2 years of the study were: 'Concord' = 7, 5; 'Cabernet franc' = 39, 35; 'Frontenac' = 38, 37; 'Norton' = 44, 44; 'Chambourcin' = 46, 43; 'Cabernet Sauvignon' = 48, 38. With its thicker epidermis, 'Concord' had markedly less injury than all the other cultivars, which were very close together in terms of defoliation. 'Concord' is in Group 3 of Langford and Cory (1948), while all of the other cultivars fall in Group 1 (preferred). Hammons et al. (2010) adjusted defoliation levels by using different pesticide regimes: carbaryl every 7 versus 14 days, or no insecticide. 'Norton' exhibited reduced vine growth, delayed synthesis of sugars, and reduced yield, while 'Concord' showed little effect. Insecticides to protect 'Concord' grapes from JB offered no benefit for vine growth or cluster yield and quality.

17.2.4 Pheromone Biology and Monitoring

Trapping for JB began soon after its establishment in the United States. Much research attention was given to this area for several decades, not only to determine phenology and to time control measures, but to follow the spread of this invasive species. In the 1930s, traps baited with a 1:10 blend of eugenol and geraniol were used with or without phenethyl alcohol (Britton and Johnson 1938).

In 1970, evidence for a sex pheromone in JB was discovered (Ladd 1970), and there were attempts to use adult females as lures (Goonewardene et al. 1973). Male extract caused greater electroantennogram response than did a female extract (Adler and Jacobson 1971). The sex pheromone was eventually described as (*R,Z*)-5-1-decenyl)-dihydro-2(3*H*)-furanone (Doolittle et al. 1980) and given the name Japonilure. The *R* enantiomer is required for attraction and contamination with small amounts of the *S* enantiomer inhibits attraction. This is part of a reproductive isolating mechanism used by the sympatric scarab *Anomala osakana* Sawada in its native Japan because this species uses *S*-japonilure as its pheromone (Leal 1998).

In more recent years, a lure containing PEP (phenethyl propionate) became standard for monitoring JB. Adding eugenol to PEP enhanced captures, as did addition of Japonilure, the sex pheromone (Ladd et al. 1981; Ladd 1986). The sex pheromone Japonilure is more effective when used with plant volatiles than when used alone (Klein et al. 1981; Allsopp et al. 1992a).

In a study to evaluate a visual component, white traps were determined to be the most attractive, followed by yellow. Shielding traps to limit emission of attractants by trapped virgin females enhanced trap captures by keeping attracted beetles from accumulating on the outside of the canister (Klein et al. 1973). While bag type traps sometimes lose efficiency because of beetles escaping through drain apertures, the larger volume of such traps is useful during periods of high beetle activity (Klostermeyer 1985). Agronomic habitats such as fields of corn and soybean increased trap catches (Hamilton et al. 2007).

Beetle captures increase if trapped beetles are removed each day before decomposition occurs. Traps are highly attractive and they may become filled with beetles quickly and may need to be serviced frequently in times of high beetle activity. However, traps do not provide control of JB, possibly because more beetles are attracted into the area than are collected. In fact, defoliation near traps is sometimes greater than where no traps are present (Gordon and Potter 1986). If traps are to be used as part of a JB management program, they should not be placed near the vines to be protected. Instead, they should be placed some distance away, upwind of the vineyard, so that attractant volatiles will drift over the crop, allowing beetles to be attracted upwind to the traps while minimizing attraction of additional beetles into the site.

Traps are most effective for monitoring, including detection of isolated populations. Use of JB traps aided the successful eradication of isolated populations of this pest in California (Alm et al. 1996). Some of the other species of *Popillia* in Asia (e.g., *P. lewisi*, *P. uchidai* and *P. indigonacea*) respond to lures for JB, although there are some specific differences (Klein and Edwards 1989; Reed et al. 1991).

17.2.5 Biological Control

Biological control of JB was reviewed by Fleming (1968). Explorations in the native range of the beetle began soon after its establishment in the eastern United States. Two entomologists searched in Japan and other parts of Asia for several years beginning in 1920 (Fleming 1968). Fleming (1968) provided a list of parasites and predators, including some that were released but were not known to have become established. The most important species will be discussed here.

Tiphia vernalis Rohwer (Tiphiidae) emerges in the spring, and overwinters in the pupal stage. *Tiphia popilliavora* Rohwer (Tiphiidae) emerges in summer and fall, and overwinters in the larval stage. Both species are specialists on JB in Japan. Frass kairomones help orient *Tiphia* to its hosts (Rogers and Potter 2002). Both species were released in New Jersey, *T. vernalis* beginning in 1921, and *T. popilliavora* in 1925. Both *Tiphia* species were released in mid-1930s in Connecticut, where they successfully established (Britton and Johnson 1938). *Tiphia vernalis* was considered to be the most effective of the introduced parasitoids. When adult wasps are active in May, the grubs are in the third instar, the primary target of ovipositing female *Tiphia*. Furthermore, there was a strong density dependent numerical response, with % parasitization increasing with high JB density. *Tiphia vernalis* is now found in every county of Connecticut, after the state introduced this natural enemy, including at two sites where it was never released, reflecting natural spread (Ramoutar and Legrand 2007).

Three geographical strains of *T. popilliavora* were released from Japan, Korea and China (Fleming 1968). The Japanese strain flies in August and September. For the first half of its flight, most grubs are in first and second instars, not preferred by the hunting wasps. Later in the flight, most JB are in second and third instars. The third instar is preferred by the parasitoid. The Korean strain is active a little later in September, when most grubs are in the third instar. Hence, this strain is more closely matched to JB phenology. Insecticides have inconsistent negative effects on parasitism by *T. vernalis*. Parasitism was greater when an insecticide was combined with the parasitoids (Oliver et al. 2005). Nevertheless, bifenthrin, chlorpyrifos, and imidacloprid lowered survival of adult *T. vernalis*, while halofenozide had a minimal effect (Oliver et al. 2006). Isophenphos and diazinon decreased predation by ants on JB immature stages, whereas imidacloprid and a halofenozide treatment had no effect (Zenger and Gibb 2001a). The *Tiphia* species are now widely established, but sporadic in occurrence.

A univoltine tachinid, *Istocheta aldrichi* (Mesnil) (formerly known as *Hyperecteina aldrichi*), parasitizes adult beetles. Eggs are laid on the pronotum of mating female JB. About 36–48 h after eggs are laid, larvae drill downward into the body cavity where internal organs are consumed, killing the beetle usually within 5 days (Fleming 1968). In Japan it is a specialist predator of JB (Fleming 1968). In the United States, it is not well synchronized with JB and only attacks the earliest-emerging adults, missing the peak of JB activity. *Istocheta aldrichi* was first released in New Jersey in

1922, and over nearly 30 years, in more than 50 sites in 12 states, with successful establishment occurring in most (Fleming 1968). Another tachinid from Japan, *Centeter cinerea* Aldrich, was released but its establishment is unknown (Britton and Johnson 1938), and it was not mentioned by Fleming (1968). This species was thought to be the most successful in northern parts of the range of JB (King 1931).

Ants may also be a source of natural mortality for JB eggs (López and Potter 2000). While Fleming (1968) believed that ant predation would usually impose insufficient mortality on JB, pesticide impact studies have shown generalist predators to be a significant source of natural mortality. Ant-induced mortality of eggs has been reported to exceed 80% (Zenger and Gibb 2001b).

Japanese beetle larvae are subject to attacks by a bacterium, *Paenibacillus* (formerly *Bacillus*) *popilliae* (Dutky), causing milky disease. After sporulation, the hemolymph turns milky white, hence the name of the disease. It was found naturally infecting JB grubs in NJ in 1933 (Fleming 1968). A second bacterium, *Paenibacillus lentimorbus* (Dutky), was also found in the grubs. These pathogens are thought to be natural mortality agents for JB only. However, other scarab species are infected by different host races of *P. popillae*. These races cause mortality mainly in the scarab species in which they were collected (Fleming 1968). There have been reports in recent years of lower efficacy of commercial preparations of *P. popilliae* (Dunbar and Beard 1975), including contamination of preparations with nonpathogenic *Bacillus* species (Stahly and Klein 1992). In fact, preparations known to contain *P. popilliae* led to incomplete control. Therefore, milky disease was considered as one of a complex of agents that could help suppress JB populations, but not as a stand-alone control tactic. Infection levels of larvae in a Connecticut survey were only 3.5% (Hanula and Andreadis 1988). In addition to quality control problems, there are also environmental variables that may slow the development of efficacious soil titers of bacteria. Soil temperatures of 21°C are needed for bacterial development, and the number of weeks above 21°C will affect the time required by the bacteria to become effective (Fleming 1968). This biological control agent can be used in grassy areas with large larval populations, but it is ineffective against adults entering the vineyard. Adults are capable of flying great distances and may invade the vineyard from untreated areas. Consequently this organism is more important in turf management of JB than in fruit systems.

Entomopathogenic nematodes attack JB (Fleming 1968; Koppenhöfer et al. 2000). One of the most important species was reported to be the entomogenous nematode *Steinernema* (Formerly *Neoaplectana*) *glaseri* (Steiner) (Fleming 1968). This species was released over wide areas, but most of these were later deemed unsuccessful, owing either to low tolerance of cold temperatures (this nematode was found only in southern New Jersey), or to the elimination of the bacterial symbionts in the rearing procedures. Symbionts are needed to overcome host defenses, and this became known only more recently (Gaugler et al. 1992). In some cases there are differences in susceptibility, but in others there is a uniform response (Koppenhöfer and Fuzy 2004). Root cues enhance infection by *S. glaseri* and *Heterorhabditis bacteriophora*

Poinar (Wang and Gaugler 1998). *Steinernema glaseri* and *H. bacteriophora* were the most effective nematodes against JB (Wang et al. 1994). The Japanese beetle showed a strong encapsulation defense against all injected nematodes except *S. glaseri*. Of the three nematode species (*Steinernema carpocapsae* (Weiser), *S. scapterisci* n. sp., and *H. bacteriophora*) that induced the encapsulation response, *H. bacteriophora* and *S. carpocapsae* were able to overcome the response, but *S. scapterisci* was not. *Steinernema glaseri* was also found to be the most effective nematode (Alm et al. 1992), although a high level of control was not consistently attained. *Steinernema glaseri* was the most common nematode collected in North Carolina (Régnière and Brooks 1978). In a study in The Azores, *S. glaseri* and *H. bacteriophora* caused complete mortality of larvae. *Steinernema carpocapsae* caused almost 60% mortality. It was also reported to be an inferior control agent for JB by Georgis and Gaugler (1991). The entomopathogenic nematodes *H. bacteriophora* HP88 and *H. marelatus* Liu & Berry performed poorly to moderately (Mannion et al. 2001). Elsewhere, *H. marelatus* has outperformed insecticides in other trials (Mannion et al. 2000). Irrigation immediately before and after application of entomopathogenic nematodes improves the level and consistency of control (Downing 1994).

Strains of nematodes with more effective host detection ability have been identified. These strains have increased ability to detect CO_2, and hence non-diapausing larvae, but their ability to find diapausing JB larvae has not improved (Gaugler and Campbell 1991). When these nematodes are widespread they may protect turf from white grub feeding and decrease population pressure. However, they are usually insufficient to protect grapevines from immigrating adult beetles.

Larvae of JB have some defense against entomopathogenic nematodes. Through grooming by rubbing with their legs or raster, and by host encapsulation, successful infection rates are decreased (Gaugler et al. 1994; Wang et al. 1995). There can also be avoidance behavior, with JB grubs moving to sections of grass plantings not treated with *H. bacteriophora* (Schroeder et al. 1993; Gaugler et al. 1994).

Other pathogens have been evaluated for JB management. In a survey for white grub pathogens in Connecticut (Hanula and Andreadis 1988), four of the seven species of scarabs encountered in 49 sites were exotics and made up 91% of the samples. Cephaline gregarines were the most (42 sites) widely distributed pathogens. The microsporidium fungus *Ovavesicula popilliae* n. g., n. sp. was found in JB from 34 sites. Overall, 25% of the larvae were infected, but prevalence was 80–90% in some locations. Described from JB (Andreadis and Hanula 1987), this pathogen lowers fecundity of JB by 50% (Hanula 1990). The fungus *Metarhizium anisopliae* (Metschnikoff) Sorokin infected about 1.2%. In a Michigan study, some parasitoids and parasites common in more eastern States were uncommon. The most common parasite was a cephaline gregarine (*Stictospora* sp.) (Cappaert and Smitley 2002), described as *S. villani* n. sp. (Hays et al. 2004). Dutky and Gooden (1952) described the rickettsia *Coxiella popilliae* (now *Ricketsiella popilliae* (Dutky and Gooden) Philip), that causes a blue disease in larvae. Though not present in all survey sites, it was common in some and thought to have potential as a microbial control agent. Koppenhöfer et al. (2000) reported *Bacillus thuringiensis* Berliner var. *japonensis* strain Buibui, to cause limited to high mortality. According to Mannion et al. (2001) this strain of *B. thuringiensis*, and *Beauveria bassiana* (Balsamo) Vuillemin caused poor to moderate mortality.

Some vertebrates also feed on JB, either in the adult or larval stages. Fleming (1968) listed several species of birds whose stomach contents contained remains of JB. Common grackle was the most important avian predator of JB adults, followed by meadowlark, European starling, northern cardinal and catbird. Among mammals, the most important predator was the skunk, which digs below the soil surface for grubs. The disruption of turf by hunting skunks poses a secondary problem resulting from larval presence, mainly in golf courses and other high-value turf. Hogs, moles and short-tailed shrews were other mammalian predators of JB larvae (Fleming 1968). Vertebrate predators are rarely able to make a significant impact on JB numbers in an area.

17.2.6 Cultural Control

In a study of ovipositional preferences, Wood et al. (2009) suggested that planting hybrid Bermudagrass may decrease JB oviposition and the resulting infestations of white grubs in high value turf. Larval densities can be reduced by planting non-grass cover crops in perennial fruit plantings (Szendrei et al. 2005; Szendrei and Isaacs 2006), but this may also lead to increased feeding by the adults. Withholding irrigation during peak JB flight may successfully reduce larval populations (Potter et al. 1996). Beetles move more slowly in strip cropped soybean (Bohlen and Barrett 1990), but in this case the dwarf sorghum plants used as interplants were about the same height as the soybean plants. Unfortunately, this would be hard to implement in vineyard settings.

The use of geranium as a companion plant to protect against JB has also been examined. Zonal geranium (*Pelargonium* x *hortorum*) was reported to cause narcotic/paralytic effects in JB by Fleming (1972). This was confirmed by Potter and Held (1999), who reported that naïve beetles prefer geranium petals over the otherwise attractive linden leaves, undergoing a temporary period of paralysis thereafter. The paralysis ensued rapidly and lasted 12–16 h. The preference for geranium petals was retained even after several bouts of paralysis resulting from their consumption. Such relative comparisons must be done for different crops. For example, geranium petals are less competitive with leaves of raspberry (Maxey et al. 2009); companion or trap planting may be less likely in the raspberry system than in grape, and this warrants further investigation.

Endophyte-infected grasses have resistance against some phytophagous insects because of toxic alkaloids. While the alkaloids in endophyte-enhanced grasses are mainly active against defoliators of grasses (alkaloids are absent in roots (Breen 1994)), there is some evidence for a negative effect of such grasses on JB grubs (Potter et al. 1992).

17.2.7 Chemical Control

17.2.7.1 Adult Control

Carbamates and organophosphates have long been employed against JB. Carbaryl has been a standard insecticide causing high mortality and rapid knockdown (Lockwood et al. 2010). However it is detrimental to beneficial arthropods and may induce secondary pest outbreaks. Phosmet is one of the few organophosphates currently registered for vineyards in the United States (Wise et al. 2009), and is highly active on JB. Pyrethroids have been reported to be more effective than carbaryl (Baumler and Potter 2007). However, they are even more toxic to predators and parasitoids, and have been linked with outbreaks of grape mealybug in vineyards, a vector of grapevine leafroll virus.

Neonicotinoids are a newer class of insecticide that are used on a wide range of cropping systems. Imidacloprid has both lethal and sublethal effects. Direct mortality is most evident when berries and leaf surface residues are high, thereafter sublethal feeding deterrent effects become evident (Wise et al. 2007). Acetamiprid is moderately toxic to JB (Williams and Fickle 2007, 2008). Thiamethoxam provides some repellent activity for JB, contributing to efficacy of a pre-mix blend sold under the trade name Voliam Flexi®, where defoliation was decreased without a significant reduction in JB numbers (Wise et al. 2009). Other new chemical classes have representatives that can be used for JB control, including indoxacarb and chlorantraniliprole (Williams and Fickle 2007, 2008).

Organic Adulticides

Particle film technology such as kaolin successfully reduced JB adults and their damage in peach (Lalancette et al. 2005). A disadvantage of this product is the high use rate recommended, i.e. 28–56 kg/ha per application. Furthermore, the label (CDMS 2010) warns that for wine grapes: 'Harvest parameters can be altered and maturity can be delayed especially in white wine varieties. Harvest parameters have to be closely monitored to determine optimal time to harvest. Changes in harvest parameters can affect final taste. Wine grapes sprayed up to veraison will have minimal adherence to berries. Applications after veraison will adhere more on grape berries.' An advantage is that this product also controls some diseases, and protects against sunburn in hot regions of production.

Natural Insecticides

Azadirachtin is an extract from the neem tree, originally from India and Africa. It has a complex mode of action and it acts as an insect growth regulator (inhibiting biosynthesis of ecdysone) and as a repellent. As an insect growth regulator azadirachtin has no effect on adult JB. Ladd et al. (1978) showed that extracts of neem seeds were highly repellent to adult JB, protecting sassafras and soybean leaves almost completely from JB feeding. Some commercial formulations of azadirachtin also performed well against adults on peach (Lalancette et al. 2005). In Florida, studies showed azadirachtin to cause low morbidity and mortality against adult JB (Vitullo and Sadof 2007a, b). Those workers found that at low JB pressure the use of repeated azadirachtin sprays exerted some control, but a single application did not. Some effect was seen by removing beetle-marked plant tissue, but this was not enough to be of importance. A commercial azadirachtin formulation (Aza-Direct®) has been recommended for JB (Lockwood et al. 2010), but it was found ineffective to control JB on primocane-bearing raspberries by Maxey et al. (2008). A second formulation (Neemix®), applied alone also did not differ from untreated plots. The second formulation, when combined with a clarified hydrophobic extract of neem oil (Trilogy®), was effective (L. M. Maxey and D. G. Pfeiffer, unpubl. data). JB was found to habituate to residues of azadirachtin (Held et al. 2001). However, those authors felt that such habituation would have little significance in the field because of the great mobility of adult beetles among a wide host range.

17.2.7.2 Larval Control

Larval control is more important in turf management of JB than in fruit systems. However, there may be a role for larval management in overall suppression of beetle populations in an area. This may be practical in an area where high cash value turf (i.e., golf courses or residential lawns) are near vineyards, but less so when pasture or range are nearby. Both types of habitats abound in the eastern US. Imidacloprid, thiamethoxam, and halofenozide performed well against white grubs in turf (Cowles et al. 1999). Soil type can also affect the relative efficacy of soil insecticides. Some insecticides can decrease the efficacy of nematode control (Cowles and Villani 1994). Conversely, synergism has been reported between neonicotinoids and entomopathogenic nematodes used together to control JB larvae (Koppenhöfer and Kaya 1998). Azadirachtin completely interrupted normal development of larvae (Ladd et al. 1984). While resistance is not often reported against JB, possibly because of the large population of unsprayed beetles with such a wide host range, resistance to cyclodiene insecticides was reported after repeated use of this class of insecticides to control larvae in turf (Niemczyk and Lawrence 1973).

17.3 Green June Beetle

17.3.1 Appearance

The adult green June beetle (GJB), *C. nitida* (Scarabaeidae), is about 25 mm long and 13 mm wide, and flat on the top. Beetles are dull velvety green above, with deep yellow to bronze margins and metallic green below (Fig. 17.2). Grubs are grayish white and considerably larger than JB grubs, less C-shaped than other white grubs, though when disturbed they will coil tightly.

Fig. 17.2 Green June beetle adult

17.3.2 Biology

The green June beetle has a similar life history to JB and although differences in injury to grapevines exist, there are some similarities such as causing mainly skeletonizing injury to foliage. Adult GJB feed on the foliage of many shrubs and trees and will attack most tree fruits and berries. There is one generation per year. Grubs overwinter up to 30 cm below the soil surface. They gradually make their way close to the surface during the spring and feed mainly on rich organic matter such as decaying plant material, and to a lesser degree on roots. Larvae may leave their protected sites and crawl on their backs to establish a new site elsewhere. By May, grubs have pupated. Adults emerge in early July and August. Females oviposit in soil with decaying vegetation. Adults feed on petioles, leaves and fruit, and a single beetle can cause significant damage. Adults are often found in groups and take large chunks from the fruit.

Adult GJB are unable to break through the skin of grape berries. However, JB or other factors such as hail, yellowjacket injury may break the berry skin and allow GJB to feed (Hammons et al. 2008). A study of head space volatiles of fermenting apples led to development of a 5-component blend that was equally or more attractive than the natural material (Johnson et al. 2009). Males and females feeding on ripe fruit emit an aggregation pheromone (Domek and Johnson 1988) that is produced by yeasts in the diet or digestive tract of the beetles (Domek and Johnson 1990). More males are attracted to feeding females than to males, but females are attracted equally to either sex. Three to six days are required after feeding before the pheromone is produced (Johnson and Vishniac 1991). Adult beetles lack yeasts at the time of emergence but they acquire the microflora as they feed (Vishniac and Johnson 1990). There is also evidence for a sex pheromone (Domek and Johnson 1987). Before coupling for mating, both sexes cast about in a zigzag pattern, until the male drops on the female, hooking front tarsi at the leading edge of the female pronotum (Patton 1956).

In a search for an attractant for this species, molasses was found to be more attractive than a variety of other candidate feeding attractants (Wylie 1969). Green June beetle adults are also attracted to isopropanol bait (Landolt 1990). Once mated, females lay eggs in rich soils (Chittenden and Fink 1922).

17.3.3 Importance of Injury

Fruit injury is more common than that caused by JB, and it is more likely to occur when populations are large. Most injury to grapes is seen in late July and August, and unlike injury from JB, it can occur on both unripe and ripening fruit.

17.3.4 Monitoring

Traps used for JB are somewhat effective for GJB, but are only used to indicate the initial adult emergence. Direct fruit counts by examining berry clusters on the vine are the most effective way of assessing damage. Since feeding may be unevenly distributed, every effort should be made to collect a representative sample before deciding on control measures. A treatment is justified if feeding exceeds 1% of clusters examined. Adults may be monitored by quietly jarring several cordons along the vine row, and counting how many beetles fly off.

17.3.5 Biological Control

In a study in Norfolk, Virginia (Chittenden and Fink 1922), two sarcophagid parasites were reared from GJB. *Sarcophaga utilis* Aldrich, was reared from adults, and *S. (Helicobia) helicis* Townsend was reared from both pupae and adults. The latter species is half the size of *S. utilis*, but more common. A digger wasp, *Discolia dubia* Say, was also collected. Several other insect predators were listed in that study. The fungal pathogen *M. anisopliae* infected GJB, and several birds were found to be predatory.

17.3.6 Chemical Control

Generally the same insecticides are recommended for GJB as for JB, although GJB may be more difficult to control. Carbaryl has been a standard material used for control of both species. Thiamethoxam, imidacloprid, chlorantraniliprole, deltamethrin, beta-cyfluthrin, clothianidin, fenpropathrin, and carbaryl all provided a high degree of control. A blend of 10% rosemary oil with 2% peppermint oil, metaflumizone, and a plant oil extract from *Chenopodium ambrosioides* (Requiem®) have all provided moderate control. The Aza-Direct formulation of azadirachtin provided inadequate control (Johnson and Lewis 2008, 2009).

17.4 Rose Chafer

17.4.1 Appearance

The adult rose chafer (RC), *M. subspinosus* (Scarabaeidae), is 13 mm long, with a straw colored body, reddish brown head and legs (Fig. 17.3). The legs bear long spines. The larva is about 19 mm long (McCleod and Williams 1990).

17.4.2 Biology

The rose chafer emerges in late May or early June in the southern part of its range, and mid-June in the north. It ranges from Canada and Minnesota to Virginia and Tennessee, west to Oklahoma and Colorado. It is most destructive from southern

Fig. 17.3 Rose chafer adults (Photo by Rufus Isaacs)

New England to the mid-Atlantic states. Rose chafer is polyphagous, but rose and grapevines are among the most vulnerable hosts, where it feeds on blossoms, leaves and berries. Adults are active from 4–6 weeks. Females lay 24–36 eggs singly, several cm below the soil surface. Eggs hatch in 2–3 weeks, feeding on grass roots until cold weather, when they descend below the frost line (Chittenden and Quaintance 1916). Eggs are laid preferentially in soils with high moisture content (Allsopp et al. 1992b).

17.4.3 Monitoring

The same attractants for JB are effective for rose chafer. However, addition of eugenol does not increase trap captures of RC (Williams and Miller 1982). Caproic and valeric acids have been reported as potential attractants (Williams et al. 1982). Monitoring should take place during grape bloom, since the adults will feed on clusters at that time and in the following few weeks.

17.4.4 Cultural Control

Site selection affects vulnerability of grapevines to RC. This species is mainly a problem in vineyards on sandy soils (McCleod and Williams 1990).

17.4.5 Chemical Control

With the regulatory demise of chlorinated hydrocarbon and organophosphate insecticides, carbaryl is now a standard recommendation (Lockwood et al. 2010). Acetamiprid is also effective and is less disruptive to biological control. Phosmet and fenpropathrin may also be recommended (Bordelon et al. 2011).

17.5 Grape Flea Beetle

17.5.1 Appearance

The adult grape flea beetle (GFB), *A. chalybea* (Chrysomelidae), is a metallic blue-green beetle and is almost 5 mm long (Fig. 17.4). Eggs are light yellow and are laid in masses. They hatch in a few days and larvae feed on grape leaves for 3–4 weeks. Larvae are brown with black spots, and reach a length of 10 mm (Fig. 17.5).

Fig. 17.4 Grape flea beetle adult

Fig. 17.5 Grape flea beetle larva and associated injury to grape foliage (Photo by Rufus Isaacs)

17.5.2 Biology

Adult grape flea beetles overwinter in debris in and near the vineyard. They become active early in spring and lay eggs in cracks in the bark, at bases of buds, between bud scales, and on leaves. After feeding on foliage, mature larvae drop to the ground and pupate in an earthen chamber. Adults emerge 1–2 weeks later in July and August. They feed on grape foliage for the rest of the summer causing little damage. In the fall they seek protected places in which to overwinter. In addition to wild and cultivated grapes, grape flea beetles feed on Virginia creeper. The grape flea beetle is more common in neglected vineyards, but some eastern commercial growers consider this species their main insect pest, especially in vineyard rows near deciduous forest.

17.5.3 Importance in Injury

Larval feeding damage consists of characteristic chain-like feeding marks on leaves. Individual leaves may become very ragged in appearance, but real effect on vines or crop is rare. The damage by adult grape flea beetles is more important. The beetles eat holes into the sides of buds and gouge out the contents as the buds swell. They also feed on the unfolding leaves. Once the young shoots have grown past 5–12 cm, they are no longer vulnerable.

This pest may escape accurate identification because the injury caused by adults can easily be mistaken for that caused by climbing cutworms. Injury by the latter is more likely to be ragged in appearance, though there is an overlap in appearance. The grape flea beetle injury is heaviest near wooded edges, whereas cutworm injury may be spread throughout the block. Injury to the leaves by larvae may be confused with that caused by adults of grape rootworm. Proper identification is of paramount importance to take appropriate management measures.

17.5.4 Chemical Control

Insecticide applications directed against grape berry moth aid in controlling GFB. However, where a history of damage is known, targeted adulticides may be needed in early season. The pyrethroids fenpropathrin and beta-cyfluthrin and the carbamate carbaryl have provided very good control of GFB (Lockwood et al. 2010). Pyrethroids are very damaging to populations of natural enemies. This undesirable effect may be less pronounced at the bud swell stage than in the summer. Although phosmet is effective, a recent lengthening of the restricted entry interval in the US to 14 days has made it impractical for many growers. In India, carbaryl and monocrotophos provided good control of another flea beetle attacking grapevines, *Scelodonta strigicollis* Mots (Rao et al. 1983). However, larvae of this species are found in the soil rather than being foliar feeders (Rao et al. 1984).

17.6 Grape Rootworm

17.6.1 Appearance

The adult grape rootworm (GRW), *F. viticida* (Chrysomelidae), is a chestnut brown beetle about 6 mm long, covered with tiny yellow-white hairs (Fig. 17.6). Creamy-white egg clusters of 20–30 eggs are laid on canes or under loose bark. Larvae are white with a brown head capsule.

Fig. 17.6 Grape rootworm adult

17.6.2 Biology

Grape rootworm was earlier reported as the most destructive grape pest in the Chautauqua-Erie grape region in New York State (Hartzell 1918), and caused the beginning of entomological research in that region (Jubb 1977). Egg clusters of 20–30 eggs are laid on canes or under loose bark, averaging about 100 eggs per female. Eggs hatch in 1–2 weeks and larvae drop to the ground, enter the soil, and feed on grape roots until cold weather. Overwintering takes place among the roots, at depths of 1–2 cm to more than 50 cm. In spring, feeding on roots is resumed. Pupation cells are formed close to the surface, usually 40–60 cm from vine bases, about the time of grape bloom. Adults appear about 2 weeks later. The grape rootworm has been most severe in the Chautauqua and Lake Erie regions. An effect on yield is difficult to quantify. Numbers of eggs deposited are the best reflection of feeding intensity. Hartzell (1918) recommended that if beetles become a problem, a spray should be applied within a week of beetles becoming active, repeating 10 days later. If populations are high, pesticide applications are recommended the day the adults appear (Hartzell 1918). A related species, *Fidia longipes* Melsheimer, was reported to have replaced GRW as a pest in Arkansas (Isely 1930).

17.6.3 Importance of Injury

Foliar injury is caused by adult feeding. Adults feed on foliage for a month or more, making chain-like feeding marks (Fig. 17.7) similar to those made by larval grape flea beetles. Larvae consume smaller roots and eat pits into larger ones. Root injury

Fig. 17.7 Adult grape rootworm feeding injury to grape foliage

has a much greater impact on the vine than the foliar feeding of adults. As a result of grape rootworm larval infestation on roots, vines become unthrifty, yield is reduced, and in cases of continued high infestation over several years, vine death may occur. Root damage by grape rootworm will be compounded by planting in poor soil.

17.6.4 Biological Control

No information is available on biological control of GRW.

17.6.5 Cultural Control

Until adults emerge in late June, intensive shallow cultivation of soil may destroy pupae. If an infested vineyard is removed and planted immediately back to grapevines, GRW that had remained in the soil can concentrate onto the young poorly developed root systems and create an immediate risk in the young vineyard.

17.6.6 Chemical Control

Few chemical control studies have been conducted because of the relatively low importance of this insect in most areas. Carbaryl is a standard recommendation (Weigle et al. 2010), targeted at adults soon after they become active, and before egg-laying begins. Early sprays of some materials directed against grape berry moth not selective for Lepidoptera (e.g., carbaryl, phosmet, fenpropathrin, cyfluthrin, beta-cyfluthrin, bifenthrin, methomyl, or diazinon) may provide control of GRW (Bordelon et al. 2011).

17.7 Conclusion

Several coleopteran foliar feeders are associated with grape in eastern North America. The most important of these is the JB, often responsible for extensive defoliation. This defoliation is very damaging to young vines. Mature vines can tolerate substantial foliar feeding by JB without affecting fruit yield or quality. Green June beetle causes similar foliar injury. The other species (grape flea beetle, rose chafer, and grape rootworm) cause minor injury to foliage but cause more important injury on other plant parts (expanding primary buds, young clusters, and roots respectively).

References

Adler VE, Jacobson M (1971) Electroantennogram responses of adult male and female Japanese beetles to their extracts. J Econ Entomol 64:1561–1562

Allsopp PG, Klein MG, McCoy EL (1992a) Effect of soil moisture and soil texture on oviposition by Japanese beetle and rose chafer (Coleoptera: Scarabaeidae). J Econ Entomol 85:2194–2200

Allsopp PG, Ladd TL, Klein MG (1992b) Sample sizes and distributions of Japanese beetles (Coleoptera: Scarabaeidae) captured in lure traps. J Econ Entomol 85:1797–1801

Alm SR, Yeh T, Hanula JL, Georgis R (1992) Biological control of Japanese, oriental, and black turfgrass *Ataenius* beetle (Coleoptera: Scarabaeidae) larvae with entomopathogenic nematodes (Nematoda: Steinernematidae, Heterorhabditidae). J Econ Entomol 85:1660–1665

Alm SR, Yeh T, Dawson CG, Klein MG (1996) Evaluation of trapped beetle repellency, trap height, and string pheromone dispensers on Japanese beetle captures (Coleoptera: Scarabaeidae). Environ Entomol 25:1274–1278

Ando Y (1986) Seasonal prevalence and outbreaks of the Japanese beetle, *Popillia japonica* Newman (Coleoptera: Scarabaeidae). Jpn J Appl Entomol Zool 30:111–116

Andreadis TG, Hanula JL (1987) Ultrastructural study and description of *Ovavesicula popilliae* n.g., n.sp. (Microsporida: Pleistophoridae) from the Japanese beetle, *Popillia japonica* (Coleoptera: Scarabaeidae). J Protozool 34:15–21

Barrows EM, Gordh G (1978) Sexual behavior in the Japanese beetle, *Popillia japonica*, and comparative notes on sexual behavior of other scarabs (Coleoptera: Scarabaeidae). Behav Biol 23: 341–354

Baumler RE, Potter DA (2007) Knockdown, residual, and antifeedant activity of pyrethroids and home landscape bioinsecticides against Japanese beetles (Coleoptera: Scarabaeidae) on linden foliage. J Econ Entomol 100:451–458

Bohlen PJ, Barrett GW (1990) Dispersal of the Japanese beetle (Coleoptera: Scarabaeidae) in strip-cropped soybean agroecosystems. Environ Entomol 19:955–960

Bordelon B, Ellis M, Welty C (2011) 2011 Midwest small fruit and grape spray guide. Purdue University, Lafayette

Boucher TJ (1986) Japanese beetle *Popillia japonica* Newman: foliar feeding on wine grapes in Virginia. Virginia Polytechnic Institute and State University, Blacksburg

Boucher TJ, Pfeiffer DG (1989) Influence of Japanese beetle (Coleoptera: Scarabaeidae) foliar feeding on 'Seyval Blanc' grapevines in Virginia. J Econ Entomol 82:220–225

Boucher TJ, Pfeiffer DG, Barden JA, Williams JM (1987) Effects of simulated insect injury on net photosynthesis of potted grapevines. Hortscience 22:927–928

Breen JP (1994) *Acremonium* endophyte interactions with enhanced plant resistance to insects. Annu Rev Entomol 39:401–423

Britton WE, Johnson JP (1938) The Japanese beetle in Connecticut. Conn Agric Exp Stn Bull 411:455–486

Cappaert DL, Smitley DR (2002) Parasitoids and pathogens of Japanese beetle (Coleoptera: Scarabaeidae) in southern Michigan. Environ Entomol 31:573–580

CDMS (2010) Surround WP crop protectant label. http://www.cdms.net/LabelsMsds/LMDefault.aspx?manuf=145&t=

Chittenden FH, Fink DE (1922) The green June beetle. U S Dep Agric Bull 891:1–52

Chittenden FH, Quaintance AL (1916) The rose-chafer: a destructive garden and vineyard pest. U S Dep Agric Farm Bull 721:1–8

Cowles RS, Villani MG (1994) Soil interactions with chemical insecticides and nematodes used for control of Japanese beetle (Coleoptera: Scarabaeidae) larvae. J Econ Entomol 87:1014–1021

Cowles RS, Alm SR, Villani MG (1999) Selective toxicity of halofenozide to exotic white grubs (Coleoptera: Scarabaeidae). J Econ Entomol 92:427–434

Domek JM, Johnson DT (1987) Evidence of a sex pheromone in the green June beetle, *Cotinus nitida* (Coleoptera: Scarabaeidae). J Entomol Sci 22:264–267

Domek JM, Johnson DT (1988) Demonstration of semiochemically induced aggregation in the green June beetle, *Cotinis nitida* (L.) (Coleoptera: Scarabaeidae). Environ Entomol 17:147–149

Domek JM, Johnson DT (1990) Inhibition of aggregation behavior in the green June beetle (Coleoptera: Scarabaeidae) by antibiotic treatment of food substrate. Environ Entomol 19: 995–1000

Doolittle RE, Tumlinson JH, Proveaux AT, Heath RR (1980) Synthesis of the sex pheromone of the Japanese beetle. J Chem Ecol 6:473–483

Downing AS (1994) Effect of irrigation and spray volume on efficacy of entomopathogenic nematodes (Rhabditida: Heterorhabditidae) against white grubs (Coleoptera: Scarabaeidae). J Econ Entomol 87:643–646

Dunbar DM, Beard RL (1975) Present status of milky disease of Japanese and oriental beetles in Connecticut. J Econ Entomol 68:453–457

Dutky SR, Gooden EL (1952) *Coxiella popilliae*, n. sp., a rickettsia causing blue disease of Japanese beetle larvae. J Bacteriol 63:743–750

Fleming WE (1968) Biological control of the Japanese beetle. U S Dep Agric Tech Bull 1383:1–78

Fleming WE (1970) The Japanese beetle in the United States. U S Dep Agric Agric Handb 236:1–30

Fleming WE (1972) Biology of the Japanese beetle. U S Dep Agric Tech Bull 1449:1–129

Gaugler R, Campbell JF (1991) Selection for enhanced host-finding of scarab larvae (Coleoptera: Scarabaeidae) in an entomopathogenic nematode. Environ Entomol 20:700–706

Gaugler R, Campbell J, Selvan M, Lewis E (1992) Large-scale inoculative releases of the entomopathogen *Steinernema glaseri*: assessment 50 years later. Biol Control 2:181–187

Gaugler R, Wang Y, Campbell J (1994) Aggressive and evasive behaviors in *Popillia japonica* (Coleoptera: Scarabaeidae) larvae: defenses against entomopathogenic nematode attack. J Invertebr Pathol 64:193–199

Georgis R, Gaugler R (1991) Predictability in biological control using entomopathogenic nematodes. J Econ Entomol 84:713–720

Goonewardene HF, Townshend BG, Bingham RG, Borton R (1973) Improved technique for field use of female Japanese beetles as lures. J Econ Entomol 66:396–397

Gordon CF, Potter DA (1986) Japanese beetle (Coleoptera: Scarabaeidae) traps: evaluation of single and multiple arrangements for reducing defoliation in urban landscape. J Econ Entomol 79:1381–1384

Gu S, Pomper GW (2008) Grape cultivar feeding preference of adult Japanese beetles. Hortscience 43:196–199

Hamilton RM, Foster RE, Gibb TJ, Sadof CS, Holland JD, Engel BA (2007) Distribution and dynamics of Japanese beetles along the Indianapolis airport perimeter and the influence of land use on trap catch. Environ Entomol 36:287–296

Hammons DL, Kurtural SK, Potter DA (2008) Japanese beetles facilitate feeding by green June beetles (Coleoptera: Scarabaeidae) on ripening grapes. Environ Entomol 37:608–614

Hammons DL, Kurtural SK, Potter DA (2010) Impact of insecticide-manipulated defoliation by Japanese beetle (*Popillia japonica*) on grapevines from vineyard establishment through production. Pest Manag Sci 66:565–571

Hanula JL (1990) Epizootiological investigations of the microsporidium *Ovavesicula popilliae* and bacterium *Bacillus popilliae* in field populations of the Japanese beetle (Coleoptera: Scarabaeidae). Environ Entomol 19:1552–1557

Hanula JL, Andreadis TG (1988) Parasitic microorganisms of Japanese beetle (Coleoptera: Scarabaeidae) and associated scarab larvae in Connecticut soils. Environ Entomol 17: 709–714

Hartzell FZ (1918) Experiments for the control of the grape root-worm. N Y Agric Exp Stn Bull 453:255–332

Hays J, Clopton RE, Cappaert DL, Smitley DR (2004) Revision of the genus *Stictospora* and description of *Stictospora villani*, n. sp. (Apicomplexa: Eugregarinida: Actinocephalidae) from larvae of the Japanese beetle, *Popillia japonica* (Coleoptera: Scarabaeidae), in Michigan. J Parasitol 90:1450–1456

Heath JJ, Williams RN, Phelan PL (2001) High light intensity: a critical factor in the wind-tunnel flight of two scarabs, the rose chafer and Japanese beetle. J Chem Ecol 27:419–429

Held DW, Potter DA (2004) Floral affinity and benefits of dietary mixing with flowers for a polyphagous scarab, *Popillia japonica* Newman. Oecologia 140:312–320

Held DW, Eaton T, Potter DA (2001) Potential for habituation to a neem-based feeding deterrent to Japanese beetles, *Popillia japonica*. Entomol Exp Appl 101:25–32

Held DW, Gonsiska P, Potter DA (2003) Evaluating companion planting and non-host masking odors for protecting roses from the Japanese beetle (Coleoptera: Scarabaeidae). J Econ Entomol 96:81–87

Isely D (1930) *Fidia longipes* as a grape pest. J Econ Entomol 23:95–97

Iwabuchi K, Takahashi J (1983) Aggregative distribution of the Japanese beetle, *Popillia japonica* Newman (Coleoptera: Scarabaeidae), and the role of former occupants in the formation of an aggregation. Appl Entomol Zool 18:324–329

Johnson DT, Lewis BA (2008) Efficacy of feeding on insecticides and dosage response of contact insecticides against green June beetle, 2007. Arthropod Manag Tests 33:C10

Johnson DT, Lewis BA (2009) Efficacy of insecticides against green June beetle, 2008. Arthropod Manag Tests 34:C13

Johnson DT, Vishniac HS (1991) The role of *Trichosporon cutaneum* in eliciting aggregation behavior in *Cotinis nitida* (Coleoptera: Scarabaeidae). Environ Entomol 20:15–21

Johnson DT, Lewis BA, Bryant RJ, Liyanage R, Lay JO, Pszczolkowski MA (2009) Attractants for the green June beetle (Coleoptera: Scarabaeidae). J Econ Entomol 102:2224–2232

Jubb GL (1977) History of entomological research on grapes and other crops in Erie County, Pennsylvania. Melsheimer Entomol Ser 22:7–11

Keathley CP, Potter DA (2008) Quantitative resistance traits and suitability of woody plant species for a polyphagous scarab, *Popillia japonica* Newman. Environ Entomol 37:1548–1557

King JL (1931) The present status of the established parasites of *Popillia japonica* Newman. J Econ Entomol 24:453–462

Klein MG, Edwards DC (1989) Captures of *Popillia lewisi* (Coleoptera: Scarabaeidae) and other scarabs on Okinawa with Japanese beetle lures. J Econ Entomol 82:101–103

Klein MG, Lawrence KO, Ladd TL (1973) Japanese beetles: shielded traps to increase captures. J Econ Entomol 66:562–563

Klein MG, Tumlinson JH, Ladd TL, Doolittle RE (1981) Japanese beetle (Coleoptera: Scarabaeidae): response to synthetic sex attractant plus phenethyl propionate: eugenol. J Chem Ecol 7:1–7

Klostermeyer LE (1985) Japanese beetle (Coleoptera: Scarabaeidae) traps: comparison of commercial and homemade traps. J Econ Entomol 78:454–459
Koppenhöfer AM, Fuzy EM (2004) Effect of white grub developmental stage on susceptibility to entomopathogenic nematodes. J Econ Entomol 97:1842–1849
Koppenhöfer AM, Kaya HK (1998) Synergism of imidacloprid and an entomopathogenic nematode: a novel approach to white grub (Coleoptera: Scarabaeidae) control in turfgrass. J Econ Entomol 91:618–623
Koppenhöfer AM, Wilson M, Brown I, Kaya HK, Gaugler R (2000) Biological control agents for white grubs (Coleoptera: Scarabaeidae) in anticipation of the establishment of the Japanese beetle in California. J Econ Entomol 93:71–80
Kruse KC, Switzer PV (2007) Physical contests for females in the Japanese beetle, *Popillia japonica*. J Insect Sci 7:1–10
Ladd TL (1966) Egg viability and longevity of Japanese beetles treated with tepa, aphodate, and metepa. J Econ Entomol 59:422–425
Ladd TL (1970) Sex attraction in the Japanese beetle. J Econ Entomol 63:905–908
Ladd TL (1986) Enhancement of lures for Japanese beetles (Coleoptera: Scarabaeidae) by eugenol and Japonilure. J Econ Entomol 79:405–409
Ladd TL (1987) Influence of food, age, and mating on production of fertile eggs by Japanese beetles (Coleoptera: Scarabaeidae). J Econ Entomol 80:93–95
Ladd TL, Jacobson M, Buriff CR (1978) Japanese beetles: extracts from neem tree seeds as feeding deterrents. J Econ Entomol 71:810–813
Ladd TL, Klein MG, Tumlinson JH (1981) Phenethyl propionate + eugenol + geraniol (3:7:3) and Japonilure: a highly effective joint lure for Japanese beetles. J Econ Entomol 74:665–667
Ladd TL, Warthen JD, Klein MG (1984) Japanese beetle (Coleoptera: Scarabaeidae): the effects of azadirachtin on the growth and development of the immature forms. J Econ Entomol 77:903–905
Lalancette N, Belding RD, Shearer PW, Frecon JL, Tietjen WH (2005) Evaluation of hydrophobic and hydrophilic kaolin particle films for peach crop, arthropod and disease management. Pest Manag Sci 61:25–39
Landolt PJ (1990) Trapping the green June beetle (Coleoptera: Scarabaeidae) with isopropanol. Fla Entomol 73:328–330
Langford GS, Cory EN (1948) Host preference in Japanese beetles with special reference to grape and apple. J Econ Entomol 41:823–824
Leal WS (1998) Chemical ecology of phytophagous scarab beetles. Annu Rev Entomol 43:39–61
Lockwood D, Sutton T, Burrack H, Pfeiffer D, Mitchem W, Bellinger B et al (2010) southeast regional bunch grape integrated management guide. Southern Region Small Fruit Consortium, Raleigh
López R, Potter DA (2000) Ant predation on eggs and larvae of the black cutworm (Lepidoptera: Noctuidae) and Japanese beetle (Coleoptera: Scarabaeidae) in turfgrass. Environ Entomol 29:116–125
Loughrin JH, Potter DA, Hamilton-Kemp TR (1995) Volatile compounds induced by herbivory act as aggregation kairomones for the Japanese beetle (*Popillia japonica* Newman). J Chem Ecol 21:1457–1467
Loughrin JH, Potter DA, Hamilton-Kemp TR, Byers ME (1996) Role of feeding-induced plant volatiles in aggregative behavior of the Japanese beetle (Coleoptera: Scarabaeidae). Environ Entomol 25:1188–1191
Mannion CM, Winkler HE, Shapiro DI, Gibb TJ (2000) Interaction between halofenozide and the entomopathogenic nematode *Heterorhabditis marelatus* for control of Japanese beetle (Coleoptera: Scarabaeidae) larvae. J Econ Entomol 93:48–53
Mannion CM, McLane W, Klein MG, Moyseenko J, Oliver JB, Cowan DC (2001) Management of early-instar Japanese beetle (Coleoptera: Scarabaeidae) in field-grown nursery crops. J Econ Entomol 94:1151–1161
Maxey LM, Laub CS, Mays RS, Pfeiffer DG (2008) Japanese beetle (*Popillia japonica*) control and varietal comparisons in primocane-bearing brambles. In: Proceedings of the 84th Cumberland-Shenandoah fruit workers conference 84:66–71

Maxey L, Laub C, Pfeiffer DG (2009) Effects of geranium exposure on Japanese beetle (*Popillia japonica*) feeding on primocane-bearing raspberries. In: Proceedings of the 85th Cumberland-Shenandoah fruit workers conference 85:26–30
McCleod MJ, Williams RN (1990) Life history and vineyard damage by rose chafer. Vinifera Wine Growers J 17:25–27
Mercader RJ, Isaacs R (2003) Damage potential of rose chafer and Japanese beetle (Coleoptera: Scarabaeidae) in Michigan vineyards. Gt Lakes Entomol 36:166–178
Moore AR, Cole WH (1921) The response of *Popillia japonica* to light and the Weber-Fechner law. J Gen Physiol 3:331–335
Niemczyk HD, Lawrence KO (1973) Japanese beetle: evidence of resistance to cyclodiene insecticides in larvae and adults in Ohio. J Econ Entomol 66:520–521
Oliver JB, Mannion CM, Klein MG, Moyseenko JJ, Bishop B (2005) Effect of insecticides on *Tiphia vernalis* (Hymenoptera: Tiphiidae) oviposition and survival of progeny to cocoon stage when parasitizing *Popillia japonica* (Coleoptera: Scarabaeidae) larvae. J Econ Entomol 98:694–703
Oliver JB, Reding ME, Moyseenko JJ, Klein MG, Mannion CM, Bishop B (2006) Survival of adult *Tiphia vernalis* (Hymenoptera: Tiphiidae) after insecticide, fungicide, and herbicide exposure in laboratory bioassays. J Econ Entomol 99:288–294
Patton CN (1956) Observations on the mating behavior of the green June beetle, *Cotinus nitida* (Linn.). Fla Entomol 39:95
Pfeiffer DG, Boucher TJ, Lachance MW, Killian JC (1990) Entomological research in Virginia (USA) vineyards. In: Bostanian NJ, Wilson LT, Dennehy TJ (eds) Monitoring and integrated management of arthropod pests of small fruit crops. Intercept, Andover, pp 45–61
Potter DA, Held DW (1999) Absence of food-aversion learning by a polyphagous scarab, *Popillia japonica*, following intoxication by geranium, *Pelargonium* x *hortorum*. Entomol Exp Appl 91:83–88
Potter DA, Held DW (2002) Biology and management of the Japanese beetle. Annu Rev Entomol 47:175–205
Potter DA, Patterson CG, Redmond CT (1992) Influence of turfgrass species and tall fescue endophyte on feeding ecology of Japanese beetle and southern masked chafer grubs (Coleoptera: Scarabaeidae). J Econ Entomol 85:900–909
Potter DA, Powell AJ, Spicer PG, Williams DW (1996) Cultural practices affect root-feeding white grubs (Coleoptera: Scarabaeidae) in turfgrass. J Econ Entomol 89:156–164
Ramoutar D, Legrand A (2007) Survey of *Tiphia vernalis* (Hymenoptera: Tiphiidae), a parasitoid wasp of *Popillia japonica* (Coleoptera: Scarabaeidae), in Connecticut. Fla Entomol 90:780–782
Rao RV, Lakshminarayana K, Subbbaratnam GV (1983) Residual toxicity of certain insecticides to grapevine flea beetle. Entomon 8:395–396
Rao RV, Lakshminarayana K, Subbaratnam GV (1984) Seasonal occurrence of grapevine flea beetle, *Scelodonta strigicollis* (Mots) around Hyderabad. Entomon 9:59–60
Reed DK, Lee MH, Kun SH, Klein MG (1991) Attraction of scarab beetle populations (Coleoptera: Scarabaeidae) to Japanese beetle lures in the Republic of Korea. Agric Ecosyst Environ 36: 163–174
Régnière J, Brooks WM (1978) Entomogenous microorganisms associated with the Japanese beetle, *Popillia japonica*, in eastern North Carolina. J Invertebr Pathol 32:226–228
Régnière J, Rabb RL, Stinner RE (1983) *Popillia japonica* (Coleoptera: Scarabaeidae): distribution and movement of the adults in heterogeneous environments. Can Entomol 115:287–294
Rogers ME, Potter DA (2002) Kairomones from scarabaeid grubs and their frass as cues in below-ground host location by the parasitoids *Tiphia vernalis* and *Tiphia pygidialis*. Entomol Exp Appl 102:307–314
Rowe WJ II, Potter DA (1996) Vertical stratification of feeding by Japanese beetles within linden tree canopies: selective foraging or height per se? Oecologia 108:459–466
Schroeder PV, Villani MG, Ferguson CS, Nyrop JP, Shields EJ (1993) Behavioral interactions between Japanese beetle (Coleoptera: Scarabaeidae) grubs and an entomopathogenic nematode (Nematoda: Heterorhabditidae) within turf microcosms. Environ Entomol 22:595–600

Stahly DP, Klein MG (1992) Problems with in vitro production of spores of *Bacillus popilliae* for use in biological control of the Japanese beetle. J Invertebr Pathol 60:283–291
Szendrei Z, Isaacs R (2006) Ground covers influence the abundance and behaviour of Japanese beetle in blueberry fields. Environ Entomol 35:789–796
Szendrei Z, Mallampalli N, Isaacs R (2005) Effect of tillage on abundance of Japanese beetle, *Popillia japonica* Newman (Col., Scarabaeidae), larvae and adults in highbush blueberry fields. J Appl Entomol 129:258–264
Vishniac HS, Johnson DT (1990) Development of a yeast flora in the adult green June beetle (*Cotinus nitida*, Scarabaeidae). Mycologia 82:471–479
Vittum PJ (1986) Biology of the Japanese beetle (Coleoptera: Scarabaeidae) in eastern Massachusetts. J Econ Entomol 79:387–391
Vitullo JM, Sadof CS (2007a) Effects of pesticide applications and cultural controls on efficacy of control for adult Japanese beetles (Coleoptera: Scarabaeidae) on roses. J Econ Entomol 100: 95–102
Vitullo JM, Sadof CS (2007b) Efficacy of soil and foliar-applied azadirachtin in combination with and in comparison to soil-applied imidacloprid and foliar-applied carbaryl against Japanese beetles on roses. HortTechnology 17:316–321
Wang Y, Gaugler R (1998) Host and penetration site location by entomopathogenic nematodes against Japanese beetle larvae. J Invertebr Pathol 72:313–318
Wang Y, Gaugler R, Cui L (1994) Variations in immune response of *Popillia japonica* and *Acheta domesticus* to *Heterorhabditis bacteriophora* and *Steinernema* species. J Nematol 26:11–18
Wang Y, Campbell JF, Gaugler R (1995) Infection of entomopathogenic nematodes *Steinernema glaseri* and *Heterorhabditis bacteriophora* against *Popillia japonica* (Coleoptera: Scarabaeidae) larvae. J Invertebr Pathol 66:178–184
Weigle TH, Muza AJ, Gardner R, Helms M, Bates TR, Reisch BI et al (2010) 2010 New York and Pennsylvania pest management guidelines for grapes. http://www.ipmguidelines.org/grapes
Williams RN, Fickle DS (2007) Chemical evaluations for control of Japanese beetle on grapes, 2006. Arthropod Manag Tests 32:C12
Williams RN, Fickle DS (2008) Chemical evaluations for control of Japanese beetle on grapes, 2007. Arthropod Manag Tests 33:C27
Williams RN, Miller KV (1982) Field assay to determine attractiveness of various aromatic compounds to rose chafer adults. J Econ Entomol 75:196–198
Williams RN, McGovern TP, Klein MG (1982) Evaluation of aromatic compounds and virgin females as attractants for rose chafer. Res Circ Ohio Agric Res Dev Cent 272:38–40
Wise JC, Vandervoort C, Isaacs R (2007) Lethal and sublethal activities of imidacloprid contribute to control of adult Japanese beetle in blueberries. J Econ Entomol 100:1596–1603
Wise JC, Poppen RV, Isaacs R (2009) Grape berry moth control, 2008. Arthropod Manag Tests 34:C21
Wood TN, Richardson M, Potter DA, Johnson DT, Wiedenmann RN, Steinkraus DC (2009) Ovipositional preferences of the Japanese beetle (Coleoptera: Scarabaeidae) among warm- and cool-season turfgrass species. J Econ Entomol 102:2192–2197
Wylie WD (1969) Attractants for green June beetle adults. Ark Farm Res 18:11
Zenger JT, Gibb TJ (2001a) Identification and impact of egg predators of *Cyclocephala lurida* Bland and *Popillia japonica* Newman (Coleoptera: Scarabaeidae) in turfgrass. Environ Entomol 30:425–430
Zenger JT, Gibb TJ (2001b) Impact of four insecticides on Japanese beetle (Coleoptera: Scarabaeidae) egg predators and white grubs in turfgrass. J Econ Entomol 94:145–149

Chapter 18
Ecological Management of Ants in Vineyards of the Cape Floristic Region Biodiversity Hotspot, South Africa

Pia Addison and Michael J. Samways

18.1 Introduction

Ants are among the most ecologically significant of all insects, as well as being important study organisms for biologists in general. However, ants have a close mutualistic relationship with insects occurring in the subfamily Sternorrhyncha (mealybugs, scales and aphids), with this mutualism impacting negatively on natural enemies, and therefore on biological control efforts. Ants utilize the nutrient-rich honeydew as a lucrative, readily-available food source, but in doing so, they deter parasitic wasps and predatory beetles from feeding on the honeydew-producing insects (Way 1963; Way and Khoo 1992). This mutualism occurs in association with the vine mealybug, *Planococcus ficus* (Signoret) (Kriegler and Whitehead 1962), a major direct pest (Joubert and Walters 1955; Walton and Pringle 2004) and vector of the grapevine leafroll virus (Engelbrecht and Kasdorf 1990).

In South Africa, the Argentine ant, *Linepithema humile* (Mayr) (Fig. 18.1), is widespread and of major concern for both agricultural and natural ecosystems, in which it has become established since its introduction in 1901. Its geographical range expansion is almost exclusively via human-mediated jump dispersal rather than by natural spread (Luruli 2007), and the invasion of natural vegetation is restricted largely to the Western Cape (De Kock and Giliomee 1989; Dean 1992; Botes et al. 2006). Indeed, the Mediterranean climate of the Cape Floristic Region (CFR), Western Cape, is ideal for *L. humile* establishment, as well as vineyard cultivation, both of which pose major threats to the CFR, a biodiversity hotspot with

P. Addison (✉) • M.J. Samways
Department of Conservation Ecology and Entomology, Faculty of AgriSciences, University of Stellenbosch, Matieland 7602, South Africa
e-mail: pia@sun.ac.za; samways@sun.ac.za

N.J. Bostanian et al. (eds.), *Arthropod Management in Vineyards: Pests, Approaches, and Future Directions*, DOI 10.1007/978-94-007-4032-7_18,

Fig. 18.1 *Linephitema humile* adults

high conservation priority and dominated by fynbos vegetation (characterized by proteoid, ericoid and restioid plants) (Myers et al. 2000). A further consideration is that several indigenous ant species in the CFR have become pests, given modified microclimates in vineyards and an over-abundance of food in the form of honeydew from *P. ficus* (Addison and Samways 2000).

The total surface area of South Africa planted to wine grape vineyards, 95% of which is in the Western Cape, has increased steadily from almost 90,000 ha in 1998 to over 102,000 ha in 2006, declining over the following 2 years by around 800 ha (Cupido and Isaacs 2008). Fairbanks et al. (2004) estimated that almost 15,000 ha of the most threatened habitats are particularly suitable for vineyard expansion, which is a further disturbance factor that could facilitate *L. humile* expansion. With new environmental initiatives being taken by wine grape producers, such as the Scheme for Integrated Production of Wine (IPW) and the Biodiversity and Wine Initiative (BWI), the aim is that even with vineyard expansion, the whole area-wide ecosystem can be managed so as to benefit both the industry and the environment (IPW 2009).

18.2 Ant Diversity in the Cape Floristic Region Biodiversity Hotspot

Several ant assemblage studies and surveys have been carried out in the CFR and Western Cape in general, most of which have focused on natural ecosystems. A summary of species occurring in vineyards versus natural habitats in the CFR is given in Table 18.1.

Table 18.1 Ant diversity, as documented from available literature (numbers refer to source, listed below), in the Cape Floristic Region of the Western Cape Province, South Africa

Species[a]	Vineyards	Reserves	Forest plantations	Natural habitat and road verges
Aenictus rotundatus Mayr		14		7, 9
Anochetus levaillanti Emery		1, 14, 15		7, 9
Anoplolepis custodiens (Smith)	10	1, 3a, 4, 5, 11, 16		2, 6, 13
Anoplolepis steingroeveri (Forel)	10	1, 4, 15, 16		2, 3a, 6, 7, 8, 9, 13
Camponotus angusticeps Emery		14		
Camponotus baynei Arnold		19	19	
Camponotus cuneiscapus Forel				9
Camponotus emarginatus Emery		14		
Camponotus fulvopilosus (DeGeer)	10	1, 14, 15		6, 7, 8, 9
Camponotus irredux Forel				17
Camponotus maculatus (Fabricius)		1, 3a, 4, 5, 14		9
Camponotus mystaceus Emery		14, 15		7, 8
Camponotus niveosetosus Mayr		3a, 5, 14, 16, 19		17
Camponotus rufoglaucus (Jerdon)		3a, 3b		7
Camponotus simulans Forel				9
Camponotus vestitus (Smith)		14, 15		9
Camponotus werthi Forel		3a, 3b		
Cardiocondyla emeryi Forel	10			9
Cardiocondyla schuckardi Forel	10			
Cerapachys peringueyi Emery		3a, 3b		17
Cerapachys wroughtoni Forel		14		17
Crematogaster (=*Acrocoelia*) *delagoensis* (Forel)		1		
Crematogaster liengmei Forel	10	5		7
Crematogaster melanogaster Emery	10	14		6, 7, 9
Crematogaster peringueyi Emery	10, 18	1, 5		18
Crematogaster transvaalensis Forel				2
Dorylus helvolus (Linnaeus)	10	1, 3a, 3b, 5, 14, 15	12	7, 8, 17
Hypoponera spei (Forel)		5		
Lepisiota (=*Acantholepis*) *capensis* (Mayr)	10, 18	1, 3a, 3b, 4, 5		6, 7, 8, 9, 17, 18
Lepisiota laevis (Santschi)	10			
Lepisiota (=*Acantholepis*) *spinosior* (Forel)		3a, 3b		
Leptogenys castanea (Mayr)	10			
Leptogenys intermedia (=*nitida*) Emery		1, 5		
Linepithema (=*Iridomyrmex*) *humile* (Mayr)	10, 18	3a, 5, 11, 16, 19	12, 17, 19	2, 13, 17, 18
Meranoplus peringueyi Emery		3a, 3b, 4, 5, 11, 16, 19		17
Messor barbarus Mayr		1, 3a		

(continued)

Table 18.1 (continued)

Species[a]	Vineyards	Reserves	Forest plantations	Natural habitat and road verges
Messor capensis (Mayr)	10	15		7, 8, 9, 17
Monomorium australe Emery				7
Monomorium braunsi Mayr		5		
Monomorium delagoense Forel		3a, 3b, 4		7
Monomorium fridae Forel		14		
Monomorium havilandi Forel	10	15, 16		7, 9
Monomorium lubricum Arnold		5		
Monomorium macrops Arnold	10	14		9
Monomorium monomorium (=*minutum*) Bolton		3a		9
Monomorium musicum Forel		5		
Monomorium nuptialis Forel		5		
Monomorium ocellatum Arnold				9
Monomorium prossae Forel		3a		
Monomorium rhopalocerum (=*leimbachi*) Emery	10	3a, 3b		
Monomorium schultzei Forel	10	3a, 3b, 5		
Monomorium springvalense Forel		5		
Monomorium subopacum (Smith)				7
Monomorium tablense Santschi			12	
Monomorium torvicte Bolton		5		
Monomorium willowmorense Bolton		5		9
Myrmicaria nigra Mayr		16, 19		
Ocymyrmex barbiger Emery	10	3a, 3b, 16		9, 17
Ocymyrmex cilliei Prins and Roux		5		6, 7, 8, 13
Pachycondyla cavernosa (Roger)		14		
Pachycondyla (=*Opthalmopone*) *hottentota* Emery				6, 7, 8, 9
Pachycondyla (=*Hagensia*) *peringueyi* (Emery)		1		
Pachycondyla pumicosa (Roger)		14, 16		
Pachycondyla (=*Euponera*) *wroughtoni* (Forel)		1		
Pheidole capensis Mayr		1, 3a, 3b, 4, 11, 16		2, 6, 7, 9, 13, 17
Pheidole foreli Mayr		5		
Pheidole tenuinodis Mayr		3a, 3b, 5		7
Plagiolepis jouberti Forel		5		
Plagiolepis pygmaea (Latreille)		3a, 3b, 5		
Rhoptromyrmex transversinodis Mayr		3a, 3b		
Solenopsis punctaticeps Mayr	10	1, 3b, 14, 19	19	
Strumigenys (=*Smithistruma*) *emarginata* (Mayr)		3a, 3b, 5		
Tapinolepis (=*Anoplolepis*) *trimenii* (Forel)				7, 8, 9
Tapinoma arnoldi Forel				7
Technomyrmex albipes (Smith)	10	1, 3a, 3b		7, 8

(continued)

Table 18.1 (continued)

Species[a]	Vineyards	Reserves	Forest plantations	Natural habitat and road verges
Technomyrmex pallipes Wetterer			17	17
Tetramorium bevisi Arnold	10			9
Tetramorium bothae Forel		5		
Tetramorium capense Mayr		3a, 3b, 5		
Tetramorium emeryi Mayr		5		
Tetramorium erectum Emery	10			17
Tetramorium flaviceps Arnold		5		
Tetramorium frigidum Arnold	10	3a	17	17
Tetramorium grassii Emery		5	12	
Tetramorium lobulicorne Santschi		5		
Tetramorium peringueyi Arnold				9
Tetramorium pusillum Emery	10	1, 3a, 3b		
Tetramorium quadrispinosum Emery	10	1, 3a, 5, 11, 14, 15, 16, 19	19	2, 7, 8, 9, 13, 17
Tetramorium regulare Bolton	10		12	
Tetramorium signatum Emery				7, 8, 9
Tetramorium simillimum (Smith)		3a, 3b		
Tetramorium solidum Emery	10			9
Tetramorium squaminode Santschi		3a, 3b		
Tetraponera clypeata (Emery)		3b		7
Total number of species: 95	30	72	10	50

[a]Species names checked for validity and adjusted where necessary (Agosti and Johnson 2005). Old name, as used in source, is included in brackets where applicable

Source: 1Prins (1967), 2Bond and Slingsby (1984), 3aDonnelly and Giliomee (1985a), 3bDonnelly and Giliomee (1985b), 4de Kock (1990), 5de Kock et al. (1992), 6Dean (1992), 7Milton et al. (1992), 8Dean and Milton (1995), 9Tshinguvho et al. (1999), 10Addison and Samways (2000), 11Christian (2001), 12Ratsirarson et al. (2002), 13Witt et al. (2004), 14Botes et al. (2006), 15Boonzaaier et al. (2007), 16Luruli (2007), 17Schoeman (2008), 18Gaigher (2008), 19Pryke (2008)

The often ecologically dominant *Pheidole megacephala* (F.) has not yet been documented in the Western Cape, although it occurs as an agricultural pest in most other parts of the country, including the Eastern Cape Province (Samways 1981). This ant appears to thrive in the humid tropics, although it has also been documented in New Zealand (Lester et al. 2003), Europe and California (McGlynn 1999). *Linepithema humile* and *P. megacephala* are highly competitive, and will exclude each other completely depending on climate, with *L. humile* dominating in warm temperate regions between 30° and 36° latitude, while *P. megacephala* dominates in the tropics (Hölldobler and Wilson 1990). In South Africa, *P. megacephala* was recorded as far south as 33°22′S in the Eastern Cape Province (Compton and Robertson 1988). However, *P. megacephala* is not necessarily a local top species, with the indigenous *Myrmicaria natalensis* (Smith) physically able to out-compete it through dumping of soil on its nest through its own nest building (Samways 1983).

Table 18.2 Dominant ant species sampled in a survey conducted in 18 vineyards throughout the Western Cape Province, South Africa, over 2 weeks using pitfall traps, as described in Addison and Samways (2000)

Vineyard no.	Dominant species	Total abundance (%)	No. of dominant individuals	Total no. of species
1	*Technomyrmex albipes*	28.7	181	12
2	*Anoplolepis steingroeveri*	58.7	478	12
3	*Anoplolepis custodiens*	69.6	1,450	9
4	*Linepithema humile*	70.2	811	6
5	*Linepithema humile*	70.2	630	10
6	*Anoplolepis steingroeveri*	70.9	1,747	8
7	*Linepithema humile*	86.5	2,606	8
8	*Anoplolepis custodiens*	90.7	1,954	10
9	*Linepithema humile*	91.7	4,045	11
10	*Anoplolepis steingroeveri*	95.6	2,038	9
11	*Anoplolepis steingroeveri*	95.8	3,875	9
12	*Linepithema humile*	97.2	2,089	7
13	*Anoplolepis custodiens*	97.3	3,656	8
14	*Linepithema humile*	97.6	1,923	6
15	*Anoplolepis custodiens*	98.9	5,323	6
16	*Anoplolepis custodiens*	98.9	5,499	10
17	*Anoplolepis custodiens*	99.4	10,880	10
18	*Anoplolepis custodiens*	99.9	16,248	5

There are five ant species associated with *P. ficus* in South African vineyards, with *L. humile* having the highest pest status (Tables 18.2 and 18.3), and *Anoplolepis custodiens* (Smith) being common throughout the country (Steyn 1954; Botes et al. 2006). *Anoplolepis custodiens* is a well known pest in citrus orchards as well as vineyards, where it attains high populations, and can exhibit extreme dominance over other ant species (Steyn 1954; Samways 1981, Fig. 18.2, Table 18.2). However, *A. custodiens*, as well as its congener *Anoplolepis steingroeveri* (Forel), are both involved in myrmechochory (plants with seeds dispersed by ants) in Cape fynbos, and are frequently out-competed by *L. humile*, which threatens seed dispersal in natural habitats (Skaife 1961; Bond and Slingsby 1984). *Crematogaster peringueyi* Emery (Fig. 18.3) is primarily arboreal and nests in vines, possibly causing deterioration of vines (Kriegler and Whitehead 1962) and agitating workers during harvest. With the nearly 95 confirmed ant species occurring in the ecologically sensitive CFR (Table 18.1), threatened by the invasive *L. humile* and vineyard encroachment, area-wide pest management in vineyards has become essential.

18.3 The Ant-Mealybug Mutualism

A deeper understanding of insect behavior is important for successful arthropod pest management. Trophobiotic relationships are poorly understood and the factors involved are not always fully appreciated. In South African vineyards, the

Table 18.3 Ant species associated with *Planococcus ficus* in vineyards, their origin and estimated pest status in South Africa

Ant species	Type location[a]	No. of vineyards present[b]	No. of vineyards dominant	Publications relating to ant control in South African vineyards	Pest status[c]
Linepithema humile	Argentina	7	6	Dürr (1953), Joubert and Walters (1955), Schwartz (1988), Addison (2002)	17
Anoplolepis custodiens	South Africa	7	7	Addison (2002), Addison and Samways (2006)	16
Anoplolepis steingroeveri	Namibia	7	4	Addison (2002)	12
Technomyrmex albipes	Sulawesi	9	1	–	10
Crematogaster peringueyi	South Africa	5	0	Kriegler and Whitehead (1962)	6

[a]Type location is according to Taylor (2006)

[b]No. of vineyards present as sampled in Addison and Samways (2000), where a total of 22 vineyards throughout the Western Cape were sampled

[c]Pest status = (No. of vineyards present) + (No. of vineyards dominant) + (No. references)

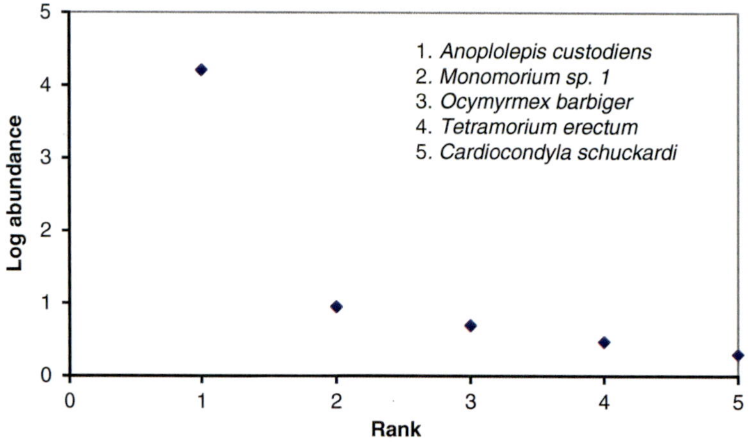

Fig. 18.2 Abundance/rank plot of the epigaeic ant assemblage in a highly infested wine grape vineyard contaminated with vine mealybug *Planococcus ficus*

Fig. 18.3 *Crematogaster peringueyi* adult

success of *P. ficus* biological control is dependent on several factors, notably the characteristics of the parasitoids, ants and mealybugs involved in this interaction. The primary vine mealybug parasitoids in South Africa are *Anagyrus* sp. near *pseudococci* (Girault) (henceforth referred to as *A. pseudococci*) and *Coccidoxenoides perminutus* (Girault) (Walton and Pringle 2004). Characteristics of the parasitoids include the stage of attack: *Coccidoxenoides perminutus* prefers the first three mealybug instars (Joyce et al. 2001), while *A. pseudococci* prefers later mealybug instars (Islam and Copland 1997). *Coccidoxenoides perminutus* has a lower threshold temperature (Walton and Pringle 2004) than *A. pseudococci* (Daane et al. 2004). In laboratory trials, *Coccidoxenoides perminutus* was found to be significantly more ant tolerant than *A. pseudococci* (Mgocheki and Addison 2009a). On the other hand, *A. pseudococci* adults were more tolerant to pesticides than *C. perminutus* (Mgocheki and Addison 2009b). Using both parasitoids together would, in theory at least, imply improved biological control of *P. ficus*. Using additional species of natural enemies was also found to be key for improving biological control of *P. ficus* in California (Gutierrez et al. 2007).

Inter and intraspecific competitions are characteristic of ants. Competition determines ant dominance and, in turn, determines the overall effect of the ants on the parasitoids and therefore the success of biological control. In South Africa, *C. peringueyi* causes significantly higher parasitoid mortality than the epigaeic ants *L. humile* and *A. steingroeveri*, and that vineyard ant generally has a significant effect on both parasitoid mortality and parasitism in the laboratory (Mgocheki and Addison 2009a). In vineyards, *C. peringueyi* and *L. humile* formed significant spatial associations with *P. ficus* and parasitoids, while *A. steingroeveri* formed no significant associations with parasitoids (Mgocheki and Addison 2010), confirming laboratory trials. Daane et al. (2007) found that *L. humile* significantly reduced *Pseudococcus viburni* (Signoret) parasitoids in vineyards, while Chong et al. (2010) found no significant effect of native ants on the grapevine scale *Parthenolecanium persicae* (F.). It is likely that ant aggression plays a major role in this dynamic interaction, and that highly aggressive and dominant ants should be the prime targets within biological control programs (Buckley and Gullan 1991).

Characteristics of *P. ficus* that lend themselves to this mutualism have been little studied in South African vineyards. Delabie (2001) summarized ant protection of trophobionts as being influenced by their accessibility, individual numbers, ability to aggregate, capacity to produce honeydew and diversity of food sources available to the ants. Developmental biology of *P. ficus* has been determined in laboratory trials on grapevine seedlings (Walton and Pringle 2005). However, the ant-*P. ficus* mutualism has not yet been explored within the greater ecological context, involving arthropod community structure and host plant characteristics (Styrsky and Eubanks 2007).

18.4 Ecological Ant Management in Vineyards

A standardized sampling system, based on monitoring 20 evenly distributed plots (with five vines per plot) per 2-ha block, has been developed for use in vineyards for monitoring key arthropod pests (Walton 2003; De Villiers and Pringle 2008). For ants and *P. ficus*, visual vine inspections using a presence-absence cluster sampling system is used. Management actions are recommended when 2% of vine stems are infested with *P. ficus* (Walton 2003; De Villiers and Pringle 2008), and when 20% of vine stems are infested with ants (Mgocheki and Addison 2009c). This information can be used by producers to reduce unnecessary and hazardous pesticide applications, forming the basis for the ecological management of *P. ficus* and attending ants.

18.4.1 Exclusion Barriers

Most ant control experiments on vines in South Africa have been directed against *L. humile* (Dürr 1953; Joubert and Walters 1955; Whitehead 1958, 1961; Schwartz 1988). Early trials showed that chlorinated hydrocarbon insecticides such as DDT, dieldrin and chlordane gave good control as soil and stem treatments, and were a good alternative to baiting, which was the standard practice fifty or so years ago (Dürr 1953; Joubert and Walters 1955; Whitehead 1957, 1961; Kriegler and Whitehead 1962). After the withdrawal of DDT and dieldrin, Schwartz (1988) tested various sticky stem barriers such as Plantex®, Formex®, Rever Ant®, and physical barriers such as Sper®. Although Schwartz (1988) achieved good control with polybutene-based sticky barriers, he stated that this treatment became expensive as a result of having to use a backing (Bidim®) to prevent possible phytotoxicity. According to Samways et al. (1981), Samways (1982), Moreno et al. (1987), and James et al. (1996), direct stem insecticide sprays acting as ant barriers are an effective, environmentally-friendly method for controlling epigaeic ants, although sticky barriers are even better.

Parasitism of *P. ficus* reaches its highest point during the months of January and February in the Western Cape Province, shortly before harvest (Walton 2003; Mgocheki 2008). Since there is seasonal movement of *P. ficus* up the vine stems to leaves and bunches as the bunches ripen (Walton 2003; Mgocheki 2008), *P. ficus* is most likely to be parasitized and eaten by predators during this time, when the mealybugs are exposed to natural enemies. Chemical stem treatments against ants are usually applied in spring and, if applied correctly and population pressure is moderate, can last over 100 days (Addison 2002). The application of current ant control treatments therefore do not coincide, either temporally or spatially, with natural *P. ficus* parasitism, as was also demonstrated for situations where augmentative releases were made (Walton and Pringle 2003). However, care must be taken when applying chemical stem treatments against ants, because currently registered products (NDA 2007) are considered highly toxic to the primary mealybug parasitoids *A. pseudococci* and *C. perminutus* (Mgocheki and Addison 2009b). These compounds are high risk treatments for integrated production of wine and subject to strict conditions of use (IPW 2009). No season-long pesticide impacts on arthropod communities have been assessed in South African vineyards. However, even the cumulative effects of moderately toxic pesticides, such as some fungicides, can have adverse effects over the longer term (Nash et al. 2010).

Directed chemical stem treatments have proven effective in reducing foraging ants in vineyards (Addison 2002; Klotz et al. 2003) and citrus canopies (Samways and Tate 1984; Moreno et al. 1987; Stevens et al. 1995; James et al. 1998). However, it is important to consider several factors when applying stem barriers. The spray must form a complete barrier around the stem above the irrigation lines, otherwise the ants will bypass the barriers. In many cases, producers do not have the correct equipment for applying stem barriers correctly (Addison 2002). Weed control is important too, as high growing weeds provide access for ants into the vine canopy. Under high ant pressure, barriers may need to be reapplied during the growing season. Lastly, barriers only target foraging workers, thereby suppressing, not eliminating, the population. However, elimination is not always the aim, as ants fulfill important ecological functions such as myrmechochory, pollination, nutrient recycling, soil improvement and predation of pest insects (Way and Khoo 1992), except for invasive pest ants such as *L. humile*. For these reasons, low toxicity baits have been investigated as an alternative to chemical stem applications. Stem barriers cannot be applied in vine nurseries or on bush vines, and low toxicity baits could therefore be a valuable alternative.

18.4.2 Low Toxicity Baits

Low toxicity baits exploit the ants' food sharing behavior (trophallaxis) by delaying death so that the workers are still able to distribute toxic baits among colony members, thereby destroying the whole colony. Effective baits have been described for controlling *L. humile* (Baker et al.1985; Blachly and Forschler 1996; Klotz et al. 1996; Daane et al. 2006), but few have reported success against *Anoplolepis* spp. (Samways 1985) and no such research has been done on *C. peringueyi*. Indeed, most bait trials have been conducted in an urban context against *L. humile* and not in agricultural locations (Baker et al. 1985; Knight and Rust 1991; Blachly and Forschler 1996; Klotz et al. 1996, 1998). Despite the possible environmental problems that could be associated with toxic baits, such as leaching of chemicals into the soil and killing non-target arthropods, a species-specific (containerized) bait with low mammalian toxicity could be of great value in vineyards, especially those infested with the invasive *L. humile* (Daane et al. 2006). The success of low toxicity baits is also dependent on the type of bait attractant, dispenser design, dispenser density in the field and target ant species. Laboratory bioassays showed that certain toxicants are effective as low toxicity baits against *L. humile* and *C. peringueyi* (Nyamukondiwa 2008). This study also found that *L. humile*, *C. peringueyi* and *A. custodiens* had a preference for wet attractants over dry ones, and were significantly attracted to 25% sugar water over protein baits, except for *A. custodiens*, which was also attracted to tuna (Baker et al. 1985; Silverman and Roulston 2001). Dispenser density is another critical aspect of bait efficacy and relates to central place foraging theory (Hölldobler and Wilson 1990). Daane et al. (2006) proposed that between 5 and 20 dispensers per ha are economically feasible in vineyards, but with such a deployment rate only a small proportion of vineyard foraging workers will be affected in South Africa (Nyamukondiwa 2008).

Research on low toxicity baits in South African vineyards is still preliminary. A practical and effective ant bait station is described by Grout (2008) for use in citrus orchards against *P. megacephala*. This could be adapted for deploying a liquid bait, although it may be more practical to investigate the use of a gel as a bait matrix (Silverman and Roulston 2001), as this would reduce evaporation and prevent ants from drowning. However, it would seem that at this stage low toxicity baits still require a substantial amount of modification before they became practical and economically viable in vineyards. A combination of barrier sprays and low toxicity baits may provide better control if hemipteran population pressure is moderate (i.e., honeydew food source) (Silverman and Brightwell 2008), and the cost of such treatments can be justified. Indeed, the food source in the form of honeydew is often the main driver regulating ant abundance in agroecosystems (Samways 1983; Addison and Samways 2006).

18.5 Improving Indigenous Ant Biodiversity in Association with Vineyards in the Cape Floristic Region

Pesticides and other agricultural disturbances can impact negatively on arthropod diversity in agro ecosystems (McLaughlin and Mineau 1995; Matson et al. 1997; Altieri and Nicholls 1999; Bengtsson et al. 2005; New 2005) and in vineyards specifically (Sharley et al. 2008; Peverieri et al. 2009; Nash et al. 2010; Gaigher and Samways 2010). In comparison with natural CFR fynbos patches, agricultural patches with pesticide applications have a negative impact on arthropods in general, including ants. This means that a more ecological approach towards pest management should involve reducing the contrast between agricultural and natural land patches, resulting in a more equitable distribution of arthropods (Witt and Samways 2004). While agricultural impacts can vary (Bruggisser et al. 2010), including that on ants (Altieri and Letourneau 1984; Chong et al. 2007; Sharley et al. 2008), increasing habitat heterogeneity for improving pest management and the restoration of natural habitat (Pryke and Samways 2009) appears to be the most sustainable way forward (Altieri et al. 2005; New 2005). It appears that ants can be relatively tolerant to altered habitats, especially dominant ants associated with mealybugs (Addison and Samways 2006). Investigations into the use of corridors of natural habitat and floral patches in and around vineyards in South Africa require further attention. This could impact positively on natural enemies (Landis et al. 2000; Nicholls et al. 2001) and mammalian predators (Hilty and Merenlender 2004), which would fit well with ecotourism and improved environmental wine production.

18.6 Conclusion

In this chapter we have compiled a species list of ants occurring in the CFR. The 95 species representing 26 genera that are documented from this 90,000 km^2 area compare very well with the 281 species representing 44 genera of California, with its

Mediterranean climate but approximately 4.5 times larger area (Ward 2005; Born et al. 2007). With vineyard production being interconnected with natural vegetation, and in many instances replacing it, those ants dominating in both habitats will always be a problem for biological control in vineyards. Current control measures which make use of chemical exclusion barriers are relatively effective if applied correctly, but it would be more desirable to reduce pesticide reliance and further promote sustainability. Unanswered questions include: (1) what abiotic factors affect ant dominance in vineyards, (2) how can the ant-*P. ficus* mutualism be explained within the greater ecological context involving both arthropod assemblages and the host plant, and (3) how can ecological engineering of the landscape benefit ant management and improve biological control? With the average cost of pesticide (insecticide and fungicide) control amounting to US$170/ha for wine grapes during 2008, this figure is by far the highest direct input cost for producers (VinPro and Winetech 2009). As most of the insecticides are applied against *P. ficus* and associated ants, more sustainable control of these pests would drastically reduce overall input costs for the producer. Alternative management strategies for *P. ficus,* which are currently being investigated, include the use of entomopathogenic nematodes (De Waal et al. 2007), but optimizing biological control with parasitoids still remains the focus of most research.

References

Addison P (2002) Chemical stem barriers for the control of ants (Hymenoptera: Formicidae) in vineyards. S Afr J Enol Vitic 23:1–8

Addison P, Samways MJ (2000) A survey of ants (Hymenoptera: Formicidae) that forage in vineyards in the Western Cape Province. Afr Entomol 8:251–260

Addison P, Samways MJ (2006) Surrogate habitats demonstrate the invasion potential of the African pugnacious ant. Biodivers Conserv 15:411–428

Agosti D, Johnson NF (2005) Antbase. World Wide Web electronic publication. http://www.antbase.org, version (05/2005)

Altieri MA, Letourneau DL (1984) Vegetation diversity and insect pest outbreaks. CRC Crit Rev Plant Sci 2:131–169

Altieri MA, Nicholls CI (1999) Biodiversity, ecosystem function, and insect pest management in agricultural systems. In: Collins WW, Qualset CO (eds) Biodiversity in agroecosystems. CRC Press, Boca Raton, pp 69–84

Altieri MA, Ponti L, Nicholls CI (2005) Manipulating vineyard biodiversity for improved insect pest management: case studies from northern California. Int J Biodivers Sci Manag 1:1–13

Baker TC, Van Vorhis Key SE, Gaston LK (1985) Bait-preference tests for the Argentine ant (Hymenoptera: Formicidae). J Econ Entomol 78:1083–1088

Bengtsson J, Ahnström J, Weibull A-C (2005) The effects of organic agriculture on biodiversity and abundance: a meta-analysis. J Appl Ecol 42:261–269

Blachly JS, Forschler BT (1996) Suppression of late-season Argentine ant (Hymenoptera: Formicidae) field populations using a perimeter treatment with containerized baits. J Econ Entomol 89:1497–1500

Bond W, Slingsby P (1984) Collapse of an ant-plant mutualism: the Argentine ant (*Iridomyrmex humilis*) and myrmecochorous proteaceae. Ecology 65:1031–1037

Boonzaaier C, McGeoch MA, Parr CL (2007) Fine-scale temporal and spatial dynamics of epigaeic ants in fynbos: sampling implications. Afr Entomol 15:1–11

Born J, Linder HP, Desmet P (2007) The greater Cape Floristic Region. J Biogeogr 34:147–162
Botes A, McGeoch MA, Robertson HG, van Niekerk A, Davids HP, Chown SL (2006) Ants, altitude and change in the Northern Cape Floristic Region. J Biogeogr 33:71–90
Bruggisser OT, Schmidt-Entling MH, Bacher S (2010) Effects of vineyard management on biodiversity at three trophic levels. Biol Conserv 143:1521–1528
Buckley R, Gullan P (1991) More aggressive ant species (Hymenoptera: Formicidae) provide better protection for soft scales and mealybugs (Homoptera: Coccidae, Pseudococcidae). Biotropica 23:282–286
Chong C-S, Hoffmann AA, Thomson LJ (2007) Commercial agrochemical applications in vineyards do not influence ant communities. Environ Entomol 36:1374–1383
Chong C-S, D'Alberto CF, Thomson LJ, Hoffmann AA (2010) Influence of native ants on arthropod communities in a vineyard. Agric For Entomol 12:223–232
Christian CE (2001) Consequences of a biological invasion reveal the importance of mutualism for plant communities. Nature 413:635–639
Compton SG, Robertson HG (1988) Complex interactions between mutualisms: ants tending homopterans protect fig seeds and pollinators. Ecology 69:1302–1305
Cupido J, Isaacs N (2008) Statistics of wine-grape vines. http://www.sawis.co.za
Daane KM, Malakar-Kuenen RD, Walton VM (2004) Temperature-dependent development of *Anagyrus pseudococci* (Hymenoptera: Encyrtidae) as a parasitoid of the vine mealybug, *Planococcus ficus* (Homoptera: Pseudococcidae). Biol Control 31:123–132
Daane KM, Sime KR, Hogg BN, Bianchi ML, Cooper ML, Rust MK, Klotz JH (2006) Effects of liquid insecticide baits on Argentine ants in California's coastal vineyards. Crop Prot 25: 592–603
Daane KM, Sime KR, Fallon J, Cooper ML (2007) Impacts of Argentine ants on mealybugs and their natural enemies in California's coastal vineyards. Ecol Entomol 32:583–596
De Kock AE (1990) Interactions between the introduced Argentine ant *Iridomyrmex humilis*, and two indigenous fynbos ant species. J Entomol Soc South Afr 53:107–111
De Kock AE, Giliomee JH (1989) A survey of the Argentine ant, *Iridomyrmex humilis* (Mayr), (Hymenoptera: Formicidae) in South African fynbos. J Entomol Soc South Afr 52:157–164
De Kock AE, Giliomee JH, Pringle KL, Majer JD (1992) The influence of fire, vegetation age and Argentine ants (*Iridomyrmex humilis*) on communities in Swartboskloof. In: Van Wilgen BW, Richardson DM, Kruger FJ, van Hensbergen HJ (eds) Fire in South African mountain fynbos. Springer, Berlin, pp 203–215
De Villiers M, Pringle KL (2008) Developing a generic sampling system for monitoring the key arthropod pests of table grapes, *Vitis vinifera* L. Int J Pest Manag 54:207–217
De Waal JY, Wohlfarter M, Malan AP (2007) Laboratory bioassays for the differential susceptibility of *Planococcus ficus* to infection by entomopathogenic nematodes (Rhabditida: Heterorhabditidae and Steinernematidae). In: Fifth international table grape symposium, extended abstracts, Somerset West, South Africa, 14–16 Nov 2007
Dean WRJ (1992) Temperatures determining activity patterns of some ant species in the southern Karoo. J Entomol Soc South Afr 55:149–156
Dean WRJ, Milton SJ (1995) Plant and invertebrate assemblages on old fields in the arid southern Karoo, South Africa. Afr J Ecol 33:1–13
Delabie JHC (2001) Trophobiosis between Formicidae and Hemiptera (Sternorrhyncha and Auchenorrhyncha): an overview. Neotrop Entomol 30:501–516
Donnelly D, Giliomee JH (1985a) Community Structure of epigaeic ants (Hymenoptera: Formicidae) in fynbos vegetation in the Jonkershoek Valley. J Entomol Soc South Afr 48:247–257
Donnelly D, Giliomee JH (1985b) Community structure of epigaeic ants in a pine plantation and in newly burnt fynbos. J Entomol Soc South Afr 48:259–265
Dürr HJR (1953) Die Argentynse mier, *Iridomyrmex humilis* (Mayr.). Farm S Af 27:429, 442
Engelbrecht DJ, Kasdorf GGF (1990) Field spread of corky bark, fleck, leafroll and Shiraz decline diseases and associated viruses in South African vineyards. Phytophylactica 22:347–354
Fairbanks DHK, Hughes CJ, Turpie JK (2004) Potential impact of viticulture expansion no habitat types in the Cape Floristic Region, South Africa. Biodivers Conserv 13:1075–1100

Gaigher R (2008) The effect of different vineyard management systems on the epigaeic arthropod assemblages in the Cape Floristic Region, South Africa. M.S. thesis, Stellenbosch University, Matieland, South Africa. http://scholar.sun.ac.za/handle/10019.1/1565
Gaigher R, Samways MJ (2010) Surface-active arthropods in organic vineyards, integrated vineyards and natural habitats in the Cape Floristic Region. J Insect Conserv 14:595–605
Grout TG (2008) An inexpensive ant bait station. S Afr Fruit J 7:15
Gutierrez AP, Daane KM, Ponti L, Walton VM, Ellis CK (2007) Prospective evaluation of the biological control of vine mealybug: refuge effects and climate. J Appl Ecol 45:524–536
Hilty JA, Merenlender AM (2004) Use of riparian corridors and vineyards by mammalian predators in northern California. Conserv Biol 18:126–135
Hölldobler B, Wilson EO (1990) The ants. Belknap, Cambridge, MA
IPW (2009) Integrated production of wine: guidelines for farms. http://www.ipw.co.za
Islam KS, Copland MJW (1997) Host preference and progeny sex ratio in a solitary koinobiont mealybug endoparasitoid, *Anagyrus pseudococci* (Girault) in response to its host stage. Biocontrol Sci Technol 7:449–456
James GD, Stevens MM, O'malley K, Heffer R (1996) Ant control strategies show promise in citrus orchards. Farmers Newsl 178:6–8
James DG, Stevens MM, O'Malley KJ (1998) Prolonged exclusion of foraging ants (Hymenoptera: Formicidae) from citrus trees using controlled-release chlorpyrifos trunk bands. Int J Pest Manag 44:65–69
Joubert CJ, Walters SS (1955) Bestryding van die Argentynse mier deur die toediening van insektemiddels in die grond. Farm S Afr 30:269–272
Joyce AL, Hoddle MS, Bellows TS, Gonzalèz D (2001) Oviposition behaviour of *Coccidoxenoides peregrinus*, a parasitoid of *Planococcus ficus*. Entomol Exp Appl 98:49–57
Klotz JH, Oi DH, Vail KM, Williams DF (1996) Laboratory evaluation of a boric acid liquid bait on colonies of *Tapinoma melanocephalum* Argentine ants and Pharaoh ants (Hymenoptera: Formicidae). J Econ Entomol 89:673–677
Klotz J, Greenberg L, Venn EC (1998) Liquid boric acid bait for control of the Argentine ant (Hymenoptera: Formicidae). J Econ Entomol 91:910–914
Klotz J, Rust MK, Gonzales D, Greenberg L, Costa H, Phillips P et al (2003) Directed sprays and liquid baits to manage ants in vineyards and citrus groves. J Agric Urban Entomol 20: 31–40
Knight RL, Rust MK (1991) Efficacy of formulated baits for the control of Argentine ants (Hymenoptera: Formicidae). J Econ Entomol 84:510–514
Kriegler PJ, Whitehead VB (1962) Notes on the biology and control of *Crematogaster peringueyi* var. *angustior* Arnold on grape vines (Hymenoptera: Formicidae). J Entomol Soc South Afr 25:287–290
Landis DA, Wratten SD, Gurr GM (2000) Habitat management to conserve natural enemies of arthropod pests in agriculture. Annu Rev Entomol 45:175–201
Lester PJ, Baring CW, Longson CG, Hartley S (2003) Argentine and other ants (Hymenoptera: Formicidae) in New Zealand horticultural ecosystems: distribution, hemipteran hosts, and review. N Z Entomol 26:79–89
Luruli NM (2007) Distribution and impact of the Argentine ant, *Linepithema humile* (Mayr.) in South Africa. M.S. thesis, Stellenbosch University, Matieland, South Africa
Matson PA, Parton WJ, Power AG, Swift MJ (1997) Agricultural intensification and ecosystem properties. Science 277:504–509
McGlynn TP (1999) The worldwide transfer of ants: geographical distribution and ecological invasions. J Biogeogr 26:535–548
McLaughlin A, Mineau P (1995) The impact of agricultural practices on biodiversity. Agric Ecosyst Environ 55:201–212
Mgocheki N (2008) The relationship between ants (Hymenoptera: Formicidae), vine mealybug (Hemiptera: Pseudococcidae) and parasitoids in vineyards of the Western Cape Province, South Africa. Ph.D. dissertation, Stellenbosch University, Matieland, South Africa. http://scholar.sun.ac.za/handle/10019.1/1488

Mgocheki N, Addison P (2009a) Interference of ants (Hymenoptera: Formicidae) with biological control of the vine mealybug *Planococcus ficus* (Signoret) (Hemiptera: Pseudococcidae). Biol Control 49:180–185

Mgocheki N, Addison P (2009b) Effect of contact pesticides on vine mealybug parasitoids, *Anagyrus* sp. near *pseudococci* (Girault) and *Coccidoxenoides perminutus* (Timberlake) (Hymenoptera: Encyrtidae). S Afr J Enol Vitic 30:110–116

Mgocheki N, Addison P (2009c) Incorporating sampling precision into an action threshold for monitoring ant (Hymenoptera: Formicidae) population levels in vineyards. Crop Prot 28: 257–263

Mgocheki N, Addison P (2010) Spatial distribution of ants (Hymenoptera: Formicidae), vine mealybugs and mealybug parasitoids in vineyards. J Appl Entomol 134:285–295

Milton SJ, Dean WRJ, Kerley GIH (1992) Tierberg Karoo Research Centre: history, physical environment, flora and fauna. Trans R Soc S Afr 48:15–46

Moreno DS, Haney PB, Luck RF (1987) Chlorpyrifos and diazinon as barriers to Argentine ant (Hymenoptera: Formicidae) foraging on citrus trees. J Econ Entomol 80:208–214

Myers N, Mittermeier RA, Mittermeier CG, da Fonseca GAB, Kent J (2000) Biodiversity hotspots for conservation priorities. Nature 403:853–858

Nash MA, Hoffmann AA, Thomson LJ (2010) Identifying signatures of pesticide applications on indigenous and invasive non-target arthropod communities from vineyards. Ecol Appl 20: 1693–1704

NDA (2007) A guide for the control of plant pests, 40th edn. Directorate Food Safety and Quality Assurance, National Department of Agriculture, Government Printer, Pretoria

New TR (2005) Invertebrate conservation and agricultural ecosystems. Cambridge University Press, Cambridge

Nicholls CI, Parrella M, Altieri MA (2001) The effects of a vegetational corridor on the abundance and dispersal of insect biodiversity within a northern California organic vineyard. Landsc Ecol 16:133–146

Nyamukondiwa C (2008) Assessment of toxic baits for the control of ants (Hymenoptera: Formicidae) in South African vineyards. M.S. thesis, Stellenbosch University, Matieland, South Africa

Peverieri GS, Simon S, Goggioli D, Liguori M, Castagnoli M (2009) Effects of variety and management practices on mite species diversity in Italian vineyards. Bull Insectol 62:53–60

Prins AJ (1967) The ants of our national parks. Koedoe 10:63–81

Pryke JS (2008) Conservation of the invertebrate fauna on the Cape Peninsula. Ph.D. dissertation, Stellenbosch University, Matieland, South Africa. http://scholar.sun.ac.za/handle/10019.1/1305

Pryke JS, Samways MJ (2009) Recovery of invertebrate diversity in a rehabilitated city landscape mosaic in the heart of a biodiversity hotspot. Landsc Urban Plann 93:54–62

Ratsirarson H, Robertson HG, Picker MD, van Noort S (2002) Indigenous forests versus exotic eucalypt and pine plantations: a comparison of leaf litter invertebrate communities. Afr Entomol 10:93–99

Samways MJ (1981) Comparison of ant community structure (Hymenoptera: Formicidae) in citrus orchards under chemical and biological control of red scale, *Aonidiella aurantii* (Marskell) (Hemiptera: Diaspididae). Bull Entomol Res 71:663–670

Samways MJ (1982) Ecologically sound and commercially acceptable control of ants in guava trees. Subtropica 3:19–20

Samways MJ (1983) Asymmetrical competition and amensalism through soil dumping by the ant, *Myrmicaria natalensis*. Ecol Entomol 8:191–194

Samways MJ (1985) Appraisal of the proprietary bait "Amdro" for control of ants in southern African citrus. Citrus Suptrop Fruit J 621:14–17

Samways MJ, Tate BA (1984) Evaluation of several trunk barriers used to prevent the movement of the pugnacious ant [*Anoplolepis custodiens* (Smith)] into citrus trees. Citrus Subtrop Fruit J 608:9–12, 20, 23, 25, 26

Samways MJ, Weaving AJS, Nel M (1981) Efficacy of chemical and stickybanding in preventing ants entering guava trees. Subtropica 2:13–15

Schoeman C (2008) Synergistic impact of invasive alien plants and the alien Argentine ant on local ant assemblages in the Western Cape. M.S. thesis, Stellenbosch University, Matieland, South Africa
Schwartz A (1988) Efficacy of trunk barriers for the control of key pests on trellised grapevines. S Afr J Enol Vitic 9:16–18
Sharley DJ, Hoffmann AA, Thomson LJ (2008) The effects of soil tillage on beneficial invertebrates within the vineyard. Agric For Entomol 10:233–243
Silverman J, Roulston TH (2001) Acceptance and intake of gel and liquid sucrose compositions by the Argentine ant (Hymenoptera: Formicidae). J Econ Entomol 94:511–515
Silverman J, Brightwell RJ (2008) The Argentine ant: challenges in managing an invasive unicolonial pest. Annu Rev Entomol 53:231–252
Skaife SH (1961) The study of ants. Longman, London
Stevens MM, James DG, O'Malley KJO (1995) Evaluation of alpha cypermethrin-treated proprietary trunk barriers for the exclusion of *Iridomyrmex* spp. (Hymenoptera: Formicidae) from young citrus trees. Int J Pest Manag 41:22–26
Steyn JJ (1954) The pugnacious ant *Anoplolepis custodiens* (Smith) and its relation to the control of the citrus scales at Letaba. Mem Entomol Soc South Afr 3:1–96
Styrsky JD, Eubanks MD (2007) Ecological consequences of interactions between ants and honeydew-producing insects. Proc Biol Sci 274:151–164
Taylor B (2006) The ants of Africa. http://www.antbase.org/ants/africa/antcover.htm
Tshinguvho TE, Dean WRJ, Robertson HG (1999) Conservation value of road verges in the semi-arid Karoo, South Africa: ants (Hymenoptera: Formicidae) as bio-indicators. Biodivers Conserv 8:1683–1695
VinPro and Winetech (2009) The cost of grape production and producer profitability. National Agricultural Marketing Council, South Africa. http://www.namc.co.za
Walton VM (2003) Development of an integrated pest management system for vine mealybug, *Plannococcus ficus* (Signoret), in vineyards in the Western Cape Province, South Africa. Unpublished M.S. thesis, University of Stellenbosch, Matieland, South Africa
Walton VM, Pringle KL (2003) Evaluating effectiveness of mass releases of the vine mealybug (*Planococcus ficus*) parasitoid *Coccidoxenoides peregrinus* in Western Cape province vineyards, South Africa. In: Proceedings of the 1st international symposium on biological control of arthropods, USDA-Forest Service FHTET, Honolulu, HI, 14–18 Jan 2002
Walton VM, Pringle KL (2004) A survey of mealybugs and associated natural enemies in vineyards in the Western Cape Province, South Africa. S Afr J Enol Vitic 25:23–25
Walton VM, Pringle KL (2005) Developmental biology of vine mealybug, *Planococcus ficus* (Signoret) (Homoptera: Pseudococcidae), and its parasitoid *Coccidoxenoides perminutus* (Timberlake) (Hymenoptera: Encyrtidae). Afr Entomol 13:143–147
Ward PS (2005) A synoptic review of the ants of California (Hymenoptera: Formicidae). Zootaxa 936:1–68
Way MJ (1963) Mutualism between ants and honeydew-producing Homoptera. Annu Rev Entomol 8:307–344
Way MJ, Khoo KC (1992) Role of ants in pest management. Annu Rev Entomol 37:479–503
Whitehead VB (1957) A study of the predators and parasites of *Planococcus citri* (Risso) on vines in the Western Cape province, South Africa. Unpublished M.S. thesis, Rhodes University, Grahamstown, South Africa
Whitehead VB (1958) Nuwe lig op Argentynse mier en witluis. Deciduous Fruit Grow (April):94–96
Whitehead VB (1961) Integrated biological and chemical control of mealybug on table grapes. Deciduous Fruit Grow (September):258–260
Witt ABR, Samways MJ (2004) Influence of agricultural land transformation and pest management practices on the arthropod diversity of a biodiversity hotspot, the Cape Floristic Region, South Africa. Afr Entomol 12:89–95
Witt ABR, Geertsema H, Giliomee JH (2004) The impact of an invasive ant, *Linepithema humile* (Mayr) (Hymenoptera: Formicidae), on the dispersal of the elaiosome-bearing seeds of six plant species. Afr Entomol 12:223–230

Chapter 19
Threatening the Harvest: The Threat from Three Invasive Insects in Late Season Vineyards

Douglas G. Pfeiffer, Tracy C. Leskey, and Hannah J. Burrack

19.1 Introduction

19.1.1 Scope

An integral goal of integrated pest management programs is to reduce the pesticide load in the cropping system. Reducing pesticide applications will generally lower pressure to develop pesticide resistance, enhance the presence of beneficial arthropods, and reduce unintended effects on beneficial arthropods, environment, farm workers, and consumers. It is generally desirable to eliminate late season applications, because such applications would lead to the highest residues at harvest. The fact that growers must observe label pre-harvest intervals (PHIs) is often a complicating factor in vineyard management. In recent years, three invasive species from Asia have become pests in North American vineyards. The purpose of this chapter is to discuss their biology, the relationship of their injury to grape harvest, and possible management approaches.

D.G. Pfeiffer (✉)
Department of Entomology, Virginia Tech., 205C Price Hall, Blacksburg, VA 24061-0319, USA
e-mail: dgpfeiff@vt.edu

T.C. Leskey
USDA-ARS, Appalachian Fruit Research Station, 2217 Wiltshire, Road, Kearneysville, WV 25430, USA
e-mail: tracy.leskey@ars.usda.gov

H.J. Burrack
Department of Entomology, North Carolina State University, Research Annex West, Ligon Road, Box 7630, Raleigh, NC 27695-7630, USA
e-mail: hannah_burrack@ncsu.edu

N.J. Bostanian et al. (eds.), *Arthropod Management in Vineyards: Pests, Approaches, and Future Directions*, DOI 10.1007/978-94-007-4032-7_19,

19.2 Brown Marmorated Stink Bug

19.2.1 Introduction

The brown marmorated stink bug (BMSB), *Halyomorpha halys* (Stål), is sometimes called yellow-brown stink bug or east Asian stink bug (Hoebeke and Carter 2003; Hamilton et al. 2008). The native range of BMSB is northeastern Asia (China, Korea, Taiwan, and Japan). In the United States, this species was first collected in Allentown, Pennsylvania, in 1996 (Hoebeke 2002; Hoebeke and Carter 2003). In 1999, it was first detected in New Jersey in a blacklight trap, though BMSB was not officially identified until 2001 based on two specimens collected near Allentown, Pennsylvania. Officially, this insect was identified as present in New Jersey in 2002, Maryland in 2003, and in West Virginia in 2004. An isolated population was found in a recreational vehicle in Maine, recently arrived from Maryland, where there was a known BMSB population (Maine Department of Agriculture 2006). Hamilton et al. (2008) reported BMSB as feeding on peach, pear, raspberry, string beans, asparagus, as well as many ornamentals. At the time, it was not clear whether this species would become a widespread pest in the US (Wermelinger et al. 2008). While widespread across Virginia since its first detection there in 2004 (Day et al. 2011), it was first reported as a nuisance in buildings in 2008–2009, and in 2010 it caused significant damage in orchards and vineyards throughout the mid-Atlantic region. It was detected in 2005 in California, in items stored by a resident that had recently arrived from Pennsylvania (CDFA 2005). It was intercepted in Florida, with no indication of establishment (Halbert 2009; Gyeltshen et al. 2010), and in 2009 it was collected in Tennessee (Jones and Lambdin 2009). There have been isolated reports from Massachusetts and Ohio as well (Welty et al. 2008). Officially, the United States Department of Agriculture-Animal and Plant Health Inspection Service (USDA-APHIS) has reported that BMSB has been detected in 35 states and the District of Columbia as of November 2010. Detection does not necessarily indicate establishment but it rather indicates a potential risk of establishment. The brown marmorated stink bug has also been introduced into Switzerland (Wermelinger et al. 2008). The taxonomy of this species has been unclear, but there is apparently only one species of *Halyomorpha* in eastern Asia (*H. halys*), and all other names from this region (e.g., *H. mista* (Uhler)) are considered synonyms (Rider 2005).

19.2.2 Appearance

The adult (Fig. 19.1) has a typical pentatomid shape, flattened, broad and shield-shaped. Females are slightly larger than males (length 15.6 mm, 13.6 mm, respectively, and width (humeral area) 9.0 mm, 7.6 mm, respectively). Adults are brown, flecked with white (hence the common name marmorated, or marbled). The edge of the abdomen has alternating dark and white spots, and there are white bands on the

Fig. 19.1 Adult brown marmorated stink bug

antennae and tibiae. The most likely insects to be confused with BMSB are stink bugs in the genus *Brochymena* (Hoebeke 2002). However, *Brochymena* spp. have a dentate or crenulate margin of the pronotum, which is smooth in BMSB. Sexes can be differentiated by the forked appearance of the last sternite of the abdomen (Niva and Takeda 2002).

Eggs are barrel-shaped, laid in groups of about 28 (Nielsen et al. 2008a), and nymphs have markings of red and white on the abdomen (Fig. 19.2). The following characters may be used to identify the instars (Hoebeke and Carter 2003): first instar – eyes not projecting; second instar – eyes spherically projecting; third instar – development of white bands on tibiae; fourth instar – anterior wing pads only; fifth instar – posterior wing pads in addition to the anterior ones (Fig. 19.3).

19.2.3 Biology

The brown marmorated stink bug has a wide host range, including many fruit, vegetable and ornamental plants, including apple, pear, grape, kidney bean, pea, and cucumber (Panizzi et al. 2000). The host list is probably about 300 species (Nielsen and Hamilton 2009a). In the review of Pentatomidae by Panizzi et al. (2000), little was known of the life history of this species, despite it being an important pest of commercial crops in Japan. In Japan, BMSB uses *Prunus* trees as reproductive hosts. Overwintered adults appear on trees in early June, and females already have eggs

Fig. 19.2 Egg mass and first-instar brown marmorated stink bug nymphs (Photo by Eric Day)

Fig. 19.3 Fifth instar nymph of brown marmorated stink bug

ready for oviposition. Trees have developing nymphs all season (Funayama 2007). After the final molt to the adult stage, a further 14 days are required for sexual maturation (Kawada and Kitamura 1983). If mated once, a female can produce eggs for about half her life span, but females commonly mate multiple times, even up to five times a day, with duration of copulations averaging 10 min. There is generally one generation in most of Asia and apparently in Pennsylvania (Hoebeke 2002; Funayama

2007) and in New Jersey (Nielsen and Hamilton 2009a). Hoebeke and Carter (2003) cite Hoffmann (1931) who reported up to six generations in the southern part of its range in China. The potential for multiple generations of BMSB in more southerly locations in North America is supported by a 2010 field cage study conducted in the eastern panhandle of West Virginia. The study showed that two distinct generations completed their development within approximately 50 days from egg to adult under field conditions (T.C. Leskey et al., unpubl. data). A modeling study in New Jersey estimated that 537.63 growing degree-days (DD) are needed for total development (egg to adult eclosion). An additional 147.65 DD are needed for the 2-week preovipositional period of the female (Nielsen et al. 2008a). Brown marmorated stink bug eggs have been found hatching in a commercial vineyard in Orange County, Virginia, in the first week of June (D. G. Pfeiffer, unpubl. data).

Kiritani (2006, 2007) pointed to the potential impact of climate change on BMSB populations with winter mortality predicted to decrease by 15% with a rise in temperature of 1°C and the potential for increase in the number of generations per year for BMSB and other bug species that attack rice and fruits.

Stock cultures have been maintained on a diet of shelled sunflower seeds for nymphs, and peanuts for adults, with water containing 0.5% ascorbic acid and 0.25% L-cysteine. Carrot has also been used to augment standard soybean-peanut diets for BMSB cultures with reportedly increased colony viability (Funayama 2006).

In New Jersey, Nielsen and Hamilton (2009a) reported that BMSB eggs were first seen on the Empress tree, *Paulownia tomentosa* (Thunberg) Steudel. Ash (*Fraxinus americana* L.) was an important mid- and late season host for adults. Nymphal abundance shifted among hosts during the season. *Paulownia tomentosa* supported high populations early in the season, while *Viburnum opulus* L. var. *americanum* Aiton was a preferred mid-season host. In the late season, highest nymphal densities were found on *Viburnum prunifolium* L. and *Rosa rugosa* Thunberg. Given the univoltine cycle established in this study, the shifting nymphal population reflects adults changing their preferred oviposition sites. Abundance of nymphs was strongly associated with the presence of maturing fruit or pods.

The defensive scent glands of BMSB have been reported to be located on the dorsal surface of the abdomen and ventral surface of the thorax (Hamilton et al. 2008), but their chemical ecology has been little studied. However, there is a pattern in related bugs of nymphs having dorsal abdominal scent glands, which decline in importance at the adult molt, when the metathoracic scent glands become functional (Aldrich 1988). The defensive odor of BMSB was reported to be *trans*-2-decenal and *trans*-2-octenal in a web site (EOL 2011), but the support for this assertion is a study completed using unidentified stink bugs before BMSB could have been in the area (Henderson et al. 2006).

In addition to their defensive odors, pentatomids use male-produced aggregation pheromones. An aggregation pheromone has not been specifically identified for BMSB. However, in Asia, it has been attracted to methyl (*E,E,Z*)-2,4,6-decatrienoate, the pheromone of another East Asian pentatomid, *Plautia stali* Scott (Aldrich et al. 2007). This compound attracts female and male adults, as well as nymphs (Khrimian et al. 2008). Using this material, BMSB has been detected in traps in Maryland,

and from 2004 to 2008 it became more commonly trapped than the green stink bug, *Acrosternum hilare* (Say), which responds to the same chemical (Aldrich et al. 2009). Diapausing BMSB likely respond to short-range chemical stimuli that result in overwintering aggregations (Toyama et al. 2006), although no specific stimuli associated with this behavioral response have been identified.

19.2.4 Injury

19.2.4.1 Feeding Injury

Little work has been done on feeding effects on grapevines. Stink bugs have traditionally not been considered to be important grape pests. This may change with the high populations of BMSB seen recently at harvest in the mid-Atlantic region of the United States. There has been a grower account of BMSB feeding on the rachis, causing abscission of clusters, with loss of several ha of grapes in 2010 (S. Dorn, pers. comm.). Collapsed berries among table grape cultivars were detected on farms in the mid-Atlantic region, and there is a concern that feeding punctures may increase incidence of fruit rots. More work has been done on tree fruit crops than on grapevines. During mid- and late season growth stages, apples and peaches are susceptible (Nielsen and Hamilton 2009b). During final pit hardening and final growth, feeding impacts are most visible. Early feeding (petal fall in apple, shuck split in peach) results in premature abscission of fruit. In 2010, severe BMSB feeding injury was detected in commercial peach and apple orchards in the mid-Atlantic region. Fruit injury ranged from 15–85% and 25–80% among commercial peach and apple orchards, respectively (T. C. Leskey, unpubl. data). One grape grower reported BMSB feeding on the rachis of grape clusters resulted in loss of berries equal to the production of 1.2–1.6 ha in 2010 (S. Dorn, pers. comm.). Not only would the resulting nymphs have easy access to the rachis, but at this time of the season, expanding berries and leaf canopies would impede spray coverage of this area.

19.2.4.2 Impact of Presence for Wine Quality

Another pressing concern regarding BMSB in vineyards is the result of bugs being collected along with clusters at harvest, and being transported in lugs or bins to the winery. If crushed with the berries, they can impart a noticeable odor or flavor, referred to as 'stink bug taint'. Although BMSB has been noted in Virginia vineyards for a few years, populations became much more severe in 2010, and the taint imparted to juice aroused the concern of vineyard/winery managers (Kelly 2010). A preliminary study showed that as few as 10 adult BMSB per lug can taint the wine (J. Fiola, pers. comm.). Further research is needed on the minimum number of bugs needed to impart a noticeable taint, and on the stability of this effect in the finished wine. This is currently an area of research in Maryland and Virginia.

19.2.4.3 Role as Plant Disease Vector

The brown marmorated stink bug is the vector of the phytoplasma that causes witches broom in *P. tomentosa* (Weintraub and Beanland 2006). Since this phytoplasma also attacks roses, and phytoplasma vectors sometimes carry more than one phytoplasma species, Jones and Lambdin (2009) speculated on potential economic impact of BMSB in North America. The relevance of this reasoning remains to be seen for vineyards.

19.2.4.4 Role as a Nuisance Pest

In its native range, BMSB congregates on buildings in autumn, entering them for the winter, and becoming a nuisance (Hoebeke 2002). This habit has created a public prominence in the eastern US as well (Day et al. 2011). Unprecedented numbers of BMSB invading buildings in 2010 elicited much public concern. During warm days in the fall, large numbers gather on the sides of buildings, especially on south- and west-facing walls. They enter buildings through cracks at doors and windows, and may enter interior rooms either directly, or later from attics and other spaces. As populations develop in grape-producing regions, this may become a significant concern for winery and tasting room managers.

19.2.5 Management

19.2.5.1 Chemical Control

Effective control of BMSB using insecticides has been difficult. The most effective classes have been the pyrethroids and the neonicotinoids. Whereas it is possible to reduce populations immediately after a treatment, it is more difficult to prevent re-infestation. In a glass-vial bioassay, bifenthrin was found to be highly toxic (Nielsen et al. 2008b). Other pyrethroids tested, with similar toxicity, were beta-cyfluthrin, cyfluthrin, fenpropathrin, and lambda-cyhalothrin. Recovery was recorded with all the pyrethroids. Neonicotinoids (dinotefuran, acetamiprid, and thiamethoxam) were also very toxic. The organophosphate phosmet had LD_{50} values almost fourfold higher than other insecticide classes tested. Nymphs were more sensitive to insecticides than adults, and females were more sensitive than males, despite being larger.

A list of insecticides ranked with a 'lethality index' (ranking materials from 0 to 100) was presented by Leskey (2011). This index reflects both immediate mortality, as well as the effect of recovery from initial paralysis. There was considerable variation in the lethality index within pesticide classes, e.g., dimethoate (93.3) and malathion (92.5) at the upper end of the range, and phosmet (20.0) near the lower end. Bifenthrin (91.5) was at the upper end of the pyrethroid class, with esfenvalerate (43.3) much lower. Insecticides on this list that are registered in the US for grape, with the associated lethality index values, are: malathion (92.5), bifenthrin (91.5), endosulfan (90.4), dinotefuran (67.3), fenpropathrin (66.7), kaolin + thiamethoxam (66.7), thiamethoxam (56.3), clothianidin (55.6), beta-cyfluthrin (54.8), zeta-cypermethrin (52.1), cyfluthrin (49.0), imidacloprid (40.0), kaolin (23.1), diazinon (20.4), phosmet (20.0), acetamiprid (18.8), abamectin (16.3), indoxacarb (11.3), spirotetramat (9.8), carbaryl (9.2), and cyantraniliprole (1.7).

Given the late season infestation seen in vineyards, and the problem of harvesting bugs along with the fruit, the PHI becomes of great importance. The following materials may be recommended and are available for BMSB control (followed by Restricted Entry Interval and PHI): **pyrethroids**: fenpropathrin (24 h, 21 days), cyfluthrin (12 h, 3 days), **neonicotinoids**: acetamiprid (12 h, 7 days), clothianidin (12 h, 0 day), dinotefuran (12 h, 1 day), imidacloprid (12 h, 0 day), **carbamates**: methomyl (7 days, 14 days), **chlorinated hydrocarbons**: endosulfan (24 h, 7 days), **botanicals**: pyrethrin (PyGanic® 1.4% or 5%) (12 h, 0 day), pyrethrin plus Canola oil (12 h, 0 day).

In a vineyard study, PyGanic® and clothianidin were both used successfully to eliminate BMSB from clusters by applying them with an airblast sprayer late in the day preceding harvest (Pfeiffer et al. 2010).

A disadvantage of the pyrethroid class is the extremely damaging effect on populations of beneficial arthropods. It is common to see induction of secondary pest outbreaks, including spider mites and mealybugs. The latter is of special interest to vineyard managers, since mealybugs are the vectors for grapevine leafroll virus. Mealybug outbreaks and subsequent infection by leafroll virus have been noted following pyrethroid application for grape berry moth, *Paralobesia viteana* (Clemens) (D. G. Pfeiffer, unpubl. data). Sometimes there are also negative impacts of neonicotinoids on beneficial species. If insecticides are needed, special attention should be paid to vineyard edges, where populations of BMSB are often higher.

19.2.5.2 Biological Control

A tachinid fly in the genus *Bogosia* was reported parasitizing BMSB in Japan. The female laid an egg on the pronotum of the bug. After entering the host, the larva consumed the reproductive system, and sterilized the host. The tachinid larva then left the host to pupate. Parasitism of nymphs has also been observed (Kawada and Kitamura 1992). Parasitization rates were reported between 6% and 7% (Kawada and Kitamura 1983) and>10% for overwintering adults (Kawada and Kitamura 1992). In an 8-week survey in China, Koppel (2010) found BMSB eggs to be parasitized by *Trissolcus halyomorphae* Yang in four different host plants in Nanjing, Kunming, and Xi'an. Parasitization rates can reach 70%, with annual mean parasitization of 50% (Yang et al. 2009). A picorna-like virus, named *P. stali* intestine virus (PSIV), was found infecting the brown-winged greenbug, *P. stali* Scott. It was also found infecting BMSB. Infected *P. stali* have an adult life span of about 13 days, compared to about a month in non-viruliferous adults (Nakashima et al. 1998).

19.2.5.3 Mechanical Control

Infestations in dwellings can be controlled by sealing and screening openings, to reduce numbers entering the dwelling. Light fixtures, exhaust fans and baseboards are sealed with caulk to keep those that have entered attics and basements from entering interior rooms (Day et al. 2011). Insecticides may be applied to the sides of the dwelling, but this should be done by a professional.

19.3 Spotted Wing Drosophila

19.3.1 Introduction

Native to eastern Asia, the spotted wing drosophila (SWD), *Drosophila suzukii* (Matsumura), is a pest of soft skinned fruit. It has recently been detected in and spread throughout North America. Markow and O'Grady (2006) recognize *D. suzukii* as a complex of the *suzukii* subgroup, within the *melanogaster* species group, with the following description of ranges: '*Drosophila mimetica* … is known from Malaysia, *D. lucipennis* … is disjunctly distributed in eastern India and Taiwan, *D. biarmipes* … is known from India and Sri Lanka to southeast Asia, and *D. pulchrella* … is found from India, China, and southeast Asia to Japan.'

The spotted wing drosophila invaded Hawaii several years ago (Walsh et al. 2011). Since Hawaii is a US State, some considered this was therefore already a US pest, complicating later detection and management efforts. Upon its detection in Florida in late 2009, USDA-APHIS declared that it was not a regulated pest, so there would be no eradication program (Lehnert 2010). The spotted wing drosophila was found in California in 2008 infesting strawberries and caneberries (Bolda et al. 2009; Lehnert 2010; Walsh et al. 2011). In 2009 it spread up the Pacific Coast to infest fruit in Oregon, Washington, and the Fraser Valley of British Columbia. The spotted wing drosophila is now found in all western counties of Washington State, and in eight eastern counties. In 2009, it was found in Florida (Anonymous 2009b; Acheampong 2010; Lehnert 2010), where it was found in feeding lure traps 4.8 km apart. In 2010 it was found in Louisiana, South Carolina, North Carolina (H. J. Burrack, unpubl. data) and Michigan (Milkovich 2010; Isaacs 2011). In 2011, SWD was detected in Georgia, Virginia, Alabama (H. J. Burrack, unpubl. data; D. G. Pfeiffer, unpubl. data), New Jersey (Rodriguez-Saona and Polk 2011), and Pennsylvania (Anonymous 2011).

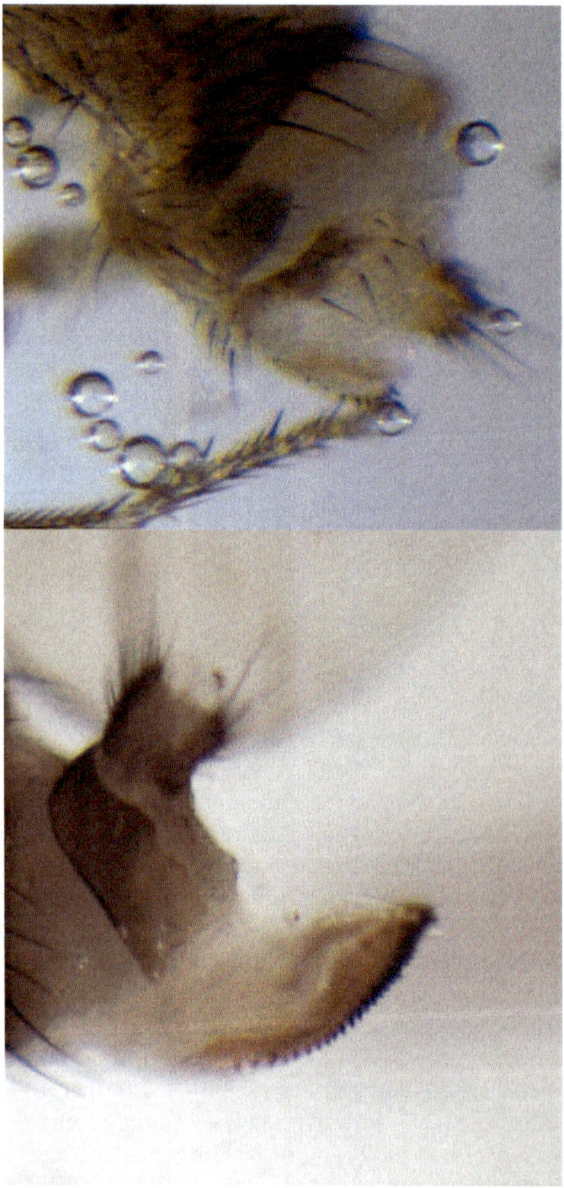

Fig. 19.4 Comparison of ovipositor of spotted wing drosophila with another *Drosophila* sp. (Photo by Hannah Burrack)

19.3.2 Appearance

Females have a relatively large serrated ovipositor (Fig. 19.4), unusual among the drosophilids. The typically modest development of female terminalia in *Drosophila* spp. was described by Demerec (1965). Males have a single spot present on the first

Fig. 19.5 Comparison of wing of male spotted wing drosophila with female (Photo by Hannah Burrack)

vein at the distal end of each wing (Fig. 19.5) and a unique sex comb arrangement on the basitarsis and first tarsomere of their forelegs. Larvae reach a length of 3.5 mm (Walsh et al. 2011).

19.3.3 Biology

The spotted wing drosophila has been often reported to attack grapes (Anonymous 2009a; Dreves et al. 2009; Anonymous 2010). Other hosts include apples, figs, hardy kiwi, apricots, persimmons, pluots, blackberries, blueberries, cherries, nectarines, peaches, pears, plums, raspberries, strawberries, their wild relatives, and at least one native weed, *Phytolacca americana* L. (Anonymous 1993; Bolda et al. 2009; Acheampong 2010; Walsh et al. 2011). Reports from Japan indicate that grape is among the most vulnerable fruit crops (Anonymous 2009b, 2010; Walsh et al. 2010), and that grape is a preferred late season host (Walsh et al. 2010). Additional work is needed to determine the crops at greatest risk in North America.

There are important gaps in our knowledge of the basic biology of this pest (Bolda et al. 2009). The spotted wing drosophila prefers warm areas, but tolerates the cold of northern Japan (Bolda et al. 2009). The range of drosophilids is limited by cold tolerance. In an effort to predict the ability of selected drosophilid species to expand their geographic ranges, Kimura (2004) measured the LT_{25}, LT_{50}, and LT_{75}

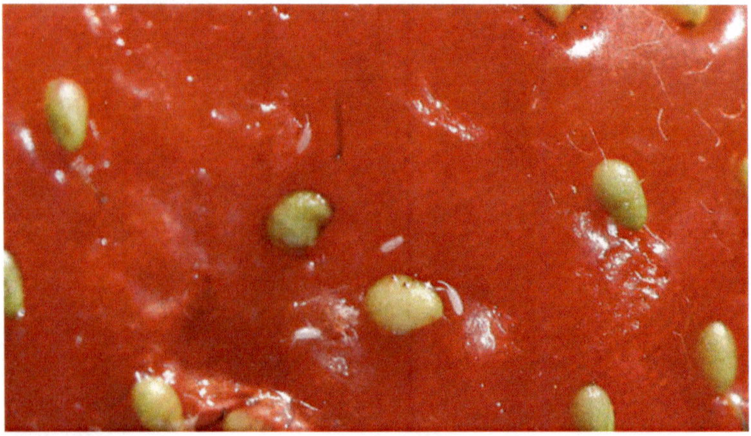

Fig. 19.6 Eggs of spotted wing drosophila in a strawberry (Photo by Hannah Burrack)

(Lethal Temperature) of northern and southern strains of several species. There was little difference in northern and southern representative strains of SWD. However, this study used non-adapted, non-diapausing individuals for testing, and their conclusions may be conservative. Although a congeneric species, *D. lutescens* Okada, cannot tolerate winter conditions in Sapporo, Japan, it is common in this region in summer and autumn (Kimura 2004). This period is the most problematic for SWD management in vineyards. All life stages may die when frozen. However, adults are more tolerant to cold temperatures and may survive short periods of freezing or sustained cool temperatures (10°C) for longer periods (Dalton et al. 2011).

In Japan, there are up to 13 generations. A life cycle can be completed in 8–14 days, but adults can live up to 9 weeks. Females use the atypically large and serrated ovipositor to lay eggs in fruits before they become overripe and soft. Eggs are inserted under the skin of ripe or ripening fruit. Each female lays 7–16 eggs per day. Eggs have prominent respiratory horns projecting from one end (Fig. 19.6). Eggs hatch in 1–3 days, and larval feeding on the flesh causes a collapse of localized tissue after another 2 days, followed by growth of fungal or bacterial organisms; yeasts may be carried on the ovipositor (Walsh et al. 2010). Larvae are slender white maggots (Fig. 19.7). Pupae are brown and seed-like, about 3 mm long, with two small respiratory horns protruding from one end.

19.3.4 Injury

Cherries were reported to have 70–80% injury by SWD, with eggs laid in sound fruit, by Kamizawa (1936) (cited by Demerec 1965). Growers of blueberries, caneberries and cherries, were reported to have experienced injury levels ranging

Fig. 19.7 Larvae of spotted wing drosophila in a strawberry (Photo by Hannah Burrack)

from 33% to 100% (Lehnert 2010). The economic significance of SWD in wine grapes remains unclear. However, in Japan seasonal activity was found to be greatest when cherries and grapes were ripening; this occurs in the fifth and sixth generation of the 10 seasonal generations observed in the Far East (Kanzawa 1939; Walsh et al. 2011). Walsh (2011) reported that while blueberry, blackberry, raspberry, marionberry, boysenberry, strawberry, cherry and peach are considered being of prime importance, grapes are also considered at risk.

19.3.5 Management

19.3.5.1 Monitoring

Traps baited with either yeast or apple cider vinegar may be used for monitoring adults (Fig. 19.8). Instructions for construction of simple plastic cup traps are presented by Walsh et al. (2010). Traps should be checked at least weekly. Most of the *Drosophila* flies collected will not be SWD, so the flies collected must be filtered from the trap fluid and poured into a white pan. Male SWD have a characteristic black spot at the tip of the wings. Females lack this spot, but are slightly larger than females of other fruit fly species, and have a larger ovipositor.

Caprile et al. (2010) mentioned the use of traps to protect cherries in home gardens. This would likely be impractical for large plantings such as commercial vineyards. Use of such traps early in the season may give early warning of the presence of SWD. There seems to be little relationship between trap captures and the degree of fruit damage. Traps are best used to determine presence and not density.

Fig. 19.8 Trap for spotted wing drosophila baited with apple cider vinegar (Photo by Hannah Burrack)

19.3.5.2 Chemical Control

Control measures are directed against the adults. There are no effective controls for larvae in the fruit. Insecticides with different modes of action should be rotated in order to delay the development of pesticide resistance. Walsh et al. (2010) recommended the following insecticides: malathion, spinosad, spinetoram, acetamiprid, imidacloprid, and thiamethoxam. In addition, fenpropathrin and zeta-cypermethrin may be effective. However, these last two products may disrupt biological control components of vineyard pests.

19.3.5.3 Cultural and Physical Control

Fruit must be harvested promptly to eliminate breeding sites. This issue should be kept in mind once SWD has established in an area, because grape growers may occasionally leave berries on the vines to allow greater development of some harvest parameters. Any overripe or rotten fruits nearby should be destroyed. If a crop is found to be infested with SWD, especially if not known to be established in the area, it should be destroyed after samples are taken for proper identification. Solarization and burying are being explored as means of destruction of infested fruit. Destruction of infested fruit will be especially important when there are nearby blocks approaching maturity (Walsh et al. 2011). As fruit approach maturity, covering plants with netting may be helpful (Walsh et al. 2011), but this may be difficult in a vineyard setting.

19.4 Multicolored Asian Lady Beetle

19.4.1 Introduction

The multicolored Asian lady beetle (MALB), *Harmonia axyridis* (Pallas), is an exotic coccinellid, and is now common in many North American and European cropping systems. It has come to dominate the coccinellid fauna (LaMana and Miller 1996; Brown and Miller 1998; Harmon et al. 2007; Lucas et al. 2007; Mizell 2007; Finlayson et al. 2008). Snyder and Evans (2006) discussed several possible ecological factors potentially involved in promoting high numbers of introduced arthropod predators, including absence of natural enemies, competition, intraguild predation, and disease transmission.

Koch (2003) reported its native range from the Altai Mountains (Siberia) to the Pacific Ocean, and southern Siberia to southern China. Many attempts to introduce MALB into North America have taken place since 1916, though it apparently did not establish until 1988 (Koch 2003). It has since spread to South America and Europe (Koch et al. 2006). The multicolored Asian lady beetle is now an important contributor to biological control. It is an important factor in reducing the soybean aphid, *Aphis glycines* Matsumura, another invasive species (Landis et al. 2004). Compounds have been evaluated for repellency to MALB (Riddick and Aldrich 2004). Camphor repelled adults but its effectiveness was too short lived. DEET was also repellent but studies on urban structures are needed. There is some potential for modifying behavior with plant products. Menthol, catnip and grapefruit seeds have been reported to cause avoidance by MALB (Riddick et al. 2000, 2008).

19.4.2 Appearance

The multicolored Asian lady beetle (Fig. 19.9) is described by several authors (Obata 1997; Nalepa et al. 2004; Pfeiffer 2008). There is a large amount of variation in color, giving rise to the common name. The background color ranges from light orange or pink to dark red, with 19 spots ranging from unapparent to heavy black patches (color forms termed *succinea* 1 and 2 by Seo et al. (2008)). More rarely the beetle is black with red spots (color forms *conspicua* and *spectabilis*). This color

Fig. 19.9 Adult multicolored Asian lady beetle

variation is affected by genetics, larval diet, and seasonal effects (Koch 2003). It is also partly affected by nonrandom mating (Seo et al. 2008). The pronotum is white with a black M- or W-shaped pattern, sometimes reduced to a pair of curved lines. Larvae are dark gray to black with orange patches, with two- or three-pronged scoli. The scoli are projections from the dorsal abdominal surface and they are branched, unlike those of other coccinellids. There are four instars, which can be distinguished by coloration: first instar – dark blackish coloration; second instar – same as first instar but with orange on the dorsolateral areas of first and/or second abdominal segment; third instar – orange color extends to dorsolateral areas of second through fifth abdominal segments; fourth instar – same coloration as third, but the scoli on fourth and fifth abdominal segments are also orange (Koch 2003).

19.4.3 Biology

LaMana and Miller (1998) determined temperature-dependent developmental rates, with 267.3 DD >11.2°C required for the development from egg to adult. At 26°C, 2.8 days were spent in the egg state, and 2.5, 1.5, 1.8, and 4.4 days in the four larval instars. The pupal stage was 4.5 days. Adults may live up to 3 years (Koch 2003). Females can lay an average of 1,642 eggs (Stathas et al. 2001), in batches of 20–30 (Takahashi 1987). However, when food is a limiting factor, larvae grow more slowly, producing smaller and less fecund adults (Agarwala et al. 2008). Nutritional requirements for development of MALB were studied by Agarwala et al. (2008). To some extent, MALB can compensate for low prey availability by feeding at extrafloral nectaries and the presence of such nectaries may interfere with biological control (Spellman et al. 2006). Presence of aphids enhances mate-finding behavior of MALB males, and receptivity to mating by females (Obata 1997).

The multicolored Asian lady beetle is bivoltine in much of its range, including North America (Koch 2003), though up to five generations have been reported (Katsoyannos et al. 1997). Adults search for overwintering sites in late October, seeking out isolated shapes on the horizon (Obata 1986). As winter approaches, their supercooling point drops to −16°C to −19°C (Watanabe 2002), and in Minnesota, even to −24°C (Carillo et al. 2004). Mating often occurs as adults leave overwintering sites (LaMana and Miller 1996), though 12–41% of females had sperm in their spermathecae in collections at North Carolina overwintering sites (Nalepa et al. 1996). Parts of its North American range exhibit winter temperatures that should be lethal. The beetle finds suitable microclimates in the fall, making winter minimum temperatures a poor indicator of potential range (Carillo et al. 2004; Labrie et al. 2008).

Intraguild predation has been raised as a factor related to the high numbers of MALB, and affecting the efficacy of other predators. Burgio et al. (2002) reported that in feeding trials with the native coccinellid, *Adalia bipunctata* (L.), intraguild predation between the two coccinellids was generally less than within-species egg cannibalism by *A. bipunctata*. In a biological control study for the hemlock woolly adelgid, *Adelges tsugae* Annand, two specialists *Laricobius nigrinus* Fender and *Sasajiscymnus* (=*Pseudoscymnus*) *tsugae* Sasaji & McClure, were examined with MALB. All species fed on eggs of the two specialists, but eggs of MALB were only fed upon by MALB, putting it at an advantage (Flowers et al. 2005). Cannibalism by siblings is a trait of MALB. Osawa (1993) concluded that sibling cannibalism was density-independent and the non-sibling cannibalism was density-dependent. A different situation exists with larval intraguild predation. When exposed to foraging MALB, larval *Coccinella septempunctata* L. were more likely than *A. bipunctata* to drop from the plant. As a result, 95% of *A. bipunctata* larvae were consumed by MALB, compared with about 54% of *C. septempunctata* larvae (Sato et al. 2005). Intraguild egg predation by MALB and lack of feeding on MALB eggs by other coccinellid species may have favored the spread and population increase of MALB in Great Britain (Snyder et al. 2000; Ware et al. 2008).

19.4.4 Injury

Adult MALB have been reported feeding on the fruit of peaches, apples, raspberries and grapes (Kovach 2004). In order to determine whether beetle presence represented a primary or secondary problem, fruit of several types were placed in screened containers: 'Gala' apples, 'Redhaven' peaches, and 'Red Flame' seedless grapes. Some fruit were injured mechanically to simulate bird-injured fruit while other fruit were left uninjured as controls. Beetles were much more likely to feed on injured than on uninjured fruit. However, uninjured fruit, mainly grapes, were also fed on by adult MALB. Moreover, injured grapes were also twice as attractive as injured apples or peaches. However, that study involved caged beetles and fruit, and only a minority of beetles fed upon grape berries. In contrast, Koch et al. (2004) reported that MALB do not cause primary feeding injury to grapes. Galvan et al. (2006a) found that MALB were more likely to be found in clusters if berries had been previously injured, and that simple presence of MALB was not a problem unless accompanied by injured fruit. It appears that, while MALB is capable of feeding on uninjured grape berries, beetles fed preferentially on previously injured fruit. Hence, vineyard managers should be most mindful of MALB in blocks where fruit exhibit splitting from rain, birds or insect injury after berries have started to accumulate sugars.

Galvan et al. (2009) found that there was a major peak of MALB flight activity between veraison and harvest, allowing greater numbers in vineyards at the most vulnerable time. Populations in vineyard surroundings peak about 10 days earlier than populations in vineyards.

19.4.4.1 Effect on Wine

Tasting panels in Ohio noted an occasional taste reminiscent of 'rancid peanut or cooked spinach odor.' This could mask varietal characteristics of wine, and judges reported that the odor was similar to MALB found in homes in the area (Kovach 2004). In a preliminary assessment of effects of beetles crushed with the berries on wine quality, Kovach (2004) crushed 100 MALB adults in 100 ml of white wine, centrifuged and filtered, and this product was used to spike various concentrations of wine. The detection limit was determined to be about 1.2 beetles per liter, about one per bottle. A field threshold was set of 12 beetles per lug. Fermenting MALB with the wine causes altered aroma and flavor profiles (Pickering et al. 2004). Pickering et al. (2005) found that 2-Isopropyl-3-methoxypyrazine (IPMP) was determined to be above a sensory threshold when fermented with MALB. The aroma and flavor profiles of aged wines were not different from fresh wines, with reduced fruit and floral traits in red and white wines. Research on potential remedial treatments for wine have shown that IPMP titers were lowered by activated charcoal in white wine, deodorized oak in red wine, and the addition of oak chips reduced the 'ladybug taint' in both red and white wines (Pickering et al. 2006).

19.4.4.2 Domestic Invasion

Huelsman et al. (2002) reported that the movement of adults to overwintering sites starts on the first day when temperatures exceed 18°C after near-freezing temperatures have been reached. Adult MALB sometimes enter buildings in large numbers, creating a nuisance. There may be a greater risk of allergic responses because of its greater proximity to humans (Yarbrough et al. 1999). Beetles prefer buildings with at least one wooded side. Caulking of cracks and other spaces is helpful but not sufficient (Huelsman and Kovach 2004). Beetles are kept out by 2 mm gaps, though most can pass through a 3 mm gap (Nalepa 2007). Beetles prefer to land on structures with high contrast patterns of light and dark. Unfortunately, many structures on buildings (drain pipes, gutters, etc.) cast contrasting shadows that may be attractive to immigrating MALB adults (Nalepa et al. 2004, 2005).

19.4.5 Management

19.4.5.1 Monitoring

Galvan et al. (2007) examined the distribution of MALB in vineyards, and developed sequential sampling programs. The distribution was determined to be random at low densities, and aggregated at higher densities. Enumerative and binomial sampling plans were compared. For management purposes, an average sample of 180 clusters were needed to determine the population at precision level of 0.25 (SE/$\bar{x}$). However, binomial sampling plans were more efficient for pest management purposes, where 19–26 clusters were needed to determine the action threshold of one beetle per cluster. Correct decisions were made in 83–96% of the time. Sampling should be initiated 2–3 weeks before harvest. In the final week, sampling should precede the PHI of the insecticide to be employed.

19.4.5.2 Chemical Control

Buprofezin is highly toxic to MALB larvae, but showed only low-moderate toxicity to MALB pupae (James 2004). In a vineyard comparison, bifenthrin, carbaryl, zeta-cypermethrin, imidacloprid and thiamethoxam reduced adult populations in clusters (Galvan et al. 2006b). In the laboratory, residues of bifenthrin, carbaryl, and thiamethoxam were all toxic to MALB 7 days following treatment. In 2006, few insecticides registered on grape had sufficiently short PHIs to allow their use to control MALB. The choice was limited to carbaryl, malathion, and imidacloprid. Indoxacarb was toxic to third instars and adults while spinosad was ineffective (Galvan et al. 2006c). Vineyard managers should check current labels for effective materials registered with the required short PHI. For example, clothianidin now has MALB on the label, with a 0-day PHI (Valent U.S.A Corporation 2010) and dinotefuran has a 1-day PHI (Cornell Cooperative Extension Publication 2010).

19.4.5.3 Biological Control

Koch (2003) reviewed the biological control literature of MALB, listing several parasitoids. The phorid *Phalacrotophora philaxyridis* Disney has been reported in Asia, but may have also followed MALB to North America. Two tachinids attack MALB adults: *Degeria lutuosa* Meingen in Korea (Park et al. 1996) and *Strongygaster triangulifera* (Loew) in North America. The braconid *Dinocampus* (=*Perilitus*) *coccinellae* (Schrank) parasitized MALB in Korea and North America. A parasitic fungus specific to Coccinellidae was found infecting MALB in Pennsylvania, with infection levels>50% (Riddick and Schaefer 2005). Nalepa et al. (1996) reported overwintering adults to be parasitized by the tachinid *S. triangulifera* at levels of 14.2% and 1.4%, respectively in 1993 and 1994. *Spiroplasma* bacteria kill only male embryos, potentially resulting in a skewed sex ratio (Nakamura et al. 2005).

19.4.5.4 Mechanical Control

Floating row covers successfully reduced the number of adults per cluster (Galvan et al. 2006b).

19.5 Prospects

It is likely that both BMSB and SWD will continue to expand their geographic ranges. Their presence will present a complication for vineyard management at harvest. However, the magnitude of this disruption is still unclear. Bolda et al. (2009) projected significant losses to Pacific Coast small fruit crops from SWD, particularly if trade barriers are erected. There will likely be greater pesticide use shortly before harvest of grapes and other fruits and vegetables. Pesticide efficacy trials are now underway in many locations and optimum choices for chemical control would be available in the near future. Currently there is cause for concern, because the pyrethroid class, providing some of the greatest mortality at this time, has extremely negative effects on vineyard IPM programs, potentially inducing secondary pest problems. Some of these pests (e.g., mealybugs, Chap. 12) are vectors of important grapevine diseases. Under a global warming scenario, it is likely that BMSB will become a greater problem, because of decreased winter mortality. Based on a Japanese study, each increase of 1°C is expected to result in approximately 13.5–16.5% decrease of winter mortality (Musolin 2007). In addition, earlier spring emergence of BMSB has been seen because of higher early spring temperatures. The wide host range of BMSB and SWD will complicate control programs (Nielsen and Hamilton 2009a). Efforts may be needed at the landscape level. MALB has been present in vineyards for a longer period of time. It now seems that this insect will pose the greatest threat not from primary injury to berries, but by concentrating in clusters that are already injured by other organisms. Hence, managers should be vigilant in order to protect wine quality.

References

Acheampong S (2010) Spotted wing drosophila (*Drosophila suzukii*), a new vinegar fly pest in British Columbia. British Columbia Ministry of Agriculture and Lands, Kelowna, BC

Agarwala BK, Yasuda H, Sato S (2008) Life history response of a predatory ladybird, *Harmonia axyridis* (Pallas) (Coleoptera: Coccinellidae), to food stress. Appl Entomol Zool 43:183–189

Aldrich JR (1988) Chemical ecology of the Heteroptera. Annu Rev Entomol 33:211–238
Aldrich JR, Khrimian A, Camp MJ (2007) Methyl 2,4,6-decatrienoates attract stink bugs and tachinid parasitoids. J Chem Ecol 33:801–815
Aldrich JR, Khrimian A, Chen X, Camp MJ (2009) Semiochemically based monitoring of the invasion of the brown marmorated stink bug and unexpected attraction of the native green stink bug (Heteroptera: Pentatomidae) in Maryland. Fla Entomol 92:483–491
Anonymous (1993) Host plants of *Drosophila suzukii*. Pest Res Rep 44. www.affrc.go.jp/ja/research/seika/data_tnaes/h05/tnaes93122
Anonymous (2009a) Fruit fly pest identified in wine grapes. Science Daily. Oregon State University, Corvallis. http://www.sciencedaily.com/releases/2009/10/091015163605.htm
Anonymous (2009b) Spotted wing drosophila, *Drosophila suzukii* (Matsumura) (Diptera: Drosophilidae), a Fruit Pest New to North America and Florida. University of Florida, Gainesville, FL
Anonymous (2010) Pest alert: spotted wing drosophila. Oregon Department of Agriculture, Corvallis. http://www.oregon.gov/ODA/PLANT/IPPM
Anonymous (2011) New fruit pest found in Pennsylvania. Pennsylvania Integrated Pest Management, http://extension.psu.edu/ipm/news/2011/new-fruit-pest-found-in-pennsylvania
Bolda MP, Goodhue RE, Zalom FG (2009) Spotted wing drosophila: potential economic impact of a newly established pest. Giannini Foundation of Agricultural Economics, University of California, Berkeley, CA
Brown MS, Miller SS (1998) Coccinellidae (Coleoptera) in apple orchards of eastern West Virginia and the impact of invasion by *Harmonia axyridis*. Entomol News 109:143–151
Burgio F, Santi F, Maini S (2002) On intra-guild predation and cannibalism in *Harmonia axyridis* (Pallas) and *Adalia bipunctata* L. Biol Control 24:110–116
Caprile JM, Flint L, Bolda MP, Coates WW, Grant JA, Zalom FG, Van Steenwyck RA, Haviland D (2010) Spotted wing drosophila, *Drosophila suzukii*: a new pest in California. http://www.ipm.ucdavis.edu/EXOTIC/drosophila.html
Carillo MA, Koch RL, Venette RC, Cannon CA, Hutchison WD (2004) Response of the multicolored Asian lady beetle (Coleoptera: Coccinellidae) to low temperatures: implications for winter survival. Am Entomol 50:157–158
CDFA (2005) New state records: *Halyomorpha halys* (Stål) (Pentatomidae), California plant pest and disease report. California Department of Food and Agriculture, Sacramento
Cornell Cooperative Extension Publication (2010) 5.3 pest management guidelines. Pest management schedules for minor insects. Cornell University and PennState, NY and PA. http://ipmguidelines.org/Grapes/content/CH05/default-3.asp
Dalton DT, Walton VM, Shearer PW, Walsh DB, Caprile J, Isaacs R (2011) Laboratory survival of *Drosophila suzukii* under simulated winter conditions of the Pacific Northwest and seasonal field trapping in five primary regions of small and stone fruit production in the United States. Pest Manag Sci 67:1368–1374
Day ER, McCoy T, Miller D, Kuhar TP, Pfeiffer DG (2011) Brown marmorated stink bug, Hemiptera, Pentatomidae: *Halyomorpha halys*. Virginia Tech, Fact Sheets 2902–1100:1–2
Demerec M (1965) Biology of drosophila. Hafner, New York, NY
Dreves AJ, Fisher G, Walton V (2009) A new pest attacking healthy ripening fruit in Oregon: spotted wing drosophila, *Drosophila suzukii* (Matsumura), regional pest alert. Oregon State University, Corvallis
EOL (2011) *Halyomorpha halys*: brown marmorated stink bug. Encyclopedia of life. http://eol.org/pages/3686128/overview
Finlayson CJ, Landry KM, Alyokhin AV (2008) Abundance of native and non-native lady beetles (Coleoptera: Coccinellidae) in different habitats in Maine. Ann Entomol Soc Am 101: 1078–1087
Flowers RW, Salom SM, Kok LT (2005) Competitive interactions among two specialist predators and a generalist predator of hemlock woolly adelgid, *Adelges tsugae* (Homoptera: Adelgidae), in the laboratory. Environ Entomol 34:664–675
Funayama K (2006) A new rearing method using carrots as food for the brown marmorated stink bug, *Halyomorpha halys* (Stål) (Heteroptera: Pentatomidae). Appl Entomol Zool 41:415–418

Funayama K (2007) Reproduction of the brown marmorated stink bug, *Halyomorpha halys* (Stål) (Heteroptera: Pentatomidae) on Japanese bird cherry trees, *Prunus grayana* Maxim. Jpn J Appl Entomol Zool 51:238–240

Galvan TL, Burkness EC, Hutchison WD (2006a) Influence of berry injury on infestations of the multicolored Asian lady beetle in wine grapes. Plant Health Prog doi:10.1094/PHP-2006-0607-01-BR

Galvan TL, Burkness EC, Hutchison WD (2006b) Efficacy of selected insecticides for management of the multicolored Asian lady beetle on wine grapes near harvest. Plant Health Prog doi:10.1094/PHP-2006-1003-01-RS

Galvan TL, Koch RL, Hutchison WD (2006c) Toxicity of indoxacarb and spinosad to the multicolored Asian lady beetle, *Harmonia axyridis* (Coleoptera: Coccinellidae), via three routes of exposure. Pest Manag Sci 62:797–804

Galvan TL, Burkness EC, Hutchison WD (2007) Enumerative and binomial sequential sampling plans for the multicolored Asian lady beetle (Coleoptera: Coccinellidae) in wine grapes. J Econ Entomol 100:1000–1010

Galvan TL, Burkness EC, Koch RL, Hutchison WD (2009) Multicolored Asian lady beetle (Coleoptera: Coccinellidae) activity and wine grape phenology: implications for pest management. Environ Entomol 38:1563–1574

Gyeltshen J, Bernon G, Hodges A (2010). Brown marmorated stink bug, *Halyomorpha halys* Stål (Insecta: Hemiptera: Pentatomidae). University of Florida IFAS Extension. http://edis.ifas.ufl.edu/in623

Halbert SE (2009) Tri-ology. Entomology section, DACS-P-00124. Florida Department of Agriculture and Consumer Services

Hamilton GC, Shearer PW, Nielsen AL (2008) Brown marmorated stink bug: a new exotic insect in New Jersey. R. U. C. Extension, New Brunswick, NJ

Harmon JP, Stephens E, Losey J (2007) The decline of native coccinellids (Coleoptera: Coccinellidae) in the United States and Canada. J Insect Conserv 11:85–94

Henderson W, Khalilian A, Han Y (2006) Detecting stink bugs/damage in cotton utilizing a portable electronic nose, paper number 061103. In: ASABE annual international meeting, 9–12 July 2006, Portland, OR, pp 1–10

Hoebeke ER (2002) Brown marmorated stink bug, *Halyomorpha halys* (Stål): Heteroptera: Pentatomidae. Entomol Circ 204:34–38

Hoebeke ER, Carter ME (2003) *Halyomorpha halys* (Stål) (Heteroptera: Pentatomidae): a polyphagous plant pest from Asia newly detected in North America. Proc Entomol Soc Wash 105:225–237

Hoffmann WE (1931) A pentatomid pest of growing beans in south China. Peking Nat Hist Bull 5:25–26

Huelsman MF, Kovach J, Jasinski J, Young C, Eisley B (2002) Multicolored Asian lady beetle (*Harmonia axyridis*) as a nuisance pest in households in Ohio. In: Jones SC, Zhai J, Robinson WH (eds) Proceedings of the 4th international conference on urban pests, 7–10 July 2002, Charleston, USA, pp 243–250

Huelsman M, Kovach J (2004) Behavior and treatment of the multicolored Asian lady beetle (*Harmonia axyridis*) in the urban environment. Am Entomol 50:163–164

Isaacs R (2011) First detection and response to the arrival of spotted wing drosophila in Michigan. Newsl Michigan Entomol Soc 56:10–12

James DG (2004) Effect of buprofezin on survival of immature stages of *Harmonia axyridis*, *Stethorus punctum picipes* (Coleoptera: Coccinellidae), *Orius tristicolor* (Hemiptera: Anthocoridae), and *Geocoris* spp. (Hemiptera: Geocoridae). J Econ Entomol 97:900–904

Jones JR, Lambdin PL (2009) New county and state records for Tennessee of an exotic pest, *Halyomorpha halys* (Hemiptera: Pentatomidae), with potential economic and ecological implications. Fla Entomol 92:177–178

Kamizawa T (1936) Studies on *Drosophila suzukii* Mats. J Plant Prot 23(66–70):127–132

Kanzawa T (1939) Studies on *Drosophila suzukii* Mats. Rev Appl Entomol 29:622

Katsoyannos P, Kontodimas DC, Stathas GJ, Tsartsalis CT (1997) Establishment of *Harmonia axyridis* on citrus and some data on its phenology in Greece. Phytoparasitica 25:183–191

Kawada H, Kitamura C (1983) The reproductive behavior of the brown marmorated stink bug, *Halyomorpha mista* Uhler (Heteroptera: Pentatomidae). I. Observation of mating behavior and multiple copulation. Appl Entomol Zool 18:234–242

Kawada H, Kitamura C (1992) The tachinid fly, *Bogosia* sp. (Diptera: Tachinidae) as a parasitoid of the brown marmorated stink bug, *Halyomorpha mista* Uhler (Heteroptera: Pentatomidae). Jpn J Environ Entomol Zool 4:65–70

Kelly T (2010) A stinky situation, Grape Press (Virginia Vineyards Association) 26(3):1, 7–8

Khrimian A, Shearer PW, Zhang A, Hamilton GC, Aldrich JR (2008) Field trapping of the invasive brown marmorated stink bug, *Halyomorpha halys*, with geometric isomers of methyl 2,4,6-decatrienoate. J Agric Food Chem 56:197–203

Kimura MT (2004) Cold and heat tolerance of drosophilid flies with reference to their latitudinal distributions. Oecologia 140:442–449

Kiritani K (2006) Predicting impacts of global warming on population dynamics and distribution of arthropods in Japan. Popul Ecol 48:5–12

Kiritani K (2007) The impact of global warming and land-use change on the pest status of rice and fruit bugs (Heteroptera) in Japan. Glob Change Biol 13:1586–1595

Koch RL (2003) The multicolored Asian lady beetle, *Harmonia axyridis*: a review of its biology, uses in biological control, and non-target impacts. J Insect Sci 3:1–16

Koch RL, Burkness EC, Wold-Burkness SJ, Hutchison WD (2004) Phytophagous preferences of the multicolored Asian lady beetle (Coleoptera: Coccinellidae) to autumn ripening fruit. J Econ Entomol 97:539–544

Koch RL, Venette RC, Hutchison WD (2006) Invasions by *Harmonia axyridis* (Pallas) (Coleoptera: Coccinellidae) in the Western Hemisphere: implications for South America. Neotrop Entomol 35:421–434

Koppel AL (2010) Stink bug egg studies in southeastern Virginia: parasitoid survey, and susceptibility and chorion permeability to insecticides. Ph.D. dissertation, Virginia Polytechnic Institute and State University, Blacksburg, VA

Kovach J (2004) Impact of multicolored Asian lady beetles as a pest of fruit and people. Am Entomol 50:159–161

Labrie G, Coderre D, Lucas É (2008) Overwintering strategy of multicolored Asian lady beetle (Coleoptera: Coccinellidae): cold-free space as a factor of invasive success. Ann Entomol Soc Am 101:860–866

LaMana ML, Miller JC (1996) Field observations on *Harmonia axyridis* Pallas (Coleoptera, Coccinellidae) in Oregon. Biol Control 6:232–237

LaMana ML, Miller JC (1998) Temperature-dependent development in an Oregon population of *Harmonia axyridis* (Coleoptera: Coccinellidae). Environ Entomol 27:1001–1005

Landis DA, Fox TB, Costamagna AC (2004) Impact of multicolored Asian lady beetle as a biological control agent. Am Entomol 50:153–155

Lehnert D (2010) New fruit fly is more than just a nuisance. Fruit Growers News 49:12

Leskey T (2011) The challenges posed by the invasive brown marmorated stink bug, *Halyomorpha halys* (Stål), to U.S. agriculture. U. S. Department of Agriculture-ARS, Kearneysville

Lucas E, Vincent C, Labrie G, Chouinard G, Fournier F, Pelletier F, Bostanian NJ et al (2007) The multicolored Asian ladybeetle *Harmonia axyridis* (Coleoptera: Coccinellidae) in Quebec agroecosystems ten years after its arrival. Eur J Entomol 104:737–744

Maine Department of Agriculture (2006) Interception of the brown marmorated stink bug (*Halyomorpha halys*) in Maine. Maine Department of Agriculture, Division of Plant Industry. http://www.maine.gov/agriculture/pi/pestsurvey/pestinfo/BMSB.htm

Markow TA, O'Grady PM (2006) *Drosophila*: a guide to species identification and use. Academic, London

Milkovich M (2010) SWD found in Michigan; team planning a response. Fruit Growers News November, pp 1–14

Mizell RF (2007) Impact of *Harmonia axyridis* (Coleoptera: Coccinellidae) on native arthropod predators in pecan and crape myrtle. Fla Entomol 90:524–536

Musolin DL (2007) Insects in a warmer world: ecological, physiological and life-history responses of true bugs (Heteroptera) to climate change. Global Change Biol 13:1565–1585

Nakamura K, Ueno H, Miura K (2005) Prevalence of inherited male-killing microorganisms in Japanese population of ladybird beetle *Harmonia axyridis* (Coleoptera: Coccinellidae). Ann Entomol Soc Am 98:96–99

Nakashima N, Sasaki J, Tsuda K, Yasunaga C, Noda H (1998) Properties of a new picorna-like virus of the brown-winged green bug, *Plautia stali*. J Invertebr Pathol 71:151–158

Nalepa CA (2007) *Harmonia axyridis* (Coleoptera: Coccinellidae) in buildings: relationship between body height and crevice size allowing entry. J Econ Entomol 100:1633–1636

Nalepa CA, Kidd KA, Ahlstrom KR (1996) Biology of *Harmonia axyridis* (Coleoptera: Coccinellidae) in winter aggregations. Ann Entomol Soc Am 89:681–685

Nalepa CA, Kennedy GG, Brownie C (2004) Orientation of multicolored Asian lady beetles to buildings. Am Entomol 50:165–166

Nalepa CA, Kennedy GG, Brownie C (2005) Role of visual contrast in the alighting behavior of *Harmonia axyridis* (Coleoptera: Coccinellidae) at overwintering sites. Environ Entomol 34:425–431

Nielsen AL, Hamilton GC (2009a) Life history of the invasive species *Halyomorpha halys* (Hemiptera: Pentatomidae) in northeastern United States. Ann Entomol Soc Am 102:608–616

Nielsen AL, Hamilton GC (2009b) Seasonal occurrence and impact of *Halyomorpha halys* (Hemiptera: Pentatomidae) in tree fruit. J Econ Entomol 102:1133–1140

Nielsen AL, Hamilton GC, Matadha D (2008a) Developmental rate estimation and life table analysis for *Halyomorpha halys* (Hemiptera: Pentatomidae). Environ Entomol 37:348–355

Nielsen AL, Shearer PW, Hamilton GC (2008b) Toxicity of insecticides to *Halyomorpha halys* (Hemiptera: Pentatomidae) using glass-vial bioassays. J Econ Entomol 101:1439–1442

Niva CC, Takeda M (2002) Color changes in *Halyomorpha brevis* (Heteroptera: Pentatomidae) correlated with distribution of pteridines: regulation by environmental and physiological factors. Comp Biochem Physiol B Biochem Mol Biol 132:653–660

Obata S (1986) Determination of hibernation site in the ladybird beetle, *Harmonia axyridis* Pallas (Coleoptera, Coccinellidae). Kontyu 54:218–223

Obata S (1997) The influence of aphids on the behavior of adults of the ladybird beetle, *Harmonia axyridis* (Col.: Coccinellidae). Entomophaga 42:103–106

Osawa N (1993) Population field studies of the aphidophagous ladybird beetle *Harmonia axyridis* (Coleoptera: Coccinellidae): life tables and key factor analysis. Res Popul Ecol 35:335–348

Panizzi AR, McPherson JE, James DG, Javahery M, McPherson BA (2000) Stink bugs (Pentatomidae). In: Schaefer CW, Panizzi AR (eds) Heteroptera of economic importance. CRC Press, New York, pp 421–474

Park H, Park YC, Hong OK, Cho SY (1996) Parasitoids of the aphidophagous ladybeetles, *Harmonia axyridis* (Pallas) (Coleoptera: Coccinellidae) in Chuncheon areas, Korea. Korean J Entomol 26:143–147

Pfeiffer DG (2008) Major insect and mite pests of grape in eastern North America. In: Wolf TK (ed) Wine grape production guide for eastern North America. Natural Resource, Agriculture, and Engineering Service (NRAES), Ithaca, NY, pp. 241–261, 307–313

Pfeiffer DG, Jordan TA, Laub CA, Mays RS (2010) Elimination of brown marmorated stink bug from winegrape clusters at harvest-2010. In: Cumberland-Shenandoah fruit workers conference, 2010, Winchester, VA

Pickering GJ, Lin Y, Riesen R, Reynolds A, Brindle I, Soleas G (2004) Influence of *Harmonia axyridis* on the sensory properties of white and red wine. Am J Enol Vitic 55:153–159

Pickering GJ, Lin Y, Reynolds A, Soleas G, Riesen R, Brindle I (2005) The influence of *Harmonia axyridis* on wine composition and aging. J Food Sci 70:128–135

Pickering GJ, Lin J, Reynolds A, Soleas G, Riesen R (2006) The evaluation of remedial treatments for wine affected by *Harmonia axyridis*. Int J Food Sci Technol 41:77–86

Riddick EW, Aldrich JR (2004) Search for repellents, attractants, and pheromones of the multicolored Asian lady beetle. Am Entomol 50:167–168

Riddick EW, Schaefer PW (2005) Occurrence, density, and distribution of parasitic fungus *Hesperomyces virescens* (Laboulbeniales: Laboulbeniaceae) on multicolored Asian lady beetle (Coleoptera: Coccinellidae). Ann Entomol Soc Am 98:615–624

Riddick EW, Aldrich JR, De Milo A, Davis JC (2000) Potential for modifying the behavior of the multicolored Asian lady beetle (Coleoptera: Coccinellidae) with plant-derived natural products. Ann Entomol Soc Am 93:1314–1321

Riddick EW, Brown AE, Chauhan KR (2008) *Harmonia axyridis* adults avoid catnip and grapefruit-derived terpenoids in laboratory bioassays. Bull Insectology 61:81–90

Rider DA (2005) *Halyomorpha halys* Stål, 1855. Pentatomoidea home page. http://www.ndsu.nodak.edu/ndsu/rider/Pentatomoidea/Species_Cappaeini/Halyomorpha_halys.htm

Rodriguez-Saona C, Polk D (2011) Spotted wing drosophila – a potential pest of New Jersey blueberries and other soft fruit. Plant Pest Advisory Rutgers Coop Ext 16(17):1–2

Sato S, Yasuda H, Evans EW (2005) Dropping behaviour of larvae of aphidophagous ladybirds and its effects on incidence of intraguild predation: interactions between the intraguild prey, *Adalia bipunctata* (L.) and *Coccinella septempunctata* (L.), and the intraguild predator, *Harmonia axyridis* Pallas. Ecol Entomol 30:220–224

Seo MJ, Kim GH, Youn YN (2008) Differences in biological and behavioural characteristics of *Harmonia axyridis* (Coleoptera: Coccinellidae) according to colour patterns of elytra. J Appl Entomol 132:239–247

Snyder WE, Evans EW (2006) Ecological effects of invasive arthropod generalist predators. Annu Rev Ecol Evol Syst 37:95–122

Snyder WE, Joseph SB, Preziosi RF, Moore AJ (2000) Nutritional benefits of cannibalism for the lady beetle *Harmonia axyridis* (Coleoptera: Coccinellidae) when prey quality is poor. Environ Entomol 29:1173–1179

Spellman B, Brown M, Mathews C (2006) Effect of floral and extrafloral resources on predation of *Aphis spiraecola* by *Harmonia axyridis* on apple. BioControl 51:715–724

Stathas GJ, Eliopoulos PA, Kontodimas DC, Giannopapas J (2001) Parameters of reproductive activity in females of *Harmonia axyridis* (Coleoptera: Coccinellidae). Eur J Entomol 98:547–549

Takahashi K (1987) Differences in oviposition initiation and sites of lady beetle, *Coccinella septempunctata bruckii* Mulsant and *Harmonia axyridis* (Pallas) (Coleoptera: Coccinellidae) in the field. Jpn J Appl Entomol Zool 31:253–254

Toyama M, Ihara F, Yaginuma K (2006) Formation of aggregations in adults of the brown marmorated stink bug *Halyomorpha halys* (Stål) (Heteroptera: Pentatomidae): the role of antennae in short-range locations. Appl Entomol Zool 41:309–315

Valent U.S.A Corporation (2010) Belay supplemental label. http://www.valent.com/agriculture/products/belay/label-msds.cfm

Walsh D, O'Neal S, Brooks T (2010) Spotted wing drosophila: what Washington State wine grape growers need to know. Washington State University Extension, Pullman. swd.hort.oregonstate.edu/files/webfm/editor/Wine_Grape_SWD_Bulletin_WSU.pdf

Walsh DB, Bolda MP, Goodhue RE, Dreves AJ, Lee J, Bruck DJ, Walton VM, O'Neal SD, Zalom FG (2011) *Drosophila suzukii* (Diptera: Drosophilidae): invasive pest of ripening soft fruit expanding its geographic range and damage potential. J Integr Pest Manag 2:1–7

Ware RL, Yguel B, Majerus MEN (2008) Effects of larval diet on female reproductive output of the European coccinellid *Adalia bipunctata* and the invasive species *Harmonia axyridis* (Coleoptera: Coccinellidae). Eur J Entomol 105:437–444

Watanabe M (2002) Cold tolerance and myo-inositol accumulation in overwintering adults of a lady beetle, *Harmonia axyridis* (Coleoptera: Coccinellidae). Eur J Entomol 99:5–9

Weintraub PG, Beanland L (2006) Insect vectors of phytoplasmas. Annu Rev Entomol 51:91–111

Welty C, Shetlar D, Hammond R, Jones S, Bloetscher B, Nielsen A (2008) Brown marmorated stink bug. Ohio State University Extension Fact Sheet. http://ohioline.osu.edu/hyg-fact/pdf/FS_3824_08.pdf

Wermelinger B, Wyniger D, Forster B (2008) First records of an invasive bug in Europe: *Halyomorpha halys* Stål (Heteroptera: Pentatomidae), a new pest on woody ornamentals and fruit trees? Bull Soc Entomol Suisse 81:1–8

Yang ZQ, Yao YX, Qiu LF, Li ZX (2009) A new species of *Trissolcus* (Hymenoptera: Scelionidae) parasitizing eggs of *Halyomorpha halys* (Heteroptera: Pentatomidae) in China with comments on its biology. Ann Entomol Soc Am 102:39–47

Yarbrough JA, Armstrong JL, Blumberg MZ, Phillips AE, McGahee E, Dolen WK (1999) Allergic rhinoconjunctivitis caused by *Harmonia axyridis* (Asian lady beetle, Japanese lady beetle, or lady bug). J Allergy Clin Immunol 104:704–705

Chapter 20
Vineyard IPM in a Changing World: Adapting to New Pests, Tactics, and Challenges

Rufus Isaacs, Charles Vincent, and Noubar J. Bostanian

Nothing that is can pause or stay;
The moon will wax, the moon will wane,
The mist and cloud will turn to rain,
The rain to mist and cloud again,
Tomorrow be today.

– Henry Wadsworth Longfellow

20.1 Introduction

As the chapters throughout this book demonstrate, the 7.5 million ha of vineyards around the globe (FAO 2009) are home to dynamic communities of insects and mites that require active management to prevent economic levels of injury to vines. Some pests cause indirect damage to the vines, attacking leaves and roots, whereas others infest berry clusters, causing direct infestation or reduced production of the harvested part of this crop. In some cases, arthropods are pests because they act as vectors of plant diseases. Other insects are natural enemies, but the relative importance of these arthropods varies with the location, type of grapes grown, and the management approach taken.

R. Isaacs (✉)
Department of Entomology, Michigan State University, East Lansing, MI 48824, USA
e-mail: isaacsr@msu.edu

C. Vincent • N.J. Bostanian
Horticultural Research and Development Centre, Agriculture and Agri-Food Canada, 430 Gouin Blvd., Saint-Jean-sur-Richelieu, QC, Canada J3B 3E6
e-mail: charles.vincent@agr.gc.ca; noubar.bostanian@agr.gc.ca

N.J. Bostanian et al. (eds.), *Arthropod Management in Vineyards: Pests, Approaches, and Future Directions*, DOI 10.1007/978-94-007-4032-7_20,

Since grapes are produced in so many regions of the world it can be dangerous to generalize. However, some important themes have emerged throughout this book, and an examination of the current status of integrated pest management (IPM) in vineyards and the future direction of viticultural entomology will be presented within this final chapter. Our aim here is to highlight some of the future challenges that will face those involved in arthropod management in vineyards and to offer our perspectives on where this field of investigation is heading.

For vineyard managers, the goal will continue to be the determination of levels of pest infestations, identifying when populations of certain key species are approaching economic thresholds, and if needed, selecting an effective and economical response to prevent crop loss and reduced vine health. To do this, they need tools that can be combined into a reliable IPM program for use in vineyards. In most regions, management of grape diseases is a dominant concern for growers. Thus, it should be noted here that arthropod pest management is often implemented within a context of actions that are also used to manage leaf, cluster, and trunk diseases. We do not expect this to change. However, there are regulatory and consumer-driven forces (Barber et al. 2009), as well as internal industry desires for reduced environmental footprints (Marshall et al. 2005), that guide grape producers and wineries to explore more environmentally conscious approaches to their businesses. This includes adoption of IPM practices for viticulture, whether for disease or arthropod management.

20.2 Integrated Pest Management for Vineyards

Long before Stern et al. (1959) articulated the Integrated Control Concept that led to the formalization of IPM, entomologists have been playing important roles in the course of viticulture. A well-known example is available in the research of C. V. Riley who worked with French plant scientist J.-E. Planchon in the late 1800s to develop resistant rootstocks that saved the French wine industry from the ravages of grape phylloxera. As described by Powell (Chap. 10), resistant rootstocks remain a foundation for management of this insect in vineyards, but changing geographic distributions and host-adapted biotypes require continued vigilance to minimize the economic impact of this pest.

For the past 50 years, applied entomologists have been developing programs for grape producers within the IPM paradigm. By taking state-of-the-art theories, approaches, and techniques into the vineyard, agricultural researchers have responded to some of the most challenging insect and mite pest problems faced by grape growers. These strategies have changed over time, as scientific advances have enabled new techniques to be applied to grape pest management.

From the late 1940s through the 1960s, neurotoxic pesticides developed during World War II replaced the arsenate insecticides that had been relied upon for insect control. The new chemicals provided very high levels of pest control and were quickly adopted by vineyard managers. However, once the effects of these chemicals on the environment, farm workers, and consumers became apparent, growers were under increasing pressure to reduce their use and to adopt non-chemical approaches where possible. The major advance that Stern et al. (1959) provided was to lay out a theoretical framework for how pest management could be based on the integration of biological (host plant resistance, biological control, etc.) and chemical (selective insecticides and acaricides) approaches. By doing so in a system that emphasized crop sampling for pests and economic thresholds (Chap. 2), the authors highlighted a science-based approach to pest management that could address many of the negative side effects caused by widespread use of broad-spectrum insecticides.

Grape producers were some of the first to adopt IPM tactics, and this industry has supported development of intensive scouting and monitoring programs coupled with adoption of cutting edge technology to aid in decision-making. The high value of grapes and their large area in some of the industrialized nations have also supported the availability of mating disruption technology for *Lobesia botrana* (Denis & Schiffermüller) and *Eupoecilia ambiguella* (Hübner) in Europe and for *Paralobesia viteana* (Clemens) in North America, as well as the relatively early registration of reduced-risk insecticides and miticides.

The grape industry has also been a pioneer in the organic movement and in sustainable agriculture and biologically-based farming (Broome and Warner 2008). This has included grower-led programs that have combined education and self-assessment to document how adoption of IPM practices has improved the level of pest control and reduced pesticide use, as well as improved the environment for beneficial arthropods (Thomson et al. 2007). These approaches to grape production have organizations that can certify vineyards and wineries. Many producers adopt such approaches without using these certifications in marketing their products, so consumers may not be aware of how the grapes were grown. Grower motivation for embracing such viticultural systems may range from the purely economical to the philosophical and may have more to do with their wine quality goals than pest management goals (Warner 2007). Whatever the reason, the growth of organic, biodynamic, or sustainable viticulture also depends on biological processes (Reeve et al. 2005) and these are more information-intensive production systems with greater reliance on IPM.

20.3 Current Situation

20.3.1 Biological Approaches

After the chemical-based vineyard pest management programs of the first half of the twentieth century, and the integrated systems of management promoted in the late twentieth century, there is now an increasing interest in biologically-based components for use in vineyard management. This is reflected in the four chapters in this book that focus on biological control (Chaps. 5, 6, 7, and 8) authored by viticultural entomologists from the United States, South Africa, New Zealand, and Australia. The enhancement of natural predators through conservation biological control is a compelling approach to increase the level of natural pest suppression, thereby reducing the likelihood that pest populations will ever reach the action thresholds. Providing appropriate habitat to improve the nutritional status and reproductive potential of natural enemies can also have aesthetic benefits that attract people to visit wine regions, and these are being promoted by wineries and tourist groups in some major wine producing regions, notably in New Zealand, where the value of such approaches is being estimated (Porter et al. 2009) and used to build support for these programs. Certain management tactics may also support beneficial organisms above and below ground, further supporting soil quality. Integrating these 'ecosystem services' (i.e., the delivery of services by natural areas with economic value to humans) into viticulture is providing a new direction for thinking about how applied pest management programs can integrate entomology, sociology, and economics for the benefit of society at large (Sandhu et al. 2010). We expect that such multidisciplinary projects will be replicated in other regions in the coming years, and it will be interesting to see the extent to which wine grape industries engage in the concept of 'natural capital' as part of their sustainability programs.

As mentioned by Walton et al. (Chap. 5), there are still unexplored agents of biological control, such as nematodes and viruses, that are now being commercialized for use in other crops and that may soon play a larger role in grape IPM, particularly in those vineyards with sustainable, organic, or biologically-based production systems. As with all such tactics, the economics of crop production imply that these new approaches must be able to compete with the generally less expensive chemical pesticides to increase their market share in conventional grape production. The higher costs of biological controls are a major impediment to grower adoption, but long term studies have repeatedly demonstrated that multi-year transition from chemically-based to biologically-based or IPM systems can lead to higher revenues. Detailed studies are needed in different viticultural regions to determine how the adoption of biologically-based pest management programs affects the arthropod community, and the implications of these changes for pest populations and grower economic return.

An interesting new area of research related to biological control in vineyards is described by Simpson et al. (Chap. 6). In this attract and reward approach, the vine's volatile defensive chemical signals are stimulated by application of elicitor chemicals, and this is combined with provision of habitat for the natural enemies to survive on when they are not attacking the pests. Testing this combination of approaches is an active area of research in multiple crop systems, and recent reports suggest that the combination of both tactics into an attract and reward system will have additive effects on pest control (e.g., Simpson et al. 2011). There is much yet to learn about optimal deployment of the rewarding plants and stimulation of the vine to produce attractant chemicals, and how best to selectively 'attract and reward' the insects needed for biological control, but these early tests of the concept are promising. We expect future development of this research to include tests in other regions and at much larger spatial scales to determine its practicality within intensive IPM programs.

Currently, biological control of pests is increasingly considered at the landscape scale (Landis et al. 2000). As Miles et al. discuss in Chap. 8, there is much yet to learn about the implications of habitat composition close to vineyards and in the landscape surrounding vineyards for the abundance and composition of pest and natural enemy communities. This is an active area of research but consistent patterns are not yet evident in the implications of landscape composition for the risk of pest infestation. Rather, the ecology and host associations of specific insects are likely to drive these patterns to affect whether simple or diverse landscapes are positive or negative for pest pressure. As this area of research develops, advanced IPM systems are increasingly expected to consider more than individual vineyard blocks to integrate whole farm mapping, monitoring, and management. This is already underway in many grape growing regions as ecologically-based farm management practices are being developed, and growers may adopt this approach independently or with support from government incentive payments that aim to meet certain resource conservation goals.

Mating disruption using pheromones is another biologically-based approach with widespread adoption in some regions and potential for greater adoption in vineyards (Witzgall et al. 2010). Used on a high proportion of vineyards in the wine-producing areas of Germany and Switzerland for control of cluster-infesting moth pests, it has nevertheless only been adopted at lower levels elsewhere. Future identification of the sex pheromones of new vineyard pest species, coupled with development of efficient delivery systems, will bring about some new opportunities for this approach. Where area-wide adoption has been stimulated by coordinated decisions to employ this approach, pest populations have been suppressed and insecticide use has declined. Recent advances in identifying the pheromone of grape mealybugs may also provide additional avenues for integrating mating disruption into vineyard IPM systems. Regions with successful adoption of mating disruption tend to have more stringent restrictions on pesticide use, and may also provide incentives for adoption of non-chemical approaches. The relative costs of application and of the product itself remain barriers to widespread adoption, but there is continual development of dispenser technology and application devices that may improve these parameters (e.g., Teixeira et al. 2010).

20.3.2 Chemical-Based Approaches

The past 10 years has seen a dramatic change in the spectrum of insecticides available for grape producers, with new modes of action and pest spectra allowing an unparalleled opportunity for growers to target specific pests for control while also minimizing the risks to non-target organisms. This has been spurred in part by the discovery of new chemical classes, thanks to advances made in molecular physiology which led to critical changes in the search as well as the synthesis of xenobiotics, and also by a more restrictive regulatory environment for registration of pest control chemicals. There has also been keen interest worldwide in the effects of pesticides on non-target arthropods that are an integral component of IPM programs. This interest has been emphasized in Chaps. 4, 5, 9, 10, 12, 13, 14, 17, and 19.

There is now increased potential for realizing integrated control, since many of the most effective new insecticides have been evaluated and shown to have relatively low impact on natural enemies. For example, registration of the insect growth regulator insecticides methoxyfenozide and diflubenzuron for use in vineyards and the recent availability of the diamide insecticides rynaxapyr and flubendiamide allow more selective and long-lasting control of lepidopteran pests without high levels of natural enemy mortality. Entomologists have developed and validated the essential degree-day models to enable accurate spray timing for such compounds that require stage-selective application for maximum efficacy, and we expect that greater adoption of these newer reduced-risk insecticides will result in more favorable environmental profile of grape pest control programs.

Acaricides have also changed from broad-spectrum to more selective chemistries. The vineyard manager now has an array of different acaricide modes of action available, many of which can selectively kill pest mites without injuring predatory species. Some of these are also systemic, thereby providing a route of exposure that further protects predators from direct contact with the acaricide. The availability of selective acaricides and insecticides is leading to a transition towards vineyard insect control programs that rely less on the neurotoxic modes of action of contact insecticides and more on selective insecticide classes that provide a range of different modes of action. This has resulted in a much more complex landscape of pesticide options that will require greater understanding of pest-plant-pesticide interactions for appropriate recommendations to be made. Major issues related to pesticides that will continue to be relevant for grape industries worldwide are the preservation of biodiversity in vineyards (Bruggisser et al. 2010), how pesticides disrupt beneficial arthropod communities (Nash et al. 2010), pre-harvest intervals and maximum residue limits (Navarro et al. 2000), and mitigation of arthropod resistance to pesticides (Whalon et al. 2008).

20.3.3 Decision Support to Facilitate Vineyard IPM

Without sampling plans and economic thresholds, implementation of complete IPM programs is not possible. To ensure their greatest chance of adoption, development of these tools requires careful research and then subsequent validation to test them under commercial vineyard conditions. Despite their importance for IPM programs, there are relatively few widely used thresholds for grape pest management (Chap. 2) and they tend to be developed and adopted only for the most economically important pests. This is in part due to the time and expense of scouting, the economic risk inherent in making an incorrect decision not to treat, and the high per hectare value of grapes. Techniques are available for incorporating these factors into thresholds, however, and there are situations where thresholds have been developed and adopted. Examples of how threshold-based decision making has reduced grower costs, improved profit, and reduced-risk should be highlighted more so that the end users of such management tools can see the value to their businesses. Nevertheless, there are instances where quantitative data are not being used by growers and the reasons merit further examination.

Delivery of decision support tools to vineyard managers through internet-based systems is increasingly common, but there remains great value in having crop advisors and extension staff that can help managers interpret model outputs, guide sampling programs, and select appropriate insecticides if needed. This requires investment in developing programs that can teach the current and future generations of agricultural professionals, but there is a trend in some regions towards greater investment in basic science and away from the applied sciences that support IPM programs. As argued by Castle and Naranjo (2009), this trend has the potential to undermine the great advances in IPM implementation, leading to an erosion of grower's ability to access impartial advice from independent crop scouts and consultants.

We hope that these changes will evolve so that vineyard managers continue to have free access to trained professionals who can interpret and deliver the results of publicly-supported research to answer grape producers' questions and guide adoption of IPM programs.

20.4 What Next for Vineyard IPM?

Grape producers have always had to respond to changing pest complexes over the years, and they have learned new techniques for vineyard management that address these novel situations. It is apparent that the pace of change in crop pest management has increased, with the converging forces of increased global trade, climate change, consumer preference, new technology and regulations coming together to make the pest management approaches of only one generation ago seem obsolete. This pace of change argues strongly for continued vigilance against new pests, development of regionally-appropriate management tactics, and investment in educational programs that focus on technology transfer to grape growers and other vineyard managers.

As stated in Chap. 1, viticulture is an ancient form of agriculture. It has developed in many production regions over hundreds or thousands of years with growers learning which cultivars worked well for the soils and climate of their regions. The changing climate will alter both rainfall patterns and temperature in many of the major viticultural regions of the world, making some regions more suitable for cultivars that require warmer growing seasons (White et al. 2006). Indeed, these trends are already being seen in increasing quality and decreasing inter-year variation in wines (Jones et al. 2005). This warming trend is also affecting where grapes can be grown. Thus, southern England now produces wines from cultivars that could not be grown since the Medieval Warm Period over 700 years ago. While the distribution and types of grape grown will be changing, these climatic changes are also expected to change suitability for insect pests. Temperature affects the growth rate, reproduction, survival of insects and mites, and the amount of feeding that plants will experience (Bale et al. 2002). Increasing temperatures will lead to higher arthropod pest pressure from the combined effects of expanding geographic ranges and increasing numbers of pest generations. This will further emphasize the need for monitoring and scouting so that appropriate responses can be made to protect crops.

In addition to the expansion of existing pest distributions, we expect invasive species to become an increasingly common feature of vineyard pest management. Grapes are grown in different continents, with strong economic pressure for movement of plant materials and vineyard machinery among these production regions. Regulatory and quarantine systems are in place to prevent the introduction of novel pests and diseases. However, history teaches us from the experiences with phylloxera, glassy wing sharpshooter (Hoddle 2004) and multicolored Asian ladybeetle (Hutchison et al. 2010) or more recently, spotted wing drosophila (Lee et al. 2011) and European grapevine moth (Gilligan et al. 2011), that there will be continued opportunities for new arthropods to become established in regions of grape production. Regulatory systems must remain well supported if the devastating consequences of these new pests are to be minimized by interception at borders or by early detection. Once a pest is found, the likely source(s) of its origin can be investigated using molecular techniques such as those employed by Downie (2002) for phylloxera. However, prevention is far easier and less costly in the long run than attempting post-invasion eradication.

Viticulture is expected to continue moving towards increasing adoption of organic, sustainable, and biodynamic management, but we expect these approaches to be implemented on a minority of vineyards around the globe for the foreseeable future. Knowledge gained in these systems will be expected to have an influence on conventional viticulture, but it is likely that chemical methods of pest control will continue to dominate grape production. With agrochemical companies under increasing societal and regulatory pressure to develop new pest control options that have low non-target impact, new classes of insecticides are providing unparalleled opportunities for insect and mite control without disrupting natural enemies. The greater complexity inherent in managing insects and mites using chemicals that have distinct and varying modes of action will necessitate continued investment in applied entomology and acarology to understand the spectrum of activity, non-target effects, mechanisms, and field performance of these pesticides.

Genetic modification of grapevines has been pursued in the past few decades by plant scientists interested in developing vines with resistance to pests (Gray et al. 2002; Vivier and Pretorius 2002). Targeted pests include insect-vectored diseases such as the economically important *Xylella fastidiosa* Wells et al. that causes Pierce's disease (Agüero et al. 2005). There is also interest in developing herbicide-resistant vines that would allow use of broad-spectrum herbicides, particularly during vineyard establishment. Insect resistance through engineering of plants to express *Bacillus thuringiensis* Berliner is common in some regions of field crop production. Advances in grapevine plant physiology in the foreseeable future may enable selective expression of *B. thuringiensis* toxins in the vegetative, but not the reproductive parts of the vine. However, there have only been some small field trials to date and there is no commercial production of engineered vines yet. Regulatory constraints on this technology will likely be tight in many regions of the world. Marketing products from engineered vines may be challenging, and consumer acceptance has to be far more accommodating than at the present time before the grape industry would adopt such technology (Vivier and Pretorius 2002). The ecological risks of such approaches have been discussed widely in the scientific and popular press (e.g., Snow and Palma 1997), and this debate is expected to be an intense one between different groups of people with diverse views on the future direction for grape production.

Integrated pest management is a well-tested approach that relies on knowledge of pest abundance and its relationship to crop damage to determine the point at which pest control is needed to prevent economic loss. The method of control may be one or a combination of tactics, and the range of options available to grape producers is expected to continue to expand as new technologies are developed. We see a significant need for researchers to develop action thresholds for key pests that can guide decision-making using the foundational concepts that initiated IPM in the 1950s. These ideas remain relevant currently (Mitchell and Hutchison 2009) and will be important well into the future as a means for growers to reduce pest risk and increase returns in a sustainable manner.

References

Agüero CB, Uratsu SL, Greve C, Powell ALT, Labavitch JM, Meredith CP, Dandekar AM (2005) Evaluation of tolerance to Pierce's disease and *Botrytis* in transgenic plants of *Vitis vinifera* L. expressing the pear PGIP gene. Mol Plant Pathol 6:43–51

Bale JS, Masters GJ, Hodkinson ID, Awmack C, Bezemer TM, Brown VK et al (2002) Herbivory in global climate change research: direct effects of rising temperature on insect herbivores. Glob Change Biol 8:1–16

Barber N, Taylor C, Strick S (2009) Wine consumers' environmental knowledge and attitudes: influence on willingness to purchase. Int J Wine Res 1:59–72

Broome JC, Warner KD (2008) Agro-environmental partnerships facilitate sustainable wine-grape production and assessment. Calif Agric 62:133–141

Bruggisser OT, Schmidt-Entling MH, Bacher S (2010) Effects of vineyard management on biodiversity at three trophic levels. Biol Conserv 143:1521–1528

Castle S, Naranjo SE (2009) Sampling plans, selective insecticides and sustainability: the case for IPM as 'informed pest management'. Pest Manag Sci 65:1321–1328

Downie DA (2002) Locating the sources of an invasive pest, grape phylloxera, using a mitochondrial DNA gene genealogy. Mol Ecol 11:2013–2026

FAO (2009) Food and Agriculture Organization agribusiness handbook – grapes and wine, Rome. http://www.fao.org/docrep/012/al176e/al176e.pdf

Gilligan TM, Epstein ME, Passoa SC, Powell JA, Sage OC, Brown JW (2011) Discovery of *Lobesia botrana* (Denis & Schiffermüller) in California: an invasive species new to North America (Lepidoptera: Tortricidae). Proc Entomol Soc Wash 113:13–40

Gray DJ, Jayasankar S, Li Z, Cordts J, Scorza R, Srinivasan C (2002) Transgenic grapevines. In: Khachatourians GG, McHughen A, Scorza R, Nip WK, Nui YH (eds) Transgenic plants and crops. Marcel Dekker, New York, pp 397–405

Hoddle MS (2004) The potential adventive geographic range of glassy-winged sharpshooter, *Homalodisca coagulata* and the grape pathogen *Xylella fastidiosa*: implications for California and other grape growing regions of the world. Crop Prot 23:691–699

Hutchison WD, Galvan TL, Burkness EC, Koch RL (2010) *Harmonia axyridis* as an economic pest of wine grapes in the U.S.: progress in developing an IPM program and potential impact in Europe. IOBC/WPRS Bull 58:47–52

Jones GJ, White MA, Cooper OR, Storchmann K (2005) Climate change and global wine quality. Clim Chang 73:319–343

Landis DA, Wratten SD, Gurr GM (2000) Habitat management to conserve natural enemies of arthropod pests in agriculture. Annu Rev Entomol 45:175–201

Lee JC, Bruck DJ, Dreves AJ, Ioriatti C, Vogt H, Baufeld P (2011) In focus: spotted wing drosophila, *Drosophila suzukii*, across perspectives. Pest Manag Sci 67:1349–1351

Marshall R, Cordano M, Silverman M (2005) Exploring individual and institutional drivers of proactive environmentalism in the US wine industry. Bus Strat Environ 14:92–109

Mitchell PD, Hutchison WD (2009) Economic risk and decision making in IPM. In: Radcliffe EB, Hutchison WD, Cancelado RE (eds) IPM: concepts, tactics, strategies, and case studies. Cambridge University Press, Cambridge, pp 35–50

Nash MA, Hoffmann AA, Thomson LJ (2010) Identifying signature of chemical applications on indigenous and invasive nontarget arthropod communities in vineyards. Ecol Appl 20: 1693–1703

Navarro S, Barba A, Navarro G, Vela N, Oliva J (2000) Multiresidue method for the rapid determination – in grape, must and wine – of fungicides frequently used on vineyards. J Chromatogr 882:221–229

Porter J, Costanza R, Sandhu H, Sigsgaard L, Wratten S (2009) The value of producing food, energy, and ecosystem services within an agro-ecosystem. Ambio 38:186–193

Reeve JR, Carpenter-Boggs L, Reganold JP, York AL, McGourty G, McCloskey LP (2005) Soil and winegrape quality in biodynamically and organically managed vineyards. Am J Enol Vitic 56:367–376

Sandhu HS, Wratten SD, Cullen R (2010) Organic agriculture and ecosystem services. Environ Sci Policy 13:1–7

Simpson M, Gurr GM, Simmons AT, Wratten SD, James DG, Leeson G, Nico HI, Orre GUS (2011) Field evaluation of the 'attract and reward' biological control approach in vineyards. Ann Appl Biol 159:69–78

Snow AA, Palma PM (1997) Commercialization of transgenic plants: potential ecological risks. Bioscience 47:86–96

Stern VM, Smith RF, van den Bosch R, Hagen KS (1959) The integrated control concept. Hilgardia 29:81–101

Teixeira LAF, Mason KS, Mafra-Neto A, Isaacs R (2010) Mechanically-applied wax matrix (SPLAT-GBM) for mating disruption of grape berry moth (Lepidoptera: Tortricidae). Crop Prot 29:1514–1520

Thomson LJ, Sharley DJ, Hoffmann AA (2007) Beneficial organisms as bioindicators for environmental sustainability in the grape industry in Australia. Aust J Exp Agric 47:404–411

Vivier VA, Pretorius IS (2002) Genetically tailored grapevines for the wine industry. Trends Biotechnol 20:472–478

Warner KD (2007) The quality of sustainability: agroecological partnerships and the geographic branding of California winegrapes. J Rural Stud 23:142–155

Whalon ME, Mota-Sanchez D, Hollingworth RM (2008) Global pesticide resistance in arthropods. CABI, Oxfordshire

White MA, Diffenbaugh NS, Jones GV, Pal JS, Giorgi F (2006) Extreme heat reduces and shifts United States premium wine production in the 21st century. Proc Natl Acad Sci USA 103:11217–11222

Witzgall P, Kirsch P, Cork A (2010) Sex pheromones and their impact on pest management. J Chem Ecol 36:80–100

General Index

N.J. Bostanian et al. (eds.), *Arthropod Management in Vineyards: Pests, Approaches, and Future Directions*, DOI 10.1007/978-94-007-4032-7,

D

Species Name Index

N.J. Bostanian et al. (eds.), *Arthropod Management in Vineyards: Pests, Approaches, and Future Directions*, DOI 10.1007/978-94-007-4032-7,